The Physical Basis of Thermodynamics

With Applications to Chemistry

The Physical Basis of Thermodynamics
With Applications to Chemistry

Pascal Richet
Institut de Physique du Globe
Paris, France

Kluwer Academic/Plenum Publishers
New York, Boston, Dordrecht, London, Moscow

Library of Congress Cataloging-in-Publication Data

Richet, Pascal.
The physical basis of thermodynamics: with applications to chemistry/by Pascal Richet.
p. cm.
Includes bibliographical references and index.
ISBN 0-306-46584-1
1. Thermodynamics. I. Title.

QC311 .R5 2001
536′.7—dc21

2001023459

ISBN 0-306-46584-1

233 Spring Street, New York, New York 10013

http://www.wkap.nl/

10 9 8 7 6 5 4 3 2 1

A C.I.P. record for this book is available from the Library of Congress

Printed in the United States of America

To Evelyne

A physical understanding is a completely unmathematical, imprecise, and inexact thing, but absolutely necessary for a physicist.

Richard Feynman

PREFACE

As indicated by its title, this book aims at explaining thermodynamic phenomena and concepts in physical terms before proceeding to the requisite mathematical aspects of the problems. Given the breadth of thermodynamics, it does not claim to be comprehensive. In a self-consistent way, it focuses on the effects of pressure, temperature and chemical composition on thermodynamic properties. Some emphasis has been put on rapidly evolving fields such as amorphous materials, metastable phases, numerical simulations of microsystems and high-pressure thermodynamics. On the other hand, it describes in less depth topics, like redox reactions, for which much literature is already available. Without requiring a background in quantum mechanics, the last three chapters will illustrate the main practical applications of statistical thermodynamics and give a microscopic interpretation of temperature, pressure, and entropy

Our main goal is therefore to show that thermodynamics is not an abstruse discipline where one readily gets lots in a maze of difficult concepts. For example, entropy usually raises many difficulties even though it is simply for thermal energy the analog of charge for electrical energy. In the same way that one would not expect to wait until the middle of a treatise on electrostatics to mention charge, we will introduce entropy in the very first chapter and stress how it is amenable to measurement. Many other concepts may look strange, such as the Gibbs-Duhem equation and the critical point of fluids. We will therefore physically explain why the former is not a mathematical trick and the latter is not a point where the liquid phase mysteriously disappears.

The approach followed seems particularly relevant as thermodynamics maintains an essentially experimental basis. The importance of statistical methods should not conceal that a great many aspects of thermodynamics remains empirical, in particular when one investigates real objects which, as such, are often complex. This is why the topics discussed are referred to well founded case studies taken from a variety of fields, whereas familiar substances like water, ice or rocksalt have been included as examples. By avoiding qualitative schemes, experimental results are thus presented in more than 100 original figures. Along with recent measurements, one will also find older results whose precision has not yet been surpassed; as a matter of fact, one often unknowingly makes use of them through current compilations of thermodynamic data. For more extensive mathematical treatments or developments on transport properties, kinetics, magnetism, etc., we refer to the excellent books listed in the bibliography where, for ease of consultation, books have been listed separately from research publications.

This text has grown from years of teaching at the graduate and undergraduate levels. Originally, it addressed the needs of student who already have a basic knowledge of thermodynamics and wish to gain a deeper understanding of the subject. Because experience has shown that newcomers in the field could also profit from this course, the basic concepts have been thoroughly discussed. Likewise, topics such as the properties of solutions are briefly introduced before being discussed more thoroughly in the following chapters. Whereas parts of an elementary nature will of course be skipped by more conversant readers, beginners will themselves skip chapters 6, 10–11 and 13–15 which refer to a higher level. As usual, sections in small type are of a more specialized character; from the negative pressures and temperatures to information theory and temperatures at the center of the earth, they will also illustrate the relevance of the topics discussed to other contexts.

Problems are appended at the end of most chapters, but many topics have been left as exercises and the reader could also check the examples discussed in the text and in the figures. In this respect a great benefit to be drawn from personal computers is the possibility of investigating problems in more depth than simply finding a solution for some particular conditions. The reader is thus advised to play with the data and the equations and visualize the results obtained over wide pressure, temperature and composition ranges in order to arrive at a real, quantitative understanding of the subject.

Following a common usage in continental Europe, the term *principles* has been preferred over *laws* to stress the fundamental difference between the two great pillars of thermodynamics, which count among the highest achievements of human thought, and innumerable laws which have only a restricted validity. Although the nomenclature is not always agreed upon, common usage has been followed. Because the same symbols happen to be used for denoting different properties, they have been redefined each time a risk of confusion is possible. Also note that *concentration* will denote any composition scale, and will not be synonymous with molarity. SI units have been used consistently, with the single exception of pressure for which bars are more convenient than Pascals (resulting for instance in a unit instead of a 10^5 standard pressure).

Finally, the author warmly thanks the numerous people without whom this book would not have been brought to completion, first of all his specially understanding family circle. Valuable information was provided by D. Andrault, M.A. Bouhifd, F. Félin, P. Jarry, H. Kéhiaian, G. Moguilny, J.-P. Petitet, M. Pichavant, A. Polian, J.-P. Simonin and L. Thiéblot. Helpful remarks on some chapters were offered by F. Chabaux, B. Guillot, G. Michard, Ph. Sarda, T. Trull and Y. Simon. Special thanks are due to Y. Bottinga, K. Bourova, S. Dellerue, J. Sipowska, A. Whittington and, particularly, to E. Richet and M. Toplis for thoughtful comments on the whole manuscript; J. Dyon for beautiful artwork; A. Glitho for skillful preparation of the camera-rady copy; and alert readers who would point out remaining errors for which the author begs pardon in advance.

CONTENTS

S.I. UNITS AND PHYSICAL CONSTANTS
1998 CODATA Recommended Values

	Symbol	Value	Unit
Universal constants			
Speed of light in a vacuum	c	2.997 924 58 10^{8}	m s^{-1}
Permittivity of vacuum	ε_0	8.854 187 817 10^{-12}	F m^{-1}
Gravitation constant	G	6.673 (10) 10^{-11}	m^3 kg^{-1} s^{-2}
Planck constant	h	6.626 068 76 (52) 10^{-34}	J s
		4.135 667 27 (16) 10^{-15}	eV s
Masses			
Electron	m_e	9.109 381 88 (72) 10^{-31}	kg
Proton	m_p	1.672 621 58 (13) 10^{-27}	kg
Neutron	m_n	1.674 927 16 (13) 10^{-27}	kg
Physical constants			
Charge of the electron	e	1.602 176 462 (63) 10^{-19}	C
Avogadro number	N_A	6.022 141 99 (47) 10^{23}	mol^{-1}
Faraday constant	F	9.648 534 15 (39) 10^{4}	C mol^{-1}
Ideal gas constant	R	8.314 472 (15)	J mol^{-1} K^{-1}

	Symbol	Value	Unit
Boltzmann constant	k	1.380 6503 (24) 10^{-23}	J K^{-1}
	k/hc	69.503 56 (12)	m^{-1} K^{-1}
Stefan-Boltzmann constant	σ	5.670 400 (40) 10^{-8}	W m^{-2} K^{-4}
Molar volume of ideal gases (at 273.15 K and 1 atm)	V_m	22.413 996 (39) 10^{-3}	m^3 mol^{-1}

As consistently made in this book, numbers in parentheses denote the uncertainties in the final decimal places. Only in the above Table reported values are by definition exact when no uncertainties are mentioned. Data from Mohr and Taylor (1999).

CONVERSION FACTORS

Electron-Volt	eV	1.602 176 462 (63) 10^{-19}	J
Standard atmosphere	atm	101.325 10^3	Pa

ELEMENTS
Atomic Numbers and 1995 Standard Atomic Weights

Calculated from the average natural isotopic abundances, see Coplen (1995). For radioactive elements, whose weights depend on the origin, the values in brackets refer to the longest lived isotope. The uncertainties on weights are indicated by the number in parenthese. They essentially result from natural variations in isotopic abundances and are included only to show the remarkable precision with which atomic weights are generally known.

Actinium	Ac	89	227.0278	Cadmium	Cd	48	112.411 (8)
Aluminum	Al	13	26.981538 (2)	Calcium	Ca	20	40.078 (4)
Americium	Am	95	[243]	Californium	Cf	98	[251]
Antimony	Sb	51	121.760 (1)	Carbon	C	6	12.0107 (8)
Argon	Ar	18	39.948 (1)	Cerium	Ce	58	140.116 (1)
Arsenic	As	33	74.92160 (2)	Cesium	Cs	55	132.90545 (2)
Astatine	At	85	[210]	Chlorine	Cl	17	35.4527 (9)
Barium	Ba	56	137.327 (7)	Chromium	Cr	24	51.9961 (6)
Berkelium	Bk	97	[247]	Cobalt	Co	27	58.933200 (9)
Beryllium	Be	4	9.012182 (3)	Copper	Cu	29	63.546 (3)
Bismuth	Bi	83	208.98038 (2)	Curium	Cm	96	[247]
Boron	B	5	10.811 (7)	Dysprosium	Dy	66	162.50 (3)
Bromine	Br	35	79.904 (1)	Einsteinium	Es	99	[252]

Erbium	Er	68	167.26 (3)
Europium	Eu	63	151.964 (1)
Fermium	Fm	100	[257]
Fluorine	F	9	18.9984032 (5)
Francium	Fr	87	[223]
Gadolinium	Gd	64	157.25 (3)
Gallium	Ga	31	69.723 (1)
Germanium	Ge	32	72.61 (2)
Gold	Au	79	196.96655 (2)
Hafnium	Hf	72	178.49 (2)
Helium	He	2	4.002602 (2)
Holmium	Ho	67	164.93032 (2)
Hydrogen	H	1	1.00794 (7)
Indium	In	49	114.818 (3)
Iodine	I	53	126.90447 (3)
Iridium	Ir	77	192.217 (3)
Iron	Fe	26	55.845 (2)
Krypton	Kr	36	83.80 (1)
Lanthanum	La	57	138.9055 (2)
Lawrencium	Lr	103	[262]
Lead	Pb	82	207.2 (1)
Lithium	Li	3	6.941 (2)
Lutetium	Lu	71	174.967 (1)
Magnesium	Mg	12	24.3050 (6)
Manganese	Mn	25	54.938049 (9)
Mendelevium	Md	101	[258]
Mercury	Hg	80	200.59 (2)
Molybdenum	Mo	42	95.94 (1)
Neodymium	Nd	60	144.24 (3)
Neon	Ne	10	20.1797 (6)
Neptunium	Np	93	237.0482
Nickel	Ni	28	58.6934 (2)
Niobium	Nb	41	92.90638 (2)
Nitrogen	N	7	14.00674 (7)
Nobelium	No	102	[259]
Osmium	Os	76	190.23 (3)
Oxygen	O	8	15.9994 (3)
Palladium	Pd	46	106.42 (1)
Phosphorus	P	15	30.973762 (4)
Platinum	Pt	78	195.078 (2)
Plutonium	Pu	94	[244]
Polonium	Po	84	[209]
Potassium	K	19	39.0983 (1)
Praseodymium	Pr	59	140.90765 (2)
Promethium	Pm	61	[145]
Protactinium	Pa	91	231.03588 (2)
Radium	Ra	88	226.0254
Radon	Rn	86	[222]
Rhenium	Re	75	186.207 (1)
Rhodium	Rh	45	102.90550 (2)
Rubidium	Rb	37	85.4678 (3)
Ruthenium	Ru	44	101.07 (2)
Samarium	Sm	62	150.36 (3)
Scandium	Sc	21	44.955910 (8)
Selenium	Se	34	78.96 (3)
Silicon	Si	14	28.0855 (3)
Silver	Ag	47	107.8682 (2)
Sodium	Na	11	22.989770 (2)
Strontium	Sr	38	87.62 (1)
Sulfur	S	16	32.066 (6)
Tantalum	Ta	73	180.9479 (1)
Technetium	Tc	43	[98]
Tellurium	Te	52	127.60 (3)
Terbium	Tb	65	158.92534 (2)
Thallium	Tl	81	204.3833 (2)
Thorium	Th	90	232.0381 (1)
Thulium	Tm	69	168.93421 (2)
Tin	Sn	50	118.710 (7)
Titanium	Ti	22	47.867 (1)
Tungsten	W	74	183.84 (1)
Uranium	U	92	238.0289 (1)
Vanadium	V	23	50.9415 (1)
Xenon	Xe	54	131.29 (2)
Ytterbium	Yb	70	173.04 (3)
Yttrium	Y	39	88.90585 (2)
Zinc	Zn	30	65.39 (2)
Zirconium	Zr	40	91.224 (2)

GREEK LETTERS

α	*A*	alpha	ι	I	iota	ρ	P	rho
β	*B*	beta	κ	K	kappa	σ	Σ	sigma
γ	Γ	gamma	λ	Λ	lambda	τ	T	tau
δ	Δ	delta	μ	M	mu	υ	Υ	upsilon
ε	E	epsilon	ν	N	nu	φ	Φ	phi
ζ	Z	zeta	ξ	Ξ	xi	χ	X	chi
η	H	eta	o	O	omicron	ψ	Ψ	psi
θ	Θ	theta	π	Π	pi	ω	Ω	omega

The Physical Basis of Thermodynamics

With Applications to Chemistry

1
ENTROPY AND PRINCIPLES

Thermodynamics would be less disconcerting if entropy did not play so prominent a role. The difficulty is that entropy is neither palpable nor directly measurable. This is why is a good part of the first chapter will be devoted to it. But we first have to introduce a few basic concepts, including phase and state function, and present the internal energy and the first principle of energy conservation. Then, a review of various forms of energy will show that entropy represents for thermal energy the analog of mass for gravitational energy, or of electrical charge for electrical energy. Associated with the concept of irreversibility, this analog has some peculiarity, however, whose physical meaning will be discussed within the general framework of the second principle and of mutual conversions of the different forms of energy.

1.1 PREAMBLE

1.1a SYSTEMS, PROPERTIES, PHASES

A *system* is simply that part of the universe to be studied: this can be a droplet of water, a few liters of gas in a bottle, an ignition engine or reactants in a test tube. It could be an animal, a gas factory, a mountain range, the Earth or the whole universe. A system can even be devoid of matter and consist solely of electromagnetic radiation. It is the problem under consideration that leads to the most convenient and fruitful way of defining a system. Once this is made, it is the problem considered that will dictate what are the relevant properties to be determined. Some properties such as the chemical composition, pressure, temperature, volume or density sound familiar.

Others, like the entropy, are less so, and one may need to consider physical properties like the refractive index or the dielectric constant that are important in other branches of physics. Regardless of this diversity, a basic distinction leads to split properties into two categories: *intensive properties*, like density, temperature and pressure, which are independent of the size of the system, and *extensive properties*, like volume or mass, which are proportional to the size of the system.

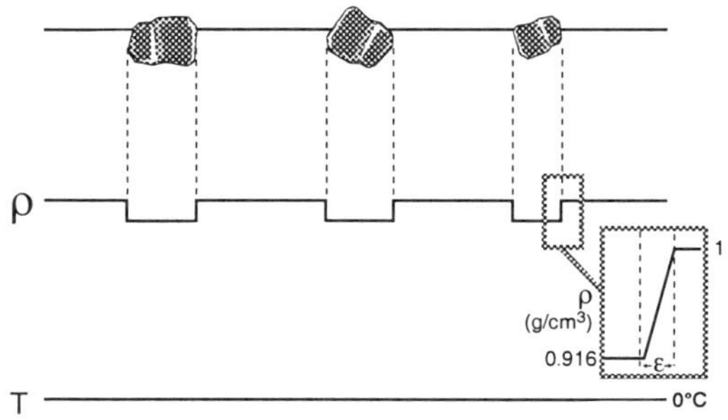

Figure 1.1. Density (ρ) and temperature profiles at the surface of a vessel containing melting ice. The interface (ε) is the zone through which the density and other physical properties vary markedly and has typically a thickness of the order of 100 angstroms.

An *homogeneous* system is made up of only one *phase*, a phase being defined as a structurally homogeneous portion of matter characterized by the same intensive properties. A phase has thus a uniform composition and its intensive properties are constant. (The pressure and density variations induced by gravity in long columns of matter will be neglected here). Like melting ice at 0°C and atmospheric pressure, an *heterogeneous* system is made up of more than one phase. Water and ice are of course two different phases but, even if present as thousands of pieces of different size, ice constitutes only one phase because all pieces have identical intensive properties. The boundary between two phases is marked by a discontinuity in at least one intensive property. The density, for instance, increases abruptly when ice melts whereas other intensive properties such as the pressure or temperature remain constant (Fig. 1.1). For many phases, the chemical composition can be varied through mixing of various components. Such phases are called *solutions*, and they can be fluid or even solid, such as alloys. Finally, we will assume here that the interface between any two phases is thin enough that the variations of intensive properties can be considered as discontinuous.

In the following we will deal only with systems in which the role of surface forces can be neglected. In general this can be done when the size of the constitutive entities of the system (grains, bubbles, etc.) is greater than a few microns. For emulsions or very finely ground solid powders, surface tension has to be accounted for. This is made through inclusion of surface energies into all the relevant energy balances.

All matter that is not part of the system (strictly speaking, the rest of the universe) constitutes the *surroundings*. A system is *isolated* when neither matter nor energy is exchanged with the surroundings. As shown by the inexorable cooling of a thermos bottle, a system cannot be completely isolated. Nevertheless, one designates as *adiabatic* any process through which heat exchange with the surroundings is so small that it can be neglected. It is easier to realize *closed* systems, which exchange no matter but only energy with the surroundings, and especially *open* systems where both matter and energy are exchanged. (Note, however, that a system is never strictly closed because no wall is totally impervious to the particles of cosmic radiation.)

1.1b STATE, STATE FUNCTIONS, TRANSFORMATIONS, ADVANCEMENT OF REACTION

The *state* of a system is characterized by the intensive and extensive properties of the constitutive phases. Two states are identical if and only if all these properties are identical. Conversely, all properties characterizing the state of the system are *state functions*. Their fundamental feature is that they depend only on the current conditions of the system, and not on its previous history. Example: let some water boil and then condense at a given temperature and pressure; the density of water then depends only on the actual pressure and temperature, and is not at all affected by the fact that water has been vaporized. The density is therefore a state function.

In other words, *the variation of a state function when the system goes from one state to another depends only on the initial and final states, and not on the actual pathway followed during the transformation.* Mathematically, this is tantamount to saying that the differentials of state functions are exact differentials (see Appendix A). By *transformation*, we denote any process leading to a change in the state of the system. A transformation can represent a slight temperature variation of an isolated system, without any change in the nature or proportion of the phases present, as well as drastic variations of all state functions in an open system. In any case, however, a transformation is correctly defined only if the initial and final sates are known.

Regardless of the physical or chemical nature of a transformation, it is often useful to describe its extent by a single parameter, the *advancement of the reaction* ξ. For convenience, this variable is defined to be 0 in the initial state and 1 in the final one. It is impossible to define ξ in a general way because it depends on the transformation considered. In the case of the *isothermal* expansion of a gas (i.e., expansion at constant temperature) between the pressures and volumes (P_0, V_0) and (P_1, V_1), for instance, one could define $\xi = (P - P_0)/(P_1 - P_0)$ or $\xi = (V - V_0)/(V_1 - V_0)$. For a chemical reaction, we will express in 7.1a ξ as a function of the numbers of moles of the reactants and products.

1.1c STATE CHARACTERIZATION, DUHEM POSTULATE

As already noted, the density of water is determined if the temperature and pressure are known. Specifying two state functions (e.g., temperature and pressure) is thus sufficient to fix a third one such as the density (Fig. 1.2). For a system with a given mass whose chemical composition is constant, experience indeed shows that two state functions only can be fixed independently. Instead of temperature and pressure, we could select entropy or volume or any other pair of state functions. All other properties could then be determined from such two parameters that would be referred to as the *independent variables*. Without waiting to define equilibrium or components, we will state in a more formal way: *regardless of the number of phases, components or chemical reactions, the equilibrium state of a closed system for which one knows the initial mass of all components is determined by only two independent variables.*

Known as the Duhem (1861-1916) postulate, this statement plays a fundamental role that must be stressed explicitly.

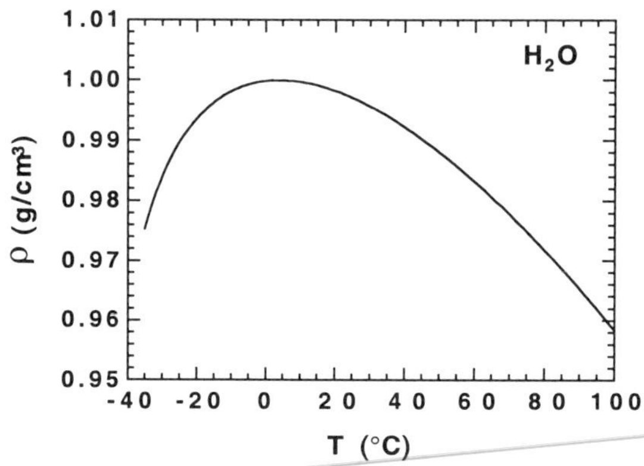

Figure 1.2. Density of water at room pressure. Note the maximum near 4°C and the rapid diminution of the density of supercooled water (i.e., water kept in the liquid state below 0°C). Because the density of water is not a single valued function of temperature, water has a thermometric interest only through the fixed points it provides. Data from Speedy (1987).

It is also worth insisting that defining the state of a system does not imply that all physical properties are known. Assume for instance that the temperature and pressure of a given mass of water are fixed. All other properties of water like the volume, entropy, heat capacity, index of refraction, viscosity, etc. are then determined in an unequivocal manner, but they are not necessarily known. They have to be determined experimentally and a feature like the density maximum of water near 4°C at room pressure (Fig. 1.2) is a peculiarity of this substance that must be determined through experimental measurements. Also note that properties like the refractive index or the viscosity cannot be related in any *exact* way to other state functions. Their investigation is thus outside the scope of classical thermodynamics.

1.2 INTERNAL ENERGY AND FIRST PRINCIPLE

1.2a TEMPERATURE AND HEAT

The "burn" caused by extreme cold illustrates how the sense of touch can be mislead. This does not prevent making a correct assessment of the difference between coldness and hotness in most situations, however, in the same way as optical illusions hardly hinder an understanding of lengths. From time immemorial, it has thus been known how to heat an initially cold body through the action of solar light, fire or of a hotter body. The actual difficulty stems instead from two different concepts, temperature and heat, being concealed behind these empirical notions. To make them explicit, we must rely on the fact that a body has differing properties depending on whether it is hot or cold. Volume expansion on heating gives a convenient means of specifying the degree of hotness or coldness reliably with *thermometers*, and it is with these instruments that the difference between temperature and heat can be clearly understood once empirical *temperature scales* have been set up.

For example, let *temperature* be the property of a system indicated by the height of mercury in a glass capillary. Experimentally, four important observations then follow. First, temperature is an intensive property since it does not change when two water masses initially at the same temperature are mixed. Second, to induce temperature changes we must add or withdraw some *heat*. Third, unlike temperature, heat is an extensive property because twice as much fuel must be burnt to produce the same temperature increase when the mass of water is doubled. (In passing, we also note that it is through heat exchange that temperatures vary; it would be absurd to assume that heat varies through temperature exchange!) Fourth, heat always flows spontaneously from a hot to a cold body until the temperatures of the two bodies become equal. This final state is called *thermal equilibrium*.

Leaving aside for the moment the question of the nature of heat, we observe that heat transfer first induces changes in the temperature of a system. Furthermore, heat results in a second kind of transformation for which H_2O gives a familiar example: when heated ice *melts* to water, and on further heating water eventually *vaporizes*. As discovered in the early 18th century by Amontons (1663-1705), the temperature of water shows the striking feature of remaining constant throughout vaporization. Such isothermal *phase transitions* allow the *fixed points* of a temperature *scale* to be defined. In the Celsius (1701-1744) scale, melting of ice and vaporization of water at room pressure correspond to 0 and 100°, respectively. With fixed points, temperatures measured with different thermometers can be compared, reproducible temperature scales can be set up over wide intervals, and conversions can be made between temperatures referred to differing temperature scales. Finally, we can check that temperature as measured with such thermometers is a state function, like the pressure or density which are also measured with suitable instruments. (For those who would be offended by this empirical approach, we note that a definition of length is a prerequisite to those of pressure and density, but that lengths are not subjected to the kind of discussions devoted to temperatures even though they cannot be measured without making reference to thermal expansion, and thus to the very concept of temperature!)

1.2b HEAT CAPACITY AND LATENT HEAT

The availability of temperature scales makes it possible to proceed to the definition of a unit of heat and to quantitative measurement of heat. Historically, the calorie was first defined arbitrarily as the amount of heat required to raise the temperature of one gram of water by one degree Celsius at room pressure. For any system, we define the *heat capacity* as the amount of heat dQ required to raise the temperature infinitesimally

$$C = dQ/dT. \tag{1.1}$$

Like heat, heat capacities are extensive variables, usually referred to one mole or a unit of mass of matter. In fact they depend on the experimental conditions. As shown in appendix A, *isobaric* heat capacities (at constant pressure, C_p) are greater than their *isochoric* counterparts (at constant volume, C_v). Furthermore, experience shows that heat capacities generally vary with temperature. Because of these effects, the calorie has been more precisely defined as the isobaric heat capacity of water in the interval 14.5 - 15.5°C.

Besides, such transformations as ice melting or water vaporization show the important fact that heating of a phase does not necessarily result in a temperature rise. As observed by Black (1728-1799) in the 18th century, a phase transition is associated with a *latent heat* and, like heat capacities, latent heats are extensive properties. One must perhaps also recall that not all such transformations are isothermal. For instance, vaporization and freezing of water no longer take place at constant T when solutes like salt have been dissolved. In such cases, part of the latent heat of transformation is spread over a finite temperature interval where it manifests itself as an anomaly in the heat capacity of the system.

Would it be possible to measure heat capacities and latent heats without knowing anything more about heat? Provided that a temperature scale and a unit of heat have been defined, the answer is yes. Let one gram of water at 14.5°C be mixed with the amount of water at 16.5°C needed to reach a final temperature of 15.5°C. If C_p were constant, one gram of water exactly would be needed. As a matter of fact, the heat capacity of water slightly diminishes when the temperature increases, so that the amount of water to be added, q, is slightly greater than one gram. Between 15.5 and 16.5°C, the heat capacity of one gram of water is thus $1/q$. It would then suffice to redo the experiment at higher or lower temperatures to determine C_p from 0 to 100°C. To measure the heat of melting of ice in a similar way, let some water at a temperature T be mixed with a given amount of ice at 0°C until the last piece of ice has melted. The heat of melting is equal to the heat released by cooling of water from T to 0°C, which is readily known from the mass of water and its heat capacity. Extending such measurements to other substances would be straightforward. This is in effect the principle of the method of mixtures, with which a number of calorimetric measurements are still performed (see 4.3a).

1.2c HEAT AS AN ENERGY: FIRST PRINCIPLE

It is time now to abandon this phenomenological approach and ponder over the real nature of heat. That heat represents the kinetic energy associated with the various kinds of atomic and molecular motions is so generally acknowledged that in retrospect it is difficult to understand the formidable problem raised by its nature. But how can one prove that heat is energy? The demonstration was by no means obvious as, at the time when Carnot (1796-1832) was laying the foundations of the second principle, heat transfer was still generally interpreted as the flow of a massless fluid, the *caloric*, which was neither created nor destroyed. The examples of the preceding section actually show that this interpretation had a sound experimental basis. As a matter of fact, the principles of calorimetry could have not been set if heat were not conserved in transformations at constant pressure involving only heat transfer. This point will be dealt with again in the next chapter.

It had long been known that heat could be produced by friction, and several kinds of caloric had to be distinguished to account for such observations. Other important observations paving the way to the rejection of the caloric theory were made by Rumford (1753-1814), who noticed that the work done to bore cannons could be estimated from the temperature rises resulting from the friction of the drills, suggesting a proportionality between the work done and the heat produced. Conversely, heat engines had long been showing how work could be produced from heat, well before new thermoelectric, electrochemical or electromagnetic effects would illustrate at the beginning of the 19th century the possibility of transformations between other known forces.

To demonstrate the equivalence of heat and work, however, it remained to be shown that the conversion factor between heat and work was constant, independent of the way transformations are performed. It was Joule (1818-1889) who eventually showed that a given amount of mechanical, gravitational or electrical energy translates into the same quantity of heat. The demonstration that heat is "energy" was then complete. Joule's experiments indicated that a calorie is worth 4.16 J, in very good agreement with the currently used conversion factor of 4.184 J. This conversion factor is called the *mechanical equivalent* of the calorie. (Although calorimeters have long been calibrated electrically, the calorie has been abandonned only in the last decades.)

The first principle extends and generalizes this conclusion by stating that energy is conserved: *for an isolated system, any decrease of a given form of energy is exactly compensated by an increase of one or several other forms.* We are so used to transforming a given form of energy into another one that the first principle probably looks obvious instead of the formidable breakthrough it actually was. In this respect, it is worth recalling that the caloric theory prevailed until the mid 19th century, and that the first principle was stated about twenty years after the second one.

As illustrated by the classic history of the neutrino, energy conservation is nowadays one of the soundest bases of physics. In 1930 Pauli assumed that an unknown particle, later called the neutrino, was emitted in order that the energy budget of β radioactivity be balanced. The existence of the neutrino was very soon acknowledged although its experimental detection was eventually made thirty years later. Historically, the first principle as stated at around 1845 by Colding (1815-1888), Mayer (1814-1878), Joule, and Helmholtz (1821-1894) in the middle of the 19th century introduced the concept of energy. Until the end of the 18th century, the word "energy" meant only the "force of a speech, of a sentence, of a word", in accordance with its Greek etymology. Hence, pre-thermodynamics physics was first the physics of forces. Before being given its modern meaning by Kelvin, energy was first introduced by Young (1773-1829) in 1807 in the case of the *vis viva* (i.e., the kinetic energy). It was the success (and the conceptual difficulty) of the first principle to unite under the same word the common consequences of apparently very different forces.

1.2d INTERNAL ENERGY

Before turning to the second principle and to entropy we must make the most of the first principle by defining a new state function, the internal energy. In this book we will generally deal with systems where surface, gravitational, electrical, magnetic or nuclear forces need not to be taken into account and where friction can be neglected. Energy will be exchanged between the system and the surroundings in the form of work, heat and matter, where work will refer only to mechanical work.

According to the first principle, a system is of constant energy when nothing is exchanged with the surroundings. We will call this energy content *internal energy* (U). For an isolated system, U is thus constant. For a transformation within a closed system, the internal energy variation is equal to the sum of the work (W) and heat (Q) exchanged with the surroundings. In the case of an infinitesimal transformation we have

$$dU = dQ + dW. \tag{1.2}$$

For an open system, the change in chemical energy associated with the exchange of matter between the system and surroundings has to be included into the energy budget (1.2). This will be made in the next chapter. In (1.2), note that the convention followed is to count as positive the energy gained by the system and negative the energy given off to the surroundings. (With this convention, heat capacities are necessarily positive. As to latent heats, they are positive when going from the low- to the high-temperature phases and negative in the other direction.)

Thanks to the equivalence of the various forms of energy, it is possible to let a system go from an initial to a final state through an infinite number of different pathways. Water, for instance, can be heated through contact with a hotter body or through friction, burning of a gas, passage of electrical current in a resistance, exposure to solar or microwave radiations, etc., and any of these mechanical, chemical, electrical or electromagnetic means can be used alone or in combination with others. Hence, knowing solely that the temperature of water rose from T_1 to T_2 at constant pressure does not give any clues as to the kinds of energy actually used. Only the total energy required for this transformation is known. This is of course the change of internal energy of water between the temperatures T_1 and T_2 at the pressure of the experiment.

More generally, consider a series of transformations through which a system exchanges various forms of energy with the surroundings. For each of them, the amount exchanged depends on the pathway actually followed but the first principle ensures that their sum is constant, regardless of any particular pathway, as long as the initial and final states are the same for all processes. Hence, the internal energy is a state function. In contrast, the heat and work exchanged are unambiguously defined in specific instances only, such as adiabatic transformations ($dQ = 0$, and thus $dW = dU$), isochoric transformations ($dW = 0$ and thus $dQ = dU$) or, as described below, isobaric transformations. In other circumstances, determining the heat (or work) exchanged requires to know beforehand the work (or heat) of the transformation along with the internal energy variation.

1.3 TENSIONS AND EXTENSITIES

In this section we will show that exchange of energy results from the transfer of extensities between different tensions. What is an *extensity*? What is a *tension*? In fact these terms represent simple concepts that are especially useful as they will allow the entropy and chemical potentials to be introduced in a natural way. To define them, we first review various forms of energy whose thermodynamic role can be usually neglected, but which have the advantage of being conceptually more familiar than thermal energy.

1.3a GRAVITATIONAL ENERGY

Consider first gravitational energy in the case of a system of two reservoirs (Fig. 1.3a) having water masses m_1 and m_2 ($m_2 < m_1$) and heights h_1 and h_2 ($h_2 > h_1$). Suppose that the gravitational energy of the low, big reservoir is smaller than that of the high, small one ($h_1g_1m_1 > h_2g_2m_2$, with g = acceleration of gravity at height h). When a sluice is opened, water flows from the high to the low reservoir, and not from the big to the small one, or from high to low energies. In other words, the relative heights determine the direction of the flow. Besides, when the high reservoir 2 loses a mass of water dm it gives off an energy

$$dE = h_2g_2dm \tag{1.3}$$

whereas the reservoir 1 similarly gains the energy h_1g_1dm.

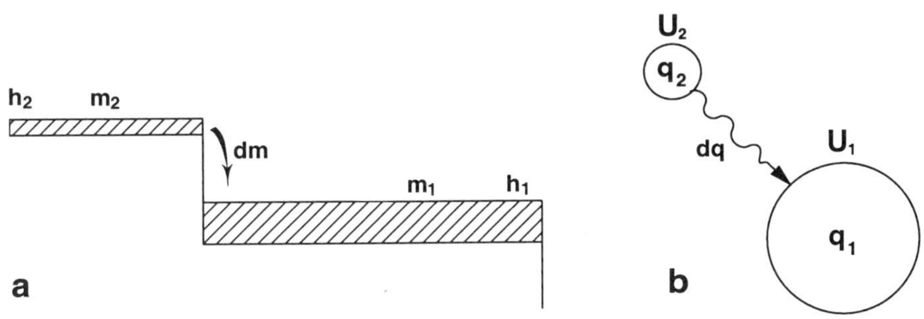

Figure 1.3. Extensity transfers for gravitational (a) and electrical energies (b)

For each reservoir, the energy change thus appears to be the product of an extensive property, the mass transferred dm, and an intensive property, actually hg (not h alone). When an infinitesimal amount of water flows from one reservoir to the other, the extensity dm is transferred from a higher tension h_2g_2 to a lower tension h_1g_1. This tension difference determines the direction of the extensity transfer and it is through this extensity transfer that gravitational energy is also exchanged. Finally, the difference between the energy variations of the two reservoirs is

$$dW = (h_2g_2 - h_1g_1)dm. \tag{1.4}$$

It represents the energy that could be transformed into work, or electrical or mechanical energy by appropriate means.

1.3b ELECTRICAL ENERGY

Consider then (Fig. 1.3b) an analogous exchange of electrical energy between spheres bearing electrical charges q_1 and q_2 ($q_1 > q_2$) brought at the potentials U_2 and U_1, respectively, with $U_2 > U_1$. Regardless of the total charge or electrical energy, electric charges move from the high to the low potential when a connection is established. The energy lost by the sphere 2 through transferring a charge dq is

$$dE = U_2\, dq, \tag{1.5}$$

and that gained by the other sphere is $U_1 dq$. Again, both variations are the product of the extensity transferred, the charge dq, and the tension of the extensity in each sphere, whereas the difference between these energy variations is the energy that could be transformed into another form if an appropriate device were inserted in the electrical circuit.

1.3c KINETIC ENERGY

Examine now the exchange of kinetic energy between two bodies moving along the same direction with different speeds. What are the tension and extensity of kinetic energy? Obviously, the speed is the tension, for kinetic energy is transferred from the faster to slower moving body. As the kinetic energy is $1/2\ mv^2$, the kinetic energy variation of each body is given by

$$dE = d(1/2\ mv^2) = mv\ dv = v\ d(mv). \tag{1.6}$$

Because the speed v is the tension, the extensity is mv, the momentum. As illustrated by the well-known example of the recoil of firearms, it is actually through exchange of momentum that kinetic energy is transferred. Note that this reasoning also holds true for relativistic conditions under which the mass would be speed dependent.

1.3d MECHANICAL ENERGY

Work is usually defined as the scalar product of a force, F, and the distance l along which the force is exerted. In thermodynamics, the relevant effects are those of the pressure and volume changes within systems. As a simple example, consider a gas cylinder divided by a frictionless piston into two compartments kept to the same temperature but initially brought to different pressures (Fig. 1.4a). When unlocked, the piston moves until the pressure becomes the same in the two compartments. The pressure is thus the tension of mechanical energy. Being transferred from one com-

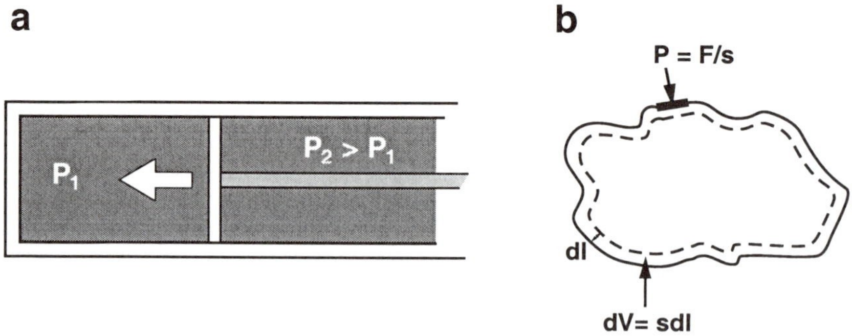

Figure 1.4. (a) Displacement (without friction) of a piston in a cylinder as a result of pressure difference ; (b) Work exchanged when some system subjected to a pressure P undergoes a volume change

partment to the other, the volume would appear to be its extensity. Contrary to other extensities, however, it would be transferred from a low to a high tension. To remove this anomaly, we will simply let $-V$, and not V, be the extensity of mechanical energy.

In a more general way (Fig. 1.4b), consider a system subjected to a hydrostatic pressure P ($= F/s$, where s is the surface). When the volume varies by a quantity $dV = sdl$ the energy changes by

$$dW = -Fdl = -PdV, \tag{1.7}$$

and we see that the "minus" sign given to the volume as the extensity of mechanical energy nicely ensures consistency with (1.7), i.e., with the sign convention used for the energy: if dV is negative, the system does gain energy. We will also emphasize that the pressure against which the system is working is that *exerted by the surroundings on the system*, and not that within the system itself. As discussed below, the difference is of considerable importance for irreversible transformations.

1.3e CONJUGATE VARIABLES

In all examples reviewed above, exchange of energy results from the transfer of an extensity between energy reservoirs having different tensions. The further the transfer proceeds, the tension of the reservoir giving off the extensity decreases whereas that of the reservoir gaining this extensity increases. The transfer ends when both tensions have equalized, defining a state of *equilibrium*. At each step of the process, energy variations are the product of the extensity transferred dx and the tension X of the reservoir

$$dE = Xdx. \tag{1.8a}$$

Like pressure and volume or speed and momentum, the tension X and the extensity x are *conjugate variables*. Alternatively, the tension can be defined through

$$X = dE/dx, \tag{1.8b}$$

and we also stress that in all preceding examples the tensions, extensities and energies are state functions.

Another noteworthy feature of conjugate variables is that tensions and extensities do not vary independently of each other. For a reservoir, for instance, there is a definite relationship between the variations of the level and those of the amount of water

$$\frac{dh}{dm} = \frac{1}{s(h)\rho(h)}, \tag{1.9}$$

where $s(h)$ and $\rho(h)$ are the surface area of the reservoir and the density of water at the height h. Denoting by $g(h)$ the gravitational constant, we find that the variations of the tension and extensity of gravitational energy are mutually related by

$$\frac{d[hg(h)]}{dm} = \frac{g(h)}{s(h)\rho(h)}. \tag{1.10}$$

Like the surface area, $d[hg(h)]/dm$ is an intrinsic property of the system, which varies when the tension $hg(h)$ changes.

Similar relationships are readily established for other energies. For kinetic energy, $dv/d(mv) = 1/m$. For electrical energy, dU/dq is a familiar property, the capacity, characterizing the way the potential of a conductor increases when it is charged electrically. For mechanical energy, the isothermal compressibility $[\beta_T = -1/V\ (\partial V/\partial P)_T]$ is introduced to account for the simultaneous variations of pressure and volume and it allows us to write $dP/dV = -1/V\ \beta_T$ at constant T. In summary, the rate of change of any tension with its extensity is an intrinsic feature of a system. Depending on the form of energy, it is either a constant or depends on the tension of the system.

1.4 ENTROPY AND SECOND PRINCIPLE

1.4a REVERSIBILITY AND IRREVERSIBILITY

We are now almost ready to discuss thermal energy in the light of tensions and extensities. Before doing so, however, a few words on reversibility and irreversibility are probably in order. Consider a gas enclosed in a cylinder to which a piston is closely fitted. If an infinitesimal excess pressure is exerted on the external side of the piston, the gas contracts infinitesimally so that pressure equalizes on both sides of the piston. Repeating the operation, we can reverse the direction of the process at any moment through opposite pressure changes. This is a *reversible* process. In the expression of reversible work associated with the volume changes of the gas, $dW_{rev} = -PdV$, the pressure is always known without any ambiguity since it is the same, to within an infinitesimal amount, on both sides of the piston. The total work done by the gas during the experiment can be obtained through integration of this expression as soon as the relevant conditions (isothermal, isobaric, etc.) have been specified so as to take into account volume changes due to possible variations of the gas temperature.

Then let us create a high vacuum on the external side of the piston. If the piston is unlocked, the gas expands violently without any possibility to interrupt the phenomenon and still less to reverse its direction. The process is clearly *irreversible*. As to the work done by the expanding gas, $dW_{irr} = -P_{ext}dV$, it is zero since the external pressure opposing the gas expansion is zero. If the experiment is eventually made under a pressure intermediate between vacuum and the initial gas pressure, the gas will again expand irreversibly. The work done is no longer zero, however, because a non zero pressure opposes the expansion. The work has in fact a value intermediate between those of the "vacuum" and reversible expansions, $-dW_{rev} < dW < 0$, but it cannot be calculated by integration of $dW = -P_{ext}dV$ because the detailed pressure-volume pathway followed by the piston is unknown.

That the work done differs in these three cases is certainly not a surprise even if the initial and final states are the same — this simply illustrates that work generally depends on the pathway of the transformation. Except in special cases like expansion in vacuum, calculation of the work done requires to know the state of the system continuously from the initial to the final state. Strictly speaking, this could be achieved only for reversible transformations that are by definition infinitely slow processes. The limiting cases toward which real transformation can tend are *quasistatic* processes, made up of a series of closely spaced equilibrium states. Even with the closest possible spacing, however, real processes have the intrinsic feature that the pathway followed between any two equilibrium states is irreversible. For a real transformation, this is why most often one must be content with specifying the initial and final states. As already noted, we have to measure the work (or heat) and calculate the heat (or work) by taking advantage of the fact that their sum is determined unambiguously if we want to know separately the heat and work exchanged.

1.4b ENTROPY AND THERMODYNAMIC TEMPERATURE

Turning to thermal energy we note that, according to common experience, the direction of heat transfer is determined by temperature alone. We will thus postulate that temperature is the tension of thermal energy. Starting from this postulate, our aim will be first to identify the extensity of this energy and link it to measurable properties. Then, we will give a firmer basis to this empirical exercise by means of the second principle.

Let a hot body liberate an amount of heat dQ from a certain temperature T. By analogy with the expressions obtained for other kinds of energy, we write the change of thermal energy in the same form as (1.8a)

$$dQ = X dx. \tag{1.11}$$

Now we have just to replace in this equation the tension X by the temperature and to call *entropy* its conjugate extensity, which will be denoted by S

$$dQ = T dS. \tag{1.12}$$

Knowing only that a system exchanges a given energy does not tell us anything about the amount of extensity transferred or about its tension. Because the same amount of heat can be exchanged in a great many different ways, a complete account of the process requires specifying separately both the entropy transferred dS and its tension T. Hence, the next step is to relate the entropy as defined by (1.12) to measurable properties. For this purpose, let us rewrite this equation as

$$dS = dQ/T. \tag{1.13}$$

Entropies appear to have the dimension of an energy:temperature ratio, i.e., they are expressed in J/K. When the transformation is isothermal, the entropy variation due to the exchange of a latent heat Q is simply

$$\Delta S = Q/T. \tag{1.14}$$

Without isothermal phase transitions, the entropy variation associated with a temperature change is obtained from heat capacities. Combining (1.1) and (1.13), we get

$$dS = C/T\, dT. \tag{1.15a}$$

Specifically, we have at constant pressure

$$dS = C_p/T\, dT, \qquad C_p = T(\partial S/\partial T)_P, \tag{1.15b}$$

and at constant volume

$$dS = C_v/T\, dT, \qquad C_v = T(\partial S/\partial T)_V. \tag{1.15c}$$

Hence, entropy is not as mysterious as it might seem because equations (1.14) and (1.15) simply relate its variations to latent heats or heat capacities, which are both readily measurable thermal properties.

According to the relations (1.15), the proportionality constant between the variations of the tension and extensity of thermal energy is the ratio C/T, which has sometimes been called *encraty*. As already mentioned, heat capacities (at constant P or V, for instance) are specific functions of the system and generally vary with temperature. As a consequence, the ratios C/T are also temperature-dependent, specific functions of the system. In this respect, they conform to the other proportionality constants between the mutual variations of conjugate tensions and extensities for other forms of energy.

At this point an important remark must be made concerning temperatures. With the empirical temperature scales that have hitherto been mentioned it appears that, as defined by (1.13), the entropy depends in a nontrivial manner on the scale used. In particular, scales like the Celsius scale are inconsistent with (1.13) since they cause this equation to diverge at their zeros which have been arbitrarily selected. In order that the entropy keeps a finite value, in (1.13) temperatures must always remain positive (or negative). Hence, the origin of a *thermodynamic* temperature scale cannot be chosen arbitrarily. This choice is described in section 5.1a. Here, we just recall that the scale used has as a unit the Kelvin (K) and is related to the Celsius scale by the linear relationship

$$T(\mathrm{K}) = T(^\circ\mathrm{C}) + 273.15. \tag{1.16}$$

For completeness, let us consider the entropy of a phase undergoing a series of phase transitions as a function of temperature. Let Q_i be the latent heat at a transition temperature T_i and C_{pi} (or C_{vi}) the isobaric (or isochoric) heat capacity of the phase between the i and $i+1$ transitions. Combining (1.14) and (1.15), we calculate any entropy increase between 0 K and a temperature T with

$$S(T) - S(0) = \int_0^{T_1} \frac{C_0}{T}\, dT + \frac{Q_1}{T_1} + \ldots + \int_{T_i}^{T_{i+1}} \frac{C_i}{T}\, dT + \frac{Q_{i+1}}{T_{i+1}} + \ldots \tag{1.17}$$

Using (1.17) to calculate entropy changes for heterogeneous systems is straightforward since the entropy is an extensive property: as such, it is the sum of the entropies of the constituting phases. From latent heats and heat capacities, however, we calculate *relative entropies*, not entropies at 0 K. What can be said about these entropies at 0 K? We will see in 4.4 that they can often be considered to be zero and describe the way *residual entropies*, as they are called when they are not zero, can be determined experimentally.

The word entropy was coined by Clausius (1822-1888) after a term meaning "turning back", an apparently deceptive word whose origin will be clearer when it will be seen that this property is conserved throughout Carnot cycles, see 5.2g. Defining the entropy by way of (1.13) has the other important consequence that heat capacities must also tend toward 0 when the temperature tends toward 0 K in order to prevent any divergence of (1.15). Such a constraint is inconsistent with classical mechanics where the continuous variations of energy levels of atomic systems result in temperature independent isochoric heat capacities (see 14.3c). Fifty years after entropy had been defined, it was one of Einstein's first successes to explain the marked decrease of heat capacities at low temperatures as a consequence of the quantization of atomic vibrational energy levels. As to the ratio C/T, one observes that it also tends toward 0 at the lowest temperatures although finite values of the entropy could nevertheless be achieved with certain kinds of divergence of C/T.

1.4c SECOND PRINCIPLE

The second principle will be stated in a simple form that generalizes, with one important restriction, the observations made for the other forms of energy: *the entropy is a state function; for reversible transformations, the entropy change of a closed system is*

$$dS = dQ/T, \tag{1.18}$$

where T *is the thermodynamic temperature of the system, and* Q *the heat exchanged with the surroundings*. Corollary: *For a reversible transformation, the entropy of an isolated system remains constant; for an irreversible transformation, it increases.*

This statement of the second principle first ensures that the entropy is a state function. This has a very important consequence: *to calculate the entropy change of a system going from a given initial to a given final state, following an irreversible pathway, it suffices to imagine a series of reversible steps, leading from the same initial state to the same final state, along which entropy variations will be given at any point by* $dS = dQ/T$.

To illustrate irreversible heat exchange, let us consider some water at room pressure in a thermostat rapidly cooled down to -10°C and kept at this temperature. Bypassing ice crystallization, water *supercools* for a certain period of time. When crystallization eventually begins the reaction proceeds very quickly until completion. The only way of terminating the process is to supply the system with a finite amount of heat in order to raise the temperature to 0°C. There is no possibility of controlling ice formation at -10°C. This process is irreversible. To describe how heat could be exchanged reversibly, let us first heat the ice up to 0°C and then give as a supplement an infinitesimal amount of heat. An infinitesimal amount of ice will melt. If the latter amount of heat is withdrawn from the system, the same amount of ice will form. For the ice-water system, the transformation of one phase into the other is reversible because its direction can be changed at will through infinitesimal exchanges of heat. In practice, however, these conditions cannot be produced since heat propagation requires a finite temperature gradient, in which case the transfer of heat is irreversible and some entropy is created, see below. To imagine a reversible heat transfer, one must postulate that the temperature difference between the system and the thermostat remains infinitely small in order to neglect the resulting entropy creation. Under such circumstances, the transfer would obviously be infinitely slow.

What is the difference between a reversible and an irreversible transformation if the entropy change of the system is the same in both cases? The answer is that the entropy change is the same for the system only, not for the surroundings. Imagine that a system is reversibly given some heat by a thermostat with which it is in thermal equilibrium. As shown in the next section, entropy is then simply transferred from the thermostat to the system and the entropy of the universe remains constant. Let us bring this system irreversibly to the same final state, for instance by heating it electrically. Then, the entropy of the surroundings will not change because no heat has been exchanged with the system; in contrast, that of the system has increased by the same amount as for the reversible transformation, and this increase represents a net creation of entropy within the system in the course of the irreversible transformation. For the surroundings, the final state and the entropy change depend on the nature of the transformation. The entropy change is zero for irreversible heating and opposite in magnitude to that of the system for reversible heating. For the system we impose by definition the same initial and final states: for it, it does not matter that the transformation be reversible or not since its entropy change does not depend on the transformation pathway.

We will conclude this section with two remarks. Because of the definitions (1.15), heat capacities are state functions like the entropy. As the integral of a state function is also a state function, the heat exchanged does not depend on the pathway for isochoric and isobaric transformations where they are given by $\int C_v dT$ and $\int C_p dT$. For an isochoric transformation, this result is not a surprise since the heat exchanged is equal to the internal energy variation. For an isobaric transformation, in contrast, this result is new and we will call *enthalpy* (H) the variable whose partial differential at constant pressure is

$$dH = C_p dT = T dS. \tag{1.19}$$

Enthalpy will be introduced in a physical way in the second chapter where some of its important properties will also be discussed.

1.4d HEAT AND ENTROPY: TWO EXAMPLES

As a first example of entropy calculations for actual processes, let us take a system made up of one mole of water crystallizing in a thermostat kept at -10°C (263.15 K) at room pressure. Because of the irreversibility of the process, we must imagine a series of transformations leading reversibly from the same initial state to the same final state:

1. Supercooled water is heated from 263 to 273.15 K (hereafter noted 263 and 273 K for short). The resulting entropy variation depends only on the temperature change, not on the manner in which heat is supplied. Hence, heat can be assumed to be transferred reversibly, a process whose entropy variation is given by (1.19). With an isobaric heat capacity of 76.4 J/mol K for supercooled water in this temperature interval, the entropy change is $\Delta S_1 =$ 76.4 ln $(273/263) = +$ 2.85 J/mol K.

2. At 273 K, ice formation is reversible. With (1.14) and a heat of crystallization of - 6.008 kJ/mol, the entropy of crystallization at 273 K is found to be ΔS_2 = - 22.00 J/mol K.

3. Ice is finally cooled reversibly from 273 to 263 K. With a heat capacity of 37.3 J/mol K, the entropy change is ΔS_3 = - 1.39 J/mol K.

For the actual irreversible process at 263 K the entropy change is the sum of these three contributions, namely, ΔS = - 20.54 J/mol K. For comparison, let us calculate in a similar way the heat (enthalpy) of crystallization at 263 K from the above mentioned latent heat and heat capacities. Heating of water from 263 to 273 K, crystallization at 273 K and cooling of ice from 273 to 263 K are associated with enthalpy changes of 0.765, - 6.008 and - 0.373 kJ/mol, respectively, for the system. For the thermostat, the total enthalpy supplied is thus 5.616 kJ/mol and the entropy change is $Q/T = 21.34$ J/mol K. Summing the entropy changes of the system and thermostat, we find an excess entropy of 0.80 J/mol K. This represents the entropy created owing to the irreversible nature of the process and given off to the thermostat.

As a second example, consider the isothermal expansion of one mole of an ideal gas from an initial volume V_1 to a final volume V_2. The three aforementioned cases will be dealt with successively: reversible expansion, irreversible expansion in a vacuum and expansion against a low but non zero pressure.

An ideal gas is characterized by the well-known relationship $PV = RT$, where R is the gas constant. Let us accept for the moment the result demonstrated in 5.2b that the internal energy of an ideal gas depends only on temperature, not on pressure or volume. Hence, we can write $dU = dQ + dW = 0$ for isothermal expansion.

1. *Reversible expansion.* The work done by the system is

$$W = -\int_{V_1}^{V_2} PdV = -\int_{V_1}^{V_2} RT/V\ dV = RT\ \ln\,(V_1/V_2). \tag{1.20}$$

The heat exchanged is $Q = -W$ since the internal energy of the gas is constant. For a reversible, isothermal expansion, the entropy change of the gas is $\Delta S = Q/T = R\ \ln(V_2/V_1)$. The heat has been supplied by the surroundings whose entropy has decreased by an amount $\Delta S = -Q/T = -R\ \ln(V_2/V_1)$. Like for any reversible process, these two entropy variations exactly cancel out, and the entropy of the universe does not change.

2. *Expansion in a vacuum.* As the work done by the gas is zero, no heat is exchanged with the surroundings ($dU = dQ + dW = 0$). The entropy change of the gas is the same as for the reversible case, however, because the initial and final states are the same as for the actual irreversible transformation. Without any heat exchanged, the entropy of the surroundings does not vary at all. Because the entropy change of the gas represents a net creation of entropy, the entropy of the universe increases by $\Delta S = R\ \ln\,(V_2/V_1)$.

3. *Expansion against a lower pressure.* Again, the work done by the gas cannot be known. We can just state that $-W_{rev} < W < 0$. Although an analogous inequality holds for the heat received by the system ($Q_{rev} < Q < 0$), the entropy change of the system is unambiguously known as $\Delta S = R \ln\,(V_2/V_1)$. In contrast, that of the surroundings as given by $-Q/T$ cannot be calculated because the heat actually exchanged is not known. This variation ranges from $-R\ \ln\,(V_2/V_1)$ to 0, whereas the entropy increase of the universe is similarly bounded by 0 and $R\ \ln\,(V_2/V_1)$.

In summary, the entropy variation of the system does not depend on whether the expansion is reversible or not as long as the initial and final states are the same. The difference between the two processes affects only the surroundings. When a gas expands reversibly, it borrows some entropy from the surroundings and causes a heat transfer (which exactly compensates the work done by the gas and allows the temperature to remain constant). For an irreversible expansion in a vacuum, no heat and work are exchanged. The state of the surroundings does not vary and, as stated by the second principle, the entropy increase represents a created entropy. The most frequent case is in reality that represented by the expansion against a pressure lower than that of the gas. As the (P,T) pathway is unknown, the work, heat and entropy exchanged are not known and neither is the final state of the surroundings. This points to an essential difference between a system and its surroundings: the state of the former is necessarily known whereas in general that of the latter cannot be specified.

1.5 ENTROPY AND THERMAL ENERGY

1.5a ENTROPY AS A NON-CONSERVATIVE EXTENSITY

At this point it is appropriate to turn to the very basic process through which entropy is created whenever heat is exchanged. For this purpose, consider the simplest case of two bodies with initially different temperatures T_1 and T_2 ($T_2 > T_1$) that are brought into contact. Because of the first principle, the energy change is zero when heat flows from the hot to the cold body

$$dQ = dQ_1 + dQ_2 = 0. \tag{1.21}$$

As the entropy of the system is the sum of the entropy of the two bodies, the entropy change is

$$dS = dS_1 + dS_2. \tag{1.22}$$

Let the volume or pressure be constant. Under these conditions, $dQ = dU$ or $dQ = dH$ and, regardless of the process, the heat exchanged is known unambiguously. Furthermore, with respect to entropy, that heat is exchanged reversibly or irreversibly does not matter as long as the initial and final states are not affected. From (1.13) and (1.21), we can write

$$dS = \frac{dQ_1}{T_1} + \frac{dQ_2}{T_2} = \frac{dQ_1}{T_1} - \frac{dQ_1}{T_2} = \frac{(T_2 - T_1)dQ_1}{T_1 T_2}. \tag{1.23}$$

The heat received by the cold body is counted as positive ($dQ_1 > 0$). As $T_2 > T_1$, dS is positive and not zero: the entropy of the system thus increases when heat is transferred from the hot to the cold body. To ensure $dS = 0$ the temperatures of the two bodies should be equal, in which case no heat would of course be exchanged. As we have implicitly assumed that the two bodies were forming an isolated system, the entropy of the surroundings does not change. It follows that the entropy of the universe increases by the quantity indicated by (1.23). In practice, heat transfer always results from finite temperature differences. As such, it is an intrinsically irreversible process necessarily causing an entropy increase in the universe.

Now, the transfer of water from one reservoir to another is associated neither with the creation nor with the destruction of matter. Like mass, electrical charge, momentum or volume are conserved throughout any process. For gravitational, electrical, kinetic and mechanical energies, all these extensities are thus *conservative*. Introduced as the extensity of thermal energy, the entropy has therefore the fundamental feature of being non conservative. Here the analogy of entropy with other extensities breaks down. Before commenting on this peculiarity of thermal energy due to entropy creation, it is worth emphasizing a few salient points pertaining to the conversion of heat into work.

1.5b CONVERSION OF HEAT INTO WORK

To transform heat into work (Fig. 1.5) one needs a *hot source* to heat a medium from a temperature T_1 to T_2 (usually water, from room temperature at least to much higher temperatures). Work is produced when the medium expands and drives a turbine, after which heat is eventually released by the medium to a *cold source* at a temperature T_1 (through discharge of the still warm vapor, or condensation of it before repeating the cycle). Let dQ_2 be the heat needed to bring from T_1 to T_2 an infinitesimal part of the medium, dW the work done, and dQ_1 the heat released at temperature T_1 to the cold source. According to the first principle, the energy balance is

$$dW + dQ_1 + dQ_2 = 0. \tag{1.24}$$

If the engine were working reversibly, there would not be any entropy creation. In this case

$$dS_1 = -dS_2. \tag{1.25}$$

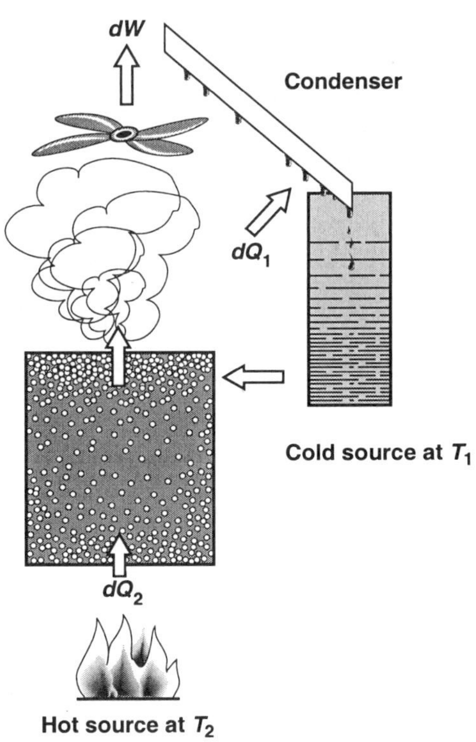

Figure 1.5. Scheme of operation of a heat engine transforming heat into work.

By combining (1.24) and (1.25) we should not be surprised to find that

$$dW = T_1 dS_1 + T_2 dS_2 = dS_2(T_2 - T_1). \tag{1.26}$$

In other words, the work done is as usual the extensity transferred dS times $T_2 - T_1$, the drop of the tension between the two heat reservoirs of the engine. In this respect, conversion of heat into work is strictly similar to the mutual conversion of other forms of energy. Likewise, in this reversible case the thermal energy exchanged is entirely converted into work.

As to the thermodynamic efficiency of the heat engine, it is given by the ratio between the net work done dW and the heat received dQ_2. According to the sign convention, dW is negative and dQ_2 positive. The efficiency r is thus

$$r = |dW/dQ_2| = -dW/dQ_2 = 1 + dQ_1/dQ_2. \tag{1.27}$$

This allows us to rewrite (1.27) in the form

$$r = 1 + T_1 dS_1/T_2 dS_2, \tag{1.28}$$

and from (1.25) we have finally

$$r = 1 - T_1/T_2. \tag{1.29a}$$

This is the well-known result that the maximum efficiency of the engine is determined by the ratio of the thermodynamic temperatures of the cold and hot sources. For an engine operated between 300 and 500 K, the efficiency would be at best 40%, and it could approach 100% only with a cold source close to 0 K or with a hot source at extremely high temperatures. For a real, irreversible engine, the efficiency would be still lower because the creation of entropy implies $dS_1 > -dS_2$.

A point worthy of attention is that the mediocre efficiency given by (1.29a) has nothing to do with irreversibility for the very reason that we have assumed the engine to be working reversibly. If given instead a work dW, such an engine could take a heat dQ_1 from the cold source and give dQ_2 to the hot source in order to restore the entropy transferred to its initial tension. This is the working principle of the refrigerator. In this way, one would go back reversibly to the initial state, in accordance with (1.25) which dictates that no entropy is created. If a heat engine must give off some heat to a cold source, it is so because it operates as any device converting a given form of energy into another one. Any energy conversion actually results from the transfer of an extensity between different tensions: mass between two reservoirs, electrical charge between two potentials, etc., and it would probably more explicit to state that the heat engine takes entropy, and not heat, from the hot source and releases it to the cold source.

More important is the fact that, contrary to a commonly made statement, the thermodynamic efficiency as given by (1.29a) is not a peculiarity of thermal energy. The maximum efficiency of any device transforming a given form of energy into another is determined solely by the ratio of the high and low tensions between which the extensity is transferred. With regard to water transfer between two reservoirs, for instance, one converts only a tiny fraction of the gravitational energy that could be tapped if there were a means of letting matter fall down to the center of the earth! The maximum efficiency is indeed

$$r = 1 - h_1 g_1 / h_2 g_2, \tag{1.29b}$$

the analog of (1.29a), as (1.26) is the analog of the equations describing the mutual conversion of other forms of energy discussed in 1.3. We thus stress again that a low thermodynamic efficiency is not indicative of irreversibility. Without the existence of coupled reservoirs, the tension of an electrical charge could not drop, the level of a mass of water could not decrease and heat could not flow out of a hot body to be transformed into another form of energy.

If neither the need for two heat sources nor the thermodynamic efficiency of conversion of heat into work are special in any respect, then what is the peculiarity of thermal energy? As already pointed out, it is the fact that its extensity is non conservative. Some entropy can thus be created without being borrowed from a preexisting reservoir. If the tension of an electrical charge dq drops spontaneously, without any attempt at recovering the electrical energy released, a thermal energy TdS is produced and the extensity dS is entirely created at the temperature T of the experiment. The same applies for any other transformation, for instance when a river flows and the gravitational energy of its waters is dissipated into heat.

In fact, the conversion efficiency (1.29a) is generally mediocre only from an implicit financial standpoint because one takes into account in the balance sheet the operating cost of the device in which heat is produced. The irreversibility does not originate so much in the process of transforming heat into work (which would be optimal if the engine were working reversibly) — it mainly results from the conversion of chemical (or nuclear) energy into heat. A much higher overall efficiency would be obtained through a direct transformation of these energies into electrical energy, for instance, which would bypass the intermediate heat production. (Or, more precisely, which would limit the amount of heat inevitably produced throughout any real process.) If the heat dQ_2 received by the engine had originally been mechanical energy, then only the small fraction$(T_2 - T_1)/T_2$ of it could be retransformed into work. This is another way of saying that the low thermodynamic efficiency of heat-work conversion is essentially due to the irreversible production of heat in the engine, and not to the conversion itself of heat into work.

1.5c THERMAL ENERGY AS A DEGRADED FORM OF ENERGY

Why is entropy created in the course of any real process, whereas other extensities like mass or electrical charge are conserved? The fact that the entropy must be made conspicuous through indirect means suggests that the answer has to be found at a microscopic, and not a macroscopic level. Before proceeding any further, however, we note that the very existence of stability in the universe shows that although all forms of energy are equal, as indicated by the first principle, one if them is "more equal than the others", as substantiated by the second principle.

Consider a ball dropped from a certain height in a vacuum. According to the first principle, its gravitational energy does not vanish during the fall; it is first transformed into kinetic energy, and then into elastic energy when the ground is impacted. If these successive transformations were total, the elastic energy would be converted back into kinetic energy, and then into gravitational energy and the whole process would repeat itself until the end of time. As a decrease in a given kind of energy is exactly offset by an increase in one or several other forms, conversion processes would run forever if all extensities were conservative. The universe would be pervaded with perpetual motions of the second kind (i.e., perpetual, but not producing more energy than being used up) during which the various energies would vary in an oscillatory way, as would the gravitational energy of a never-ending rebounding ball.

Now, experience shows that the ball eventually stands still after a limited number of rebounds. In the absence of any outside intervention to lock the system in a given state, by itself this fact points to the existence of a non conservative extensity. It is of course that of thermal energy. For the ball, each rebound is lower than the preceding one because part of kinetic energy is not transformed into elastic energy, but into heat. According to a well-known saying, heat is the degraded form of energy. As already noted, it is heat that is inevitably produced in all the cases discussed in section 1.3 if no device is used to convert into another form the energy change associated with the transfer of an extensity between two reservoirs.

1.5d MICROSCOPIC SKETCH

Let us first sketch a microscopic interpretation of heat production and entropy from the above example of the falling ball. As a simplifying feature, let us assume that both the ball and the ground are made of the same crystalline material and that there is initially a point contact between them (Fig. 1.6). The elastic energy then represents inter-atomic potential energy due to the fact that the atoms of the ball and of the ground are pushed back toward their closest neighbors. Owing to these interatomic forces, however, these neighbors in turn will be displaced toward their own neighbors, and so on, till exhaustion of the kinetic energy supplied by the ball. The relevant feature of these displacements is that their initially uniaxial directionality is progressively lost as a result of the three-dimensional nature of interatomic bonding. When the ball rebounds, only the energy associated with the vertical component of the displacements can be transformed back into kinetic energy. The rest is the fraction irreversibly converted into heat in the ball and the ground, which causes entropy to be created. As illus-

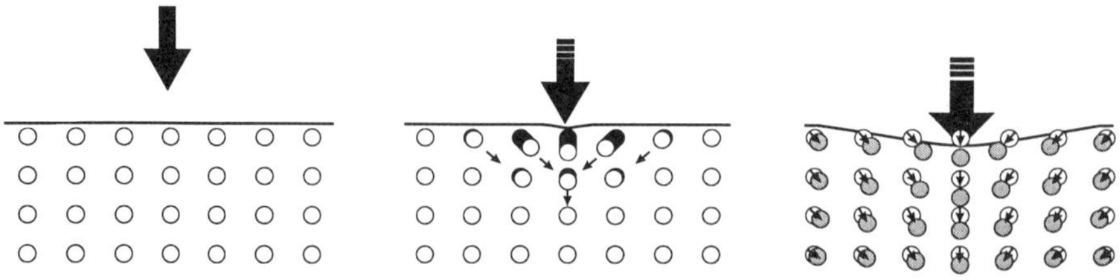

Figure 1.6. First step of the transformation of kinetic into thermal energy when a ball rebounds on a solid ground. The irreversilbility is due to the loss of the original directionality of kinetic energy.

trated by the most familiar example of friction, such an entropy creation is inevitable whenever forces are exerted between particles.

We have assumed a crystalline atomic arrangement for the ball and the ground. In such a case, a cannonball falling on a steel plate exemplifies extremely poor rebound and thus the quasi-instantaneous conversion of kinetic energy into heat. A high rebound is in contrast obtained with air balloons or rubber balls. In a balloon, the lack of cohesion allows the gas molecules to be compressed in a relatively uniaxial way, and to limit lateral dissipation of energy. For rubber balls, the high rebound is due to the existence of long molecular chains which deform elastically in the direction of the ball displacement.

The damped rebounds of a ball illustrate the general mechanisms through which heat is produced from another form of energy. In a more subtle way, this example also throws some light on the microscopic significance of entropy. The creation of entropy due to irreversible heat production results from a profound modification of the manner in which energy is distributed within the system. Initially concentrated along a vertical direction, energy appears to be progressively redistributed along other directions. The transformation is complete when atomic vibrations have equally distributed this energy throughout all directions of space. The irreversibility of entropy creation is due to the impossibility of restoring the original unidirectional energy distribution.

We draw an analogous conclusion when examining the creation of entropy in an gas expanding irreversibly at constant temperature. In this case, the difference is that a distribution of matter has to be dealt with. After having expanded, the gas molecules will not spontaneously concentrate into the fraction of space they were occupying initially. Likewise, two different gases or liquids mixed at constant total volume occupy the volume initially reserved to the other substance. By doing so, their distribution in space widens and the entropy increases irreversibly. We will finally point out that it is

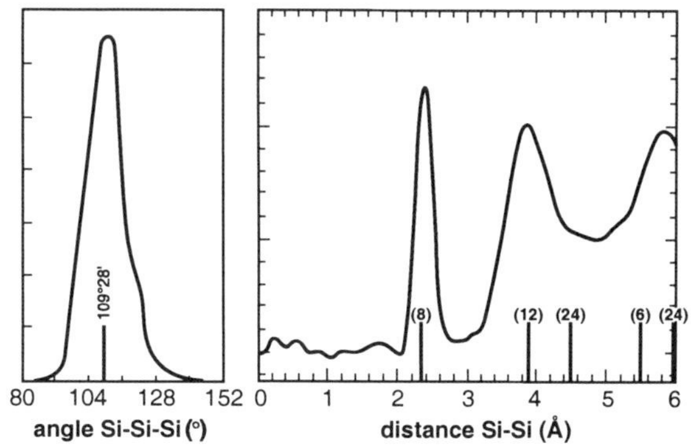

Figure 1.7. Histogram of distribution for the angles Si-Si-Si and the distances Si-Si in amorphous silicon films (Fortner and Lannin, 1989). The discrete values of these parameters for crystalline Si are included as vertical bars, the digits in parentheses indicating the numbers of atoms found from a given atom at the represented distance. In the crystalline state, silicon has the same tetrahedral structure as diamond (see Fig. 5.15).

sufficient to replace in this reasoning the word "matter" by "energy" to figure out the entropy increase caused by heat transfer between a hot and a cold body: in the same way as two gases do not spontaneously unmix, the thermal energy cannot partition preferentially into part of a system to restore a temperature difference within it.

For the time being, we will propose in a very qualitative manner that entropy is a measure of the way in which both matter and energy are distributed in space. At this point it is useful to define a *configuration* as designating any microscopic arrangement of matter consistent with a given macroscopic state of the system. We can then state that these energetic and configurational aspects of entropy are in general intricately linked, for distributions of matter depend on temperature. As will be shown later, however, this does not prevent separate identification of these two aspects and to describe them quantitatively in terms of *thermal* and *configurational* entropies, respectively.

As a further illustration of the importance of configurational entropy, let us compare the crystalline form and an amorphous phase of the same substance with the case of silicon (Fig. 1.7). In a perfect crystal each kind of atom occupies one or several well-defined crystallographic sites. For each site, the interatomic distances and the bond angles between nearest neighbors take on characteristic values. Without any distribution of these values in a crystal there is no configurational entropy. In contrast, an amorphous material is deprived of long-range order. Since the local environment varies from one place to another for each kind of atom, interatomic distances and bond angles do not have discrete values but show broad distributions which most often widen progressively when the temperature increases. Configurational entropy is therefore a fundamental feature of amorphous phases (liquids, glasses, etc.).

Problems

See at the end of chapter 2.

2
ENERGIES AND EVOLUTION CRITERIA

The second principle accounts for the stability of systems, but its most natural statement in terms of entropy increases for isolated systems is too general to be practically useful. This leads to the definition of thermodynamic potentials, such as the Gibbs free energy, whose variations dictate the feasibility of transformations under given specific conditions. These potentials will form the main theme dealt with in this chapter. In preamble, the role of changes in chemical composition, due either to chemical reactions within the system or to exchange of matter with the surroundings, will be considered in order to arrive at the most general framework for the evolution criteria that will be presented. Finally, we will discuss the various ways the state of a system can be stable, paying particular attention to the notion of metastability.

2.1 INTERNAL ENERGY OF AN OPEN SYSTEM

2.1a INTERNAL ENERGY AND COMPOSITION

Consider a system made up of one mole of methane and two moles of oxygen. With a small spark or an adequate catalyst, let us start the combustion of methane according to the familiar reaction

$$CH_4 + 2\,O_2 \rightleftharpoons CO_2 + 2\,H_2O. \tag{2.1}$$

What is the variation of the internal energy of the system if the combustion of methane is complete?

In fact, this question cannot be given any answer as long as the nature of the system has not been specified. For an isolated system, there is no exchange of matter or energy with the surroundings so that the internal energy remains constant. The chemical energy liberated by the combustion is transformed into heat but, since the volume is constant, the resulting temperature rise is associated with a pressure increase. Thermal and mechanical energies thus increase at the expense of chemical energy while the internal energy does not change.

If the combustion takes place in a closed system, the internal energy variation is the sum of the work and heat exchanged with the surroundings. This was the case considered in the first chapter, where it was not necessary to make the composition dependence of the internal energy explicit because we were not interested in the purely chemical effects of transformations. To address these effects now, we must consider not only the changes in internal energy due to chemical reactions, but also those resulting from exchange of matter with the surroundings. Let us imagine that a fraction of the gases produced through the combustion of methane is removed from the system. Thus, to establish the energy balance of the transformation, we also need to know the way in which the internal energy of an open system depends on chemical composition. This will be the first point treated in this chapter.

2.1b COMPONENTS OF A SYSTEM

Various kinds of variables can be used to describe the chemical composition of a system (see 4.1a). Here we will use the n_i numbers of moles of the N components. By *component*, we mean any chemical entity contributing to the constitution of the phases of a system. A mixture of methane and oxygen can be expressed in terms of the number of moles of methane and oxygen, but also as a function of the number of moles of elemental carbon, hydrogen and oxygen. The only constraint exerted concerns the number of *independent components*, namely, the smallest number of components required to describe the composition of the system. With elements, for instance, there are only two independent components, C and O, or H and O, because the CH_4 formula dictates that the abundances of carbon and hydrogen satisfy a 1:4 ratio.

Likewise, one usually considers the single component H_2O for water. It would be possible to chose instead H and O atoms, or H^+ and OH^- ions. To yield the H_2O stoichiometry, however, the abundances of the H and O components must be in a 2:1 ratio and those of H^+ and OH^- ions in a 1:1 ratio. In both cases the abundance of one component automatically determines that of the other. If we wanted to use a variety of subatomic entities as components, the same kind of constraints would reduce the number of independent components of water to one each time. Hence, for reasons of simplicity one usually considers water to be formed of the single component H_2O.

2.1c CHEMICAL POTENTIALS

In the first chapter we have reviewed the internal energy variations due to the exchange of entropy (and heat), and of volume (and work). To examine the influence of chemical composition, let us modify the state of a system by changing the abundance of only one of its components. As long as the variation dn_i of the number of moles of this component i is infinitesimal, we can write a proportionality relationship between dn_i and the internal energy variation

$$dU = \mu_i dn_i, \tag{2.2}$$

where μ_i denotes the proportionality constant. Caused by a change in chemical composition, the variation of internal energy is as usual the product of two terms. The latter, dn_i, is an extensity. The former, μ_i, must therefore be a tension, that of chemical energy associated with the component i of the system. For chemical energy, this tension is the analog of temperature for thermal energy or of pressure for mechanical energy, whence its name of *chemical potential*.

For simultaneous, infinitesimal changes of the numbers of moles of all components, the change in internal energy is the sum of the N individual variations (2.2)

$$dU = \sum \mu_i dn_i. \tag{2.3}$$

Because atomic interactions are specific, the chemical potentials of the various components are different and vary with the composition of the system. This complexity is at the roots of the practical difficulty of chemical potential determinations. Above all, it has the fundamental consequence that the chemical potential of a given component should not be the same in different systems. When two systems are gathered together, *chemical equilibrium* will be reached only if components are exchanged between the phases or, alternatively, if phase changes take place in order that the chemical potential of every component becomes constant throughout the whole new system.

In the more general case of simultaneous exchanges of heat, work and matter, the internal energy variation of a system is given by

$$dU = TdS - PdV + \sum \mu_i dn_i. \tag{2.4}$$

As a differential expression for the internal energy, (2.4) appears to be an explicit function of the variables S, V and n_i. These are therefore the natural variables of which the internal energy is a function. On the other hand, this equation could seem to hold true for reversible transformations only. Actually (2.4) is also valid for irreversible processes but justifying it would represent here too long a digression. This demonstration is thus deferred to section 2.3a.

2.1d INTERNAL ENERGY

Let us take advantage of the presence in equation (2.4) of intensive (T, P and μ_i) and extensive (S, V and dn_i) properties to calculate the internal energy variation of a system due to a change in its size, without any modification of the relative proportions of its components. By definition, none of the intensive properties change whereas all extensive properties vary by the same factor x as the size of the system. Let U_1 and U_2 denote the internal energy of the system before and after the size change. From the fact that $U_2 = xU_1$, it follows that

$$U_2 - U_1 = xU_1 - U_1 = (x-1)U_1. \tag{2.5}$$

With this equation and the analogous expressions obtained for the volume, entropy and the number of moles ($V_2 = xV_1$, etc.), we obtain from (2.4)

$$dU = (x-1)U_1 = T(x-1)S_1 - P(x-1)V_1 + \sum \mu_i (x-1)n_i. \tag{2.6}$$

By dividing both sides of this equation by x - 1, we derive the integrated expression of the internal energy

$$U = TS - PV + \sum n_i \mu_i. \tag{2.7}$$

In this expression, the absolute values of the variables T, P, V, and n_i are known, but not those of the entropy and chemical potentials whose variations can only be determined (see 4.4). Only changes in internal energy between different states can thus be determined.

One can arrive directly at (2.7) by noting that extensive variables are homogeneous functions of first order of the numbers of moles of the components (see appendix A). Besides, note that the absolute value of the internal energy could be derived from the relativistic equation between mass and energy, $E = mc^2$, where c is the speed of light. For one gram of matter, this energy would be about $9\ 10^{13}$ J, but the energy changes involved in the physical or chemical transformations considered here are so small with respect to numbers of this magnitude that they cannot be calculated from relativistic theory with the practically needed accuracy.

Any energy represents the product of an extensity and a tension, $E = xX$, as exemplified by gravitational ($E = mgh$), electrostatic ($E = qV$) or kinetic ($E = 1/2\ [mv]v$) energies. In equation (2.7) the same applies to the terms TS, $-PV$, and $\sum n_i \mu_i$, each representing the product of an extensity and its conjugate tension. The internal energy thus appears as the sum of three terms, namely, thermal (TS), mechanical ($-PV$) and chemical ($\sum n_i \mu_i$) energies. Being the product of two state functions, each of these energies is also a state function. Regardless of the pathway followed by a system during a transformation from a state 1 to a state 2, the thermal energy varies from T_1S_1 to T_2S_2, the mechanical energy from $-P_1V_1$ to $-P_2V_2$ and the chemical energy from $\sum n_{i_1} \mu_{i_1}$ to $\sum n_{i_2} \mu_{i_2}$. As the first principle ensures the mutual convertibility of the various kinds of energy, the difference $T_2S_2 - T_1S_1$ can be released in the form of work or of chemical energy instead of being liberated as heat. Likewise, a variation of mechanical energy $-P_2V_2 + P_1V_1$ does not necessarily translate into work, but can manifest itself as heat or chemical energy. It is in fact a direct consequence of the first principle that the amounts exchanged for each form of energy depend on the pathway of the transformation.

2.1e GIBBS-DUHEM EQUATION

Let us examine in another manner the mutual convertibility of energies within a system. Consider a gas mixture enclosed into a inflatable balloon. The internal energy of the gas is

$$U = TS - PV + \sum n_i \mu_i. \tag{2.8}$$

Let an infinitesimal amount of gas escape out of the balloon. The pressure decreases by dP, the volume by dV and the number of moles of the ith component by dn_i. In the most general case where the transformation is neither isothermal nor adiabatic, it does not matter here to know the direction of the changes in temperature, entropy and chemical potentials. The relevant fact is just that none of these variables should be constant. The total variation of internal energy is thus

$$dU = TdS + SdT - PdV - VdP + \sum \mu_i dn_i + \sum n_i d\mu_i. \tag{2.9}$$

Of these six terms, which ones are representing energies actually exchanged with the surroundings? Because they are associated with the transferred extensities dS, dV, and dn_i, these are TdS, $-PdV$, and $\sum \mu_i dn_i$, respectively. Their sum constitutes the actual internal energy variation of the system

$$dU = TdS - PdV + \sum \mu_i dn_i. \tag{2.10}$$

The three other terms also represent variations of thermal, mechanical and chemical energies. Contrary to the preceding ones, however, they do not affect at all the surroundings because they originate in tension changes within the system itself. As such, they do not contribute to the actual change in internal energy of the system, and their sum must be zero

$$SdT - VdP + \sum n_i d\mu_i = 0. \tag{2.11}$$

As no peculiar features of the system considered here have been used to derive (2.11), this expression has a general validity. Known as the Gibbs-Duhem equation, it states that the non-exchangeable thermal (SdT), mechanical ($-VdP$) and chemical ($\sum n_i d\mu_i$) energies exactly cancel out within the N-component system. Mathematically, (2.11) expresses the fact that the variations of the $N+2$ intensive properties of a system are not mutually independent. If the variations are known for of $N+1$ of them (for instance P, T and $N-1$ chemical potentials), then that of the Nth component is automatically determined. This is why (2.11) has so great a practical importance.

2.2 ENERGIES

2.2a HEATS OF TRANSFORMATION, ENTHALPY

We have already noted that $dU = dQ$ when no work is exchanged with the surroundings ($dW = 0$). At constant volume, a heat of transformation thus represents an internal energy variation. But precise measurements are generally difficult to make at constant volume, and they are even impossible for condensed phases (solids and liquids) because of the considerable pressures that would be needed to counteract their dilation. As introduced in 1.4c, the enthalpy is important in that it is much easier to measure heats at constant pressure, and especially so at room pressure. It remains, however, to relate enthalpy to internal energy variations.

To relate the heat of transformation at constant pressure to an internal energy variation, let us split a process leading from (T, P, V) to $(T+dT, P, V+dV)$ into two steps. (N.B. For a latent heat, this would correspond to the special case $dT = 0$.) If we first change the temperature at constant volume in order to go from (T, P, V) to $(T+dT, P+dP, V)$, the heat exchanged is by definition $dQ = dU$. According to the results of section 2.1c, the energy then exchanged through simultaneous changes in pressure and volume leading from $(T+dT, P+dP, V)$ to $(T+dT, P, V+dV)$ is $d(-PV)$. As the net internal energy variation sums to

$$dU = dQ - d(PV), \tag{2.12}$$

the heat exchanged during a transformation at constant pressure is given by

$$dQ = dU + d(PV) = d(U + PV) = dH. \tag{2.13}$$

By H, we of course denote the enthalpy, the name of which derives from *enthalpein* = reheat. The integrated expressions of the enthalpy are readily obtained from (2.13) and (2.7)

$$H = U + PV = TS + \sum n_i \mu_i. \tag{2.14}$$

For the differential expression of the enthalpy, (2.4) and (2.13) yield directly

$$dH = TdS + VdP + \sum \mu_i dn_i \tag{2.15}$$

and show that S, P and n_i are the natural variables of the enthalpy function.

In summary, heats of transformations at constant pressure and volume are enthalpy and internal energy variations, respectively. When they are positive, the transformation is *exothermic*. When negative, the process is *endothermic*. Besides, complementing (1.15), these heats of transformation allow the isobaric and isochoric heat capacities to be defined in another way

$$C_p = (\partial H/\partial T)_P = T(\partial S/\partial T)_P, \tag{2.16}$$
$$C_v = (\partial U/\partial T)V = T(\partial S/\partial T)_V. \tag{2.17}$$

For systems characterized by weak molecular interactions (i.e., gases) the volume increases strongly with temperature and work must be done against the external pressure to ensure dilation. At the other end, condensed phases usually expand slightly on heating. For them, the work against external pressure is small (or even negative as for water, which contracts on heating below 4°C), but considerable work has to be done within the system itself to oppose the strong atomic interactions. Hence, as discussed further in 4.3a, the difference between C_p and C_v is always positive.

2.2b HELMHOLTZ FREE ENERGY

In the first chapter, the production of work from heat alone was discussed. To include now the possible contribution of changes in chemical energy to work production, consider again the most favorable case of a reversible conversion. With a heat $dQ = TdS$ gained by the system, the maximum work done is

$$dW = dU - dQ = dU - TdS = dF, \tag{2.18}$$

where F is the function defined by

$$F = U - TS = -PV + \sum n_i \mu_i. \tag{2.19}$$

Since it represents that part of the internal energy that can be used as work, this function F is called the *Helmholtz free energy*, after the German polymath H. von Helmholtz (1821-1894). From (2.4) and (2.19), its differential expression is

$$dF = -SdT - PdV + \sum \mu_i dn_i. \tag{2.20}$$

According to this differential expression T, V and n_i are thus the natural variables of the Helmholtz free energy.

2.2c GIBBS FREE ENERGY

The internal energy is made up of the following three terms

$$U = TS - PV + \sum n_i \mu_i. \tag{2.21}$$

As long as chemical stability is concerned it will not be a surprise that the last one is especially relevant. We will give it a symbol of its own, G, and call it *Gibbs free energy*, in honor of J.W. Gibbs (1839-1903). By definition we have

$$G = \sum n_i \mu_i \tag{2.22a}$$

$$= U + PV - TS = H - TS. \tag{2.22b}$$

For a system composed of only one mole of a single component, $G = \mu$. A chemical potential is thus a molar Gibbs free energy. To avoid any confusion, however, we will use the term chemical potential when dealing with the components of a system and the term Gibbs free energy when dealing with a phase or a system.

Because the internal energy or the Gibbs free energy of a system cannot depend on the way its composition is expressed, the relations (2.21) and (2.22a) are valid for any set of components. Taking H_2O, H and O atoms, or H^+ and OH^- ions as components for water, we can write

$$G = n_{H_2O}\mu_{H_2O} = n_H\mu_H + n_O\mu_O = n_{H^+}\mu_{H^+} + n_{OH^-}\mu_{OH^-}. \tag{2.23}$$

With H and O atoms or with H^+ and OH^- ions as components, we have $n_{H_2O} = 1/2\ n_H = n_O = n_{H^+} = n_{OH^-}$. This implies that the chemical potential of a water is an additive function of the chemical potentials of its components

$$\mu_{H_2O} = 2\mu_H + \mu_O = \mu_{H^+} + \mu_{OH^-}. \tag{2.24}$$

Similar expressions would of course apply for any other set of components. As will be discussed in 7.2 when chemical equilibrium is dealt with comprehensively, it is important to note that all chemical potentials refer to a particular component in a given phase: unless water and methane are in chemical equilibrium, the chemical potential of hydrogen is different in these two species.

From (2.22b) and (2.4), we then derive the differential expression of the Gibbs free energy

$$dG = -SdT + VdP + \sum \mu_i dn_i. \tag{2.25}$$

The Gibbs free energy thus appears to be a function of the natural variables T, P and n_i. Consider again the production of work. For a system kept at constant pressure and temperature, we write

$$dG = dU + PdV - TdS. \tag{2.26}$$

Under the most favorable case of a reversible heat exchange with the surroundings, we have

$$dG = dU + PdV - TdS = dW + PdV = dW'. \tag{2.27}$$

The Gibbs free energy thus represents the difference between the maximum amount of work produced, dW, and the work $-PdV$ that has be exchanged to account for the volume variation of the system. From the definition of G, this difference is also simply related to the change in chemical energy

$$dW' = \sum \mu_i dn_i, \tag{2.28}$$

but only a small fraction of chemical energy is actually convertible directly into work for practical reasons discussed in 12.4d.

Note finally that the principle of conservation applies only to the internal energy, and not to the enthalpy or to the Helmholtz and Gibbs free energies because PV, TS, and $\sum n_i\mu_i$ are in general not conserved individually during a transformation.

2.3 EVOLUTION CRITERIA AND THERMODYNAMIC POTENTIALS

As will be shown in this section, the enthalpy ($U + PV = TS + \sum n_i\mu_i$), the Helmholtz ($U - TS = -PV + \sum n_i\mu_i$) and Gibbs ($U + PV - TS = \sum n_i\mu_i$) free energies are three other *thermodynamic potentials* defined from the internal energy. Why would additional functions like $U - \sum n_i\mu_i$, $U - \sum n_i\mu_i - PV$, etc. not also be defined? As will be shown in 2.3f, the answer is that such potentials would have no or little practical utility. In contrast, the enthalpy is especially useful as a heat of transformation at constant pressure and the importance of Helmholtz and Gibbs free energies mainly stems from the evolution criteria that will be discussed now.

2.3a INTERNAL ENERGY CRITERION

Let us consider a reversible transformation in a closed system (Fig. 2.1). At any moment the heat exchanged with the surroundings is $dQ = TdS$ and the work $dW = -PdV$. The internal energy variation of the system is thus

$$dU = dQ + dW = TdS - PdV, \tag{2.29}$$

and that of the surroundings is $-dU$ because the energy of the universe is constant. Besides, the entropy of the system varies by $dS = dQ/T$, that of the surroundings changes by $dS_E = -dS$, and the total entropy also remains constant.

Let us then follow another, irreversible pathway from the same initial to the same final state. The internal energy variation of the system is the same as in the reversible case because the internal energy is a state function. Hence, equation (2.29) is still

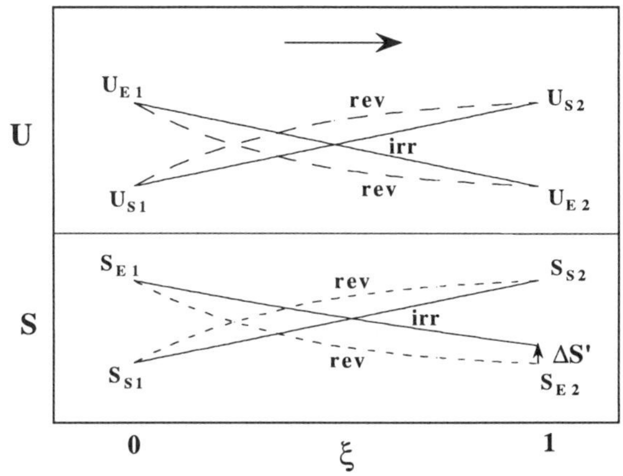

Figure 2.1. Variations of the internal energy and entropy for a system (U_S, S_S) and its surroundings (U_E, S_E) during a transformation along a reversible and an irreversible pathways. The entropy created along the latter in the surroundings is shown as $\Delta S'$.

valid, but the important difference between the reversible and irreversible pathways is that for the former only $dQ = TdS$ and $dW = -P_{rev}dV$; for an irreversible transformation $dQ < TdS$ and $dW > -P_{rev}dV$. Furthermore, the entropy is also a state function. As long as the initial and final states are the same in both cases, it does not matter for the entropy change of the system that the transformation be reversible or not.

At this point two limiting cases must be distinguished for irreversible transformations. In the first one, the entropy created within the system is entirely transferred to the surroundings; along the irreversible pathway, the entropy of the surroundings is increasingly greater than along a reversible one, with a final difference shown as $\Delta S'$ in Fig. 2.1. In the second limiting case, the entropy created is kept by the system instead of being transferred to the surroundings. As examples of the latter, we can take the isothermal, irreversible expansion of a gas, or mixing of two different gases in an adiabatic enclosure (for which the fact that the gases will not separate spontaneously witnesses the irreversibility of the process and the production of entropy).

Because the entropy increases throughout any real process ($dS > dQ/T$), it is in any case advantageous to make the entropy dS' created by the system explicit. For a closed system the entropy variation is then written

$$dS = dQ/T + dS', \tag{2.30}$$

and the internal energy variation is

$$dU = dQ + dW = TdS - TdS' - PdV. \tag{2.31}$$

In the important case of chemical reactions we will see later on how dS' can be related to the composition of the system. Inasmuch as we are interested for the moment in the physical aspects of the phenomena, we will consider here the simplest case of isochoric transformations for which the work exchanged is zero. Since TdS' is ≥ 0, a process taking place at constant V and S necessarily requires

$$dU = -TdS' \leq 0. \tag{2.32}$$

It therefore appears that the internal energy of a closed system must decrease for any spontaneous process. The inequality (2.32) thus constitutes an evolution criterion. When the internal energy is minimum, no transformation is possible. The system has achieved *stability*. It has reached a state of equilibrium.

2.3b FUNDAMENTAL REPRESENTATIONS

Since the Duhem postulate dictates that the state of a system of constant composition is determined by two variables, one could ponder over the meaning of a transformation at constant volume and entropy. In fact, the criterion of entropy production given by the second principle concerns isolated systems only, not the closed systems considered above. Let us split an isolated system into two parts, the particular system of interest and its surroundings. The entropy of the system can in fact decrease during an irreversible process as long as such a decrease is more than offset by the entropy increase of the surroundings. Now the criterion $dU \leq 0$ applies purposely to a closed system, and not to the surroundings whose state is in general not or poorly specified. Although entropy is created during an irreversible transformation, the entropy of the system can remain constant if enough heat is given off to the surroundings. Through the heat transferred, the internal energy of the system decreases and, among all the states of the system compatible with a given volume and entropy, the stable one thus is that with the lowest internal energy. This is of course what is stated by the criterion $dU \leq 0$.

By plotting in Fig. 2.2 the internal energy against the entropy, let us represent the equivalence of the two evolution criteria, $dU \leq 0$ for a closed system and $dS \geq 0$ for an isolated system. For an isolated system, the internal energy is constant and the stable state is that with the greatest entropy. In Fig. 2.2, this state is represented by the point A, the maximum of the curve given by the intersection of the surface $U = f(S,V)$ with the plane $U =$ constant. But this point also represents the minimum of internal energy in the plane that characterizes isentropic transformations. For a system whose state is described by U, S and V, defining the stable state as that with the greatest entropy, at constant internal energy, or with the smallest internal energy, at constant entropy, is a matter of convenience only. Together, the two functions, internal energy $U = f(S,V,n_i)$ and entropy $S = f(U,V,n_i)$, constitute the two *fundamental representations* of a system. From them we will now derive other, more practically convenient representations.

2.3c ENTHALPY CRITERION

In the same manner as the internal energy yields an evolution criterion for processes at constant S and V, we will show that the enthalpy determines the feasibility of transformations at constant pressure and entropy. From the definition (2.14) of the enthalpy and the differential expression (2.31) of the internal energy, we write

$$dH = dU + VdP + PdV = TdS - TdS' + VdP. \tag{2.33}$$

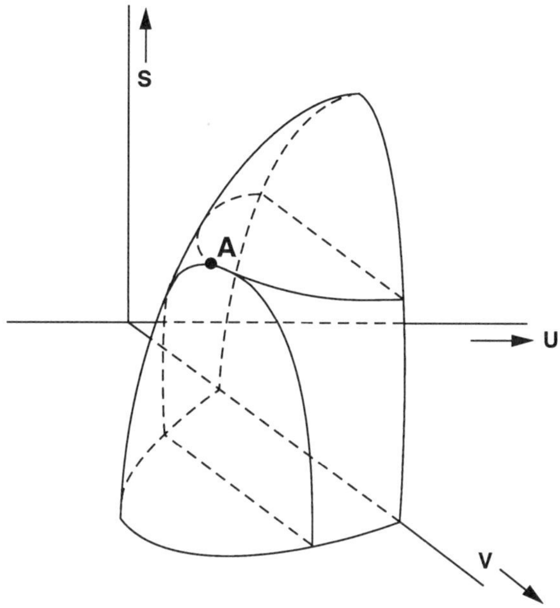

Figure 2.2. Equivalencies of the minimum of internal energy, $U = f(V,S)$, and maximum of entropy, $S = f(U,V)$ for representing the stable state of a system. Simplified from the book by Callen (1985).

If P and S are constant, we have

$$dH = -TdS' \leq 0. \tag{2.34}$$

In this case, the evolution criterion for a closed system is a decrease of the enthalpy and the stable state is that with the lowest enthalpy.

The enthalpy and internal energy decreases for spontaneous transformations at constant S and P and at constant S and V, respectively, are two equivalent statements of the second principle, but they are not much more convenient than the original one because there is no direct means of measuring enthalpy or internal energy variations as a function of entropy. In addition, entropy is a property that is especially difficult to control experimentally because most real transformations are markedly irreversible. In contrast, it is generally not very difficult to control the conjugate variable of entropy, temperature. Evolution criteria for isothermal transformations are thus more useful, whence the usual selection of T in place of S as a variable to specify the state of a system.

2.3d GIBBS FREE ENERGY CRITERION

Let us thus consider transformations at constant temperature. Through an exchange dQ of heat a closed system can give to the surroundings a part $-dQ/T$ of its entropy. Consistent with the second principle, this transfer can proceed as long as the entropy given off is smaller than the entropy created. Mathematically, this reasoning translates to the expression (2.30) which can be rewriten as

$$TdS - dQ = TdS' \geq 0. \tag{2.35}$$

If the transformation is also isobaric, $dQ = dH$, and this expression reduces to

$$dH - TdS = -TdS' \leq 0. \tag{2.36}$$

For an isobaric, isothermal transformation, the left-hand side of (2.36) is the differential of the Gibbs free energy $G = H - TS$

$$dG = -TdS' \leq 0. \tag{2.37}$$

The evolution criterion for such transformations is a decrease of the Gibbs free energy, and stability is achieved when G is minimum. We have already mentioned that it is much easier to control tensions such as P (and T) than extensities such as V (and S). In practice, the criterion $dG \leq 0$ is generally the most useful.

2.3e HELMHOLTZ FREE ENERGY CRITERION

To complete the series of criteria, let us consider finally isochoric transformations. In this case $dQ = dU$ and (2.35) leads to

$$dU - TdS = -TdS' \leq 0. \tag{2.38}$$

We recognize the differential expression at constant T of the Helmholtz free energy, $F = U - TS$. With

$$dF = -TdS' \leq 0. \tag{2.39}$$

the evolution criterion for isochoric, isothermal processes is a decrease of the Helmholtz free energy and the stability criterion is that this function be minimum.

2.3f LEGENDRE TRANSFORMATIONS AND EVOLUTION CRITERIA

The state of a system with a fixed composition is specified by only two variables, the choice of which depends on the problem considered. Here our aim will be to describe the mathematical transformations underlying the variable changes made in the preceding sections to introduce H, F and G from the internal energy. This will allow us to deal with any kind of transformation, without being limited to conditions for which the decreases of U, H, F and G constitute the relevant evolution criteria.

As already noted, the differential expression (2.4) of the internal energy

$$dU = TdS - PdV + \sum \mu_i dn_i$$

considers S and V, in addition to n_i, as the independent variables of the internal energy. Now, it is for transformations at constant S and V that the equilibrium criterion is an internal energy minimum. Likewise, the differential expression (2.15) of the enthalpy is a function of the independent variables S and P, and it is for transformations at constant S and P that a minimum enthalpy represents the equilibrium criterion. From (2.20) and (2.25), the same observations are made with T and V for the Helmholtz free energy, and with T and P for the Gibbs free energy.

Obviously, these are not coincidences. Going from $U(S,V)$ to $H(S,P)$, for example, represents the replacement of V by P as the second independent variable selected to describe the state of the system. When S is kept as the first variable, this replacement of V by P must take into account the fact that V and P are not mutually independent since only two variables can be so. From (2.4) and the definition of an exact differential, at constant entropy the pressure is in fact related to the volume by

$$-P = (\partial U/\partial V)_{S,n_i}. \tag{2.40}$$

Notwithstanding the minus sign in (2.40), the pressure appears to be the partial derivative of U with respect to V when S and n_i are kept constant. It follows that the enthalpy and internal energy are related by

$$H = U + PV = U - V(\partial U/\partial V)_{S,n_i}. \tag{2.41}$$

The expression (2.41) has a well-known mathematical form: H is the Legendre transform of U with respect to V. The remainder of this section shows how Legendre transformations enable one to determine the relevant thermodynamic potential once a given pair of independent state functions has been selected to describe the state of a system.

The properties of Legendre transformations are discussed in the book by Callen which made them popular in the thermodynamic literature. Referring to this book for a detailed treatment, we will restrict ourselves to depicting in Fig. 2.3 the geometrical interpretation of these transformations. We also note that, more generally, the Legendre transform Z of a function Y of n variables with respect to k other variables $(X_1, X_2, ..., X_k)$ is the sum of the transformations performed separately with respect to each of the k variables. For the case of

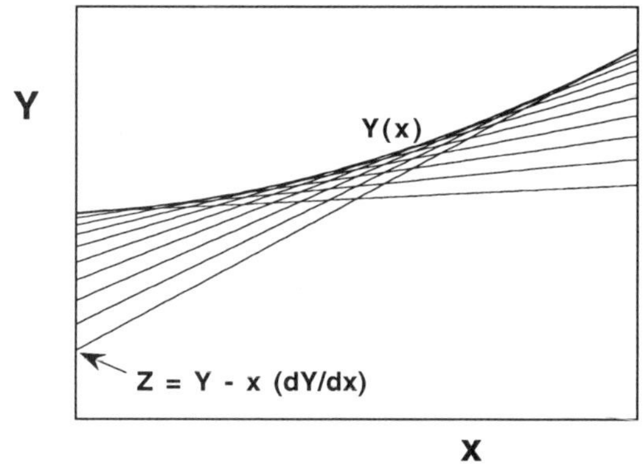

Figure 2.3. Geometrical interpretation of a Legendre transform. A function Y of a single variable x can be defined as being the envelope of the whole set of its tangents of slope $p = dY/dx$. The Legendre transform of Y with respect to x, $Z = Y - px$, represents the intercept on the Y axis of the tangent to the function Y at the point x. Knowing $Y(x)$, we determine in an univocal way this new function Z of the variable p. Conversely, we obtain immediately the function $Y(x)$ from $Z(p)$ by eliminating p and Z with the help of the relationships $x = -dZ/dp$ and $Y = px + Z$. The two representations $Y(x)$ and $Z(p)$ are thus equivalent thanks to their one to one correspondence. For functions of two or more variables, there is an analogous geometrical interpretation in terms of planes (or hyperplanes) tangent to the surface (or the hypersurface) Y. For the internal energy-enthalpy transformation, the function Y thus represents a section at constant S and n_i of $U(S,V,n_i)$, Z represents $H = U - V(\partial U/\partial V)_{S,n_i}$, which is the intercept on the Y axis of the section of the tangent to U at the point V, whereas the slope p of this tangent plane is $-P = (\partial U/\partial V)_{S,n_i}$.

two variables, in which we are interested, we have

$$Z\left[\left(\frac{\partial Y}{\partial X_1}\right)_{X_2},\left(\frac{\partial Y}{\partial X_2}\right)_{X_1}\right]=Y(X_1,X_2)-X_1\left(\frac{\partial Y}{\partial X_1}\right)_{X_2}-X_2\left(\frac{\partial Y}{\partial X_2}\right)_{X_1}. \tag{2.42}$$

Let this equation be used to determine the function Z that is the Legendre transform of the internal energy with respect to the volume and entropy

$$Z\left[\left(\frac{\partial U}{\partial S}\right)_{V,n_i},\left(\frac{\partial U}{\partial V}\right)_{S,n_i}\right]=U(S,V,n_i)-S\left(\frac{\partial U}{\partial S}\right)_{V,n_i}-V\left(\frac{\partial U}{\partial V}\right)_{S,n_i}. \tag{2.43}$$

On the right-hand side of this equation, the second partial derivative is given by (2.40). As to the first partial derivative, it is obtained in the same fashion from the differential (2.4) of the internal energy

$$T=(\partial U/\partial S)_{V,ni}. \tag{2.44}$$

We thus have

$$Z=U-TS+PV, \tag{2.45}$$

and conclude that the thermodynamic potential Z is the Gibbs free energy. This example illustrates the fact that the equilibrium state is determined by the minimum of the appropriate potential describing the state of the system as a function of the selected pair of variables.

Consider now the less common case of a system maintained at constant pressure and enthalpy. To determine what is the appropriate potential, let us find first the function of which the enthalpy is the partial derivative at constant pressure. In fact, the Gibbs-Helmholtz equation (3.4b) indicates that

$$H=\left[\frac{\partial(G/T)}{\partial(1/T)}\right]_{P,n_i}. \tag{2.46}$$

The enthalpy thus appears as the partial differential of the function G/T with respect to $1/T$ at constant pressure. From (2.25), furthermore, we know that P, the other independent variable sought after, is itself one of the two natural variables of which G is a function. Hence, we must take the Legendre transform of G/T with respect to the single variable $1/T$. Using (2.42), we find

$$Z(H,P,n_i)=G/T-H/T=-S. \tag{2.47}$$

As given by the minimum of Z, the stability criterion is that of an entropy maximum when the pressure, enthalpy and composition are maintained constant.

As a last example, let us determine the thermodynamic potential for transformations at constant volume and enthalpy. From (2.25), we conclude immediately

$$\frac{V}{T}=\frac{1}{T}\left(\frac{\partial G}{\partial P}\right)_{T,n_i}=\left[\frac{\partial(G/T)}{\partial P}\right]_{T,n_i}. \tag{2.48}$$

The desired potential Z is therefore the Legendre transform of G/T with respect to $1/T$ and V

$$Z(H,V,n_i)=G/T-H/T-PV/T=-S-VP/T. \tag{2.49}$$

The need has not been felt to give a special name to this potential simply because transformations at constant V and H are not common. For any other transformation, the pertinent potential will be determined from (2.42) in a similar manner.

2.4 STABILITIES AND INTERNAL EQUILIBRIUM

2.4a INTERNAL EQUILIBRIUM

The aforementioned evolution criteria have been obtained for transformations proceeding while two state functions are kept constant. At constant P and T, for instance, the existence of chemical reactions provides a simple means of lowering the Gibbs free energy. When no chemical reaction apparently takes place, however, what kind of process could lead to a decrease of the Gibbs free energy or of other relevant energy functions? Before answering this question, let us note that any state of a system that satisfies the Duhem postulate represents a state of *internal equilibrium*. For a given chemical composition, these states are thus characterized by only two variables.

By contrast, what kinds of states are out of internal equilibrium? The most obvious ones prevail during irreversible processes. When a gas expands irreversibly, a great many parameters would be required to describe the turbulent nature of the process before the molecular configuration has adjusted itself to satisfy the equilibrium requirements set by the particular conditions of the experiment. In the same way, internal equilibrium will not prevail as long as the temperature distribution throughout a heated body has not been equalized by heat conduction. Such transitory states are thus characterized by markedly time-dependent properties.

In a more subtle manner, time-dependent properties can also be observed in systems where the pressure and temperature are actually constant. As will be seen in 6.3b, this is the case of viscous liquids whose structure can slowly change in response to newly imposed P or T conditions. In a general way, the rate at which equilibrium is reached can be characterized by *relaxation time*. These vary from less than 10^{-14} s, for energy transfer involving atomic vibrations, to a thousand years or more for complex molecular rearrangements in very viscous liquids. When relaxation times become much longer than the time scale of an experiment, time-independent properties can again be observed. The configuration of the system has then been frozen in in a nonequilibrium state: internal disequilibrium has become permanent. Hence, the time independence of properties does not imply that a system is in equilibrium. As described below, glasses and other amorphous substances are probably the most common examples of such apparently well-behaved phases, which do not follow the Duhem postulate.

2.4b STABILITY

Consider systems in internal equilibrium, at constant P and T, but not having the lowest possible Gibbs free energy because the reactions which could lower their energy are hindered kinetically. As an example, let us mix two moles of hydrogen and one mole of oxygen at room pressure and temperature. The gaseous mix is in internal equilibrium, but equilibrium is not stable because a single spark or introduction of a catalyst results in the rapid reaction of hydrogen with oxygen to form water. With a lower Gibbs free energy, water is the *stable* phase whereas the gaseous mix is in a *metastable* equilibrium, not transforming to the stable one without any outside intervention. At temperatures ranging from -40 to 0°C, supercooled water constitutes another example of a metastable phase. One can keep it for long periods of time, especially when the temperature is close to 0°C, as long as ice seeds or impurities are not introduced.

As applied to a phase, a state or an equilibrium, these notions of stability and metastability are illustrated by the classic representation of mechanical equilibria plotted in Fig. 2.4. For both stable and metastable states the energy is minimum, but stability is associated with the lowest possible minimum. Since going from a metastable to a stable state requires overcoming energy barriers, an *unstable* state is necessarily encountered along the pathway. Unlike an equilibrium state, an unstable state is characterized by the fact that the slightest fluctuation of a single variable suffices to make the state topple over to either a metastable or stable state, depending on the direction of the fluctuation. And, as exemplified by the equilibrium of a marble on a horizontal plane, a *neutral equilibrium* is insensitive to any state change.

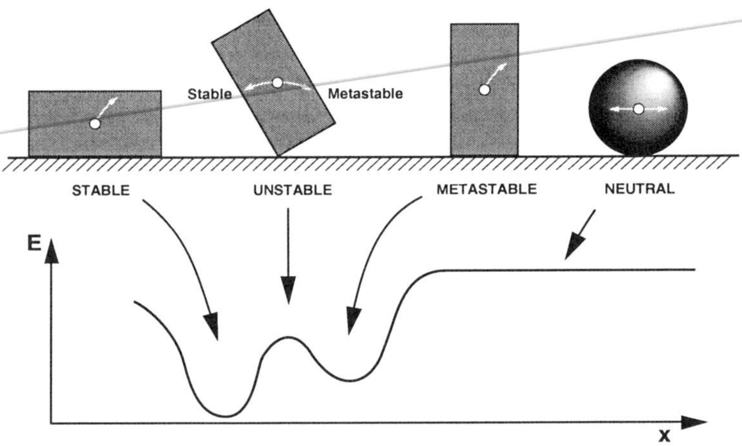

Figure 2.4. Mechanical examples of stable, metastable, unstable, and neutral equilibria. The white arrows indicate the energy changes due to a variation of the coordinate x.

For a thermodynamic equilibrium, Fig. 2.4 would represent a plot of the relevant potential (U, H, or F, etc.) as a function of one out of the $N+2$ variables of the system. A complete representation of the stability of an equilibrium should be made in an $N+3$ dimensional space. For a system whose state is characterized by only two variables, the energy is represented by a surface in the familiar Euclidean space. Unstable equilibria then correspond to the summits of this surface, neutral equilibria to its plateaus, and metastable and stable equilibria to the bottoms of its basins.

As a matter of fact, identifying which kind of equilibrium prevails is not always easy. A metastable state can be hardly distinguished from a stable state when their energy difference is small or when some peculiar reaction mechanisms lead preferentially to a metastable phase. The distinction between a metastable and an unstable state becomes blurred when energy barriers vary with temperature or any other state function. At room pressure, the Gibbs free energy of diamond is higher than that of graphite, stability at any temperature but the kinetics of the phase transition between diamond and graphite, which is zero at ambient temperature, becomes significant near 1000 K. Graphite is then the stable form of carbon whereas diamond, which is clearly metastable at lower temperatures, thus tend to become unstable at higher temperatures.

Chemical metastability is always kinetic in origin. Either the reaction producing the metastable phase is faster than that leading to the stable phase, or the metastable phase was initially in equilibrium but the kinetics of the transformation to the stable phase under the newly imposed conditions is too slow. An important difficulty in this respect stems from the fact that reaction rates depend on impurity contents or on the presence of catalysts. Example: at 1 bar, the crystalline form of germanium oxide with a hexagonal symmetry is stable in a narrow temperature interval ranging from 1320 K to the melting point of 1380 K. It is nevertheless this form that is generally produced at lower temperatures (through crystallization of the melt) in place of the actual stable tetragonal form which requires the presence of catalysts (such as Li^+ ions) or the application of kbar pressures to be made. Not only reaction rates are transformation dependent but their calculation is outside the scope of classical thermodynamics. We emphasize, however, that Gibbs free energy differences between the products and reactants constitute the driving force for transformations. The decrease of the Gibbs free energy difference between ice and water with increasing temperatures thus accounts for the fact that supercooling of water is easier when the temperature is closer to 0°C. Besides, crystallization begins with ordering of water molecules to form the nucleus of an ice crystallite that will grow and eventually yield a macroscopic crystal. Because the probability of formation of these nuclei is proportional to the amount of water, supercooling is limited to about -6°C for volumes of the order of 100 cm^3 whereas it is possible down to -40°C for droplets having a diameter of 10 μm.

2.4c STATES OF MATTER

As a conclusion to this discussion it will be useful to review briefly the stability of equilibria prevailing for the various states of matter, which are not restricted to solids, liquids and gases. In fact, the liquid and gas states do not raise much ambiguity because internal equilibrium is generally achieved rapidly with respect to experimental time scales. Here we just recall that the distinction between these two states is clear only below the critical point (see 5.3b), in relatively limited pressure and temperature domains, beyond which a single *supercritical* fluid state exists. In passing we must mention *plasmas*, i.e., gases of electrons and ionized atoms, which actually constitute 99% of the universe! They form spontaneously at temperatures of a few tens of thousand degrees, at which internal equilibrium is reached instantaneously. The local heterogeneities due to the coexistence of two kinds of particles having opposite charges result in strong electromagnetic effects that distinguish them from ordinary gases and justify their particular name.

Plasmas can alternatively be produced through exposure to ionizing radiation at lower temperatures. Under these conditions, their lifetime is significant only at very low matter densities. To be effective this process requires that the collision rates between ions (through which neutral species recombine) be very small in comparison with the rate of ion production. In nature, such conditions prevail only in interstellar space.

The crystalline state is often taken as synonymous of the solid state, but a more correct classification of the states of matter would distinguish at least two categories of solids, namely, crystals and amorphous solids. The common feature shared by solids, which distinguishes them from fluids, is a very low atomic mobility. When the structure of a solid is affected by some thermal or mechanical perturbations, the slowness of atomic diffusion possibly results in very long relaxation times to return to the equilibrium structural state. In such a case, the properties of the solid depend on its previous history and not only on two state functions. The point to be briefly reviewed here is the extent to which the resulting lack of internal equilibrium is significant.

Crystals are characterized by the existence of long-range order. Contrary to "perfect" crystals, "real" crystals have defects (point defects, dislocations, etc.) produced at high stresses or temperatures which exert a considerable influence on mechanical or transport properties (e.g., plasticity, element diffusivity). Strictly speaking, these phases are not in internal equilibrium when the nature and abundances of defects depend on their thermal and stress history. Except for crystals subjected to very severe stress conditions, however, the abundance of defects is generally too small to have a significant influence on thermodynamic properties. In contrast, departure from internal equilibrium cannot be neglected when nonequilibrium distributions of elements or molecular groups occur on crystallographic sites. These distributions can have important entropy consequences. As will be seen in 6.4f, they have a kinetic origin and are analogous to the disequilibria that characterize the bulk structure of glasses, which will be mentioned now.

Sharing a common lack of long-range order, organic and inorganic glasses are the most important members of amorphous solids. As described in 6.4 in more detail, glasses are usually formed through quenching of supercooled liquids when a very high viscosity opposes crystallization. From the simple fact that the density or refractive index at room P and T depends on the cooling rate at which a glass is cooled, one concludes that internal equilibrium is lost when the structure of the liquid is frozen in, i.e., when atomic mobility almost vanishes. Note, however, that the boundary between crystals and amorphous substances is not necessarily clear cut, as glasses can be partially crystallized at a very small scale and some solids precipitate in a poorly crystalline form.

As to aerosols, colloids, gels, suspensions or emulsions, their names are suggestive of unclear states which have long been neglected by thermodynamicists fond of simpler systems. Under these forms, matter is partitioned between solid, liquid and gaseous phases, more or less well defined and having generally a small size which can be only a few tens of angstroms. Surface tension generally plays a fundamental role and can result in time-dependent properties. (For emulsions, the sad fate of a home-made mayonnaise illustrates the changes induced by the coalescence of the initially thoroughly dispersed oil droplets.) Many of these states frequently found in the human environment are thus at best metastable, and often out of internal equilibrium.

Depending on the way they are produced, amorphous substances are themselves split into different categories. Amorphous substances closely related structurally to glasses can be formed through annealing of gels. With vapor deposition, one obtains amorphous films of diverse materials such as silicon or ice. Other amorphous substances are made when enough defects have been produced in an initially crystalline material, as a result of a long exposure to beams of neutrons, electrons or ions. Amorphization can also develop when a crystal is thoroughly ground or compressed statically or dynamically to pressures of tens or hundreds of kbar.

Finally, the considerable practical importance of supercooling must be emphasized. Organic polymer melts are often very reluctant to crystallize because of the difficulty of ordering their long molecular chains. In fact, it is this feature that makes polymers useful as "plastic" substances, readily molded when the melt viscosity is rather high, and keeping their shape when transforming to a glass on cooling to room temperature. In the same way, the existence of silicate glasses as produced industrially or naturally (by volcanoes) mainly results from the two- or three-dimensional frameworks that are present in SiO_2-rich melts and do not order readily on cooling.

Problems

1. Check whether $df = (2xy - y)dx + (x^2 - x + 4y^3)dy$ and $df = x^3ydx + xy^3dy$ are exact differentials. In both cases, calculate the variation of the function f between two points, along two different paths if df is not an axact differential.

2. The pressure of a gas is related to the temperature and volume by the expression $P = RT/(V-b) - a/V^2$, where a and b are constant parameters. Show that the differential of P as a function of T and V is exact.

3. Show that (1.6) and Einstein's relation $E = mc^2$, where the velocity of light c is constant, imply that the mass m of a moving body varies with the speed v according to the relation $m = m_0/(1 - v^2/c^2)^{1/2}$, with m_0 = mass at rest.

4. Are the seemingly perpetual revolutions of the planets around the Sun consistent with the second principle? What interactions between the Sun and the planets cause transformation of kinetic energy into heat? Same questions for the ceaseless motion of the molecules in a gas.

5. In an adiabatic enclosure at constant pressure, a 1-kg platinum bar at 900 K is laid on a 1-kg marble block (composition: $CaCO_3$) initially at 300 K. What are the equilibrium temperature and the entropy change of the system after the transformation? The heat capacities are C_{pPt} = 24.41 + 5.20 10^{-3} T J/mol K and C_{pMa} = 52.4 + 0.104 T J/mol K. (Atomic weights listed at the beginning of this book.)

6. Calculate the enthalpy, entropy and internal energy variations for the following transitions: ice (260 K, 1 bar) $\rightarrow$ water (300 K, 1 bar); water (300 K, 1 bar) $\rightarrow$ steam (500 K, 1 bar); ice (260 K, 1 bar) $\rightarrow$ steam (500 K, 1 bar). Enthalpy of fusion = 6.0 kJ/mol; enthalpy of vaporization = 40.9 kJ/mol. To simplify the calculation, assume that steam is an ideal gas and that the C_p's are constant and equal to 37.3, 75.3 and 36.0 J/mol K for ice, water and steam, respectively.

7. A 1-liter vessel is entirely filled with water at 373 K and 1 bar. With a valve the vessel can be connected to another vessel which is evacuated and kept in a heat bath at 373 K. When the valve is open all water vaporizes and one measures a final pressure of 0.2 bar. What are the heat and entropy exchanged by the water, the thermostat and the universe? How could the same final state be reached for the water through a reversible process? How would the final state of the universe differ depending on whether the transformation is reversible or irreversible? The volume of the liquid will is negligible compared to that of the vapor. The enthalpy of vaporisation and density of water at 373 K are ΔH_{vap} = 40.9 kJ/mol and ρ = 0.958 g/cm^3, respectively.

8. One wants to transfer some heat Q to a system kept at constant volume. Show that a minimum work W is needed for this purpose, such that $W/Q = 1 - T_{su}/T_s$ where T_{su} and T_s are the temperatures of the surroundings and system, respectively.

3 AUXILIARY RELATIONS

Before passing to some more practical aspects of thermodynamics, we will complete the catalog of the most useful mathematical expressions. To derive Maxwell's and other important relations, extensive use will be made of the fact that differentials of state functions are exact differentials. With partial molar properties, an additional class of properties will also be defined, and the variations of the internal energy and entropy with pressure will be related to the temperature, pressure and volume changes through the thermodynamic equations of state.

3.1 DIFFERENTIAL EXPRESSIONS

3.1a OPEN SYSTEMS

When the composition varies the resulting changes in the properties of a system must of course be taken into account. For an N-component system, the exact differentials of extensive variables are functions of $N+2$ variables. For the internal energy, for instance, we have

$$dU = (\partial U/\partial S)_{V,n_i}dS + (\partial U/\partial V)_{S,n_i}dV + \sum_i(\partial U/\partial n_i)_{V,S,n_j}dn_i, \tag{3.1a}$$

where the summation is made with respect to the number of moles n_i of the N components, and where the subscript j of n_j indicates the whole set of $N-1$ components whose numbers of moles are kept constant in the different partial differentials.

Writing analogous expressions for the enthalpy (with S and P as independent variables), the Helmholtz free energy (with T and V) and the Gibbs free energy (with T and P) we find

$$dH = (\partial H/\partial S)_{P,n_i}dS + (\partial H/\partial P)_{S,n_i}dP + \sum(\partial H/\partial n_i)_{P,S,n_j}dn_i, \quad (3.1b)$$
$$dF = (\partial F/\partial T)_{V,n_i}dT + (\partial F/\partial V)_{T,n_i}dV + \sum(\partial F/\partial n_i)_{V,T,n_j}dn_i, \quad (3.1c)$$
$$dG = (\partial G/\partial T)_{P,n_i}dT + (\partial G/\partial P)_{T,n_i}dP + \sum(\partial G/\partial n_i)_{P,T,n_j}dn_i. \quad (3.1d)$$

On the other hand, the following differential expressions have already been obtained in 2.1c and 2.2a-2.2c for U, H, F and G

$$dU = TdS - PdV + \sum \mu_i dn_i, \quad (3.2a)$$
$$dH = TdS + VdP + \sum \mu_i dn_i, \quad (3.2b)$$
$$dF = -SdT - PdV + \sum \mu_i dn_i, \quad (3.2c)$$
$$dG = -SdT + VdP + \sum \mu_i dn_i. \quad (3.2d)$$

Through identification of equations (3.1) and (3.2), the temperature, pressure, volume, entropy and chemical potentials are thus defined thermodynamically as

$$T = (\partial U/\partial S)_{V,n_i} = (\partial H/\partial S)_{P,n_i}, \quad (3.3a)$$
$$P = -(\partial U/\partial V)_{S,n_i} = -(\partial F/\partial V)_{T,n_i}, \quad (3.3b)$$
$$V = (\partial H/\partial P)_{S,n_i} = (\partial G/\partial P)_{T,n_i}, \quad (3.3c)$$
$$S = -(\partial F/\partial T)_{V,n_i} = -(\partial G/\partial T)_{P,n_i}, \quad (3.3d)$$
$$\mu_i = (\partial U/\partial n_i)_{S,V,n_j} = (\partial H/\partial n_i)_{S,P,n_j} = (\partial F/\partial n_i)_{T,V,n_j} = (\partial G/\partial n_i)_{T,P,n_j}. \quad (3.3e)$$

3.1b GIBBS-HELMHOLTZ EQUATIONS

In passing, we will derive another useful set of relations known as Gibbs-Helmholtz equations. With (3.3), the enthalpy is related to the Gibbs free energy by

$$H = G + TS = G - T(\partial G/\partial T)_{P,n_i} \quad (3.4a)$$
$$= -T^2[\partial(G/T)/\partial T]_{P,n_i} = [\partial(G/T)/\partial(1/T)]_{P,n_i}. \quad (3.4b)$$

If the Gibbs free energy is known as a function of temperature, the enthalpy can be obtained from (3.4b). From G and H, we could in addition calculate the entropy with $S = (H - G)/T$, but the experimental errors on both G and H are generally too large for S to be derived accurately in this way. Rather, (3.4) gives a test of the internal consistency of independent enthalpy, entropy and Gibbs free energy measurements. Besides, by rewriting (3.4b) as

$$[\partial(G/T)/\partial T]_{P,n_i} = -H/T^2, \quad (3.4c)$$

we arrive at a relation whose main usefulness will be to give the temperature dependence of chemical equilibrium constants.

In the same way, the second set of Gibbs-Helmholtz equations is found from (3.3d) and the definition of F

$$U = F + TS = F - T(\partial F/\partial T)_{V,n_i} \tag{3.5a}$$

$$= -T^2[\partial(F/T)/\partial T]_{V,n_i} = [\partial(F/T)/\partial(1/T)]_{V,n_i}, \tag{3.5b}$$

whence

$$[\partial(F/T)/\partial T]_{V,n_i} = -U/T^2. \tag{3.5c}$$

3.1c MAXWELL RELATIONS

Maxwell (1831-1879) relations express the fact that the second-order derivatives of state functions do not depend on the order of differentiation (Appendix A). To derive the first one, consider the internal energy. As applied to (3.3a-b), the condition

$$\left(\frac{\partial^2 U}{\partial V \partial S}\right)_{n_i} = \left(\frac{\partial^2 U}{\partial S \partial V}\right)_{n_i}$$

imposes that

$$(\partial P/\partial S)_{V,n_i} = -(\partial T/\partial V)_{S,n_i}. \tag{3.6a}$$

Likewise, three other relations are obtained from (3.3b-d)

$$(\partial V/\partial S)_{P,n_i} = (\partial T/\partial P)_{S,n_i}, \tag{3.6b}$$

$$(\partial S/\partial V)_{T,n_i} = (\partial P/\partial T)_{V,n_i} = \alpha/\beta_T, \tag{3.6c}$$

$$(\partial S/\partial P)_{T,n_i} = -(\partial V/\partial T)_{P,ni} = -\alpha V, \tag{3.6d}$$

where α is the thermal expansion coefficient and β_T the isothermal compressibility. Because α is usually positive and β_T is always so, (3.6c) and (3.6d) indicate that, at constant T, the entropy generally increases with the volume and decreases when the pressure increases. As already noted, the variations of the entropy with temperature are given by calorimetric measurements. Hence, the main usefulness of the Maxwell relations is to give the variations of the entropy with pressure and volume from P, V, T measurements.

Strictly speaking, the Maxwell relations are the four equations relating the mutual variations of T, S, P and V as obtained from (3.3a-d). Under the same name, however, we will include the eight additional equations dealing with chemical potentials obtained from (3.3e). With respect to extensive properties, a first set of equations is drawn from eqns (3.2a)-(3.2c), namely

$$(\partial\mu_i/\partial S)_{V,n_i} = (\partial T/\partial n_i)_{S,V,n_j}, \tag{3.7a}$$

$$(\partial\mu_i/\partial V)_{S,n_i} = -(\partial P/\partial n_i)_{S,V,n_j}, \tag{3.7b}$$

$$(\partial\mu_i/\partial S)_{P,n_i} = (\partial T/\partial n_i)_{S,P,n_j}, \tag{3.7c}$$

$$(\partial\mu_i/\partial V)_{T,n_i} = -(\partial P/\partial n_i)_{T,V,n_j}. \tag{3.7d}$$

The last four equations express derivatives of chemical potentials with respect to intensive properties; for reasons that will appear obvious, they are listed below in the section devoted to partial molar properties. Note that, except for the sign, the partial derivative of the entropy with respect to P or V is equal to the partial derivative of the temperature, its conjugate variable, with respect to V or P. A similar rule applies to the eight other equations that deal with the derivatives of the chemical potentials.

3.1d ORDER OF VARIABLES

The order of a variable designates that of the partial derivative of the Gibbs free energy that defines it. Being first-order derivatives of the Gibbs free energy through (3.3c), (3.3d) and (3.4c), respectively, volume, entropy and enthalpy are referred to as first-order thermodynamic variables. Because they are themselves defined as partial derivatives of first-order properties, the heat capacity, the thermal expansion coefficient and the compressibility are second-order thermodynamic properties and their own derivatives with respect to temperature or pressure are thus third-order properties.

3.2 PARTIAL MOLAR PROPERTIES

3.2a DEFINITION

The properties of an open system depend on the number of moles of its components. With (3.3e), the chemical potential of component i has been defined as the change of G (or F) with its number of moles dn_i at constant P, T (or V, T) and n_j. The fact that a chemical potential changes when the composition of the system varies has already been mentioned. But thermodynamics does not give any clue as to the manner in which μ_i depends on composition. These variations must be determined experimentally for the chemical potential as well as for the volume, the entropy or the heat capacity. As an example, Fig. 3.1 shows how the volume of sulfuric acid solutions varies as a function of concentration. This variation is nonlinear, the actual volume being smaller than the weighted mean of the volumes of pure water and sulfuric acid.

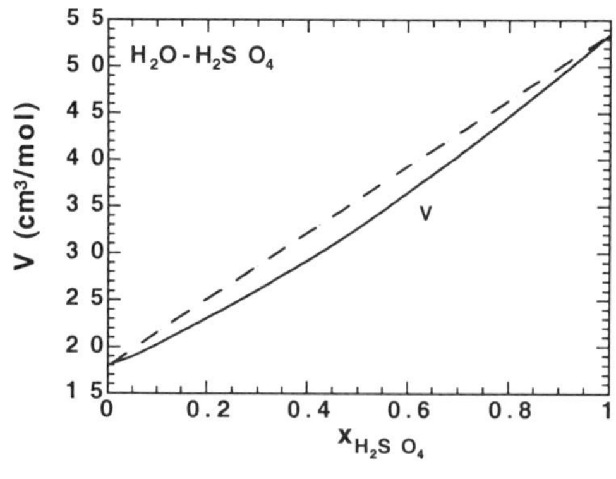

Figure 3.1. Molar volumes of sulfuric acid solutions (solid curve). For comparison, the dashed line indicates a linear variation of V with composition. Data from the Handbook of Chemistry and Physics).

To specify the possibly complex variation of a given property with composition, it is useful to define a *partial molar property*. For instance, the *partial molar volume* of component i

$$\overline{V}_i = (\partial V/\partial n_i)_{P,T,n_j} \tag{3.8}$$

represents the volume variation of the system when the number of moles of i is changed by the amount dn_i at constant T, P and numbers of moles of the other components. Contrary to a volume, $\overline{V}_i$ can be negative. This is a relatively common case, of which magnesium and calcium sulphates in dilute aqueous solutions are classic examples. In a general way, for any property Y we define

$$\overline{Y}_i = (\partial Y/\partial n_i)_{P,T,n_j} \tag{3.9}$$

as the partial derivative of Y with respect to n_i at constant P, T and n_j. The particular importance of the partial molar Gibbs free energy is emphasized by its special name (chemical potential) and symbol (μ). Also note that partial molar properties are not restricted to extensive properties. With the exception of temperature and pressure, which cannot vary with composition in an equilibrium system, it is possible to define intensive partial molar properties. Thermal expansion coefficients and compressibilities are cases in point.

Making use of (3.9), we finally complete the catalog of Maxwell relations. From (3.2c) and (3.2d) we derive the important relations

$$(\partial\mu_i/\partial T)_{P,n_i} = -(\partial S/\partial n_i)_{T,P,n_j} = -\overline{S}_i, \tag{3.10a}$$

$$(\partial\mu_i/\partial P)_{T,n_i} = (\partial V/\partial n_i)_{T,P,n_j} = \overline{V}_i. \tag{3.10b}$$

As to the last two relations

$$(\partial\mu_i/\partial T)_{V,n_i} = -(\partial S/\partial n_i)_{T,V,n_j}, \tag{3.10c}$$

$$(\partial\mu_i/\partial P)_{V,n_i} = (\partial V/\partial n_i)_{V,P,n_j}, \tag{3.10d}$$

their practical usefulness is very limited.

Equations (3.10a-b) illustrate that the relations established between properties are generally valid for their partial molar counterparts. To name only a few, we have

$$\overline{V}_i = (\partial\overline{H}_i/\partial P)_{S,n_j} = (\partial\mu_i/\partial P)_{T,n_j}, \tag{3.11a}$$

$$(\partial\overline{V}_i/\partial T)_{P,n_j} = -(\partial\overline{S}_i/\partial P)_{T,n_j}, \tag{3.11b}$$

$$\overline{H}_i = -T^2[\partial(\mu_i/T)/\partial T]_{P,n_j}, = [\partial(\mu_i/T)/\partial(1/T)]_{P,n_j} \tag{3.11c}$$

$$\overline{C_{P_i}} = (\partial\overline{H}_i/\partial T)_{P,n_j}. \tag{3.11d}$$

Note, however, that such equations are not valid when they involve intensive partial molar properties. For instance, the celebrated relation $C_p - C_v = TV\alpha^2/\beta_T$ has no counterpart for the partial molar properties of the five variables (C_p, C_v, V, α, and β_T) it involves. The article by Reis (1982) gives a detailed discussion of the various partial molar properties.

3.2b INTEGRATION OF PARTIAL MOLAR PROPERTIES

Let us consider an extensive variable, the volume, for instance. As a function of pressure, temperature and composition, its differential is

$$dV = (\partial V/\partial T)_{P,n_i} dT + (\partial V/\partial P)_{T,n_i} dP + \sum (\partial V/\partial n_i)_{T,P,n_j} dn_i. \tag{3.12}$$

At constant pressure and temperature, this expression reduces to

$$dV = \sum \overline{V}_i dn_i. \tag{3.13}$$

With the reasoning already followed in 2.1d to calculate the Gibbs-free energy variation due to a change in the size of the system, we find that the volume of the system is the sum, weighted by the number of moles, of the partial molar volumes of the components for the given composition

$$V = \sum n_i \overline{V}_i. \tag{3.14}$$

By identifying the differential of (3.14) with (3.13), we derive similarly the equivalent of the Gibbs-Duhem equation

$$\sum n_i d\overline{V}_i = 0. \tag{3.15}$$

For any extensive property Y, these two results can be written in a general way

$$Y = \sum n_i \overline{Y}_i, \tag{3.16}$$

$$\sum n_i d\overline{Y}_i = 0, \tag{3.17}$$

for conditions of constant P and T. Mathematically, the relation (3.16) simply expresses the fact that extensive variables are homogeneous functions of first degree with respect to the composition (Appendix A). Furthermore, like chemical potentials, partial molar properties vary with the composition of the system.

We stress that no hypothesis regarding the physical nature of the components has been made to derive (3.16). As already pointed out in 2.2c, the Gibbs free energy of water, for instance, can be written as any of the following expressions

$$G_{H_2O} = \mu^w_{H_2O} = 2n_H \mu^w_H + n_O \mu^w_O = n_{H^+} \mu^w_{H^+} + n_{OH^-} \mu^w_{OH^-} = \ldots \tag{3.18}$$

where the numbers of moles of the various components are fixed by the amount of water in the system. But, as will be discussed in chapter 7, the chemical potential of H, for example, is the same in water and in another hydrogen-bearing species only if both species are in chemical equilibrium. When a risk of confusion arises, it is thus preferable to indicate clearly the host phase for the component whose chemical potential is written. This has been made above by using the superscript w (for water).

3.2c DETERMINATION OF PARTIAL MOLAR PROPERTIES

Partial molar properties have been defined by (3.9) as $\overline{Y}_i = (\partial Y/\partial n_i)_{P,T,n_j}$. In practice, however, it is more convenient to express chemical compositions in terms of mole fractions x_i rather than in terms of numbers of moles. A slight difficulty then stems from the fact that in (3.9) the numbers of moles cannot be replaced by mole fractions because the conservation equation $\sum x_i = 1$ prevents the mole fraction of a given component from being changed independently of the others. In contrast, we can use mole fractions instead of numbers of moles in (3.16), so that, with modern computing means, partial molar properties are in fact more conveniently determined from multivariable fits. Equations expressing adequately the property of interest Y in terms of the $\overline{Y}_i$ and mole fractions of the components are then directly adjusted to the experimental results.

One is often interested in *binary systems* (made up of two components only) and even for complex systems it is useful to examine how properties vary along *binary joins* to determine what kind of thermodynamic model is relevant. Under these circumstances, the partial molar properties of both components are simply related to the mole fractions (Fig. 3.2a). Consider again the volume. From (3.16), the volume of a binary system is

$$V = x_1\overline{V}_1 + x_2\overline{V}_2. \tag{3.19}$$

A second expression involving these variables is needed to determine $\overline{V}_1$ and $\overline{V}_2$ from V and the mole fractions. Let us use (3.17)

$$x_1 d\overline{V}_1 + x_2 d\overline{V}_2 = 0. \tag{3.20}$$

From the expressions of $\overline{V}_2$ derived from each of these equations, we find, by noting that $dx_1 = -dx_2$

$$\overline{V}_1 - \frac{dV}{dx_1} = \frac{V}{x_2} - \frac{x_1\overline{V}_1}{x_2}. \tag{3.21}$$

Since $x_1 + x_2 = 1$, this expression is rewritten as

$$\overline{V}_1 = V + (1 - x_1)\frac{dV}{dx_1}. \tag{3.22a}$$

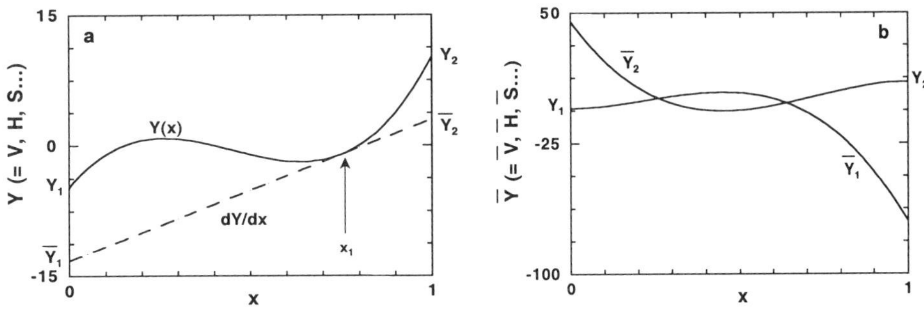

Figure 3.2. (a) Partial molar properties $\overline{Y}_1$ and $\overline{Y}_2$ of the two components of a binary system, for a given composition x_1, as given by the intercepts of the tangent to the curve $Y(x)$ on the x = 0 and 1 composition axes; (b) Variation with composition of these partial molar properties.

And we obtain in a similar way for $\overline{V}_2$

$$\overline{V}_2 = V - x_1 dV/dx_1. \tag{3.22b}$$

For a binary system, partial molar properties have a simple geometrical interpretation. If the tangent to the curve $Y = f(x)$ is drawn at a given composition x_1 (Fig. 3.2a), the partial molar properties of components 1 and 2 for this particular composition are the intercepts of the tangent with the axes $x_1 = 1$ and $x_2 = 1$. Their variations with x are shown in Fig. 3.2b.

3.3 THERMODYNAMIC EQUATIONS OF STATE

3.3a *INTERNAL ENERGY AND VOLUME*

Consider again conditions of constant composition. Through (2.4), the internal energy has until now been taken as a function of S and V only

$$dU = (\partial U/\partial S)_V dS + (\partial U/\partial V)_S dV = TdS - PdV. \tag{3.23}$$

Practically, however, it is more convenient to write its differential as a function of T and V

$$dU = (\partial U/\partial T)_V dT + (\partial U/\partial V)_T dV. \tag{3.24}$$

The first term of (3.24) is $C_V dT$. For expressing the second term as a function of $(\partial U/\partial V)_S$ which, according to (3.3b), is equal to $-P$, let us make use of (A.17) to obtain

$$(\partial U/\partial V)_T = (\partial U/\partial V)_S + (\partial U/\partial S)_V (\partial S/\partial V)_T. \tag{3.25}$$

With (3.3a) and the Maxwell relation (3.6c), we derive

$$(\partial U/\partial V)_T = -P + T(\partial P/\partial T)_V. \tag{3.26}$$

Then, (3.24) can be rewritten in the desired form where the entropy has been eliminated

$$dU = C_v dT - [P - T(\partial P/\partial T)_V]\, dV. \tag{3.27}$$

3.3b *ENTHALPY AND PRESSURE*

In the same manner, let us express dH as a function of T and P rather than as a function of T and S:

$$dH = (\partial H/\partial T)_P dT + (\partial H/\partial P)_T dP. \tag{3.28}$$

From

$$(\partial H/\partial P)_T = (\partial H/\partial P)_S + (\partial H/\partial S)_P (\partial S/\partial P)_T,$$

and (3.3c), (3.3a) and (3.6d), we obtain

$$(\partial H/\partial P)_T = V - T(\partial V/\partial T)_P. \tag{3.29}$$

We thus find

$$dH = C_p dT + [V - T(\partial V/\partial T)_P]dP. \tag{3.30}$$

The variables P, V and T are rather simply related to U and H with (3.27) and (3.30). To complement these equations, we write the variations of U with pressure and of H with volume, at constant temperature

$$\left(\frac{\partial U}{\partial P}\right)_T = \frac{(\partial U/\partial V)_T}{(\partial V/\partial P)_T}, \tag{3.31}$$

$$\left(\frac{\partial H}{\partial V}\right)_T = \frac{(\partial H/\partial P)_T}{(\partial P/\partial V)_T}. \tag{3.32}$$

An important conclusion follows: if the internal energy and the enthalpy do not depend on pressure, they are also independent of volume. From the definition of the enthalpy, we finally relate the mutual variations of U and H with the volume or the pressure. For the volume, for instance, we have

$$(\partial H/\partial V)_T = (\partial U/\partial V)_T + P + V(\partial P/\partial V)_T. \tag{3.33}$$

3.3c EQUATIONS OF STATE

The relations between P, V and T have so great an importance that they are called *equations of state*. The equations (3.26) and (3.29) can be rewritten as

$$P = T(\partial P/\partial T)_V - (\partial U/\partial V)_T, \tag{3.34}$$

$$V = T(\partial V/\partial T)_P + (\partial H/\partial P)_T. \tag{3.35}$$

These equations represent the only constraints imposed by thermodynamics on the mutual relations between P, V and T. Because they are valid for any kind of system, they are termed *thermodynamic equations of state*. Note, however, that knowledge of the equation of state does not provide one with a complete thermodynamic description of a system. The set of 4 fundamental equations (3.3) involving 8 variables, three of them remain undetermined once the equation of state is known. Although two variables have to be fixed (P, V or T, for example), there still remains one, the entropy for instance, that is not specified. In this respect, note again that Maxwell relations give only the variations of the entropy with volume and pressure, not its absolute value or its temperature dependence, which must be determined experimentally.

In (3.34), the pressure P is called the *external pressure*. It is useful to express it in terms of a *thermal* pressure, P_{th}, and an *internal* pressure, P_{int}

$$P = P_{th} - P_{int}, \tag{3.36}$$

with

$$P_{th} = T(\partial P/\partial T)_V, \qquad P_{int} = (\partial U/\partial V)_T. \tag{3.37}$$

At zero temperature, $P_{th} = 0$ and $P_{int} = -P_{ext}$. Without thermal energy at $T = 0$ K the internal pressure would oppose the pressure exerted on the system. Equilibrium would be reached when the volume would be such that $P_{ext} = -(\partial U/\partial V)_T$. If the temperature increases at constant volume, the pressure then varies by an amount that is given by the thermal pressure. The internal pressure is thus associated with the "static" part of atomic interactions, whereas the thermal pressure is related to their "dynamic" part. We can show this balance between different kinds of forces in another way, by writing (3.36) as $P_{th} = P_{ext} + P_{int}$. A compressibility is necessarily positive. Because of (3.6c),

$$(\partial P/\partial T)_V = \alpha/\beta_T, \tag{3.38}$$

the thermal pressure has the same sign as the thermal expansion coefficient α, which is generally positive. Thermal expansion can therefore be viewed as a way of relaxing the thermal pressure that would build up if the temperature were increased at constant volume.

3.4 SUMMARY OF THERMODYNAMIC RELATIONSHIPS

3.4a PARTIAL DERIVATIVES

Taking account of only the 8 most important variables (P, V, T, S, U, H, F, and G), there are 8 x 7 x 6 = 336 first partial derivatives of the form $(\partial P/\partial V)_T$ and (336 x 335 x 334 x 333) / (1 x 2 x 3 x 4) $\approx 5\ 10^8$ relations between these first derivatives! Even though almost all such relationships would be of little use, these huge numbers call for a more convenient means of evaluating the partial derivatives of interest for solving any specific problem. As noted by the physicist P.W. Bridgman (1882-1961) at the beginning of the 20th century, the 336 first derivatives can be split into 8 classes, each of them being characterized by the variable kept constant in the differentiation. In a way evoking the approach that will be followed in the next chapter to define formation properties, we can reduce considerably the number of relationships to be considered. Within each class, note that differentiation of the 7 remaining variables can be performed with respect to the same dummy variable t

$$\left(\frac{\partial x_1}{\partial x_2}\right)_{x_3} = \frac{(\partial x_1/\partial t)_{x_3}}{(\partial x_2/\partial t)_{x_3}}. \tag{3.39}$$

It then suffices to take the appropriate ratios between the 7 relations such as $(\partial x_i/\partial t)_{x_k}$ to obtain 42 relations of the kind $(\partial x_i/\partial x_j)_{x_k}$. The original 336 derivatives thus reduce to 56, and the simplicity of the method is accrued by the fact that the variable t does not even need to be made explicit. It has just to be selected in such a way that, within each class, any $(\partial x_i/\partial x_j)_{x_k}$ is written as the ratio between two factors that are as simple as possible. From a review of thermodynamic formulae, Bridgman derived a consistent set of such factors in terms of the heat capacity, compressibility and thermal expansion coefficient. He arbitrarily set $(\partial T)_P = 1$ and noted that the number of such independent "partial derivatives" is further reduced by a factor of two because eqn (A.18) implies $(\partial x_i)_{x_k} = -(\partial x_k)_{x_i}$. Instead of reporting Bridgman's original formula, which also included factors for second derivatives, we list in Table 3.1 a somewhat more convenient set of data where $(\partial T)_P$, $(\partial T)_V$ and $(\partial T)_G$ are all unity. On the right-hand side of the table, the variables T, S, P and V indicate which variable is kept constant

Table 3.1. Differentials of thermodynamic variables (after the book by Stacey, 1992)[a]

	T	S	P	V
∂T		γT	1	1
∂S	$\alpha K_T = \gamma C_v/V$		C_p/T	C_v/T
∂P	$-K_T/V$	K_S		$\alpha K_T = \gamma C_v/V$
∂V	1	$-V$	αV	
∂U	$\alpha K_T - P = \gamma\rho C_v T - P$	PV	$C_p - \alpha VP$	C_v
∂H	$-K_T(1-\alpha T)$	$K_S V$	C_p	$C_v(1+\gamma)$
∂F	$-P$	$PV - \gamma TS$	$-S - \alpha VP$	$-S$
∂G	$-K_T$	$K_S V - \gamma TS$	$-S$	$-S + \alpha K_T V = -S + \gamma C_v$
	U	H	F	G
∂T	$P - \alpha T K_T = P - \gamma C_v T/V$	$1 - \alpha T$	P	1
∂S	$C_v P/T$	C_p/T	$C_v(P/T - \gamma S/V)$	$C_p/T - \alpha S$
∂P	$-C_v(K_S - \gamma P)/V$	$-C_p/V$	$K_T(S/V + \alpha P)$	S/V
∂V	C_v	$\alpha V(1 + 1/\gamma)$	$-S$	$\alpha V - S/K_T = (1/K_T)(\gamma C_v - S)$
∂U		$C_p - PV\alpha(1 + 1/\gamma)$ $= \alpha V[K_S/\gamma - P(1 + 1/\gamma)]$	$C_v P - S\gamma C_v T/V + SP$	$C_p - \alpha TS - P\alpha V - PS/K_T$
∂H	$C_v[P(1+\gamma) - K_S]$		$SK_T(1-\alpha T) + C_v P(1+\gamma)$	$C_p + S(1 - \alpha T)$
∂F	$C_v(\gamma TS/V - P) - PS$	$-S(1-\alpha T) - PV\alpha(1 + 1/\gamma)$		$-S(1 - P/K_T) - P\alpha V$
∂G	$C_v(\gamma TS/V + \alpha P - K_S) - PS$	$-S(1-\alpha T) - Cp$	$S(K_T - P) + PV\alpha K_T$	

[a] $\alpha = \frac{1}{V}\left(\frac{\partial V}{\partial T}\right)_P$; $\gamma = \frac{\alpha K_T V}{C_v} = \frac{\alpha K_S M}{\rho C_p} = \frac{\alpha K_S V}{C_p}$, with M = molecular weight ; $K_T = \frac{1}{\beta_T} = -V\left(\frac{\partial P}{\partial V}\right)_T$; $K_S = \frac{1}{\beta_S} = -V\left(\frac{\partial P}{\partial V}\right)_S$.

in the derivative. To calculate $(\partial T/\partial V)_S$, for instance, it suffices to write the ratio of the expressions given for ∂T and ∂V under the heading S. We arrive at once at

$$(\partial T/\partial V)_S = -\alpha K_S T/C_p. \tag{3.40}$$

In the same way we find

$$(\partial T/\partial P)_S = \alpha VT/C_p. \tag{3.41}$$

3.4b JACOBIANS AND PARTIAL DERIVATIVES

With Jacobians, it is possible to give a more rigorous mathematical basis to the partial derivatives calculated with the data of Table 3.1. As will be shown with an example, this procedure is generally much less straightforward. We will nonetheless briefly present it because of its general validity and occasional use in the literature.

If the variables x and y are explicit functions of two other variables u and v, their Jacobian is by definition the determinant of their partial derivatives

$$\frac{J(x,y)}{J(u,v)} = -\frac{J(y,x)}{J(u,v)} = \left(\frac{\partial x}{\partial u}\right)_v \left(\frac{\partial y}{\partial v}\right)_u - \left(\frac{\partial x}{\partial v}\right)_u \left(\frac{\partial y}{\partial u}\right)_v. \tag{3.42}$$

As such a Jacobian has all the properties of a determinant, in particular :

$$\frac{J(x,y)}{J(x,y)} = 1, \quad \frac{J(x,y)}{J(u,v)}\frac{J(u,v)}{J(z,t)} = \frac{J(x,y)}{J(z,t)}, \quad \frac{J(x,y)}{J(u,y)} = \left(\frac{\partial x}{\partial u}\right)_y. \tag{3.43}$$

Let us take again the example of the temperature change due to adiabatic compression to illustrate the use of Jacobians. We write

$$\left(\frac{\partial T}{\partial P}\right)_S = \frac{J(T,S)}{J(P,S)} = \frac{J(T,S)/J(P,T)}{J(P,S)/J(P,T)} = -\frac{(\partial S/\partial P)_T}{(\partial S/\partial T)_P} = \alpha VT/C_p, \tag{3.44}$$

where use had to be made of the Maxwell relationship (3.6d) to express $(\partial S/\partial P)_T$. As the variables u, v are always the same, we could have more simply written

$$J(x,y) = \frac{J(x,y)}{J(u,v)}, \tag{3.45}$$

and remarked that $(\partial y)_t = J(y,t)$ thanks to the relation

$$\left(\frac{\partial y}{\partial x}\right)_t = \frac{J(y,t)}{J(x,t)}. \tag{3.46}$$

Problems

See at the end of chapter 4.

4
OBSERVABLE PROPERTIES

In this chapter emphasis is deliberately put on the empirical aspects of thermodynamics. How can one measure the variables introduced in the previous chapters? What are the auxiliary properties that have to be defined in the whole range of relevant physical conditions? Without any claim to comprehensiveness, we will describe some of the main methods used to measure thermodynamic properties. Of course, an important point concerns the precision with which a property is known. Hence, we will give rough estimates of experimental uncertainties and indicate briefly under which conditions measurements can be made. Additional relations between state variables will also be derived. We will then discuss the case of adiabatic processes, introduce the important concepts of "absolute" and residual entropies with reference to Nernst's law and conclude with a presentation of formation properties.

4.1 INTENSIVE PROPERTIES

4.1a CHEMICAL COMPOSITION

If investigated with instruments such as thermometers or barometers, the isothermal, isobaric mixing of two ideal gases would not seem to induce any change in the system. It is through chemical analysis that gas mixing can be demonstrated. As an irreversible process, mixing is accompanied by an entropy increase that could be otherwise overlooked. The role of chemical composition is thus so important that it is generally required to know the chemical composition and abundance of each phase of a system.

No attempt will be made here at reviewing the innumerable available means for determining chemical compositions. Suffice it to say that, with the current variety of analytical methods, not only major elements but even elements in trace amounts can be analyzed in quantities of matter as small as the nanogram (10^{-9} g) or less. Gas phase chromatography and mass spectrometry are very sensitive techniques in this respect. In liquids, chemical equilibrium is often reached so fast that speciation has to be determined in situ (by chemical titration, electrical conductivity or electrochemical measurements, etc.). In solids, it is at a scale of a few tens of angstroms (1 Å = 10^{-10} m) that one can determine chemical compositions, for elements heavier than sodium, with an analytical electron microscope. For a crystalline solid, note that knowledge of the bulk chemical composition is insufficient when several elements occupy the same crystallographic sites. The distribution of these elements between the various sites should also be known, as determined by various methods (neutron diffusion, X-ray diffraction, nuclear magnetic resonance spectroscopy, etc).

The unit of chemical composition is of course the mole, which can be used for any kind of entity: molecules, atoms, ions, electrons, protons, etc. By definition, a mole is made up of a number N_A of any of these entities, N_A being the Avogadro (1776-1856) number which is arbitrarily defined as the number of carbon atoms present in 12 g of isotopically pure ^{12}C. The mass of one mole of a substance is termed the *molar mass* when expressed in kg, and the *molecular weight*, or *formula weight*, when expressed in g/mol. In an attempt to reduce possible confusion, use of the dimensionless *relative molecular mass* has been recommended for the latter, but, in conformity with common usage, we will keep with the more concise molecular weights.

A problem raised by chemical compositions is the number of different ways in which these can be expressed. The thermodynamically relevant variables are generally the *mole fractions*, which we have already used

$$x_i = n_i / \sum n_i, \tag{4.1}$$

where n_i is the number of moles of component i. They have the interest of being intensive properties and must often be calculated from *weight fractions* given by chemical analyses. These are

$$w_i = n_i M_i / \sum n_i M_i \tag{4.2a}$$

where M_i is the molecular weight of component i. Not making any reference to the microscopic composition of the system, weight fractions are especially useful for complex systems (of biological, geological or industrial interest, among others). In the form of *parts per million* (ppm, i.e., $10^6 w_i$) or *parts per billion* (ppb, $10^9 w_i$), they are also the variables of choice for substances in trace amounts.

For solutions, we will designate by *concentration* any intensive compositional variable. For fluids, the numbers of moles per unit of volume (liter in general)

$$c_i = n_i / V, \tag{4.2b}$$

are termed *molarities*. Inconveniently, they vary with temperature and pressure through the effects of T and P on volume which are often not so accurately known.

In case important concentration differences lead to distinguishing solutes from the solvent, the *molality* of solute j is the number of moles of j per kg of solvent

$$m_j = 1000n_j/n_sM_s = n_j/n_sM_s^*, \tag{4.2c}$$

where M_s and M_s^* are the molecular weight (in g) and molar mass (in Kg) of the solvent, respectively, and n_s its number of moles in 1 kg. Molalities have the advantage of not depending on either temperature or pressure, and of making easy preparation of solutions with accurately known concentrations. They are especially useful when dealing with electrolyte solutions.

For convenience we list in Table 4.1 the formulae with which a composition variable can be converted into another. Also included in this table are the less commonly used *formalities* (F_i), the numbers of moles of a substance in 1 kg of solution. With the normalization condition (4.1), the composition of an N-component system may be described by one extensive variable (mass or total number of moles, for instance), and N- 1 intensive variables such as mole or weight fractions. Whenever the size of the system is irrelevant, N - 1 variables will thus suffice to specify the composition of each phase. Finally, an important feature of these concentration scales is that they tend to become proportional to each other at high dilution. Between mole fractions and molalities, we have for instance

$$x_j = \frac{m_j}{\sum m_j + 1000/M_s} \tag{4.2d}$$

where the summation is made on the solutes only. For very dilute solutions, $\sum m_j$ tends to 0 and (4.2d) indicates that m_j becomes proportional to the mole fraction. The reader may care to check that a similar conclusion applies to the molarities.

4.1b TEMPERATURE

The need for an absolute scale of temperature has already been noted. The absolute scale, whose unit is the Kelvin (K), is linked to the Celsius and Fahrenheit scales by T (K) = T (°C) + 273.15 and T (K) = 5/9 T (°F) + 255.37, respectively. For practical reasons, a *standard temperature* of 298.15 K (25°C) has been chosen as a reference temperature for the thermodynamic properties of a phase. For the sake of simplicity, we will often denote it as 298 K.

Resistive heating constitutes the most convenient means of obtaining and controlling temperatures higher than ambient in the laboratory. The resistive elements used depend on the temperature and atmosphere conditions (under oxidizing conditions, 1400 K can be reached with common Ni-Cr alloys, 2000 K with $MoSi_2$; under vacuum or a neutral atmosphere, 2500 K with tungsten, 3000 K with graphite). Temperatures as high as 5000 or 6000 K can be reached with infrared lasers, but only for samples with a size of few tens of microns. On the other hand, low temperatures are obtained with *cryostats* cooled by cryogenic liquids like boiling nitrogen (77.3 K) or helium (4.2 K). Through electrical heating of the instrument immersed in the cryostat, the temperature can then be adjusted to the required value. Still lower temperatures are reached with *adiabatic demagnetization* methods. A salt with a high magnetic susceptibility (initially gadolinium sulfate, nowadays cerium magnesium nitrate)

Table 4.1. Conversions between composition variables[a]

	n_i (mol)	x_i	c_i (mol/l)	m_j (mol/kg solvent)	F_i (mol/kg solution)	w_i
x_i	$\frac{n_i}{\sum n_i}$		$\frac{c_i}{\sum c_i}$	$\frac{m_j}{\sum m_j + 1000/M_s}$	$\frac{F_i}{\sum F_i}$	$\frac{w_i/M_i}{\sum w_i/M_i}$
c_i	$\frac{1000\,\rho n_i}{\sum n_i M_i}$	$\frac{1000\,\rho x_i}{\sum x_i M_i}$		$\frac{1000\,\rho m_j}{1000 + \sum m_j M_j}$	ρF_i	$\frac{1000\,\rho w_i}{M_i}$
m_j	$\frac{1000\,n_j}{n_s M_s}$	$\frac{1000\,x_j}{x_s M_s}$	$\frac{1000\,c_j}{1000\,\rho - \sum c_j M_j}$		$\frac{1000\,F_j}{1000 - \sum F_j M_j}$	$\frac{1000\,w_j}{M_j(1 - \sum_{j \neq s} w_j)}$
F_i	$\frac{1000\,n_i}{\sum n_i M_i}$	$\frac{1000\,x_i}{\sum x_i M_i}$	$\frac{c_i}{\rho}$	$\frac{1000\,m_j}{1000 + \sum m_j M_j}$		$\frac{1000\,w_i}{M_i}$
w_i	$\frac{n_i M_i}{\sum n_i M_i}$	$\frac{x_i M_i}{\sum x_i M_i}$	$\frac{c_i M_i}{1000\,\rho}$	$\frac{m_j M_j}{1000 + \sum m_j M_j}$	$\frac{F_i M_i}{1000}$	

[a] ρ = density (kg/l or g/cm^3), M_i = molecular weight, in g/mol ($i = s$ for solvent), subscript i running for all species (solvent included), j running for solutes only.

is magnetized at liquid helium temperatures. When the field is removed under adiabatic conditions, the transition of the salt to a disordered spin configuration requires some energy that is supplied by the crystal lattice. This ensures a temperature decrease and it suffices to repeat the process for reaching lower and lower temperatures, the record low ones being currently of the order of 1 μK (10^{-6} K). With nuclear magnetism, temperatures of the order of 10^{-8} K have been reached for nuclei in the same manner. But is it possible to reach 0 K? This temperature has the special feature that any process would be, as concluded in 4.2e from dP/dV comparisons, not only isothermal but adiabatic. Since the 0 K isotherm could not be intersected by any other adiabat, it follows that it is impossible to reach 0 K by a series of adiabatic processes. More generally, one shows that this temperature cannot be attained through a finite number of any kind of processes.

In metrology, absolute temperatures are based on gas thermometry, as discussed in 5.1a, and then on optical pyrometry, which relies on the fact that the thermal emission spectrum of a black body shifts to shorter wavelengths when the temperature increases (see 13.6f). An International Practical Temperature Scale, which is revised about every 20 years, has been constructed from a series of fixed points measured as accurately as possible. For the 1990 scale (see McGlashan, 1990), the fixed points are the triple points of H_2 (13.8033 K), Ne (24.5561 K), O_2 (54.3584 K), Ar (83.8058 K), Hg (234.3156 K) and H_2O (273.16 K), along with the the 1 atm meltings of Ga (302.9146 K), In (429.7485 K), Sn (505.078 K), Zn (692.677 K), Al (933.473 K), Ag (1234.93 K), Au (1337.33 K) and Cu (1357.77 K). From this primary scale, secondary scales are devised on the basis of the variation with temperature of certain physical properties. A most familiar example is the dilation of mercury or alcohol in capillary thermometers. At around room temperature, such thermometers can be precise to 0.01 K, but their use is restricted because of their small temperature intervals and mediocre long-term stability. The variation with temperature of the resistivity of an electric conductor is a much more sensitive probe. Platinum resistance thermometers are the reference instruments because they can accurate to almost 0.001 K from the vicinity of 0 K to more than 1000 K. Especially at higher temperatures, thermocouples are used. Their working principle is that of the Seebeck effect, whereby an electromotive force develops between the two ends of a conductor when these are brought to different temperatures. (This effect thus represents a direct production of electricity from heat with, however, an extremely small efficiency.) There exist a great many different thermocouples, the most accurate being made up of platinum alloys with which temperatures can be measured to within 0.1 K from room temperature to 2000 K. Note finally that optical pyrometry is a very accurate method in metrology, but that, as a result of numerous practical difficulties, it gives temperatures to within 10 K at best when used in the laboratory for measurements of physical properties.

4.1c PRESSURE

Pressure, the tension of mechanical energy, represents the ratio between a force and the surface on which the force is exerted. Once the units of length, mass and time have been selected, the unit of pressure is thus fixed. The *Système International* unit of pressure, the Pascal (1 Pa = 1 N/m^2), is unfortunately very small, for standard atmospheric pressure (at sea level) is 1.01325 10^5 Pa. Not to deal with too large numbers, one has long used the atmosphere or the bar (= 10^5 Pa) as pressure units. Heights of mercury (1 atm = 760 mm of mercury) are now discarded, whereas tors (1 tor = 1 mm of mercury, or 1.333 10^{-3} bar) are still used by the vacuum industry.

The gauges used to measure pressures in fluids have as a working principle the relation between the elastic deformation of a body (e.g., a membrane) and the pressure exerted.

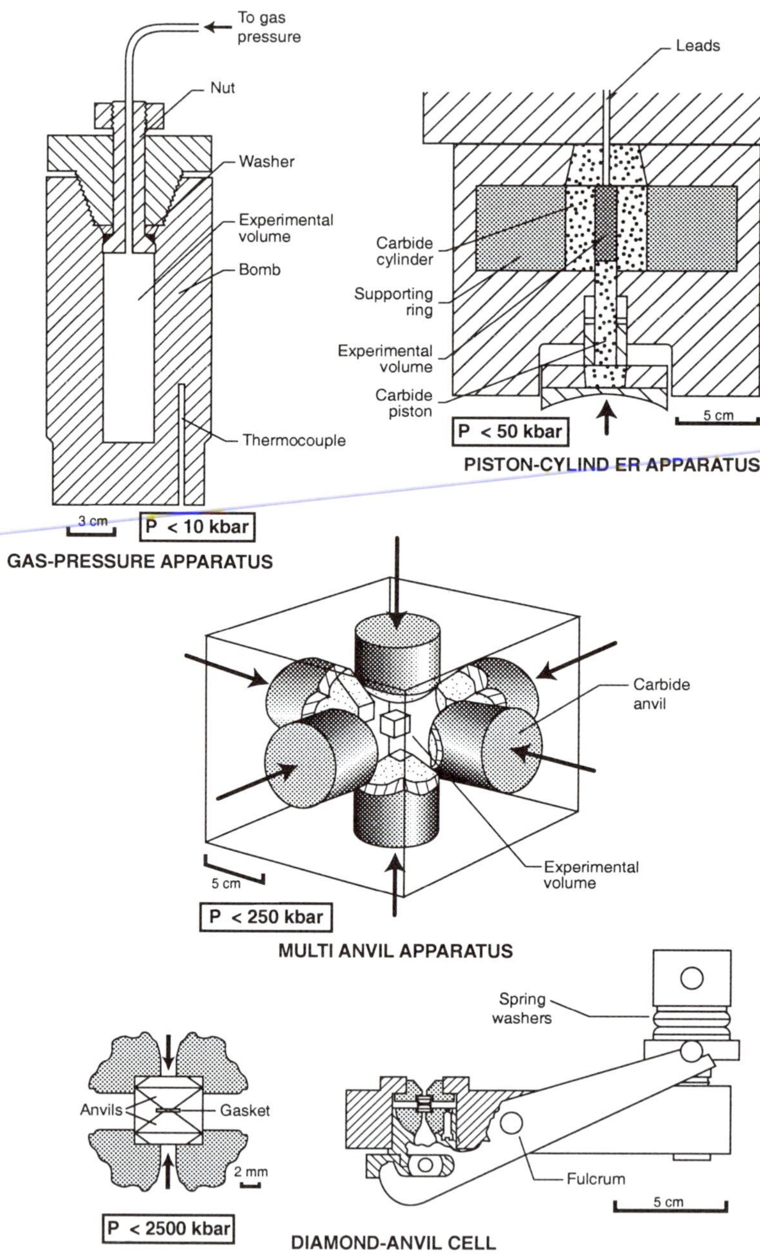

Figure 4.1. A few kinds of high-pressure apparatuses. Note that pressure intensification is greatest in diamond-anvil cells, where the ratio between the surface areas of the large and small diamond faces (see the left-hand side enlargement) typically ranges from 100 to 1000.

Pressure measurements in solid media are more difficult and less precise. Owing to the intrinsic friction developed in any apparatus, the pressure differs from the ratio between a force and a surface area. Measurements are thus made from the pressure dependence of some physical property such as the electrical resistance of a conductor (i.e., manganin resistance gauge) or the wavelength of an electromagnetic radiation (i.e., ruby fluorescence with which million atmosphere pressures can be measured to within 5%).

On the low-pressure side, pressures as low as 10^{-12} bar can be reached with various kinds of pumps. For high pressures, the general principle at work is that of intensification whereby an elevated pressure is produced when transmitting to a small surface the great force exerted over a large surface. As shown in Fig. 4.1, there are many different ways of applying this principle. The price to be paid to get to higher pressures is a marked decrease of the sample size. By compressing a fluid in a thick-walled vessel, one barely goes beyond 10 kbar. Higher pressures require the use of very hard solid pressure-transmitting media. In *piston-cylinders*, pressures higher than 50 kbar can be produced through uniaxial sliding of a piston in a cylinder, both made up of tungsten carbide. With a *multianvil* apparatus, anvils are moved by a powerful press along the three space directions toward the sample. They can ensure pressures of about 250 and 350 kbar when made of tungsten carbide and sintered diamonds, respectively, for mg samples. Much higher pressures are reached in *diamond-anvil cells* where two diamonds are opposed in a nutcracker-like fashion. With the 0.1 - 0.5 mm culet of 1/3 carat brilliant-type diamonds, pressures of up to the record 500 GPa (= 5 10^6 bar) have been reached, but the size of samples inserted in the hole of a gasket between the diamonds is of the order of 10^{-8} gram. Still higher pressures can be achieved dynamically for microsccond periods in shock waves produced by chemical or nuclear explosives and by the impact of supersonic projectiles (see appendix C) or high-power laser beams.

Regardless of the high-pressure technique, heating of samples is made difficult by the deterioration of mechanical properties in hot materials. By placing a bomb made up of the best alloys in a furnace, one goes hardly beyond 1200 K and fluid pressures of a few kbar. By inserting a furnace in an autoclave whose walls are kept cold one can reach 10 kbar and 1800 K for gram size samples. Inserting samples in small graphite furnaces allows temperatures of up to 2000 K to be reached in piston-cylinders and multianvil apparatus. Diamond graphitization at low pressures limit to about 1000 K the temperatures that can be reached with externally heated diamond-anvil cells. But good-quality diamonds are transparent for a wide range of electromagnetic radiation wavelengths so that heating of samples compressed in diamond-anvil cells is possible up to a few thousand degrees with the beam of an infrared laser. (Incidentally, this is how the P, T conditions of the deepest recesses of the earth are now reproduced.)

4.2 VOLUME PROPERTIES

4.2a VOLUME AND DENSITY

The primary SI unit of volume, the m^3, is not always used because molar volumes are generally of the order of a few tens of cm^3. As volumes often appear as a product with pressure (to give a work whose SI unit is the Joule), it can be convenient to express them in J/bar or J/Pa. Note that 1 J/bar = 10 cm^3 and 1 J/Pa = 10^{-4} cm^3. For gases, many volume measurements have been published in the scale devised by Amagat (1841-1915), whose unit is the molar volume of the gas considered at 1 bar and 273.15 K. By the same token, the Amagat unit for PV is the product of these two variables for the gas at 1 bar and 273.15 K.

The specific volume, the ratio between volume and mass, or the density represent other common ways to express volumes as intensive variables. Strictly speaking, the density is the ratio between the specific masses of the investigated body and water at 4°C. Densities are thus dimensionless quantities even though they are generally confused in the literature with specific masses. Numerically, both quantities are identical when the latter are reported in g/cm^3. The risk of confusion is slight, however, because of the units that are appended to these so-called "densities".

The volume of a given amount of gas or liquid is readily determined from that of the container. Such *piezometers* are split in two categories depending on whether pressure or volume changes are measured (at constant temperature). It goes without saying that the accuracy of the measurements decreases when working at high temperatures and pressures. Whereas errors are much lower than 1‰ near ambient conditions, they reach 1% at around 10 kbar and 1000 K, above which very few static measurements on fluids have been performed. For liquids, many different methods are being used to measure the density. The simplest one is Archimedes' whereby a solid of known volume (a bob) is immersed in the liquid. The weight difference between the bob in air and in the liquid is equal to the weight of the displaced liquid. We have thus $\rho_l = \rho_b w_a/(w_a - w_l)$, where ρ_l and ρ_b are the densities of the liquid and bob, respectively, and w_a and w_l are the weights of the bob in air and in the liquid. The main advantage of this method is to rely on mass measurements, which can be routinely made with a relative accuracy of 10^{-6}. (Only masses are so readily measured to such an accuracy.) Of course, this method can also be used at high temperatures, but the relative errors increase from 10^{-4} - 10^{-5} near ambient conditions to 10^{-2} - 10^{-3} beyond 1500 K. In contrast, direct measurements of liquid densities are difficult at high pressures. Densities are in fact generally calculated from compressibilities measured as described below.

For a solid, the density is most simply determined from the volume and weight. As this can be made only for big samples having a regular shape, the Archimedean method with a nonreactive wetting liquid of known density is in fact better suited. By dilatometry, the change in the length of a sample (oriented along some specific direction in the case of crystals) is measured as a function of temperature. For crystals, X-ray diffraction methods are used to determine unit-cell parameters and volumes to much better than 1‰ under ambient conditions. The latter methods are not biased by the intrinsic porosity of polycrystals and they can be performed not only on samples as small as 10^{-7} g but also at high temperatures and pressures (with, however, greater errors, of the order of 0.5%). For very small amounts of matter or extreme P, T conditions, the measurements can now benefit from the remarkable brightness of the X-ray beams delivered by synchrotron sources. Because microscopic defects (vacancies, dislocations, etc.) are not seen in standard X-ray diffraction experiments, the difference between the results obtained from dilatometry and X-ray diffraction gives their contribution to the volume (provided the dilatometric measurements are of course made on a samples of zero porosity). Crystal volumes as determined from X-ray diffraction experiments are generally correct because the influence of microscopic defects becomes significant only in the vicinity of the melting point. A classic experiment represented in Fig. 4.2 shows that a small part of aluminum dilation is due to the rapid increase of the concentration of Schottky vacancies (i.e., crystallographic sites becoming empty through atom migration to the surface of the crystal).

Finally, volumes can also be measured for gases, liquids and solids starting materials with the aforementioned shock-wave techniques up to several million atmospheres (see Appendix C). The main drawbacks of these dynamical methods is that the temperature and pressure

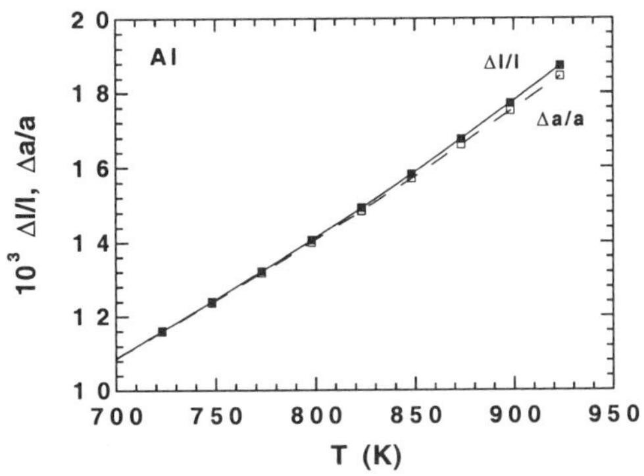

Figure 4.2. Dilation of aluminum ($\Delta l/l$), as measured by dilatometry, and relative change in lattice parameter ($\Delta a/a$), as measured from X-ray diffraction, up to the vicinity of the melting point at 933 K. The difference between the two curves is due to Schottky defects. Experimental data of Simmons and Balluffi (1960).

cannot be varied independently and that their operating cost is so high that they have long been the preserve of military research.

4.2b THERMAL EXPANSION COEFFICIENT

As defined by

$$\alpha = 1/V\ (\partial V/\partial T)_P, \tag{4.3}$$

thermal expansion coefficients have units of K^{-1}. In thermodynamics, linear thermal expansion coefficients, $\alpha_l = 1/l\ (\partial l/\partial T)_P$ where l is the length, are less important than volume coefficients. For an isotropic body (gas, liquid, crystals with a cubic symmetry), $\alpha = 3\alpha_l$. This does not hold true for anisotropic bodies for which up to six coefficients have to be defined in the case of crystals with a triclinic symmetry (three for the unit-cell parameters and three for the unit-cell angles).

There are no direct means for measuring thermal expansion coefficients. These have to be determined from fits made to volume data. Above room-temperature, their temperature dependence can be accounted for by empirical equations such as

$$\alpha = \alpha_0 + \alpha_1 T + \alpha_2 T^2. \tag{4.4}$$

If the volume is measured between temperatures T_0 and T, the α_i coefficients are determined from fits made with the equation obtained through integration of (4.3) and (4.4)

$$V = V_0 \exp[\alpha_0(T - T_0) + \alpha_1/2\ (T^2 - T_0^2) + \alpha_2/3\ (T^3 - T_0^3)]. \tag{4.5}$$

The low-temperature limit of α can be found from a simple argument. As will be seen in 4.4b, the entropy of a phase tends to a constant value at 0 K, which is independent of pressure. Hence $(\partial S/\partial P)_T = 0$ at T = 0 K. Now, relation (3.6d) gives $(\partial V/\partial T)_P = -(\partial S/\partial P)_T$, which imposes that α tends to 0 when T tends to 0 K. In contrast, thermodynamics gives no constraint on the sign of α. In general, a phase expands when the temperature increases, but water below 4°C is a well-known counter-example of a phase with a negative thermal expansion coefficient (see Fig. 1.2).

For an ideal gas, α is equal to $1/T$ (see 5.2a). At temperatures higher than 1000 K, α becomes similar to expansion coefficients of liquids which are typically of the order of $10^{-3} K^{-1}$. Solids expand still less, having expansion coefficients of the order of 10^{-4} - 10^{-5}

K^{-1} above room temperature. An unfortunate consequence of the lack of direct measurements is that the accuracy is generally one order of magnitude lower for α than that for the measured volume: for many crystals or liquids, the uncertainties reach 10 to 30% or more above 1000 K. Note finally that the definition of α as an intensive parameter results in a number of complications. For instance, this makes it difficult to calculate partial molar thermal expansion coefficients and leads to the awkward conclusion that for solutions both the volume and thermal expansion coefficient cannot be linear functions of composition at any temperature. An extensive coefficient $\alpha' = (\partial V/\partial T)_P$ would not raise these problems and would even simplify important expressions. Instead of being given by $-\alpha V$ and $V[\alpha^2 + (\partial\alpha/\partial T)_P]$, respectively, the partial derivatives $(\partial S/\partial P)_T = -(\partial V/\partial T)_P$ and $(\partial^2 V/\partial T^2)_P$ would reduce to $-\alpha'$ and $(\partial\alpha'/\partial T)_P$. But the price to be paid would be that the coefficient α' of anisotropic substances would no longer be the appropriate sum of the linear expansion coefficients.

4.2c COMPRESSIBILITIES

In addition to the isothermal compressibility already mentioned,

$$\beta_T = -1/V\ (\partial V/\partial P)_T = -1/V\ (\partial^2 G/\partial P^2)_T, \tag{4.6a}$$

we define an *adiabatic* compressibility

$$\beta_S = -1/V\ (\partial V/\partial P)_S = -1/V\ (\partial^2 H/\partial P^2)_S, \tag{4.6b}$$

where the enthalpy shows up in the second equation thanks to (3.3c). The isothermal compressibility, which we will often term for short the compressibility, is the most important one but, like the thermal expansion coefficient, it cannot be measured directly. Although the adiabatic compressibility is relevant for the more special case of processes at constant entropy, its usefulness mainly lies in its possible accurate determination from elastic constant measurements (see below). Especially for condensed phases, one often prefers to use the reciprocal of these compressibilities, which are termed the isothermal and adiabatic *bulk moduli*

$$K_T = 1/\beta_T = -V\ (\partial P/\partial V)_T, \tag{4.7a}$$
$$K_S = 1/\beta_S = -V\ (\partial P/\partial V)_S. \tag{4.7b}$$

Compressibilities are generally expressed in Pa^{-1} or bar^{-1}, bulk moduli in Pa or bar.

Through the minus sign in (4.6), β cannot be negative since such values would imply that a substance expands when compressed. (At constant temperature, the Gibbs free energy change of a phase is $dG = VdP = -dV/\beta$; with $\beta < 0$, the Gibbs free energy would decrease on compression, and the equilibrium state would be reached for a vanishing volume!) Besides the fact that it is not zero, nothing can be said a priori on the low-temperature limit of β. We refer to chapter 5 for a discussion of the pressure and temperature dependences of the compressibilities of gases, liquids and solids.

Is the difference between β_T and β_S significant? The ratio between the two compressibilities is

$$\frac{\beta_T}{\beta_S} = \frac{(\partial V/\partial P)_T}{(\partial V/\partial P)_S}. \tag{4.8}$$

Making use of (A.18), we obtain

$$\frac{\beta_T}{\beta_S} = \frac{(\partial T/\partial P)_V(\partial V/\partial T)_P}{(\partial S/\partial P)_V(\partial V/\partial S)_P} = \frac{(\partial S/\partial T)_P}{(\partial S/\partial T)_V}. \tag{4.9}$$

From the definitions of bulk moduli and heat capacities at constant pressure and volume, this ratio is

$$\beta_T/\beta_S = K_S/K_T = C_p/C_v > 1, \tag{4.10}$$

but discussion of its value is deferred to section 4.3a that is devoted to heat capacities.

One determines β_T by differentiation of volume measurements made as a function of pressure. This is the most commonly used method for pressures higher than 10 kbar. When possible, however, it is preferable to measure β_S through sound velocity experiments. For a liquid one has most simply (see Appendix B)

$$\beta_S = 1/\rho c^2, \tag{4.11}$$

where c and ρ are the sound velocity and density of the medium investigated. Errors can be as small as 1 ‰ at room temperature and pressure, and limited to 1% at higher temperatures.

Compressibilities increase when going from solids to liquids and from liquids to gases. For gases at low pressures, β_T is $1/P$. It decreases considerably with increasing pressures to become similar to those of liquids, which are typically of the order of 10^{-9} - 10^{-10} Pa^{-1}. Solids are the least compressible phases, with β_T of the order of 10^{-11} Pa^{-1} (i.e., $K_T \approx$100 GPa = 1000 kbar). For the sake of comprehensiveness, let us finally mention linear compressibilities (β_l). Like linear thermal expansion coefficients, these must be defined along each relevant direction for an anisotropic substance. For an isotropic substance, $\beta = -(\partial \ln V/\partial P) = 3\beta_l$. As to the remarks made above about the inconvenience of having defined α as an intensive variable, they also apply to the compressibilities.

The difference between the isothermal and adiabatic compressibilities gives an opportunity to stress that measurements of physical properties are never direct in the sense of using the old reference iridium-platinum bar to measure the length of a body. Especially with modern instrumentation, any measurement relies on a number of complex operations and the boundary between a direct measurement and an indirect determination is thus blurred. Arbitrarily, we will state that the adiabatic compressibility is directly measurable because it is obtained from density and sound velocities, which can be known accurately. In contrast, the isothermal compressibility has to be determined indirectly through the damaging operation, from a precision standpoint, of differentiation.

4.2d THERMAL EXPANSION COEFFICIENT VS. COMPRESSIBILITY

Thermodynamics tell us nothing concerning the variations of α with P and those of β_T with T. But volume is a state function whose differential as a function of P and T is

$$dV = (\partial V/\partial T)_P dT + (\partial V/\partial P)_T dP.$$

As dV is an exact differential

$$(\partial^2 V/\partial P \partial T) = (\partial^2 V/\partial T \partial P),$$

we obtain a simple relation between the variations of α with P and of β_T with T

$$(\partial\alpha/\partial P)_T = -(\partial\beta_T/\partial T)_P. \tag{4.12}$$

Actually, β_T generally increases with temperature and α decreases when the pressure increases.

4.2e GRÜNEISEN PARAMETER AND ADIABATIC PROCESSES

We will conclude this discussion of volume properties with other kinds of adiabatic processes. We first note that, contrary to adiabatic compressibility, an adiabatic thermal expansion coefficient is not measurable directly. With (3.40), however, it would be possible to express it in terms of other variables

$$1/V\ (\partial V/\partial T)_S = -C_p/\alpha V K_S T = -1/T\gamma, \tag{4.13}$$

where γ is an intensive, dimensionless variable which can be defined in several ways through (A.16) and (4.10)

$$\gamma = \alpha V K_S/C_p = \alpha V K_T/C_v \tag{4.14a}$$
$$= TV\alpha^2 K_T/\alpha T C_v = (C_p/C_v - 1)/\alpha T. \tag{4.14b}$$

This coefficient γ is the *thermal Grüneisen parameter*. The symbol γ being often used to designate C_p/C_v for gases, we will denote the latter ratio by γ_c to limit confusion.

Rewriting (4.13) we derive

$$(\partial \ln T/\partial \ln V)_S = -\gamma. \tag{4.15a}$$

The Grüneisen parameter thus accounts for temperature-volume changes within a system during adiabatic transformations. If γ can be considered a constant, we obtain by integration of (4.15a) between the volumes V_0 and V

$$TV^\gamma = T_0 V_0^\gamma. \tag{4.15b}$$

Since γ has the same sign as α, adiabatic expansion generally causes the temperature to decrease.

The temperature and pressure changes during adiabatic transformations are given by (3.41), which we rewrite

$$(\partial T/\partial P)_S = \alpha V T/C_p = \gamma T/K_S. \tag{4.16a}$$

Now, measuring these mutual changes is not an experimental challenge. The equation (4.16a) has thus been used to determine heat capacities of gases whose thermal expansion was known, and thermal expansion coefficients (or Grüneisen parameters) of condensed phases whose heat capacity (or bulk modulus) was known. From (4.13) and (4.16a) we then obtain the P, V relation for adiabatic processes

$$(\partial P/\partial V)_S = \alpha VT/C_p = -K_S/V. \tag{4.16b}$$

Comparing this derivative with the isothermal value $(\partial P/\partial V)_T = -K_T/V$, we find that the absolute value of the dP/dV slope of an adiabat is necessarily steeper than that of an isotherm because (4.10) states that $K_S > K_T$ for $T > 0$. This slope difference could thus vanish only at 0 K where $K_S = K_T$.

Finally note that α and K_S generally have opposite variations with both temperature and pressure. On the other hand, the variations with temperature of heat capacities are usually not so pronounced above room temperature. Hence, the Grüneisen parameter varies much less with temperature and pressure than V, α, K_S, K_T, C_p, or C_p individually. In general, it ranges between 1 and 1.5 for solids, decreases slightly when the temperature increases and, as a first approximation, can thus be estimated from its value at room temperature and pressure.

4.3 HEAT CAPACITY AND ENTHALPY

4.3a HEAT CAPACITY

We have seen in 2.1d that the reference value of the internal energy is fixed arbitrarily. The same applies to the enthalpy. Between two temperatures, the variations of both functions are calculated through integrations of the heat capacities at constant volume or pressure

$$C_p = (\partial H/\partial T)_P = T(\partial S/\partial T)_P = -T(\partial^2 G/\partial T^2)_P, \tag{4.17a}$$
$$C_v = (\partial U/\partial T)_V = T(\partial S/\partial T)_V = -T(\partial^2 F/\partial T^2)_V. \tag{4.17b}$$

The difference between these two heat capacities

$$C_p - C_v = TV\alpha^2/\beta_T, \tag{4.18}$$

is calculated in Appendix A, and we have seen in the previous section that their ratio is equal to that of the isothermal and adiabatic compressibilities, $C_p/C_v = \beta_T/\beta_S$.

If needed, heat capacities can be defined for any other kind of transformation. At constant enthalpy, the heat capacity would be $C_H = T(\partial S/\partial T)_H$. This quantity is not measurable, but the reader may check that it is equal to $C_p/(1-\alpha T)$. More important is the fact that, for condensed phases, the measured heat capacities (C_{sat}) are actually not isobaric but refer to the variable saturation pressure of the vapor. Denoting by $(\partial P/\partial T)_{sat}$ the variation with temperature of this pressure, and remarking that

$$(\partial H/\partial T)_{sat} = (\partial H/\partial T)_p + (\partial H/\partial P)_T(\partial P/\partial T)_{sat}, \tag{4.19}$$

we find that

$$C_{sat} = (dQ/dT)_{sat} = (\partial H/\partial T)_{sat} - V(\partial P/\partial T)_{sat}$$

$$= C_p + [(\partial H/\partial P)_T - V](\partial P/\partial T)_{sat}. \tag{4.20}$$

Unless the vapor pressure is much higher than 1 atm, however, (4.20) indicates that the difference between C_p and C_{sat} is negligible. For water, it amounts to 0.03 % of C_p at 373 K.

Heat capacities are usually expressed in J/g K or J/mol K. The 4.184 J per calorie conversion factor corresponds to the temperature rise of one gram of water from 14.5 to 15.5°C, but one finds, especially in the older literature, other calories worth from 4.182 to 4.186 J, which thus refer to other temperature intervals than 14.5-15.5°C. Calorimeters having long been calibrated by electrical measurements, the calorie should have only an historical interest.

How do the differences between C_p and C_v, or between β_T and β_S, vary as a function of temperature? For an ideal gas, the well-known result $C_p - C_v = R$ (gas constant) is discussed in 5.2a. For a condensed phase, α tends to 0 when the temperature tends to 0 K, and, as discussed above, β_T tends to a finite limit. Hence, C_p tends to C_v and β_T to β_S in the vicinity of 0 K. The differences between β_T and β_S and between C_p and C_v thus vanish at 0 K. For solids, they remain generally smaller than 1% up to room temperature. Because of the increase of α with temperature and the presence of a T term in the expression of $C_p - C_v$, this difference then increases rapidly to reach from 5 to 10% of C_p near 1500 K. For liquids, no general conclusion can be drawn except that the difference between C_p and C_v is in principle significant.

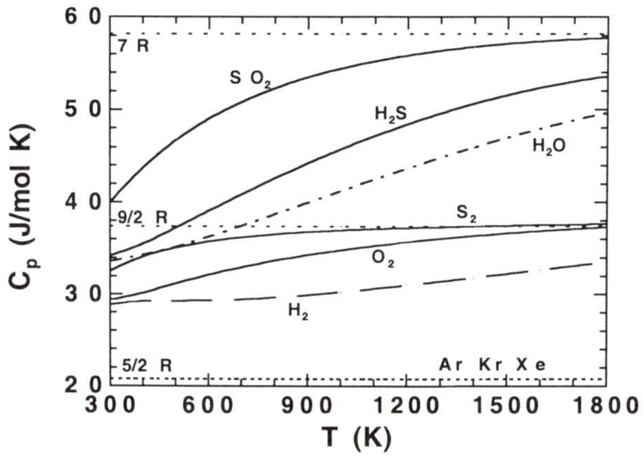

Figure 4.3. Heat capacities of a few ideal gases above room temperature. The high-temperature limits have been included for comparison as dashed lines. Data from the compilation of Robie et al. (1979)

The heat capacities of a few ideal gases are plotted in Fig. 4.3. For monoatomic gases (e.g., the noble gases), C_p is constant and equal to 5/2 R. For diatomic and polyatomic gases, C_p increases with temperature and the mass of the atoms constituting the molecule. At high temperature, it becomes close to $(3n - 3/2)R$ for a linear molecule made up of n atoms, and to $(3n - 2)R$ for a nonlinear molecule (see 14.2).

For solids, it has been noted in 1.4b that C_p tends to 0 when the temperature tends to 0 K. The data for sapphire (the so-called α crystalline form of alumina, Al_2O_3, which is a basic calorimetric standard) plotted in Fig. 4.4a show the typical s-shaped C_p curve imposed by this limit. On the high-temperature side, these data also show the important C_p increase that usually accompanies the enthalpy discontinuity at the

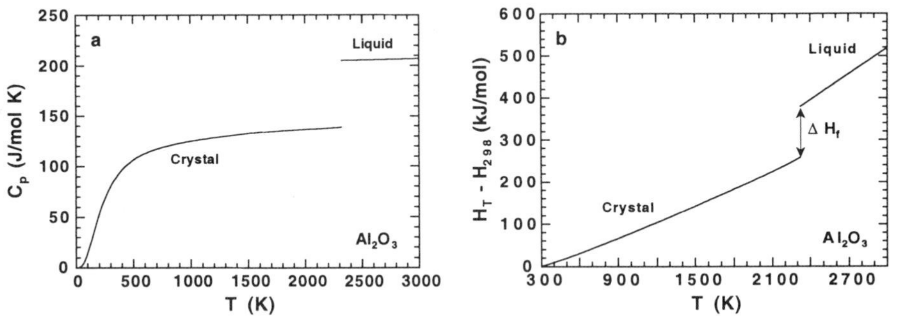

Figure 4.4. (a) Heat capacity of sapphire (α-Al_2O_3) and liquid alumina. (b) Relative enthalpy of sapphire and liquid alumina against temperature; reference state at 298 K: crystal. Data of Ditmars et al. (1982) and Barkhatov et al. (1973).

melting temperature (Fig. 4.4b). This effect is more clearly revealed by the measurement on selenium which, thanks to a much lower melting temperature, could be made very accurately during melting (Fig. 4.5).

Besides, selenium is an element that readily forms a glass through cooling of a liquid. Its various phases have been subjected to calorimetric measurements from the vicinity of 0 K up to the melting point. As shown in Fig. 4.5, the C_p difference

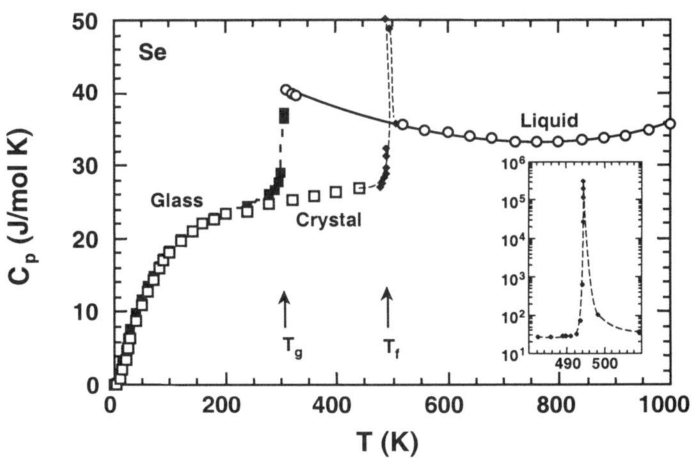

Figure 4.5. Heat capacity of crystalline, vitreous and liquid selenium. T_g and T_f are the glass transition and melting temperatures, respectively. Note the extremely high C_p values measured in the melting interval (see the logarithmic scale of the inset). Also note that the data for the liquid below the temperature of fusion refer to the supercooled liquid, which is metastable with respect to the crystal but remains in internal equilibrium. Data of Chang and Bestul (1974) and Grønvold (1973).

between the glass and crystal is small whereas the change from the liquid to glass, near 305 K, is revealed by an important C_p decrease. This is the *glass transition* where the disordered atomic configuration of the liquid is frozen in in an amorphous solid (see 6.4). Hence, both Figs 4.4a and 4.5 illustrate the very general feature that heat capacity generally increases when going from the solid (crystalline or glassy) to the liquid state.

Although statistical thermodynamics can account for the temperature dependence of heat capacities, it does so analytically in a complicated manner (see chapter 14). One thus prefers to use simple empirical equations above room temperature, and an apparent confusion could result from the numerous expressions that have been proposed. As examples, we will mention only two kinds of commonly used equations, viz.

$$C_p = a + bT + c/T^2 + d/T^{0.5} + eT^2, \tag{4.21}$$

$$C_p = k_0 + k_1/T + k_2/T^2 + k_3/T^3, \tag{4.22}$$

where the constants a, b, k_1, k_2, etc., are determined through fits to the calorimetric measurements. Often truncated forms of such expressions reproduce the data satisfactorily. The three first terms of (4.21) constitute the well-known Maier-Kelley equation.

Even though it is not too difficult to keep the volume of a gas constant, measuring C_v is not so easy a task. For an ideal gas, it is generally preferred to calculate C_v by the statistical mechanical methods described in chapter 14. For a real gas, such calculations represent a starting point in that the corrections to be made to the ideal gas values are determined from the equations of state, with eqn (5.31), and C_p is finally derived with (4.18) from C_v. It is nevertheless possible to measure C_p by various means, the simplest of which is the flux method: the gas temperature is measured at both ends of an apparatus in which a given amount of heat is transferred to the gas. This method has the additional advantage of being applicable, with some difficulties, to pressures of up to 1 kbar. Other methods are too numerous to be mentioned here.

For a condensed phase C_v measurements are impossible because extremely high pressures would be required to keep the volume constant when the temperature increases. Measurements are thus made at constant pressure, most often atmospheric pressure. Again, the simplest means consists of measuring the temperature rise of a sample to which a known amount of heat is supplied. If both quantities are small enough then their ratio directly yields C_p. This is the principle of *adiabatic* calorimetry, whereby any heat exchange between the sample and the surroundings is avoided thanks to a shield constantly kept at the same temperature as the sample (whence its name of adiabatic shield). Because the current and resistance of the electrical circuit producing the heat can be very accurately measured, errors on C_p are typically lower than a few % below 10 K and less than 1‰ between 20 and 400 K.

Between 300 and 1000 K, *differential scanning* calorimetry (DSC) has recently become popular. Schematically, a standard and a sample are heated under the same conditions and the temperature difference measured between them is assumed to be proportional to their heat capacity difference. An advantage over adiabatic calorimetry is the much smaller sample size (e.g., a few tens of mg vs. a few tens of grams), but typical errors of 1% are one order of magnitude higher. With alternative current calorimetry, some heat is supplied in a discontinuous and periodic way (by electrical or radiative means) to a sample initially kept at a constant temperature. The amplitude of the fluctuations measured with respect to the baseline temperature is inversely proportional to the heat capacity of the sample. Like any dynamical method, this one suffers from errors due to the time lag characterizing heat exchange between the sample and the surroundings.

Maintaining adiabatic conditions becomes increasingly difficult at higher temperatures where radiative transfer sets in. Beyond 1000 K, the heat exchanged between the sample and the surroundings can no longer be controlled adequately. It is then preferable to heat a sample to a high temperature T before transferring it to a calorimeter where it cools down to a temperature T_0, usually close to 298 K, in which the released enthalpy $H(T) - H(T_0)$ is measured accurately with the method of mixture mentioned in 1.2b. This is the principle of *drop* calorimetry whereby eqn (4.23), given below, is fitted with an empirical equation for C_p to the enthalpies measured for gram-size samples. With an ice calorimeter one measures the amount of ice that melts when the sample cools down, which is proportional to the enthalpy released. In a copper block calorimeter, it is the temperature rise of a block of copper (whose C_p is well known), which is measured.

4.3b RELATIVE ENTHALPY

Through integration of the heat capacity between T and T_0, the *relative enthalpy* or *heat content* of a phase is obtained between any two temperatures

$$H(T) - H(T_0) = \int_{T_0}^{T} C_p dT. \tag{4.23}$$

With equations (4.21) or (4.22), we find

$$\begin{aligned} H(T) - H(T_0) &= a\,(T - T_0) + b/2\,(T^2 - T_0^2) - c\,(1/T - 1/T_0) \\ &\quad + 2d\,(T^{1/2} - T_0^{1/2}) + e/3\,(T^3 - T_0^3), \end{aligned} \tag{4.24}$$

or

$$\begin{aligned} H(T) - H(T_0) &= k_0\,(T - T_0) + k_1 \ln(T/T_0) - k_2\,(1/T - 1/T_0) \\ &\quad - k_3/2\,(T^{-2} - T_0^{-2}). \end{aligned} \tag{4.25}$$

Conversely, the heat capacity can be determined through differentiation (with respect to T) of an analytical expression fitted to the relative enthalpy.

In the case of phase transitions occurring between T and T_0, integration of (4.17) must be split in a series of segments possibly separated by enthalpies of transition. Denoting by C_{p_i} the heat capacity of phase i and ΔH_i the enthalpy of transition at temperature T_{i+1} between the phases i and $i+1$, we write

$$H(T) - H(T_0) = \int_{T_0}^{T_1} C_{p_0} dT + \Delta H_1 + \ldots + \int_{T_i}^{T_{i+1}} C_{p_i} dT + \Delta H_{i+1} + \ldots \tag{4.26}$$

As already pointed out, a *first-order* phase transition manifests itself as a discontinuity in relative enthalpy. For sapphire, measurements give directly an enthalpy of fusion of 107 (5) kJ/mol at the melting temperature of 2290 K (Fig. 4.4b). Alternatively, one can determine enthalpies of transition by integrating with (4.23) heat capacities measured in the actual temperature interval of the transition, which can be very narrow. The very accurate enthalpy of fusion of selenium of 6159 (4) J/mol has been derived in this manner from the area below the sharp C_p peak centered on 494.3 K (Fig. 4.5).

4.4 ENTROPY AND NERNST'S LAW

4.4a RELATIVE ENTROPY

Below room temperature, C_p measurements cannot be reproduced as a function of temperature with conveniently simple equations. For systems above room temperature, this is not a serious problem because the low-temperature C_p data are just needed to calculate the entropy through integration of $dS = C_p/T\ dT$. This integration is then split into two parts: the first one, from 0 to 298 K, is once and for all integrated numerically to give $S_{298} - S_0$, which is called the *standard entropy*; and the second one is integrated analytically in each problem to derive the *relative entropy* $S_T - S_{298}$ between 298 K and the temperature T considered.

For a substance not undergoing any phase transition, we have

$$S(T) = S_{298} + \int_{298}^{T} C_p/T\ dT, \tag{4.27}$$

where the integration is performed with an adequate, empirical C_p expression. With (4.21) or (4.22), we obtain for any reference temperature T_0

$$\begin{aligned} S(T) &= S(T_0) + a\ln(T/T_0) + b\,(T - T_0) - c/2\,(T^{-2} - T_0^{-2}) \\ &\quad -2d\,(T^{-1/2} - T_0^{-1/2}) + e/2\,(T^2 - T_0^2), \end{aligned} \tag{4.28a}$$

or

$$\begin{aligned} S(T) &= S(T_0) + k_0\ln(T/T_0) - k_1\,(1/T - 1/T_0) \\ &\quad -k_2/2\,(T^{-2} - T_0^{-2}) - k_3/3\,(T^{-3} - T_0^{-3}). \end{aligned} \tag{4.28b}$$

With phase transitions between 298 K and T, integration is again split in series of segments possibly separated by entropies of transition

$$S(T) = S(T_0) + \int_{T_0}^{T_1} \frac{C_{p0}}{T}\,dT + \Delta S_1 + \ldots + \int_{T_i}^{T_{i+1}} \frac{C_{pi}}{T} dT + \Delta S_{i+1} + \ldots \tag{4.29}$$

At its equilibrium temperature T_i, the phase transition is by definition reversible. In (4.2a), this allows one to use

$$\Delta S_i = \Delta H_i / T_i, \tag{4.30}$$

where ΔH_i is the enthalpy of transition.

The ratio C_p/T becomes negligible at temperatures below 10 - 20 K only. Hence, it is necessary to begin C_p measurements at very low temperatures to determine entropies accurately. Also note that compounds including elements such as Fe, Mn, Co, etc., undergo low-temperature magnetic transitions whose entropy effects can represent nearly half of $S_{298} - S_0$, see 6.1b. The standard entropy difference between MnO, 59.7 (4) J/mol K and

MgO, 26.9 (2) J/mol K, thus mainly stems from a C_p anomaly observed for MnO, which is an antiferromagnetic substance, around its Néel temperature of 118 K.

4.4b NERNST'S LAW

From C_p measurements, only variations in entropy can be known. Would it be possible to determine absolute entropy values? In other words, what is the entropy of a system at 0 K? To answer meaningfully this question, we must first stress that the entropy of a system is determined by distributions of matter and energy at all scales, including subatomic, in all phases. The elementary constituents of atoms, whose list is not yet complete, thus also contribute to the entropy of a substance. As long as all these constituents are not known, the real absolute entropy of a system will remain undetermined. This is the actual reason why the entropy is known apart from a constant value.

Under well-defined circumstances, the simplest choice that can be made is to consider that entropy is zero at 0 K. There are several ways of stating this choice. That of Lewis (1875-1946) and Randall will be adopted here: *if the entropy of each element in some crystalline state be taken as zero at the absolute zero of temperature, every substance has a finite positive entropy; but at the absolute zero of temperature the entropy may become zero, and does so become in the case of perfect crystalline substances.* This statement is rigorous but its spirit detracts from that of the two principles for which the details of the real world are irrelevant. Rather than calling it a principle, we prefer to call it Nernst's (1864-1941) law and will discuss its very frequent breakdown in 6.3. Phases not obeying it include amorphous and poorly crystallized materials, as well as partially or totally disordered crystals, which actually constitute the bulk of the natural or synthetic phases that surround us.

Chemical reactions leave the subatomic properties of matter unchanged. As we are in fact interested in entropy differences only, the possible subatomic entropy effects can be neglected since they compensate exactly through chemical reactions. In contrast, the entropy resulting from isotope mixing should also be taken into account because it is not constant as the isotopic composition of species varies with temperature (see chapter 15). But either only one isotope has a high abundance (H vs. D and T for hydrogen, ^{12}C vs. ^{13}C and ^{14}C for carbon, or ^{16}O vs. ^{17}O and ^{18}O for oxygen, etc.), or the observed changes in isotope compositions are very small (for ^{6}Li and ^{7}Li, ^{35}Cl and ^{37}Cl for example). In both instances, the entropy effects of these variations are thus usually negligible.

Nernst's name is associated with this law because of his celebrated heat theorem. At a time when low-temperature calorimetric measurements were nonexistent, the now standard way of calculating G through the difference $H - TS$ could not be used. The Gibbs-free energy had instead to be calculated from integration of the Gibbs-Helmholtz equation (3.4c), but the problem was that the integration constant could not be determined. With his theorem Nernst stated that this constant is zero, which implies that entropies of reaction tend to 0 at 0 K, and that the entropies of the various phases participating in the reaction thus also tend to the same zero limit. As a consequence, the enthalpy and Gibbs free energy of a reaction involving phases all following Nernst's law become mutually tangential at 0 K. This is shown in Fig. 4.6 for the transition between quartz and cristobalite, two low-pressure modifications of SiO_2. Like for any glass (see below), the entropy of SiO_2 glass in contrast does not vanish

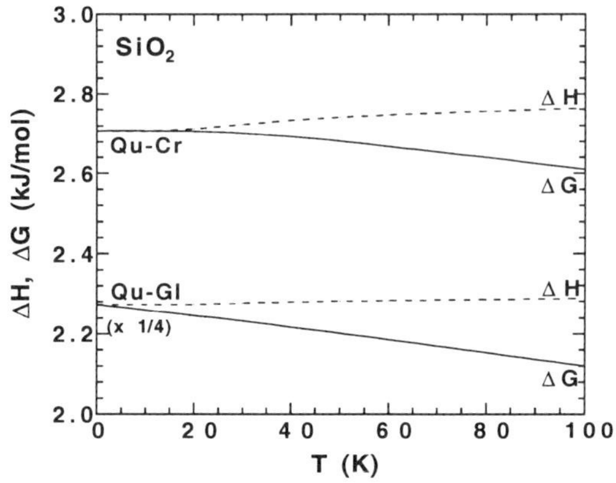

Figure 4.6. Nernst's law and differences in enthalpy and Gibbs free at low temperatures between quartz and cristobalite (Qu-Cr) and between quartz and SiO_2 glass (Qu-Gl). Data from Richet et al. (1982), divided by 4 for Qu-Gl. Note the different way the 0 K temperature is approached depending on whether Nernst's law is followed or not.

at 0 K. Although converging, the ΔH and ΔG curves for the transition between quartz and SiO_2 glass do not become tangential at 0 K (Fig. 4.6).

4.4c RESIDUAL ENTROPY OF GLASSES AND DISORDERED CRYSTALS

As another example of a phase not following Nernst's law, let us consider selenium glass whose residual entropy can be determined from the simple entropy cycle pictured in Fig. 4.7. Up to the melting point of 494.3 K the entropy of crystalline Se is calculated from its zero value at 0 K and the C_p measurements summarized in Fig. 4.5. Since these data also yield the entropy of fusion, the entropy of the liquid at the melting point is then known. From C_p measurements for the supercooled liquid and finally for the glass, a residual entropy of 3.95 J/mol K is eventually found at 0 K (Fig. 4.7). This is of course the thermodynamic measure of the configurational disorder that is quenched when the liquid structure is frozen in at the glass transition.

As exemplified by glasses, all substances having some kind of configurational disorder do not obey Nernst's law. This is why the crystalline state in itself does not constitute a sufficient criterion to ensure a zero entropy, the entropy vanishing at 0 K only "for perfect crystalline substances". Specifically, it is also required that the vari-

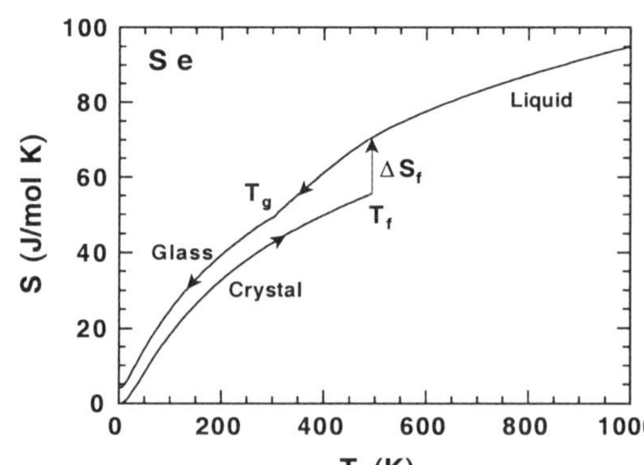

Figure 4.7. Entropy cycle for determining the residual entropy of Se glass from C_p measurements on the crystal, liquid and glass. ΔS_f is the entropy of fusion. Same data and notation as in Fig. 4.5.

ous kinds of defects be in negligible amounts and that all elements that can substitute for each other among the crystallographic sites of the structure rest on specific sites at 0 K. When a site distribution is preserved at 0 K, there remains some configurational entropy and the total entropy cannot be zero. As described in (6.4f), ice is the best-known example of this very common case.

How can residual entropies of phases such as ice be determined? This is made through a combination of calorimetric measurements and phase equilibria data. The principle of the method will thus be described in 9.3f once chemical equilibrium has been treated. But how can we assume whether or not a crystal follows Nernst's law? In general, knowledge of the structure and chemical composition gives some clues as to the possibility of residual entropies. Tin and sulfur are generally chosen to illustrate the validity of this law because of the existence of two crystal structures for both of them. The presence of a single element in the crystal structures (orthorhombic and monoclinic forms for sulfur; cubic and tetragonal forms for tin) indeed leaves no room for disorder and little for defects so that the entropy differences between the polymorphs of these elements vanish at 0 K. In contrast innumerable phases not following Nernst's law are found among solid solutions of industrial or natural interest. A first example is given by high-temperature superconductor ceramics, which are actually superconducting over narrow composition ranges. One such phase has the apparently simple stoichiometry $Bi_2Sr_2CaCu_2O_8$, which conceals a more probable formula $[Bi_{1.90}][Sr_{1.70}Ca_{0.10}Bi_{0.05}][Ca_{0.80}Sr_{0.14}Bi_{0.06}]Cu_2O_8$, where the brackets group together the elements occupying the same crystallographic sites. In addition, note that this stoichiometry indicates the existence of numerous vacancies which also contribute to the configurational entropy. Another example will be drawn from mineralogy because most minerals display very wide composition ranges. Amphiboles, for instance, are important minerals in igneous rocks and have as general formula, where bold faces indicate the naturally most abundant elements, $[\mathbf{Ca},\mathbf{Na},\mathrm{K},\mathrm{Mn}]_{2-3}[\mathbf{Mg},\mathbf{Fe},\mathbf{Al},\mathrm{Ti},\mathrm{Mn},\mathrm{Cr},\mathrm{Li},\mathrm{Zn}]_5[\mathbf{Si},\mathrm{Al}]_8\mathbf{O}_{22}[\mathbf{OH},\mathrm{F},\mathrm{Cl}]_2$. The difficulty of interpreting the configurational entropy of such phases need not be stressed as this would require knowledge of the distribution of all elements over the various crystallographic sites of the structure.

4.5 HEATS OF TRANSFORMATION

4.5a HEAT OF REACTION, CYCLE

What is the heat given off or absorbed through the transformation of diamond (d) into graphite (g), for instance, both compounds being made up of pure carbon? For a transformation at constant pressure, let us note ΔH the heat of the reaction

$$\mathrm{C\ (d)} \rightleftharpoons \mathrm{C\ (g)}. \qquad \Delta H$$

This enthalpy cannot be measured directly at low or moderate temperatures where the kinetics of the transformation is much too sluggish. In contrast, it is possible to measure the enthalpy of combustion of both forms

$$\mathrm{C\ (g)} + O_2 \rightleftharpoons CO_2, \qquad \Delta H_g$$
$$\mathrm{C\ (d)} + O_2 \rightleftharpoons CO_2, \qquad \Delta H_d$$

where $\Delta H_g = H_{CO_2} - H_{C(g)} - H_{O_2}$ and $\Delta H_d = H_{CO_2} - H_{C(d)} - H_{O_2}$. Let us then take advantage of the fact that enthalpy is a state function to imagine the following transformations

$$\begin{array}{ccc} C(d)+O_2 & \longrightarrow & C(g)+O_2 \\ \downarrow & & \uparrow \\ CO_2 & \longrightarrow & CO_2 \end{array}$$

For the enthalpy difference between diamond and graphite, it does not matter that the transformation be direct, or proceed through an intermediate CO_2 step, or even that the intermediate steps take place at a temperature differing from that of the diamond-graphite reaction in which we are interested. The enthalpy difference $\Delta H = H_{C(g)} - H_{C(d)}$ is thus equal to the difference $\Delta H_g - \Delta H_d$.

This and the following examples illustrate the notion of *cycle*, which rests on the fact that the variation of a state function depends only on the initial and final states, and not on the path actually followed. This use of enthalpy cycles is known as Hess' (1802-1850) law, but cycles can of course be used for any other state function. The main difficulty with cycles is the risk of sign errors. To evaluate any difference (Δ), the sign convention is to subtract the property for the initial state (on the left-hand side) from that of the final state (on the right-hand side).

To measure such enthalpies of combustion, let us introduce at a temperature T_1 (298 K, for instance) oxygen and carbon in a calorimetric bomb in which the reaction is initiated by an appropriate means. It does not matter that the temperature varies during the combustion as long as the reaction is complete and that the final temperature T_2 of the CO_2 formed is known. For graphite, for example, the measured heat ΔH_m represents the enthalpy of the reaction C (g,T_1) + O_2 (T_1) $\rightarrow$ CO_2 (T_2). To calculate the enthalpy of combustion at temperature T_1, let us complete the cycle by adding a last step to bring the CO_2 from T_2 to T_1. If C_{p_c} is the heat capacity of CO_2, we finally obtain from (4.23)

$$\Delta H_g = \Delta H_m - \int_{T_1}^{T_2} C_{p_c} dT.$$

At 298 K, we find in this way $\Delta H_g = -393.51$ and $\Delta H_d = -391.61$ kJ/mol, whence $\Delta H = -1.90$ kJ/mol. The impossibility of measuring ΔH directly thus does not prevent this enthalpy to be determined at any temperature.

Some methods for determining the enthalpy of rapid reactions have already been described. Many transformations are so sluggish, however, that their thermal effects cannot be measured. As illustrated by the diamond-graphite transition, this is the general case of solid-solid transformations. In such instances, *solution* calorimetry proves very useful: after having measured the enthalpy released or supplied when the relevant phases are dissolved in an appropriate solvent, one writes the enthalpy balance of a cycle analogous to that considered above for the combustion of carbon. As an example, consider the enthalpy of the transformation of silica glass, SiO_2 (g), into quartz at room temperature. Since both phases dissolve rapidly in hydrofluoric acid solutions, it suffices to measure the enthalpies of solution of the glass, ΔH_G, and quartz, ΔH_Q, and to write down the following cycle where SiO_2 (d) designates silica dissolved in the acid solution

$$\begin{array}{ccc} SiO_2(g) & \longrightarrow & \text{Quartz} \\ \Delta H_G \downarrow & & \uparrow -\Delta H_Q \\ SiO_2(d) & \longrightarrow & SiO_2(d) \end{array}$$

Because dissolved silica does not "remember" its former state, the enthalpy of the last reaction is zero and the desired enthalpy is $\Delta H = \Delta H_G - \Delta H_Q$. From the experimental results $\Delta H_G = 148.45$ (8) and $\Delta H_Q = 139.33$ (17) kJ/mol at 298 K, we find $\Delta H = 9.12$ (25) kJ/mol.

In solution calorimetry small amounts of matter are usually dissolved in the solvent to achieve a high dilution and avoid problems associated with possible enthalpies of mixing in concentrated solutions. This allows the enthalpies of reactions involving several phases to be determined from the algebraic sum of the enthalpies of solution of the individual phases. On the other hand, an enthalpy of reaction is usually a small difference between large numbers. Solvents are therefore to be selected with care in order to yield thermal effects not too great or too small, and thus ensuring errors on the enthalpy of reaction of the order of 1%. Acid solutions are commonly used near room temperature. Compounds most reluctant to dissolution eventually yield to molten salts (e.g., $2PbO.B_2O_3$), which can be used up to temperatures of about 1000 K.

4.5b ENTHALPY OF FORMATION

Let us determine the enthalpy of a reaction, for instance

$$4\,H_2 + CO_2 \rightleftharpoons CH_4 + 2\,H_2O. \tag{4.31}$$

which we will rewrite

$$CH_4 + 2\,H_2O - 4\,H_2 - CO_2 = 0,$$

to introducuce the more general notation

$$\sum \nu_i M_i = 0. \tag{4.32}$$

The ν_i are the stoichiometric coefficients of the M_i species participating in the reaction, namely, 1 for CH_4, 2 for H_2O, -4 for H_2 and -1 for CO_2.

The enthalpy of this reaction can be obtained in an apparently more complicated, but actually much more convenient way. Consider the following reactions whose ΔH are noted on the right-hand side, and where C (g) again designates carbon in graphite

$$\begin{aligned} &C\,(g) + O_2 \rightleftharpoons CO_2, && \Delta H^\circ_{fc} \\ &C\,(g) + 2\,H_2 \rightleftharpoons CH_4, && \Delta H^\circ_{fm} \\ &H_2 + 1/2\,O_2 \rightleftharpoons H_2O. && \Delta H^\circ_{fw} \end{aligned}$$

Rather than dealing with the enthalpy of reaction (4.31), let us consider instead the reactions of formation from the elements of all participating species. Then it is sufficient to determine once and for all the *enthalpies of formation* (ΔH°_f), and to sum them up appropriately to calculate ΔH for each particular reaction of interest

$$\Delta H = \sum \nu_i \Delta H^\circ_{fi}. \tag{4.33}$$

Beware of the sign! *The enthalpy of a reaction is the difference between the enthalpies of formation of the products and those of the reactants.* The enthalpy of reaction (4.31) is thus $\Delta H = \Delta H^\circ_{fm} + 2\Delta H^\circ_{fw} - \Delta H^\circ_{fc}$.

The enthalpies of formation of methane, water vapor and carbon dioxide are - 74.81 (33), - 241.81 (4) and - 393.51 (13) kJ/mol, respectively, at 298 K and 1 bar. Under these conditions, the heat supplied at this temperature by reaction (4.31) is -164.92 (50) kJ. (Because this enthalpy is negative, the heat is thus given off, and not absorbed.) To determine this enthalpy at another temperature, two equivalent paths may be followed. Either we calculate with (4.26) the variation with temperature of the enthalpies of formation from the heat capacities of all species and elements, or we obtain this variation from the C_p difference of the products and reactants.

For ionic crystals, the question naturally arises as to the energy the ions would give off when forming the actual crystal. This *cohesive energy* of the crystal (E_c) can be measured through the so-called Born-Haber cycle:

$$
\begin{array}{ccc}
\mathrm{Na^+(g)} \;+\; \mathrm{Cl^-(g)} & \xrightarrow{E_c} & \mathrm{NaCl(c)} \\
-e^-\Big\uparrow I \qquad +e^-\Big\uparrow -A & & \Big\downarrow -\Delta H_f^\circ \\
\mathrm{Na(g)} \;+\; \mathrm{Cl(g)} & \xleftarrow[\Delta H_s + 1/2\,\Delta H_d]{} & \mathrm{Na(c)} + 1/2\,\mathrm{Cl_2(g)}
\end{array}
$$

In this cycle, I, the ionization energy of Na, and A, the electron affinity of Cl^-, are determined from spectroscopic observations whereas ΔH_s, the enthalpy of sublimation of Na, ΔH_d, the enthalpy of dissociation of Cl_2, and ΔH_f°, the enthalpy of formation of NaCl, are given by calorimetric measurements. The crystal energy, E_c, is given by

$$E_c = -\Delta H_f^\circ + \Delta H_s + 1/2\,\Delta H_d - A + I.$$

For alkali halides, these energies range from 755 (= 411 + 109 + 1/2 x 226 - 368 + 490) kJ/mol for NaCl, to 644 kJ/mol for RbI and are reproduced to within 1-5% from simple theoretical calculations. This indicates that only a small part of the cohesive energy results from nonionic interactions.

4.5c STANDARD STATE OF THE ELEMENTS

By convention, the reference state selected for the elements is the form stable at room pressure (nowadays 1 bar) and the temperature considered. At 298 K, it is for instance graphite for carbon, the liquid for mercury and the gaseous forms for hydrogen or oxygen, with the additional convention that these are taken as ideal gases in the standard state. The enthalpies of formation of elements are thus zero. Like the internal energy, the enthalpy is known apart from an arbitrary constant. The convention adopted for enthalpies of formation thus fixes arbitrary zeros for their respective scales.

Note finally that, in some contexts, it can be advantageous to deal with formation from other species than the elements. The stable form of oxides at room pressure and the temperature of interest can be a convenient reference state for oxide compounds. But, of course, one should not mix formation from elements and from other substances in any calculation as the same convention must be used consistently.

4.6 GIBBS FREE ENERGY

4.6a GIBBS FREE ENERGY AT 1 BAR

The Gibbs free energy is

$$G = H - TS. \tag{4.34}$$

In enthalpy and entropy terms it is rewritten as

$$\begin{aligned} G(T) &= H(T) - TS(T) \\ &= H(T_0) + \int_{T_0}^{T} C_p dT - T\left[S(T_0) + \int_{T_0}^{T} C_p/T\, dT\right], \end{aligned} \tag{4.35}$$

where the enthalpies and entropies are calculated as a function of temperature as described in 4.3b and 4.4a. Note, however, that (4.35) is valid at the single pressure at which the calorimetric measurements have been made. Unless otherwise specified, this is atmospheric pressure but the differences with 1 bar values are negligible for condensed phases and readily dealt with for gases.

4.6b GIBBS FREE ENERGY OF FORMATION

Entropies (ΔS_f°) and Gibbs free energies (ΔG_f°) of formation are defined in the same way as enthalpies of formation. To calculate the Gibbs free energy of a reaction, we just subtract the sum of the Gibbs free energies of formation of the reactants from those of the products, viz.

$$\Delta G = \sum \nu_i \Delta G_{fi}^\circ = \sum \nu_i (\Delta H_{fi}^\circ - T\Delta S_{fi}^\circ), \tag{4.36}$$

where ΔS_f° designates the entropy of formation of the species i that participates in the reaction. Of course, the Gibbs free energies of formation of all elements are zero. (If formation from the oxides is considered, it is of course the Gibbs free energies of formation of oxides that are alternatively zero.)

4.6c GIBBS FREE ENERGY AT HIGH PRESSURE AND TEMPERATURE

It just remains to evaluate the variation of the Gibbs free energy with P to calculate G at any pressure and temperature. From (3.3c), we now that

$$(\partial G/\partial P)_{T,n_i} = V. \tag{4.37}$$

The variations of V as a function of P at a given temperature T thus determine those of the Gibbs free energy. These are given by the equations of state, to which the next chapter is devoted.

In summary, determining the Gibbs free energy of a phase at any pressure and temperature requires knowledge of: (i) its enthalpy of formation at a given temperature; (ii) its entropy at a given temperature (0 K, in general); (iii) its heat capacity as a function of temperature; (iv) its equation of state. In most cases, the first three variables are measured only at atmospheric pressure.

Table 4.2. Thermodynamic properties of water

Molecular weight = 18.015 g
Liquid from 235 to 372.8 K. Ideal gas from 372.8 to 1800 K

K	kJ/mol	J/mol K			Formation from the elements kJ/mol		
T	$\frac{H^\circ_T - H^\circ_{298}}{T}$	S°_T	$\frac{-(G^\circ_T - H^\circ_{298})}{T}$	C°_p	ΔH°_f	ΔG°_f	$\log K_f$
298.15	0.	69.95	69.95	75.21	-285.830	-237.141	41.546
Uncert.		0.08	0.08		0.042	0.084	0.015
372.8	15.216	86.76	71.63	75.86	-283.703	-225.106	31.541
372.8	124.836	196.38	71.63	34.09	-242.437	-225.106	31.541
400	118.672	198.78	80.11	34.28	-242.836	-223.882	29.236
500	101.882	206.53	104.65	35.18	-243.820	-219.035	22.883
600	90.853	213.04	122.19	36.26	-244.758	-213.987	18.629
700	83.139	218.71	135.57	37.45	-245.634	-208.786	15.580
800	77.506	223.80	146.29	38.70	-246.444	-203.473	13.285
900	73.266	228.43	155.16	39.97	-247.185	-198.045	11.494
1000	69.999	232.71	162.71	41.24	-247.859	-192.559	10.058
1100	67.442	236.69	169.25	42.48	-248.464	-186.985	8.879
1200	65.412	240.44	175.03	43.69	-249.008	-181.371	7.895
1300	63.786	243.99	180.20	44.85	-249.492	-175.730	7.061
1400	62.473	247.35	184.88	45.95	-249.923	-170.040	6.344
1500	61.406	250.56	189.15	46.98	-250.303	-164.322	5.722
1600	60.535	253.62	193.08	47.96	-250.640	-158.569	5.177
1700	59.822	256.55	196.73	48.85	-250.937	-152.805	4.695
1800	59.236	259.37	200.13	49.68	-251.201	-147.035	4.267

Temp. of fusion	273.15 K	Temp. of vaporization:	372.8 K
Enthalpy of fusion	6.008 kJ	Enthalpy of vaporization	40.866 kJ
H°_{298}(wat.) -H°_0(ice)	13.293 kJ	Molar volume at 298 K	1.8069 J/bar
			18.069 cm^3

C°_p equations (J/mol K):

$$C^\circ_p = 25.952 + 128.281\,\varepsilon - 221.405\,\varepsilon^2 + 196.894\,\varepsilon^3 - 64.812\,\varepsilon^4 + 14.2\varepsilon^{1/2},$$

with $\varepsilon = T/227.1 - 1$ (liquid water, from 235 to 323.1 K)

$$C^\circ_p = 7.3680 + 27.468 10^{-3}\mathrm{T} - 2.2316 10^5/\mathrm{T}^2 + 361.74/\mathrm{T}^{0.5} - 4.8117 10^{-6}\mathrm{T}^2$$

(water vapor, ideal gas, from 372.8 to 1800 K)

N.B. Temperature of vaporization of water: 373.15 K at 1 atm, and 372.8 K at 1 bar, the reference pressure in thermodynamic tables. Data from the compilation of Robie et al. (1979), and from Speedy (1987) for supercooled liquid water down to 235 K, the lowest temperature reached.

4.7 THERMODYNAMIC DATA

4.7a THERMODYNAMIC TABLES

Thermodynamic tables are very useful data sources since calculations of reaction properties are straightforward when thermochemical information is listed for all products and reactants. These tables, of which a short list is included in the bibliography of this textbook, are built from critical evaluations of literature measurements. The properties of water listed as an example in Table 4 do not warrant additional comments. The only new variable is the constant k_f as defined by

$$\log k_f = -\Delta G_f^\circ / RT, \tag{4.38}$$

whose usefulness in chemical equilibria calculations will become apparent in chapter 8. As to the function $-(G_T^\circ - H_{298}^\circ)/T$, it is often included for interpolation purposes because it varies more linearly with temperature than G.

We will add the superscript ° to some of these properties to prevent possible confusion and stress that they refer to the 1-bar standard state. For standard entropies as well as for the enthalpies and Gibbs free energies of formation, the experimental uncertainties are in general reported explicitly. In contrast, it would be inconvenient to give uncertainties for the coefficients of the empirical C_p equations and one prefers instead to indicate the relative uncertainty of the actual C_p in the temperature interval of the measurements. Reaction properties generally represent small differences between great numbers. To avoid round-off errors, all the figures after the decimal point must be kept even though they are seldom warranted by the accuracy of the input measurements. For the same reason, it is advisable not to mix data from different compilations, unless it has been checked that they are mutually consistent, because apparently slight differences between formation properties for a given species can translate into large discrepancies in reaction properties. In general, the temperature intervals of the tabulated data are small enough that linear interpolations are possible. If not, then more precise calculations can be made since C_p equations are also included. Finally, the usefulness of computer codes for performing routinely thermochemical calculations needs not be stressed.

4.7b EMPIRICAL ESTIMATIONS OF THERMODYNAMIC PROPERTIES

Empirical estimations of thermodynamic properties are almost as old as thermodynamics itself for the simple reason that it takes much less time to pose a problem than to make the measurements needed to solve it! In addition, there are many instances where measurements cannot be made, or made under too restricted conditions, or when it is difficult to apply the theoretical methods of calculation described in chapter 14. For more than 150 years, numerous empirical rules have thus been proposed to obviate lack of experimental data. For fluids, many of them can be found in the books by Hildebrand and Scott, and Reid et al. The most ancient rule is probably that of Neumann-Kopp, according to which the heat capacity of a substance is an additive function of its chemical composition

$$C_p = \sum x_i C_{p_i}^*, \tag{4.39}$$

where x_i and $C_{p_i}^*$ are the mole fractions and heat capacities of the components of the substance, respectively. This rule often gives good results above room temperature. Below

Table 4.3. Bond energies for single and multiple bonds (kJ/mol), from the book by Pauling (1988)

C–C	344	C–O	350	O–O	143	C–H	415
C = C	615	C = O	725	O = O	402	N–H	391
C ≡ C	812	C–S	259	N–O	175	O–H	463
C–N	292	P–O	360	N–N	159	S–H	368
C = N	615	P–P	217	N = N	418	P–H	322
C ≡N	890	P≡P	490	N≡N	946		

298 K it must be used, cautiously, only for structurally related substances otherwise standard entropies derived from (4.39) will generally be incorrect.

To give useful results, such empirical rules must indeed involve components structurally similar to the species or phases considered. Hydrocarbons, for instance, can be split into a great many kinds of groups (i.e., —CH3, =CH2, >C=CH2, —OH, —SH, —C≡N, etc.), whose contributions to the heat capacity and entropy are largely additive (see Ruzicka and Domalski, 1992; Domalski and Hearing, 1996). For inorganic solids, low-temperature heat capacities mainly depend on atom coordination (nature and number of nearest neighbors to which a given atom is bonded). This makes entropy estimations possible if atoms have the same coordination in the selected components as in the solid of interest. Thanks to the great progress made in terms of number and quality of measurements, such procedures have largely replaced universal correlations such as the famous Latimer rule, according to which the standard entropy can be approximated by $S^{\circ}_{298} = \Sigma x_i(3/2\, R \ln M_i - 3.93)$ J/mol K, where M_i is the atomic weight of element i.

Regarding enthalpies of formation, estimates can be obtained through summation of bond energies. According to Pauling, the data of Table 4.3 that can be used for this purpose yield results reliable to about 3 kJ/mol per bond. Such estimates are useless for predicting enthalpies of transition, however, because they apply without distinguishing the various isomers or polymorphs of a substance. Unlike the heat capacity and entropy, which are more readily related to the structure of a phase (because they depend only on the shape of the interatomic potentials), enthalpies of transition are much more difficult to estimate because they represent the "height" differences between the minima of these potentials. As a result, additive methods as simple as (4.39) do not exist to predict enthalpies of formation. The century old Trouton rule states that the entropy of vaporization of a nonpolar liquid is 88 J/mol K at the 1-atm boiling temperature. Alternatively, this entropy was supposed constant when referred to temperatures at which the vapor volume was the same (i.e., 85 J/mol K for V_v= 22.4 l/mol). Another rule claims that the entropies of fusion of the elements are close to R. In fact there exist marked deviations from such trends and, with the wealth of data currently available, these empirical correlations have lost much of their appeal. In any case, it is for isostructural substances that they really make sense. In ionic crystals, for example, the volume, the entropy, or even the volume derivatives (bulk modulus) often vary linearly with simple parameters such as the ratio between the ionic radius and the electrical charge of the ions. Note finally that all such rules are of course incorrect over temperature intervals where an ongoing phase transition is observed.

Problems

1. Under which conditions can we write $dG = VdP - SdT$ for a phase?
2. Making use of cross exchange of conjugate variables, write down the Maxwell relations without performing any calculation. For each relation, the sign will be determined by reference to the transformation of an ideal gas under the relevant conditions.
3. The components A and B have the mole fractions x_A and x_B in a binary mixture. The volume of the solution is represented by $V = x_A V_A + x_B V_B + C x_A x_B$. With V_A= 20 and V_B = 30 cm^3/mol, plot the partial molar volumes of A and B for C = -10.
4. The thermal expansion coefficient and the isothermal compressibility of water are -14.09 10^{-4} K^{-1} and 8.077 10^{-4} bar^{-1} at 243 K, respectively, and 5.834 10^{-4} K^{-1} and 4.520 10^{-4} bar^{-1} at 343 K (all data for 1 bar). Calculate the internal pressure at the two temperatures and comment on the difference between the results obtained.
5. An aqueous solution of NaCl is prepared from 10 g of salt and 90 g of water. At 20° C, the density of the solution is 1.073 g/cm^3. Calculate the mole fraction and molarity of both components and the molality of the ions Na^+ and Cl^-.
6. A hole is drilled at the center of a disk made of an isotropic material with a positive thermal expansion coefficient. Does the diameter of the hole increase or decrease when the disk is heated?
7. Show that the isothermal pressure variation of an ideal gas as a function of height h is given by the *barometric* formula, $P = \exp(-mgh/RT)$, with m = molecular weight and g = gravitation constant. For a temperature of 300 K and an atmospheric pressure at ground level of 1 atm, calculate the pressure at the top of the New York World Trade Center (400 m) and Mount Whitney (4300 m). Derive modified equations accounting for either the variation of g with the distance to the center of the earth, or with that of temperature with elevation (which will be assumed to be linear).
8. Air is considered as a mixture of 0.8 mole of N_2 for 0.2 mole of O_2. The heat capacity of the two gases and the sound velocity in air are listed below for three temperatures. Calculate from these data the density and isothermal compressibility of air and compare the results with the ideal gas values. Are the relative deviations from the ideal gas values larger for the density or the compressibility?

T (K)	300	500	1000
$C_{pN_2} (J/mol K)$	29.12	29.60	32.65
C_{pO_2} (J/mol K)	29.37	31.16	34.82
c_{air} (m/s)	1.161	0.696	0.348

9. Show that, if the bulk modulus K_T of a condensed phase is assumed to be constant, then the density ρ of the phase varies with pressure, at constant T, according to $\rho \sim \rho_0(P/K_T + 1)$.
10. Show that the heat capacity at constant enthalpy C_H is equal to $C_p/(1-\alpha T)$, where α is the thermal expansion coefficient.
11. The compound $Ca_2MgSi_2O_7$ is an important constituent of slags. It melts congruently at 1727 K. Proks et al. (1977) have measured the relative enthalpy $H_T - H_{298}$ of the crystalline and liquid phases of the material, and the heat of dissolution (ΔH_s), in a HF solution at 298 K, of the samples recovered after the relative-enthalpy measurements. The results are as follows (in kJ/mol), where ΔH_T designates $H_T - H_{298}$.

T (K)	ΔH_T	ΔH_s	T (K)	ΔH_T	ΔH_s	T (K)	ΔH_T	ΔH_s
1469	327.5	775.3	1669	396.4	775.3	1773	508.1	821.1
1521	344.9	775.3	1696	404.6	775.3	1798	523.1	819.9
1577	362.7	775.3	1741	514.0	804.4	1840	554.9	802.4
1630	380.7	775.3	1762	508.3	824.0	1867	556.3	812.4

Inspection of the samples quenched after the relative enthalpy measurements showed that they had partially crystallized on cooling from T to 298 K. Why does the scatter apparent in both $H_T - H_{298}$ and ΔH_s measurements disappear when one makes the sum of these measurements? Calculate graphically the enthalpy of fusion and the heat capacity of the liquid and phases.

12. The 1-bar enthalpy of formation of diamond is 2.90 kJ/mol at 298 K. Between 298 and 1000 K, the heat capacities of graphite and diamond are represented by $C_p = -4.3176 + 35.4\ 10^{-3}T - 6.591\ 10^{-6}T^2$ and - $6.0617 + 0.0285T - 2.427\ 10^{-6}T^2$, respectively. Their entropies at 298 K are 5.74 and 2.38 J/mol K. What are the enthalpies, entropies and Gibbs free energies of the transformation of graphite into diamond at 298 and 1000 K? Comment on the change in the Gibbs free energy of the reaction between these temperatures. From the volume data of Table 5.1, estimate the pressure at which the transformation would become thermodynamically possible at 1000 K. Same question for $T = 2000$ K.

13. In a crater lake the lake has a temperature of 1400 K, a density of 2.60 g/cm^3, a thermal expansion coefficient of 1.3 10^{-4} K^{-1} and an isobaric heat capacity of 1.54 J/g. If the lava went up adiabatically to the surface, what was its temperature at its source at a depth of 50 km ?

5
EQUATIONS OF STATE

The equation of state relates the pressure, temperature and volume of a system. Since the volume of a system is summed over those of the constituting phases, we will restrict ourselves in this chapter to phases with a fixed composition. We will begin with ideal gases which constitute a good starting point for real gases at low pressure and high temperatures. These will also allow Carnot cycles to be introduced in a simple way as entropy pumps. Then we will describe the main drawbacks of the ideal gas equation of state and, with special attention to the existence of the critical point and to gas-liquid phase equilibria, the manner in which these flaws are corrected by empirical expressions such as the van der Waals equation. Condensed phases will be eventually considered. After a brief presentation of equations of state that can be used up to the highest pressures experimentally accessible, we will discuss the relationship between thermal convection and adiabatic gradient.

5.1 IDEAL GASES

Ideal gases play a fundamental role in thermodynamics for a variety of reasons. First, they closely approximate real gases as long as the pressure remains lower than 1 bar and the temperature higher than 300 K. Second, the ideal gas equation of state leads to an expression for the Gibbs free energy that is not only simple, but extremely convenient for the chemical equilibria calculations described in chapter 8. Third, ideal gases are at the basis of the practical definition of the thermodynamic temperature scale with which we will begin.

5.1a IDEAL GAS LAW AND ABSOLUTE TEMPERATURES

Long before their chemical constitution could be determined, gases were subjected to intense physical scrutiny. From experiments made by Boyle (1627-1691), Mariotte (~1620-1684) or Towneley (1629-1707), the most striking conclusion was that, at constant temperature, the pressure and volume of a gas obey the simple relationship

$$PV = constant. \tag{5.1}$$

Known as Boyle's law, this equation is in fact valid over limited pressure and density intervals. To illustrate its failure, let us forget for a moment atomic theory and refer this product PV to the same mass of gas, one gram for instance, and not to one mole (Fig. 5.1). Between very low densities, for which (5.1) actually applies, and densities 520 times higher than at room pressure, comparisons between the measurements plotted in Fig. 5.1a and 5.1b indeed show that PV does not remain constant but approximately doubles at constant T. Another important feature of Fig. 5.1a is that, at any pressure, PV increases with T, in a linear way at low enough pressures and densities when temperatures are referred to the empirical Celsius scale. But of special interest is the fact that all such linear relationships intersect the T axis at a common point of − 273°C. In other words, PV would become negative at this temperature where either the pressure or the volume would themselves become negative.

With reference to Fig. 5.1b, note that negative pressures are not a paradox in themselves, for they simply characterize a system under extension and not under compression. By "pulling" isotropically a solid, one subjects it to such a negative pressure. There even exists an ingenious means, described in 5.3e, whereby negative pressures of a few hundred bars can be exerted on liquids which are then in a state of metastable equilibrium. But these states can exist only if the substance has a cohesion that allows it to transmit the tension exerted. Of course, this is not the case of an ideal gas in static equilibrium, in which pressure can only be positive.

This paradox of negative PV values is not a consequence of an unwarranted low-temperature extrapolation of the data, for the law (5.1) is the most closely followed at the lowest pressures, i.e., when PV is small. Because the intercept of − 273°C is common to all gases under the conditions where (5.1) is obeyed, it is tempting to assume that this point is the lowest temperature that could be reached, and that it constitutes the zero of an absolute temperature scale. Denoting by T' the temperatures expressed in this scale, we can represent the straight lines of Fig. 5.1a by the equation

$$PV = cT'. \tag{5.2}$$

What can we say about the c coefficient? Because V is an extensive variable, c is also an extensive parameter. It can be expressed as the product of the number of moles of gas, n, and another constant, which will be noted R and could seem to vary from one gas to another. Let us now turn to the atomic theory and refer PV to one mole of gas. In Fig. 5.1a, all straight lines will then merge. Experience therefore shows that the constant R is in fact the same for all gases, whence its name of *ideal gas* constant. In one mole of gas, we have by definition N_A (Avogadro number) molecules and the c constant of (5.2) can be expressed as the product of the number N of gas molecules

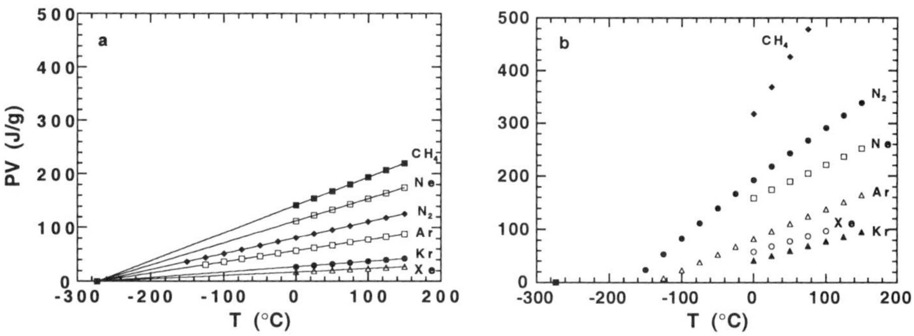

Figure 5.1. Product PV for various gases against temperature in the low-density limit (a) and for a density of 520 Amagat (b), all volumes being referred to one gram of gas. (N.B. One Amagat represents the value of PV for the gas considered at 1 bar and 273.15 K). For elevated densities (b), note that PV varies nonlinearly and, especially, that the various gases do not plot in the order of their molecular weights. Experimental data of Michels et al. (1960, 1949, 1958, 1954) for Ne, N_2, Ar and Xe, respectively; and Trappeniers et al. (1956, 1979) for Kr and CH_4.

and yet another constant, the Boltzmann (1844-1906) constant k. Hence the R and k constants are related by

$$k = R/N_A, \tag{5.3}$$

and we write

$$PV = nRT = NkT'. \tag{5.4}$$

Relating the pressure, temperature and volume, (5.4) is the equation of state of an ideal gas and it provides us with an objective temperature scale since pressures and volumes are quantities that can be measured rigorously. But it remains to show that temperatures drawn from the ideal gas law consitute a thermodynamic scale, i.e., a scale consistent with the requirements of the second principle. To demonstrate that T', the temperature of a *gas thermometer* as given by (5.4) is akin to the thermodynamic temperature T defined along with the entropy in 1.4c, let us consider the absorption of heat by an ideal gas during a reversible expansion. Because the internal energy of an ideal gas depends on temperature only, (see 5.2b), we have

$$dQ = dU + PdV = C_v dT + RT' dV/V. \tag{5.5}$$

The entropy variation is

$$dS = \frac{dQ}{T} = \frac{C_v}{T} dT + R\left(\frac{T'}{T}\right)\frac{dV}{V}. \tag{5.6}$$

As entropy is a state function, its second derivatives with respect to T and V do not depend on the order of the differentiation

$$\left[\frac{\partial(C_v/T)}{\partial V}\right]_T = \left[\frac{\partial(RT'/TV)}{\partial T}\right]_V. \tag{5.7}$$

Because the internal energy of an ideal gas is independent of volume, C_v also does not depend on V and the left-hand side of (5.7) is zero. On the left-hand side, R and V are (or are taken) as constant. If $[\partial(T'/T)/\partial T]_v$ is zero, then it follows that T and T' differ at most by a constant factor that is tied to the choice of a particular temperature unit. Therefore the temperature T' is thermodynamic and the gas thermometer actually constitutes a primary means of measuring temperatures. In practice, the constants R and k have been chosen so as to ensure a simple correspondence with the Celsius scale. The equilibrium temperature between water, ice and water vapor has been fixed at 273.16 K. (It is the water triple point, which, in contrast to the ice-water equilibrium, has the advantage of being reached at a single pressure.) With this choice, the zero of the Celsius scale (the melting temperature of ice) is 273.15 K at 1 atm and the constants R and k have the values 8.31451 J /mol K and 1.38065 10^{-23} J/K, respectively.

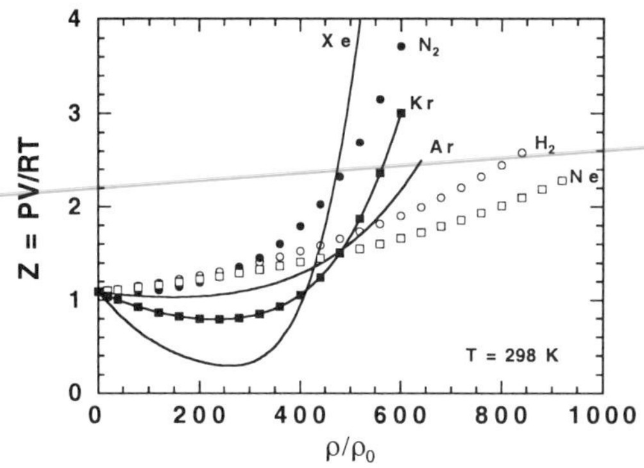

Figure 5.2. Compressibility factors of a few gases at 298 K against the density ratio ρ/ρ_0 where ρ_0 is the density at 1 bar and 298 K. Experimental data of Michels et al. (1959) for H_2; for the other gases, see the references quoted in the caption of Fig. 5.1.

5.1b VALIDITY OF THE IDEAL GAS LAW

In practice, what is the range of validity of (5.4) in the P, T plane? To answer this question, let us first define the *compressibility factor*

$$Z = PV/RT, \tag{5.8}$$

which is by definition unity for one mole of ideal gas. With the compressibility factors plotted in Fig. 5.2 against the density, one finds that the ideal gas law is generally correct at pressures lower than one atmosphere at room temperature. In passing, note the specific behavior shown by the various gases, including noble gases whose compressibility factors vary in a complex way as the atomic number increases.

With the compressibility factor of CO_2 as an example, let us examine how these deviations from an ideal gas behavior vary over wide temperature and pressure intervals (Fig. 5.3). At 400 K, Z has a minimum of 0.7 near 200 bar, similar to those shown by xenon and krypton in Fig. 5.2, and then increases rapidly to 15 for a pressure of 20 kbar. At higher temperatures, the compressibility factor minimum tends to disappear: Z remains close to 1 at lower pressures and the onset of its eventual increase at higher pressures is the highest at the highest temperatures (Fig. 5.3). At 1000 K, for instance,

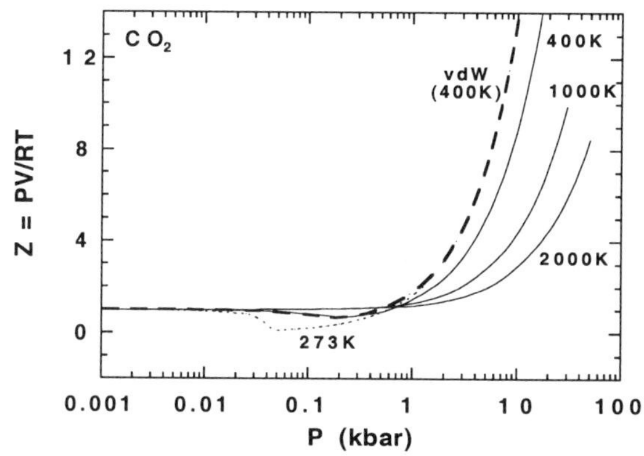

Figure 5.3. Compressibility factor $Z = PV/RT$ of CO_2 at the temperatures indicated. Note the similarities with Fig. 5.2, where the differences between 298 K and the critical temperatures of the various gases represent the analog of the differing temperatures considered here for CO_2. Data of Kennedy (1954) and Holste et al. (1987) at 273.15 K, showing the complicated variations of Z observed near the critical point, and values drawn from a Redlich-Kwong equation of state at the other temperatures (Bottinga and Richet, 1981). The dashed curve represents the compressibility factors given at 400 K by the van der Waals equation of Table 5.1.

pressures of 0.1 and 30 kbar are needed for the compressibility factor of CO_2 to depart from unity by 2 and 10%, respectively. In other words, temperature and pressure have opposite effects, and the errors made when using the ideal gas law are the greatest when the temperature is low and the pressure is high.

The main feature of the ideal gas law is that it does not depend on any specific parameter of a given gas. At a molecular scale, the validity of this law relies on the lack of any intermolecular interactions, except perfectly elastic shocks between colliding molecules. Intermolecular forces do exist, however, but their magnitude decreases when the intermolecular distances increase (i.e., when the density decreases). Hence the best results are achieved with (5.4) at low pressures. As to temperature, which is a measure of the degree of agitation of matter, it represents in the case of an ideal gas the average quadratic velocity (average of the square of the velocities) of molecules (see 13.1c). When T increases, the kinetic energy increases accordingly and progressively overcomes molecular interactions. This is why the ideal gas law is better followed at high than at low temperatures.

5.2 PROPERTIES AND TRANSFORMATIONS OF AN IDEAL GAS

Thanks to the inherent simplicity of the ideal gas law (5.4), a great many problems of practical interest can be solved readily. Illustrating nicely thermodynamic methods, calculations will be made in this section for a few important reversible and irreversible processes.

5.2a COEFFICIENTS

For an ideal gas, $\alpha = 1/T$, $\beta_T = 1/P$ and $K_T = P$. The thermal expansion coefficient is independent of pressure and, consistent with (4.12), the isothermal compressibility does not depend on temperature. In the Celsius scale, α is $1/(T+273)$ for any ideal gas. Known as the Gay-Lussac (1778-1850) and Charles (1746-1823) law, this feature did not escape the sagacity of early experimentalists who also pointed out that temperatures lower than -273°C were raising the paradox of negative thermal expansion coefficients for ideal gases.

Depending specifically on which kind of gas is considered, the values of C_p and C_v cannot be predicted from classical thermodynamics. Nevertheless, we derive from (A.16) the Mayer relation, $C_p - C_v = R$, and find that, as defined by (4.14), the Grüneisen parameter reduces to $\gamma = R/C_v = (C_p/C_v - 1)$ and is always positive. The lowest possible value of C_v is that of monoatomic gases, namely $3R/2$ per mole. The maximum value of γ is thus 2/3.

The quantity $C_p - C_v$ represents the difference between the heats required to raise the temperature of a gas by one degree pressure and at constant volume. This difference had already been measured when Mayer recognized that it represented the work done by the gas on the surroundings during its expansion, which was also known. From this, Mayer derived the mechanical equivalent of the calorie (in modern units, 3.58 J/cal) that he reported in his statement of the first principle. In the ideal gas case, this is the same reasoning as held in 2.2a to introduce the enthalpy from the internal energy.

5.2b INTERNAL ENERGY AND ENTHALPY

As already stressed several times, an important feature of ideal gases is that their internal energy and enthalpy do not depend on volume and pressure. This feature derives directly from (3.26) and (3.29). A consequence is that C_p and C_v also depend on temperature alone. Conversely, one readily demonstrates that a substance whose internal energy and enthalpy depend only on temperature follows the ideal gas equation of state.

5.2c REVERSIBLE ISOTHERMAL EXPANSION

The variations of the main properties of an ideal gas whose volume increases reversibly from V_1 to V_2 at constant T have already been calculated in 1.4d. The temperature remaining constant, the internal energy and the enthalpy do not change either. The final pressure is $P_2 = P_1 V_1/V_2$, whereas the work done is

$$W = -\int_{V_1}^{V_2} P dV = -\int_{V_1}^{V_2} RT/V dV = RT \ln(V_1/V_2). \tag{5.9}$$

The heat received is $Q = -W$, and the entropy gained $\Delta S = Q/T = R\ln(V_2/V_1)$. As noted in 1.4d, the gas entropy increases although the temperature remains constant. This result can alternatively be obtained directly through integration of the Maxwell relation (3.6c), which may be written

$$(\partial S/\partial V)_T = (\partial P/\partial T)_V = \alpha/\beta_T = P/T = R/V. \tag{5.10}$$

As to the Gibbs free energy variation, it reduces to $T\Delta S$, which, apart from the sign, is equal to the work done.

5.2d IRREVERSIBLE ISOTHERMAL EXPANSION

Provided that the initial and final volumes are the same, the state functions of the gas (U, H, S, F...) vary by the same amount as in a reversible expansion. In contrast, the final state of the surroundings is not known since the heat given to the system and the entropy exchanged are undetermined. As the surroundings (strictly speaking, the universe less the system) is in practice very large, if not infinite, direct measurements of slight changes in its properties are impossible. In fact, it is the very possibility of not having to deal explicitly with the properties of the surroundings that makes this two-part division of the universe so powerful.

5.2e REVERSIBLE ADIABATIC EXPANSION

For an adiabatic process, the variations of temperature and volume are given by (4.15). Rather than applying directly this equation, we will demonstrate it in another manner in the special case of ideal gases. We have $dU = dQ + dW = dW$, but the temperature diminishes since the internal energy of the system decreases as a result of the work done. From (5.4), the internal energy variation is

$$dU = C_v dT = dW = -PdV = -RTdV/V. \tag{5.11}$$

From (5.11), we obtain the differential equation

$$C_v dT/T = -RdV/V, \tag{5.12}$$

which is integrated directly with the approximation that R/C_v is constant. Defining $\gamma_c = C_p/C_v$, we find for a perfect gas

$$T_1 V_1^{R/C_v} = T_2 V_2^{R/C_v} = TV^{\gamma_c - 1} = TV^{\gamma} = constant. \tag{5.13a}$$

The last equality stands because $R/C_v = C_p/C_v - 1$, which is also equal to the Grüneisen parameter $\gamma = \alpha K_T/\rho C_v$. For a monoatomic gas, both γ and γ_c are actually constant whereas their variations can be neglected for other kinds of gases as long as the temperature interval of interest is not too large. From the equation of state (5.4), we finally arrive at the relation between P and V during the expansion

$$PV^{1+\gamma} = PV^{\gamma_c} = constant. \tag{5.13b}$$

From (5.11), the work done is

$$W = C_v(T_2 - T_1) = C_v/R\ (P_2V_2 - P_1V_1). \tag{5.13c}$$

The entropy does not change ($dS = dQ/T = 0$) although the temperature varies. The reason is that the temperature-induced entropy decrease is exactly compensated by the entropy rise caused by the volume increase. The Gibbs free energy variation is equal to the enthalpy change, which is itself readily derived from the preceding results.

5.2f IRREVERSIBLE ADIABATIC EXPANSION

The reasoning is similar to that developed in 5.2d. Depending on the pressure acting against the gas during its expansion, the work ranges between 0 (vacuum expansion) and the result given in 5.2e. Because $dQ = 0$, the internal energy variation ($\Delta U = W$) is also undetermined. The temperature decrease is between 0 and the value calculated for the reversible expansion. This temperature constancy of a gas expanding in a vacuum was demonstrated by Joule in the famous experiment that showed that the internal energy of an ideal gas depends only on temperature. At constant internal energy, the temperature therefore does not depend on volume, i.e., $(\partial T/\partial V)_U = 0$.

For expansion in a vacuum, the entropy change is then obtained from (A.17) and the Maxwell relation (3.6c)

$$\left(\frac{\partial S}{\partial V}\right)_U = \left(\frac{\partial S}{\partial V}\right)_T + \left(\frac{\partial S}{\partial T}\right)_V \left(\frac{\partial T}{\partial V}\right)_U = \left(\frac{\partial P}{\partial T}\right)_V = \frac{R}{V}. \tag{5.14}$$

In the intermediate case of expansion against a weak, but not zero pressure, the upper bound to the entropy variation is $\Delta S = R\ln(V_2/V_1)$, which is an entirely created entropy. The lower bound is 0, the value for the reversible expansion. The upper and lower bounds to the changes in the other state functions are determined in the same way.

Why is such an uncertainty not prevailing in the isothermal expansion against a weak pressure, since in both isothermal and adiabatic cases only the initial and final volumes are fixed along with a single constraint (ΔT or $\Delta S = 0$)? The reason is that these two constraints do not belong to the same category. The constraint $\Delta T = 0$ fixes the temperature which, together with the volume, allows the state of the system to be determined. For the adiabatic expansion, in contrast, the volume is the only known variable. Heat is not a state function, and, if the transformation is irreversible, the constraint $dQ = 0$ fixes the second variable that is needed to specify the state of the system only if the work is also zero. This is the extreme case of a vacuum expansion, not that of an expansion against a weak pressure where the state of the system cannot be specified by the single variable V.

5.2g CARNOT CYCLES

These cycles, which we have until now avoided, must be discussed here because of their considerable practical and historical importance in the study of heat engines, temperature scales and many other fundamental aspects of thermodynamics. Such a cycle is a four-stroke process originally imagined to optimize the conversion of heat into work in a heat engine (Fig. 5.4). Interestingly, its clearest description is perhaps that given by Carnot himself:

1) A gas enclosed in a cylinder with a piston is brought into contact with a hot source, kept at a temperature T_1, which gives to it a heat Q_1; the gas initially compressed expands isothermally from A to B (Fig. 5.4a) while doing the work W_1.

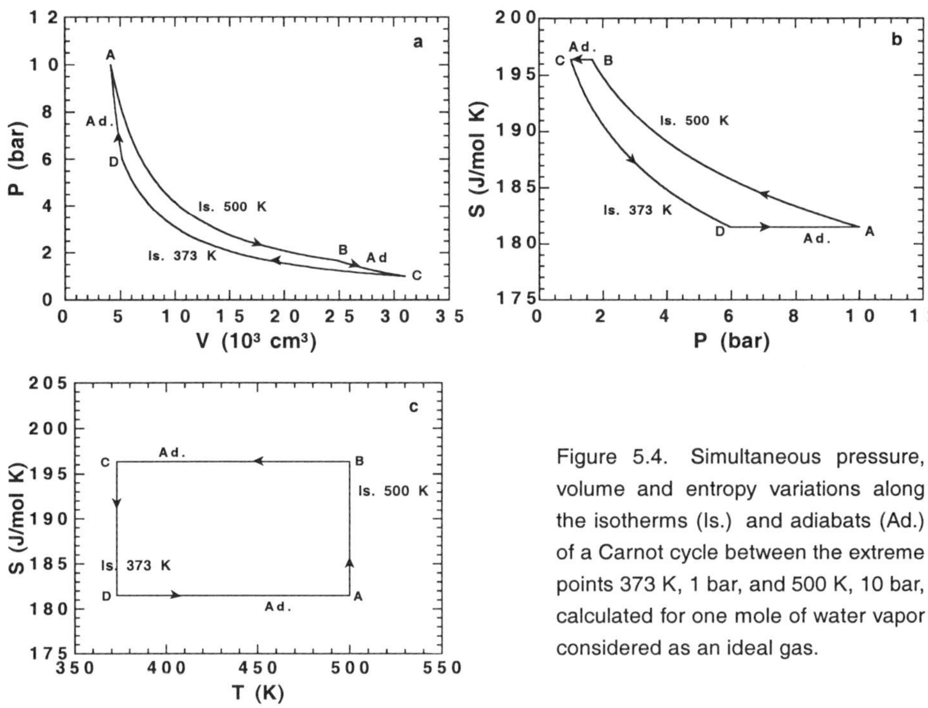

Figure 5.4. Simultaneous pressure, volume and entropy variations along the isotherms (Is.) and adiabats (Ad.) of a Carnot cycle between the extreme points 373 K, 1 bar, and 500 K, 10 bar, calculated for one mole of water vapor considered as an ideal gas.

2) Abruptly, heating is put to a stop; the gas keeps expanding for a while, however, before the piston comes to a halt; this second expansion from B to C is adiabatic.

3) To bring the piston back to its original position in order to resume work production, the gas is brought into contact with a cold source, kept at a temperature T_3; to go from C to D, the gas gives off a heat Q_3 and is given a work W_3.

4) The cold source is eventually withdrawn instantly; as in 2), the gas keeps compressing, adiabatically from D to A, allowing a new cycle to begin.

From the work production standpoint, consider the most favorable case of a reversible cycle. For the four strokes, the variations of P with V, and of S with P and T have already been calculated in 5.2c and 5.2e. They are shown in Fig. 5.4. In passing, we note in (Fig. 5.4a) the result already found in 4.2e that, in the P,V plane, the slope of an isotherm $(\partial P/\partial V)_T = -P/V$ is less steep than that that of the adiabat, $(\partial P/\partial V)_S = -K_S/V = -\gamma_c P/V$, at their crossing point.

Along the isotherm AB, the gas gains an entropy ΔS the tension of which is that of the hot source (Fig. 5.4b,c). The entropy remains constant along the adiabat BC, but its tension drops while the gas cools down to the temperature of the cold source. The entropy ΔS gained between A and B is then given back to the cold source along the isotherm CD and the rest of the gas entropy is eventually raised again to the tension of the hot source between D and A. In summary, a Carnot cycle represents an "entropy pump" transforming thermal into mechanical energy through the strictly distinct roles of the adiabats and isotherms. Since no heat is exchanged the entropy is constant along the adiabats whose role is to change the tension of the entropy, whereas the role of isotherms is to exchange entropy as much as possible with the surroundings.

In Fig. 5.4a, the work effectively done across the cycle, $\int -PdV$, is equal to the surface area ABCD. It is the greatest when the the two adiabats lie the farthest apart, i.e., when the temperature difference between the hot and cold sources is the greatest (in other words, when the entropy drop is the highest). In practice, this temperature difference is limited by the properties of the fluid and of the materials used to build the heat engine. To make this conclusion quantitative, let us calculate the maximum efficiency r of the engine, the ratio of the net work done and the heat taken to the hot source. Because the work exchanged along an adiabat is $W = C_v(T_2 - T_1)$, the net work done along the two adiabats is zero and the efficiency is solely determined by the heat exchanged along the two isotherms. Denoting by V_1, V_2, V_3 and V_4 the volumes at the points A B, C and D of thc cycle, respectively, we find from the results of 5.2c :

$$r = |\frac{(W_1 + W_3)}{Q_1}| = 1 + \frac{W_3}{W_1} = 1 + \frac{T_3 \ln(V_4/V_3)}{T_1 \ln(V_2/V_1)}. \tag{5.15a}$$

From (5.13a), we conclude that $T_1 V_2^{\gamma_c - 1} = T_3 V_3^{\gamma_c - 1}$ and $T_1 V_1^{\gamma_c - 1} = T_3 V_4^{\gamma_c - 1}$. The ratio between these two equalities indicate that $V_2/V_1 = V_3/V_4$, which allows (5.15a) to be rewritten as

$$r = 1 - T_3/T_1. \tag{5.15b}$$

Because the efficiency of an engine cannot depend on a particular temperature scale, W. Thomson (1824-1907, Lord Kelvin of Largs) relied on (5.15b) to define the absolute scale that now bears his name. And since (5.15b) is the result already found in 1.5b for the maximum efficiency, it follows that one cannot design a more efficient way than Carnot cycles to convert heat into work. To calculate this maximum efficiency in 1.5b, we made use of the fact that the entropy is conserved during a reversible process. Historically, another importance of Carnot cycles stems from that the concept of entropy has actually not been considered to derive (5.15). Now, the ratio between the heat given off to the cold source and that supplied by the hot source is

$$\frac{Q_3}{Q_1} = \frac{RT_3 \ln(V_4/V_3)}{RT_1 \ln(V_2/V_1)} = \frac{T_3}{T_1}, \tag{5.16}$$

and Clausius remarked that $Q_3/T_3 = Q_1/T_1$. Hence, the quantity Q/T does not change throughout a cycle of reversible transformations. It is itself a state function, which was called entropy by Clausius, and whose etymology (turning back) is then more understandable.

Denoting by ΔS the entropy alternatively exchanged by the gas with the two heat sources, we can more simply express the work done as

$$W = R\ln(V_2/V_1)(T_3 - T_1) = \Delta S(T_3 - T_1). \tag{5.17}$$

This equation is the same as (1.26), a result that had been obtained without making use of any specific fluid properties. With (1.26), calculation of the maximum possible work done by a Carnot engine is easy as long as the entropy change of the working fluid is known between the two extreme points of the isotherms. This is the reason for the basic importance of the graphical representations of the entropy against both temperature and pressure that have been established for most fluids of industrial interest.

One calculates in a similar way the pressure, volume, temperature and entropy changes for one mole of ideal gas subjected to a Diesel cycle, which is successively made up of an approximately adiabatic compression (of the air-fuel mix), of an isobaric expansion (due to the combustion), of an adiabatic expansion and finally of an isochoric cooling. The standard four-stroke engine follows the cycle designed by Beau de Rochas and Otto, which can be approximated as an adiabatic compression, an isochoric heating, an adiabatic expansion and an isochoric cooling. Finally, we have already noted that refrigerating engines are heat engines working the reverse way, receiving work in order to make them transfer heat from a cold to hot source. Hence, they work as described in this section.

5.3 REAL FLUIDS

5.3a VAPOR AND LIQUID PHASES

The essential limitation of the ideal gas equation of state is not so much its lack of precision as its basic inaptitude to account for the important phenomenon of condensation of a gas into a liquid. Let us fill a cylinder with water vapor and compress it isothermally at room temperature. Not departing much from that of an ideal gas, the pressure first decreases until a first droplet of water appears at P= 0.03 bar. With a molar volume of 815 10^3 cm^3, the vapor is 45 000 times less dense than the liquid whose molar volume is only 18.1 cm^3. The further the volume diminishes, the proportion of water increases whereas the pressure does not change as long as the vapor has not entirely transformed to the liquid. Then, P increases very rapidly.

If the experiment is redone at 372.8 K, the first droplet of liquid appears at a pressure of 1 bar. The pressure again remains constant while the liquid and vapor coexist, but the volume difference between the two phases has considerably reduced: the vapor and liquid molar volumes are 30.6 10^3 cm^3/mol and 18.8 cm^3/mol, respectively. The volume difference between the gas and liquid keeps decreasing if the temperature is further increased and the pressure at which the two phases coexist correlatively increases (Fig. 5.5). For a temperature T_c = 647.5 K the molar volumes of the two fluids are eventually equal, V_c = 55.4 cm^3, at a point where the pressure P_c is 215.6 bar. In the P, V plane, this *critical point* represents the maximum of a curve, the *saturation curve* or Andrews' (1813-1885) curve, shown as a dashed line in Fig. 5.5, which separates the region where the liquid and vapor coexist from the stability domains of each of these fluids at lower pressure. Below the critical point, vaporization and condensation are the hallmark of the volume, entropy, etc. discontinuities between the two fluids. Beyond the critical point, in contrast, the transition from the vapor to the liquid becomes continuous: there exists only a single *supercritical* fluid.

From the above description, the pressure at which the gas and liquid coexist should be called the condensation pressure of the vapor (at the temperature of the experiment). In fact, it has been given its name of *vapor pressure* from the reverse process of vaporization. Consider a liquid of volume V at fixed P and T that is transferred to another, larger vessel at the same temperature, initially kept under vacuum. Not to leave any vacuum and ensure instead the constancy of the pressure throughout the vessel, some liquid must vaporize. The vapor pressure is that pressure at which the two fluids are in equilibrium.

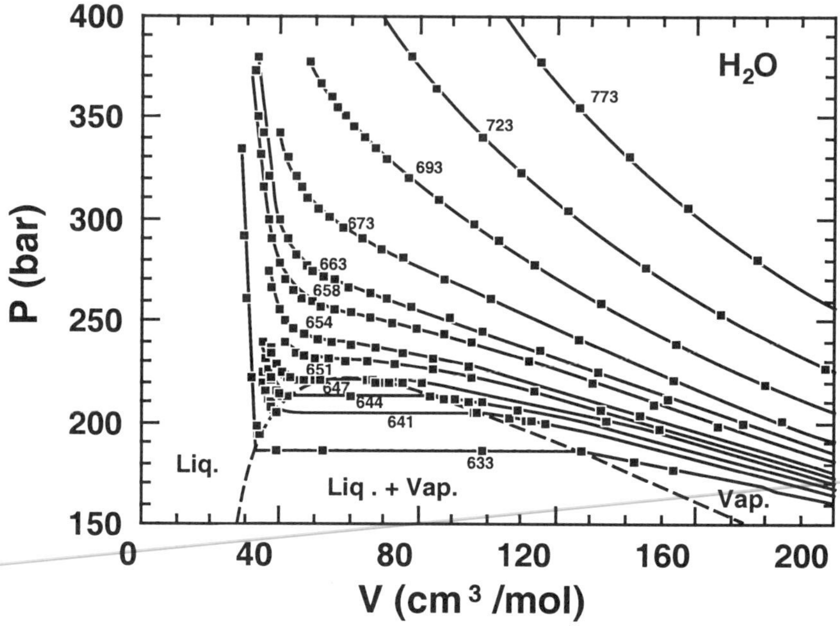

Figure 5.5. Isotherms of water at the indicated temperatures and saturation curves of the vapor and liquid. Experimental data of Rivkin and Akhundov (1962, 1963) and Rivkin and Troyanovskaya (1964). For the sake of clarity, only part of the investigated isotherms has been plotted.

5.3b CRITICAL POINT

How can an equation of state account for the critical point? When the temperature decreases, the isotherms $P = f(V)$ of the fluid phase reveal an inflexion that becomes more and more pronounced. Eventually, the critical point is reached when the isotherm has an horizontal tangent at its inflexion point. Mathematically, the critical point is thus characterized by

$$\left(\frac{\partial P}{\partial V}\right)_T = 0, \tag{5.18a}$$

$$\left(\frac{\partial^2 P}{\partial V^2}\right)_T = 0. \tag{5.18b}$$

When using these criteria to calculate critical parameters from the ideal gas equation of state, however, we obtain the absurd solutions $-RT/V^2 = 0$ and $2RT/V^3 = 0$. This is a most blatant failure of the ideal gas model. As a matter of fact, the solutions of (5.18) should differ for different gases because the critical parameters listed in Table 5.1 show a very strong dependence on composition. Contrary to the universal ideal gas law, the equation of state of a real gas must be function of specific parameters in order to account for the existence and variability of the critical point.

Table 5.1. Critical parameters, and van der Waals cohesions and covolumes for various gases (molar values)

Gas	T_c (K)	V_c (cm^3)	P_c (bar)	Z_c	a (bar cm^6)	b (cm^3)
He	5.19	57.4	2.27	0.302	34.6 10^3	23.8
Ne	44.4	41.6	27.6	0.311	0.208 10^6	16.7
Ar	150.8	74.9	48.7	0.291	1.355 10^6	32.0
Xe	289.7	118.4	58.4	0.287	4.192 10^6	51.6
Kr	209.4	91.2	55.0	0.288	5.193 10^6	10.6
Hg	1765.	42.7	1510	0.439	7.989 10^6	17.0
H_2	33.2	65.1	13.0	0.306	0.2452 10^6	26.5
N_2	126.2	89.8	33.9	0.290	1.370 10^6	38.7
O_2	154.6	73.4	50.4	0.288	1.382 10^6	31.9
H_2O	647.3	57.1	221.2	0.235	5.537 10^6	30.5
CO_2	304.1	93.9	73.8	0.274	3.658 10^6	42.9
NO_2	431.0	167.8	101	0.473	5.360 10^6	44.3
NH_3	405.5	72.5	113.5	0.244	4.225 10^6	37.1
CH_4	190.4	99.2	46.0	0.288	2.303 10^6	43.1
C_2H_6	305.4	148.3	48.8	0.285	5.580 10^6	65.1
C_3H_8	369.8	203	42.5	0.281	9.390 10^6	90.5
n-C_4H_{10}	425.2	255	38.0	0.274	13.89 10^6	116.4

[a] Critical parameters and van der Waals cohesions and covolumes from Reid et al. (1987). The data refer to ^{4}He in the case of helium. (Note the convenient conversion factor 10 cm^3 = 1 J/bar, which allows Z_c to be calculated without having to deal with elevated powers of 10.)

It is important to note that, as embodied by (5.18a), the isothermal compressibility of a fluid tends to infinity at the critical point. Along with the equality

$$\left(\frac{\partial V}{\partial T}\right)_P = -\frac{(\partial P/\partial T)_V}{(\partial V/\partial P)_T}, \tag{5.19}$$

this divergence of the compressibility implies that the thermal expansion coefficient of the fluid also diverges at T_c because the partial derivative $(\partial T/\partial P)_V$ remains finite at the critical point (see Fig. 5.5). In addition, we have from (A.11)

$$C_p - C_v = T\left(\frac{\partial P}{\partial T}\right)_V \left(\frac{\partial V}{\partial T}\right)_P. \tag{5.20}$$

Because of the $(\partial V/\partial T)_P$ term, this heat capacity difference diverges through a divergence of C_p. Although the limit of C_v remains undetermined by this equation, the calorimetric measurements available for gases do show that it is also infinite.

To examine the physical significance of these divergences, consider how the heat capacity of supercritical water varies with temperature along several isobars (Fig. 5.6). At high pressure, this variation is monotonous. On decompression, in contrast, a slight

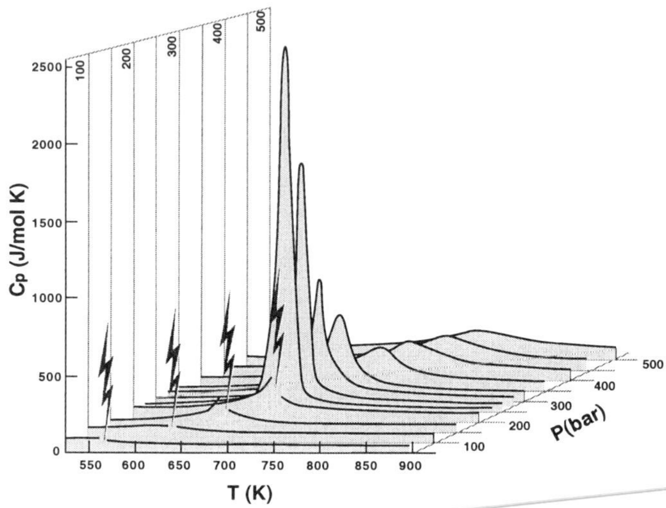

Figure 5.6. Isobaric heat capacity of H_2O against temperature at the indicated pressures. The discontinuous enthalpy and C_p changes below T_c are indicated by the sparkle-like symbols. Data of Sirota and Maltsev (1962) and Sirota et al. (1963).

bump shows up in the C_p curve at temperatures of about 800 K near 1000 bar. This bump then grows markedly, shifting toward lower temperatures and becoming sharper and sharper when the pressure becomes closer to the critical pressure. These C_p maxima thus distinguish more and more clearly a dense fluid from a less dense fluid between which, however, there still exists a continuous pathway. At the critical point, the temperature interval of the C_p anomaly has reduced to the point that the heat capacity of the latter fluid (the vapor) diverges, that of the dense fluid (the liquid) remaining finite. At pressures below P_c, the passage from one phase to the other is marked by an enthalpy discontinuity, at constant temperature. At this temperature, water and its vapor are in equilibrium at a single pressure, the vapor pressure. In the P, T plane, the liquid-vapor equilibrium can thus be represented by a curve that terminates at the critical point (Fig. 5.7).

What are the configurational changes that take place when passing from one fluid to the other? To answer this question, note first that the existence of dense, liquid phases, gives by itself direct evidence for the essentially attractive nature of intermolecular forces. Some energy is needed for opposing these cohesive forces, fragmenting the dense molecular arrangements of the liquid, and freeing the gas molecules, which are entirely independent at the high temperatures where the gas is perfect. As long as thermal energy is much smaller than cohesive energy the phase remains condensed. Now, this contrast between liquid and vapor is strongest at low pressures. Under these conditions, the Gibbs free energy is minimized either by the compact arrangement of the liquid, in which molecules are strongly bonded together, or by that of the gas, where molecules are quasi independent. But the effects of pressure are much weaker on the liquid than on the gas, where the influence of intermolecular forces gets much stronger when the density difference with the liquid decreases. Well before the critical point, association of the gas molecules can even be observed. For instance, through hydrogen bonding water molecules form pairs (i.e., *dimers*) whose fraction is already greater than 1 ‰ at 1 bar and 400 K.

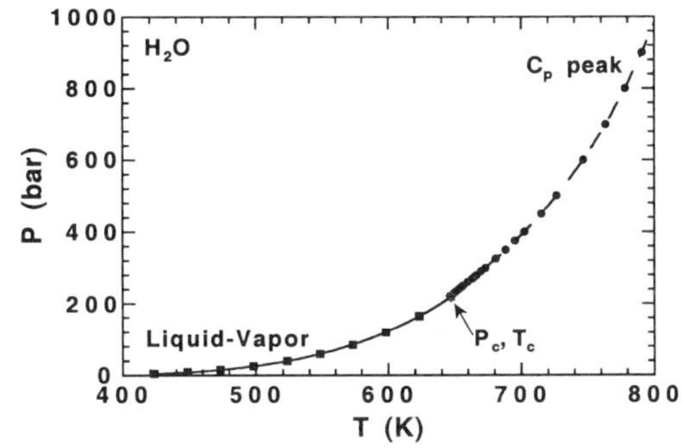

Figure 5.7. Vaporization curve of water and locus of the C_p peak maximum of supercritical water in the P, T plane meeting at the critical point P_c, T_c. Experimental data of Kell et al. (1985) and Sirota et al. (1963).

That the difference between liquid and gas has vanished at the critical point does not mean that the two fluids have become identical in all respects, however. On the contrary, the C_p anomalies observed beyond the critical point are a reminder of the barrier separating two different fluid domains since the cooperative rearrangements of the configuration of the fluid blur gradually and disappear only near 1 kbar in the case of water. A very clear illustration of this feature is provided by Fig. 5.7 where the locus of the maximum of the C_p peak in the P, T plane (Fig. 5.6) represents the extension of the vaporization curve. Hence, the critical point is not the point where any distinction between two kinds of fluids, liquid and gas, has disappeared, but that where the kinship of the gas with the liquid has become close enough that there exists a continuous pathway between the two phases. But the transformation is still cooperative, as will be discussed in more detail in 6.2, with a significant enthalpy effect that is given by the surface area under the C_p anomaly.

It is actually difficult to measure the variations of α, β_T, and C_p in the vicinity of the critical point where, announcing the divergence of the compressibility, a density gradient develops under the effect of gravity. At the critical point, the critical pressure is reached only at a certain level of the fluid column. Inasmuch as measurements are necessarily made over a finite portion of matter, they represent averages over the region investigated. In addition, near T_c there exist liquid-like fluctuations within the vapor and vapor-like fluctuations within the liquid that become large enough to scatter significant amounts of light and give rise to the so-called critical opalescence which is typical of the critical point. Under terrestrial conditions, gravity eventually causes the two phases to separate. Without gravity, phase separation would result from surface tension effects with the walls of the vessel enclosing the fluid.

5.3c VAN DER WAALS EQUATION

Intuitively, it appears that not taking into account the volume of gas atoms or molecules is a basic drawback of the ideal gas model. If this model applies to helium, the smallest monoatomic gas, as well as to big hydrocarbon molecules, it can be so only under conditions where the volume of the system is much larger than the intrinsic volume of the gas molecules. With increasing pressure, this assumption is no longer valid so that we must modify the ideal gas law by writing, for one mole

$$P(V - b) = RT, \tag{5.21}$$

where the b parameter is called the *covolume* of the gas. Although this volume correction allows the P,V,T properties of a real gas to be reproduced somewhat better, it does not result in any improvement with regard to the critical point because an absurd root $P = 0$ is again found for the conditions (5.18).

Actually, it is not surprising that this correction alone does not dispose of the drawbacks of the ideal gas law since it does not address the intermolecular attractions that are the very cause of condensation. To arrive most simply at a real gas equation of state, let us follow a phenomenological approach. The interaction term that must be included represents a force, i.e., apart from a constant factor, a pressure so that it can be added to the pressure in (5.21). On the other hand, that a real gas tends to behave as an ideal gas at low pressure suggests that this term should be a function of volume. Combining these volume and pressure corrections, we write

$$(P + aV^m)(V - b) = RT, \tag{5.22a}$$

where m is not only negative, but smaller than -1 to ensure that (5.22a) tends to the ideal gas law in the low density limit. The a constant is the *cohesion* of the gas, for it represents essentially attractive intermolecular forces that depend specifically on the gas, differing for instance for weakly interacting noble gas atoms and for molecules with a strong dipole moment like H_2O.

To calculate critical point parameters from (5.22a), let us rewrite this equation of state as

$$P = RT/(V - b) - aV^m. \tag{5.22b}$$

The conditions (5.18) give

$$(\partial P/\partial V)_T = -RT/(V - b)^2 - maV^{m-1} = 0, \tag{5.23a}$$

$$(\partial^2 P/\partial V^2)_T = 2RT/(V - b)^3 - m(m-1)aV^{m-2} = 0. \tag{5.23b}$$

Along with (5.22), these equations constitute a system of three equations from which the three parameters of the critical point can be determined. By separating the two terms in both (5.23a) and (5.23b) before taking the ratio of these equations, we first derive the critical volume

$$V_c = b(m-1)/(m+1). \tag{5.24a}$$

In passing, note that the condition $m < -1$, imposes that both $(m+1)$ and $(m-1)$ be negative. Consistent with its meaning, the covolume b must be positive in order that V_c be also positive. By replacing V by (5.24a) in (5.23a), we find

$$T_c = -4maV_c^{m+1}/[R(m-1)]^2 \tag{5.24b}$$

$$= -4ma(m-1)^{m-1}b^{m+1}/(m+1)^{m+1}. \tag{5.24c}$$

From (5.24b) we conclude that a must be positive for the critical temperature to be positive. This condition is consistent with the postulated attractive nature of intermolecular forces. It has the consequence that the pressure of a real gas is lower than that of an ideal gas at the same temperature and volume, and we will also note that the a/V^2 term represents the internal pressure previously defined by (3.37). In summary,

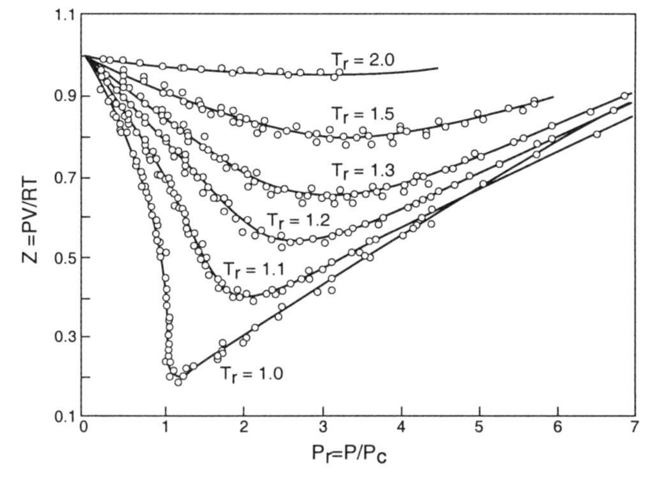

Figure 5.8. Compressibility factor of a series of gases (N_2, H_2O, CO_2, CH_4, C_2H_6, C_2H_4, C_3H_8, etc.) against reduced pressure at several reduced temperatures (from Su and Chang, 1946). Note that it is on the "vapor" side that the proximity of the critical point is strongly felt, Z decreasing extremely rapidly before reaching a common value of about 0.28 at P_r = 1.

the main features of the critical point are remarkably well explained by (5.22), which is a function of parameters whose physical meaning is also simple. With the particular value $m = -2$

$$P = \frac{RT}{V-b} - \frac{a}{V^2}, \tag{5.25}$$

we derive the equation successfully introduced by van der Waals (1837-1923).

5.3d PRINCIPLE OF CORRESPONDING STATES

From (5.24) and (5.25), we find that the critical parameters of a van der Waals gas are

$$V_c = 3b, \qquad T_c = 8a/27bR, \qquad P_c = a/27b^2. \tag{5.26}$$

When using the *reduced variables* $V_r = V/V_c$, $T_r = T/T_c$ and $P_r = P/P_c$, for representing the P, V, T measurements for differing gases, we observe that, even at some distance from the critical point, the data almost plot on the same master curve (Fig. 5.8). Another reason why van der Waals equation was originally acclaimed is that it leads to such a "universal" equation of state. These relations between the critical parameters indeed lead to elimination of the constants a and b from (5.25) and yield in terms of reduced variables

$$P_r = \frac{8T_r}{3V_r - 1} - \frac{3}{V_r^2}. \tag{5.27}$$

This equation is universal in that it is not an analytical function of specific parameters of a gas. This is why two gases having the same reduced variables are said to be in a *corresponding state*. From this standpoint, another interesting property of the van der Waals equation is that its three critical parameters depend on two constants only. This allows us to check at once the general validity of (5.27) by taking two critical parameters from Table 5.1, e.g., V_c and T_c, to calculate the third one with (5.26). We do not find perfect agreement with the experimental value, however, and could even make this test without knowing the parameters a and b because the van der Waals equation predicts that the compressibility factor at the critical point is

$$Z_c = P_c V_c / RT_c = 3/8. \tag{5.28}$$

The significant deviations of experimental Z_c values from 3/8 (= 0.376) shows that, although the van der Waals equation represents a considerable improvement over the ideal gas law, it does not allow all experimental data to be reproduced with the desired accuracy. Note finally that the notion of corresponding state is not an intrinsic feature of van der Waals' theory. As illustrated below, barring analytical problems one can write in a reduced form any three-parameter equation of state. In spite of their possible limitations, such equations are nonetheless useful to make P,V,T predictions from the sole critical parameters when experimental data are sketchy.

5.3e CONDENSATION

Important aspects of the critical point can be described with a single equation of state. This suggests that there should not exist fundamental differences between the equation of state of a liquid and that of its vapor or, in other words, that the intermolecular forces at work are essentially the same in both kinds of phases. This is what the van der Waals equation does demonstrate. Its reduced isotherms, plotted in Fig. 5.9, reproduce in effect (qualitatively, at least) the data for the liquid at the lowest volumes and clearly show that the vapor stability field expands to lower pressures at lower temperatures.

More interesting is the fact that the same pressure can correspond to three different volumes for temperatures lower than critical. This is not surprising since the van der Waals equation can be written as a third-degree polynomial in V

$$V^3 - (RT/P + b)V^2 + aV/P - ab/P = 0. \tag{5.29}$$

As a function of P, the volume curve therefore displays two extrema, shown as M and O in Fig. 5.9, which delineate three domains of liquid-vapor equilibria. Note in this respect that the Gibbs free energy varies by the amount $\int V dP$ when following an isotherm. On the $T_r = 2$ isotherm of Fig. 5.9, this variation between the points E and F is represented by the surface area between the EF segment of the isotherm and the projections of these points E and F on the pressure axis. Starting from very low volumes, let us now move along an isotherm below the critical point ($T_r = 0.8$ in Fig. 5.9, for instance). As described previously, the Gibbs free energy of the vapor begins to increase when the pressure increases. Between the points M and O, a peculiar feature is observed in that the volume expands with increasing pressure: the compressibility has become negative, and the Gibbs free energy decreases in response to a pressure rise. This segment of the isotherm thus corresponds to an unstable equilibrium.

What transformation will take place to ensure a stable state? Of course, part of the vapor condenses. To demonstrate this with a reasoning due to Maxwell, let us start again from the point P in Fig. 5.9 and go to the point N. Between these two points, the Gibbs free energy of the vapor *increases* by an amount proportional to the dashed surface area PON. Let us start from the point L, on the liquid side, to go to the same point N. The Gibbs free energy of the liquid *decreases* by an amount proportional to the surface area LMN. At equilibrium, a liquid and its vapor have the same pressure, temperature and chemical potential. The locus of the points where the liquid and gas are in equilibrium is an horizontal isotherm, along which the Gibbs free energy change is zero when going from one phase to the other. The pressure at which the vapor begins to condense is thus determined by equating the surface areas PON and LMN, where the points P and L must correspond to the same pressure (the vapor pressure) at the temperature considered.

In summary, partial condensation of the gas or partial vaporization of the liquid are the two processes through which the Gibbs free energy of the system is minimized between the points P and L. Inasmuch as the interval OM represents an unstable equilibrium, the interval LM represents the metastable domain of the liquid whereas the *oversaturated* vapor is metastable between the points O and P. The internal stability of each of these states is ensured by a normal volume decrease for a pressure increase, associated with a Gibbs-free energy increase, but the pure vapor and liq-

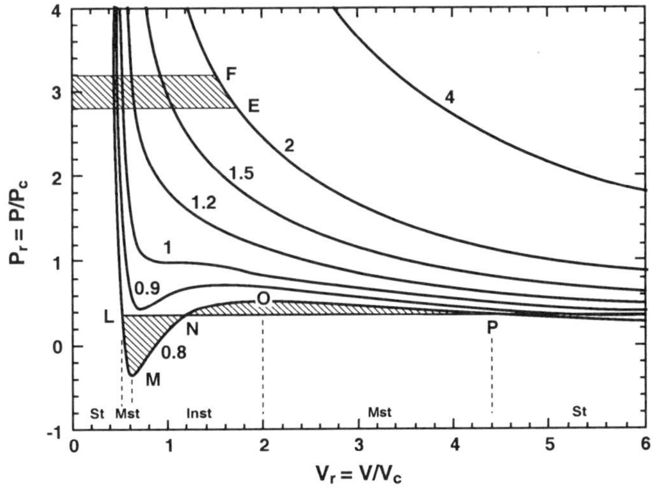

Figure 5.9. Reduced isotherms of the van der Waals equation of state for the indicated reduced temperatures. The coexistence line of the liquid and vapor is represented only for T_r = 0.8. The saturation curve of the liquid is the locus of point L, meeting at the critical point with the saturation curve of the vapor as given by the locus of point P. The domains of stability (St), metastability (Mst) and instability (Inst) are separated by the dashed vertical lines.

uid are unstable with regard to partial condensation and vaporization, respectively. As $P = -(\partial F/\partial V)_T$, the field of instability is characterized by $(\partial^2 F/\partial^2 V)_T < 0$, the metastability field by $(\partial^2 F/\partial^2 V)_T > 0$, and the boundary between these two kinds of domains by

$$-(\partial P/\partial V)_T = (\partial^2 F/\partial V^2)_T = 0. \tag{5.30}$$

These boundaries are also shown in Fig. 5.9.

When metastable, with respect to vaporization, water was feared by locomotive engine stokers because of the violence with which partial vaporization could take place. On the other hand, overvaporization would be very common in the atmosphere if it were not so easily suppressed through fog formation thanks to the nucleating agents that allow water vapor to condense. The Gibbs free energy barriers to be overcome for condensation or vaporization can indeed be lowered in many different ways (through dust or catalysts, such as silver iodide sometimes dispersed in clouds; pollutants, like SO_2 in the bygone London smog; ionized elementary particles like in the condensation chambers in the good old days of particle physics, or nowadays in bubble chambers).

Note finally that the metastability field of the liquid shown in Fig. 5.9 extends to the negative pressure side. This region does exist, and it is Berthelot (1827-1907) who discovered in 1850 a simple means to observe a liquid under a negative pressure. The investigated liquid is sealed in a glass capillary whose volume is slightly larger than that of the liquid at the P, T conditions of the experiment. The rest of the capillary is thus occupied by an air or vapor bubble (depending whether the capillary has been filled under air or a vacuum). The capillary is then heated until dilation makes the liquid absorb the bubble and fill the whole space. On cooling, one subsequently observes that the liquid keeps filling the capillary until "the least shock" eventually makes the initial gaseous phase "instantaneously reappear, with a kind of boiling, a slight noise and a more or less notable jolt", according to Berthelot's own words. It is through surface tension that negative pressures of several hundred bars can be exerted on a liquid by the glass vessel. Water, for instance, has been subjected in this way to -800 bar, and one can even reach negative pressures at temperatures lower than the freezing temperature. The liquid is then metastable with respect to both vapor and ice.

5.4 REAL GASES

5.4a SOME OTHER PROPERTIES

The simplicity of the van der Waals equation and the calculations made with it have been the main reason for choosing it to present some of the properties of real gases. This equation is no longer used when accurate calculations are required because, as already noted, it does not reproduce well the measurements above a few hundred bars (Fig. 5.3). High-pressure measurements indicate for instance that the volume of the gas becomes smaller than the covolume, as determined from low-pressure measurements, and that the critical volume is about four times the covolume, and not three as predicted from (5.27).

The incorrectly predicted volume dependence of C_v represents another limitation. This variation is given by

$$\left(\frac{\partial C_v}{\partial V}\right)_T = \frac{\partial}{\partial V}\left[T\left(\frac{\partial S}{\partial T}\right)_V\right]_T = T\frac{\partial}{\partial T}\left[\left(\frac{\partial S}{\partial V}\right)_T\right]_V$$

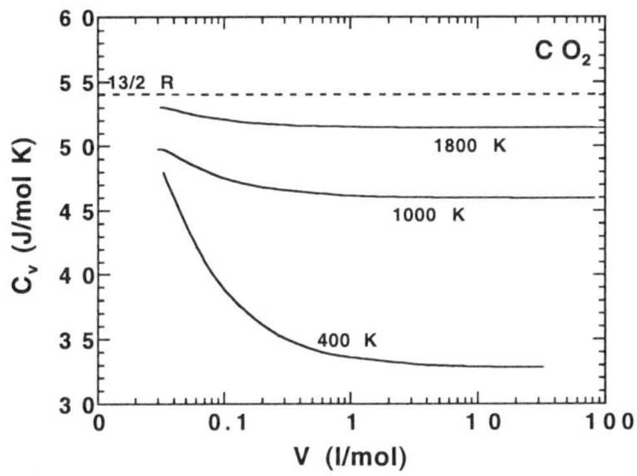

Figure 5.10. Heat capacity of CO_2 at constant volume at the three indicated temperatures. The dashed line represents the high-temperature limit of C_v, namely, 13/2 R per mole. Data of Bottinga and Richet (1981).

$$= T\frac{\partial}{\partial T}\left[\left(\frac{\partial P}{\partial T}\right)_V\right]_V = T\left(\frac{\partial^2 P}{\partial T^2}\right)_V. \tag{5.31}$$

For a van der Waals gas, we have therefore

$$(\partial C_v/\partial V)_T = 0. \tag{5.32}$$

This result clearly contradicts the volume dependence of C_v shown in Fig. 5.10 for CO_2 in temperature and pressure intervals purposely chosen far away from the critical point. At 400 K, for instance, C_v increases by 50% when the volume decreases by a factor of 1000 between 1 bar and 5 kbar. As indicated by (5.31), any equation of state where the pressure is a linear function of temperature will not account for such variations. Empirically, one could fix it by making the a or b parameters of (5.22) temperature dependent. It is preferable, however, to use other equations of state, having the same number of adjustable parameters, but which give better results and lend themselves to reliable extrapolations. We finally stress that some of the problems presented in 5.2a-2g are more difficult to solve than in the case of ideal gases because the internal energy and enthalpy of real gases do not depend on temperature only.

The various kinds of calculations performed in the preceding sections can nonetheless be repeated with any equation of state. The main difference is that, in contrast to the case of an ideal gas, the internal energy and enthalpy of real gases depend on volume and pressure. For a van der Waals gas, we find $(\partial U/\partial V)_T = a/V^2$ and $(\partial H/\partial P)_T = V - RT/[(P + a/V^2 - 2a(V-b)/V^3]$ by inserting (5.25) in the thermodynamic equations of state (3.26) and (3.29). Thanks to such relationships, changes in thermodynamic variables can be calculated when the simplifying condition $(\partial U/\partial V)_T = 0$ does not hold. For the isothermal expansion of a van der Waals gas, we calculate $\Delta U = a(1/V_1 - 1/V_2)$. The work is $W = RT\ln[(V_1 - b)/(V_2 - b)] + a(1/V_1 - 1/V_2)$, and the heat and entropy exchanged are $Q = RT\ln[(V_2 - b)/(V_1 - b)]$ and $\Delta S = R\ln[(V_2 - b)/(V_1 - b)]$, respectively.

5.4b A FEW EQUATIONS OF STATE

Webb-Benedict-Rubin Equation

One finds in the literature a wealth of equations of state, which are especially complicated when they aim at reproducing accurately a great many different properties. In a domain ranging from ice melting to 1273 K and 250 kbar, an equation proposed for water has 58 empirical parameters, of which twenty or so are needed just to account for the abrupt variations of thermodynamic properties near the critical point (Saul and Wagner, 1989). But water is the best known substance and, as extensively discussed in the books by Prausnitz and Reid *et al.*, much simpler equations must generally be used for other fluids. Most of them are actually empirical, even though they can often be derived from simple statistical mechanical models. To give a flavor of them, we will restrict our attention to three well-known different kinds of equations. We begin with the so-called Webb-Benedict-Rubin equation of state

$$P = RT\rho + (B_0RT - A_0 - C_0/T^2)\rho^2 + (bRT - a)\rho^3 + a\alpha\rho^6 + c\rho^3/T^2\,(1+\gamma r^2)\exp(-\gamma r^2), \quad (5.33)$$

where ρ is the fluid density. Thanks to its 8 adjustable parameters, this equation behaves rather successfully in the neighborhood of the critical point so that it is extensively used for calculating liquid-vapor equilibria.

Redlich-Kwong Equation

Simpler equations can give good results when getting farther from the critical point. Especially effective from this standpoint is the Redlich-Kwong equation

$$P = \frac{RT}{V-b} - \frac{a}{V(V+b)T^{1/2}}. \quad (5.34a)$$

Like the van der Waals equation, it is a third-order function of the volume. Its a and b parameters, which also represent the cohesion and covolume, are related to the critical parameters by

$$a = \frac{R^2T_c^{5/2}}{9P_c(2^{1/3}-1)}, \qquad b = \frac{(2^{1/3}-1)RT_c}{3P_c}. \quad (5.34b)$$

The Redlich-Kwong equation thus keeps all the useful features of van der Waals', but it gives a more accurate representation of real gases. This equation was in fact constructed so as to account for the empirical observation that the volume of many gases at high pressure approaches a limiting value which is practically independent of temperature and close to $0.26V_c$. This condition is satisfied by (5.34a) because the volume tends to b at high-pressure and $b = 0.26V_c$. It can also be seen that the critical compressibility factor drawn from (5.34), namely, 0.333, differs less from the data of Table 5.1 than the value of 0.376 yielded by the relation (5.28) deduced from the van der Waals equation.

As shown by deviations of the critical compressibility factors from 0.333, however, the Redlich-Kwong itself is not the answer to all problems. To reproduce correctly the P, V, T data over wide intervals, the parameters a and b are generally assumed to depend on temperature, volume, or on another characteristic parameter of the gas such as its *acentric factor* (see below). The compressibility factors of CO_2 plotted in Fig. 5.3 actually represent the values given by (5.34a) where volume-dependent coefficients (5.34b) enable experimental data to be adequately reproduced over a very wide range of molar volumes from 30 cm^3 to 10^5 cm^3. The improvement over the van der Waals equation is obvious. As to the C_v values shown in Fig. 5.10, they have been calculated from the same Redlich-Kwong equation and account correctly for the available observations even relatively close to the critical point.

On the other hand, most equations of state are deficient in that they do not give an analytical expression of V as a function of P. With modern computer means, this is no longer too serious a problem because the volume is readily calculated numerically. In addition, the lack of such an analytical expression for the volume does not prevent integrating $-PdV$, since it suffices to replace P by (5.34a) or any other expression and, as noted in 8.4a, VdP can be evaluated without much difficulty through integrations by parts.

Introduced by Pitzer (1914-1997), a third useful parameter is the acentric factor, $\omega = -\log(P_s/P_c)$ -1, where P_s is the vapor pressure at an arbitrary temperature $T_r = 0.7$, which has been chosen to be distant from the critical point. This factor owes its usefulness to the fact that the slope of the reduced vapor pressure curve is a very sensitive probe of interatomic forces upon which an additional parameter can be based. It is zero for simple fluids such as big noble gases whose intermolecular potential has a spherical symmetry, whence its name. It can be introduced in an equation of state in two different ways: either it is an additional adjustable parameter, independent of the others; or some equation of state parameters (like a and b in the Redlich-Kwong equation) are assumed to be an explicit function of it.

Virial Equations

Let us finally mention the so-called *virial equations*. These equations exist in several forms, viz.

$$Z = PV/RT\ (1 + b/V + c/V^2 + d/V^3 + ...), \tag{5.35a}$$

$$Z = PV/RT\ (1 + bP + cP^2 + dP^3 + ...). \tag{5.35b}$$

The parameters b, c, etc. of these equations are called the second, third, etc. virial coefficients. They depend only on temperature and are determined simply from the low-pressure limits of $(\partial Z/\partial Y)_T$, $1/2\ (\partial^2 Z/\partial Y^2)_T$, etc., respectively, where $Y = \rho$ or P, as given by P, V, T measurements. Converging very slowly, virial equations have too many coefficients to be really practically useful. With the first three coefficients, for instance, they generally work well up to only half the critical density and, contrary to the Redlich-Kwong equation, they do not predict that the volume tends to the co-volume b at very high pressures. On such counts virial equations compare badly with empirical two-parameters equations of state. In fact, their interest is mainly theoretical in that their coefficients are directly related to intermolecular potentials. The equation (5.35a) has the additional advantage that its coefficients can be rigorously calculated for gaseous mixtures.

Consider a system made up of N particles of masses m_i. Let $\vec{F}_i$ be the force exerted on the particle i whose vector coordinate is $\vec{r}_i$. As will be seen in 15.4a, the *virial* of the system is the sum $-1/2\ \Sigma m_i \vec{r}_i \vec{F}_i$, and a theorem demonstrated by Clausius shows that, over a long enough period, the average of the virial is equal to the kinetic energy of the system. Application of this theorem to a gas for which interatomic forces are known as a function of interatomic distances then yields the equation of state. This is why the virial equation has long been popular among theoreticians to determine intermolecular potentials from experimental equations of state. In the simple case of spherical molecules, the second virial coefficient is given by

$$b = 2\pi N_A \int_0^\infty \left\{1 - \exp[-\phi(r)/kT]\, r^2 dr\right\} \tag{5.36}$$

where N_A is Avogadro number, k, Boltzmann constant and $\phi(r)$ the potential between two molecules separated by a distance r.

5.4c MIXING LAWS

Until this point we have discussed equations of state of pure gases. What about the equation of state and critical point of a gas mixture? With regard to equations of state, a change in the number of moles in the ideal gas expression (5.4) suffices when this law is valid. For real gases, in contrast, it is necessary to determine how the empirical parameters of the equation of state depend on chemical composition. Critical parameters of mixtures are not additive functions of the pure components parameters, and another practical difficulty stems from the rapidly increasing topological complexity of the diagrams used to represent liquid-vapor equilibria. Whereas a great many pure gases have been studied, P, V, T measurements, which are tedious and difficult to make, are necessarily scarce with respect to mixtures whose diversity is intrinsically unlimited.

To obviate the general lack of experimental data, use has to be made of mixing laws whose analytical form can at best be justified from theoretical arguments. For a mixture of Redlich-Kwong (or van der Waals) gases, the following approximations are commonly used for the cohesion and the covolume

$$a = \sum\sum a_{ij} x_i x_j, \qquad b = \sum b_i x_i, \tag{5.37a}$$

where x_i is the mol fraction of gas i in the mixture, and a_{ii} and b_i the parameters of the pure gases. Representing the limiting volume of the gas at the highest pressures, the covolume can be assumed to vary linearly with the mixture composition, whereas the mixing law (5.37a) for the cohesion is inspired by considerations on interatomic potentials. Finally, the interaction parameters a_{ij} between the gas i and j of the mixture can be approximated by

$$a_{ij} = (a_{ii} a_{jj})^{1/2}. \tag{5.37b}$$

This approximation gives for the cohesion of a binary gas mixture

$$a = a_{11} x_1^2 + 2(a_{11} a_{22})^{1/2} x_1 x_2 + a_{22} x_2^2. \tag{5.37c}$$

The actual usefulness of such mixing laws rests on their ability to reproduce P, V, T measurements and phase equilibria data. As illustrated by H_2-CH_4 mixtures (Fig. 5.11), the nonlinear variations of the volume against composition, at constant T and P, are generally not only asymmetrical, but also more pronounced at low temperatures and low pressures. Such variations are embodied by (5.37a,b) and it is remarkable that these simple mixing laws can work so well for real fluid mixtures.

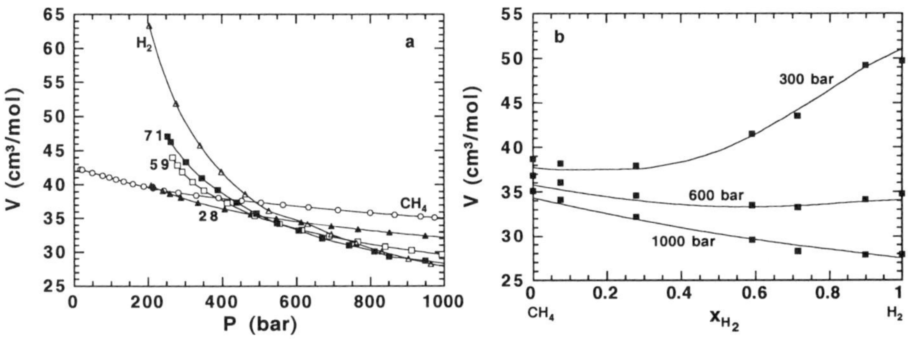

Figure 5.11. Molar volumes of hydrogen and methane mixtures at 140 K: (a) against pressure for the mole fractions of H_2 indicated; (b) against the mole fraction of H_2 for the pressures indicated. Experimental data of Machado et al. (1988) and, in (b), values calculated for the intermediate compositions with the mixing laws (5.37) and the molar Redlich-Kwong coefficients reported by these authors for the endmembers, namely, a = 17 986 bar^2/J K$^{0.5}$ and b = 1.7158 J/bar for H_2; and a = 309 793 bar^2/J K$^{0.5}$ and b = 2.9057 J/bar for CH_4.

The data plotted in Fig. 5.11 for H_2-CH_4 mixtures show that the mixing laws (5.37) can even be used for two-phase systems. Because the selected temperature and pressure are intermediate between the critical points of methane and hydrogen (see Table 5.1), the former substance is a liquid, whereas the latter and hydrogen-rich mixtures are supercritical fluids. The marginally less good agreement between calculations and observations near pure CH_4 therefore originates in the fact that the Redlich-Kwong equation of methane is slightly less appropriate for the liquid than for the gaseous phase.

5.4d JOULE-THOMSON EXPANSION

With W. Thomson, Joule performed another famous experiment for the initial purpose of checking that the internal energy of an ideal gas depends on temperature only. The experiment consists of the passage of gas through a porous wall dividing an adiabatic enclosure in two compartments between which a pressure difference is kept constant (Fig. 5.12). The observation made is that the gas temperature generally decreases by an amount that depends on the imposed pressure difference.

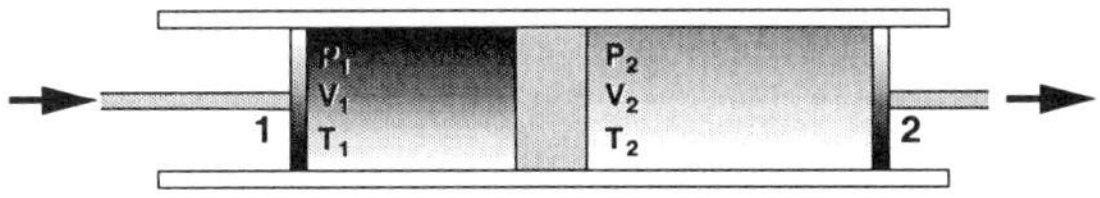

Figure 5.12. Joule-Thomson expansion of a gas. The pistons 1 and 2 are displaced at a speed such that the pressures P_1 and P_2 remain constant.

To interpret this result, it suffices to note that, without any heat exchanged, the internal energy variation is equal to the difference between the final and initial mechanical energies that represents here the net work done

$$\Delta U = U_2 - U_1 = P_1V_1 - P_2V_2. \tag{5.38}$$

From this energy balance, we conclude immediately that $H_2 = H_1$. The observed temperature change is thus characterized by $(\partial T/\partial P)_H$, a differential denoted by μ and known as the Joule-Thomson, coefficient

$$\mu = \left(\frac{\partial T}{\partial P}\right)_H = -\frac{(\partial H/\partial P)_T}{(\partial H/\partial T)_P} = -\frac{(\partial H/\partial P)_T}{C_p}. \tag{5.39}$$

For most real gases, the Joule-Thomson coefficient is positive, with values of the order of 1 K/bar. An expansion of the order of only 100 bar can result in considerable temperature decreases. Through expansion, it is thus possible to cool down gases until they liquefy. Gases such as hydrogen and helium have a negative Joule-Thomson coefficient above room temperature, which makes them become warmer on expansion. Because μ generally increases when the temperature decreases (and decreases when the pressure increases), however, these gases behave "normally" at low enough temperatures from which they can liquefy.

The generally nonzero values of μ observed demonstrate that the enthalpy of a real gas depends on pressure. To relate μ to the variation of the internal energy with the volume, let us rewrite (3.26) as

$$\begin{aligned}\left(\frac{\partial U}{\partial V}\right)_T &= \left(\frac{\partial H}{\partial V}\right)_T - P - V\left(\frac{\partial P}{\partial V}\right)_T = \left(\frac{\partial P}{\partial V}\right)_T \left[\left(\frac{\partial H}{\partial P}\right)_T - V\right] - P \\ &= -\left(\frac{\partial P}{\partial V}\right)_T [\mu C_p + V] - P.\end{aligned} \tag{5.40}$$

Since there is no reason why this partial derivative should be zero, we verify that the enthalpy of a real gas also depends on volume. What happens when the pressure vanishes, a condition for which real gases tend to behave like ideal gases? In the limiting case of the ideal gas law, we readily find that $(\partial P/\partial V)_T$ tends to $-P/V$ and $(\partial U/\partial V)_T$ to $P\mu C_p/V$ which, as expected, also tends to zero.

5.5 CONDENSED PHASES

Redlich-Kwong and other equations of state can give good results at elevated pressures and low temperatures at which gases tend to be as dense as liquids. When dealing with condensed phases over wide P, T intervals it is preferable to use simpler equations of state. These account directly for the fact that liquids have compressibilities and thermal expansion coefficients so much smaller than gases that water and other fluids have sometimes been called "incompressible". In fact, the use of specific equations of state is especially justified for solids which have even smaller compressibilities and thermal expansions and are stable phases over much wider pressure intervals than liquids. (Note that the volume of a condensed phase varies little between 0 and 1 bar. This is an essential difference between gases and condensed phases so that no significant error will be made in the following by using 0 instead of 1 bar.)

5.5a EMPIRICAL EQUATIONS

Once again, we will not review the numerous empirical or semi-empirical equations of state proposed for condensed phases. We will instead restrict ourselves to a few of them.

Simplified Equation of State

Over large temperature intervals and restricted pressure ranges, the simplest approximation consists of neglecting the slight pressure and temperature dependences of the compressibility. From the volume V_0 at given T_0 and P_0, the polynomial expansion (4.4) for α gives for the volume at the relevant P and T :

$$V = V_0 \exp[\alpha_0(T - T_0) + \frac{\alpha_1}{2}(T^2 - T_0^2) + \frac{\alpha_2}{3}(T^3 - T_0^3)] \exp[-\beta_T(P - P_0)]. \quad (5.41)$$

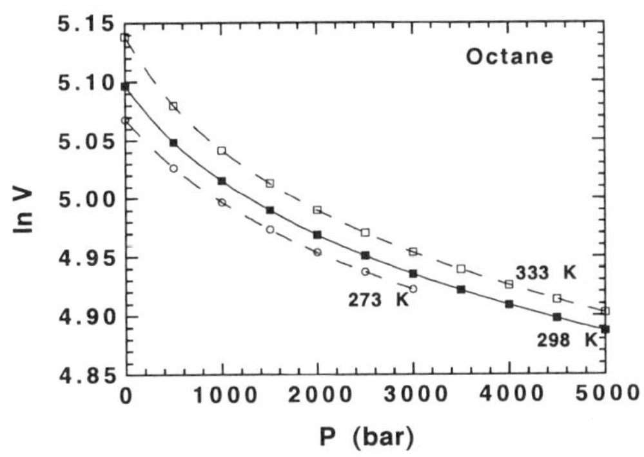

Figure 5.13. Logarithm of octane volume against pressure at different temperatures. Note the steeper slope of the curves at the highest temperatures. Experimental data of Eduljee et al. (1951).

As defined by $\beta_T = (\partial \ln V/\partial P)_T$, the compressibility is constant if there exists a linear relationship between the pressure and the logarithm of the volume. The example of octane in Fig. 5.13 shows that such a linear law is actually observed over limited pressure ranges. By extrapolating the low-pressure data to only 5000 bar, we would underestimate the volume of octane by nearly 50%. As a matter of fact, the approximation of a constant compressibility is generally satisfactory over intervals of a few hundred bars at most for liquids whereas for solids, which have a much lower compressibility, this interval is of the order of 10 kbar. In Fig. 5.13, note also that the initial volume decrease with pressure is stronger at high than at low temperatures, indicating that the compressibility increases with temperature. Besides, the influence of temperature on compressibility is the strongest at low pressure. These two observations generally hold true and they will be expressed in a different form at the end of this chapter where the effects of temperature on the volume of condensed phases will be discussed.

Murnaghan Equation

Work by Bridgman and other investigators has long demonstrated that the decrease of β_T cannot be neglected at high pressures. Nowadays, this need is particularly acute as condensed matter physics deal with the Mbar pressures that are routinely attained in static or dynamic experiments. This compressibility decrease is shown in Fig. 5.14a for a variety of crystals ranging from hydrogen to diamond. This decrease is such that the bulk modulus generally increases in a quasi linear manner (Fig. 5.14b). As a first-order approximation, we write for constant T

$$K_T = K_0 + K_0' P, \tag{5.42a}$$

where K_0 is the 1 Bar bulk modulus, and K_0' the 1-bar value of the pressure derivative of K_T. From the definition (4.7a) of K_T and (5.42a), we obtain an equation whose variables can be separated

$$-\frac{dV}{V} = \frac{dP}{K_0 + K_0' P}, \tag{5.42b}$$

and readily integrated. With $P_0 = 0$, we derive

$$V = \frac{V_0}{[1 + (K_0'/K_0)P]^{1/K_0'}}. \tag{5.42c}$$

Among the great many equations proposed for condensed phases, this equation of state is one of the few that give an analytical expression for both volume and pressure, with

$$P = \frac{K_0}{K_0'}\left[\left(\frac{V_0}{V}\right)^{K_0'} - 1\right]. \tag{5.42d}$$

These expressions (5.42) constitute the isothermal Murnaghan equation of state (which is named after the physicist who derived them, not in the empirical manner used here, but from arguments of continuum mechanics.) Correct results are generally obtained over pressure intervals of a few hundred kbar, beyond which other equations must be preferred.

Vinet Equation
It is from general considerations of the analytical form of interatomic potentials that this more complicated isothermal equation has been derived

$$P = \frac{3K_0[1 - (V/V_0)^{1/3}]\,\exp\{\eta[1 - (V/V_0)^{1/3}]\}}{(V/V_0)^{2/3}}, \tag{5.43a}$$

where $\eta = 3/2\,(K_0' - 1)$ and K_0' is a constant. By differentiating (5.43a), we find

$$K = \frac{K_0\,[2 + (\eta - 1)(V/V_0)^{1/3} - \eta(V/V_0)^{2/3}]\,\exp\{\eta[1 - (V/V_0)^{1/3}]\}}{(V/V_0)^{2/3}}. \tag{5.43b}$$

Ever since its inception in 1987, this equation has rapidly been gaining popularity. For phases brought to extreme pressures or others, like hydrogen, which show a very strong increase of the bulk modulus, it seems to reproduce best the volumes measured

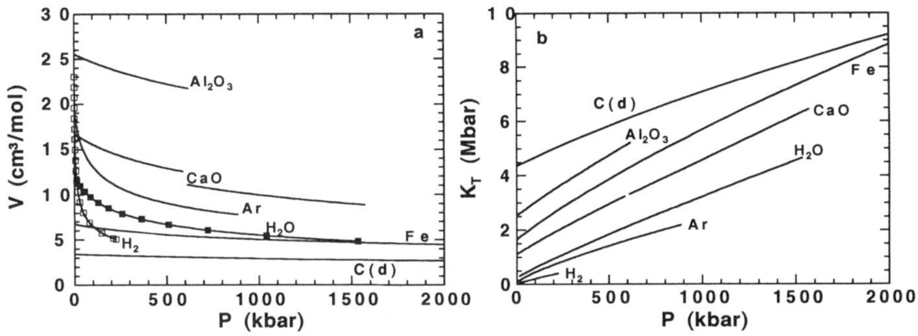

Figure 5.14. Volume (a) and bulk modulus (b) against pressure for a few crystals. Note the extraordinary strong compression of H_2, whose molar volume decreases from 23 to 10 cm^3 between 0 and 20 kbar, which contrasts with the remarkable incompressibility of diamond, C(d). Same references as below in Table 5.2.

at high-pressures as well as the bulk moduli determined at room and high pressures from elastic constant measurements.

5.5b FINITE STRAIN EQUATIONS

From a theoretical standpoint, the major difficulty experienced in equation of state calculations for condensed phases stems from the complex nature of atomic interactions. Rather than meeting this challenge, we can alternatively adopt a purely phenomenological approach within the framework of continuum mechanics by ignoring completely the nature of the phase considered. We assume that it is possible to expand the Helmholtz free energy as a function of a variable f, called the *strain*

$$F = a_0 + a_1 f + a_2 f^2 + a_3 f^3 + \dots \tag{5.44}$$

Thanks to $P = -(\partial F/\partial V)_T$ it is possible to determine the pressure as a function of the volume from such a series expansion. Before doing so, note that the Helmholtz free energy is known apart from an arbitrary constant, a_0, which can be taken as zero, and that a zero reference pressure P_0 implies $a_1 = 0$.

The main problem is to find how the strain must be defined in terms of the volume. A simple way of doing it would be to take the compression, V/V_0, in a virial-like manner, but convergence of (5.44) would then be much too slow. As a rigorous discussion of this point is beyond the scope of this book, we will just note that mechanical arguments favor the use of the so-called *Eulerian* strain

$$f = 1/2\,[(V_0/V)^{2/3} - 1]. \tag{5.45}$$

By keeping only the first three terms of (5.44), inserting (5.45) in (5.44), differentiating F with respect to V at constant T, and using the limiting conditions to eliminate the a_i

coefficients, we obtain

$$P = \frac{3}{2}K_0\left[\left(\frac{V_0}{V}\right)^{7/3} - \left(\frac{V_0}{V}\right)^{5/3}\right]\left\{1+\frac{3}{4}(K_0'-4)\left[\left(\frac{V_0}{V}\right)^{2/3} - 1\right]\right\}, \quad (5.46a)$$

$$K = K_0\left(\frac{V_0}{V}\right)^{5/3}\left\{1+\left(\frac{3}{2}K_0'-\frac{5}{2}\right)\left[\left(\frac{V_0}{V}\right)^{2/3} - 1\right]\right\}, \quad (5.46b)$$

$$K' = (\partial K/\partial P)_T = K_0'. \quad (5.46c)$$

These equations are the most extensively used ones. They are known as the isothermal Birch-Murnaghan equations, or third-order finite strain equations because they are derived from third-order expansion of F as a function of f. As a matter of fact, the expansion (5.44) is used only as an intermediate step to derive the equations (5.46) and, as such, is generally not evaluated. One could go one step farther and make a fourth-order calculation to account for the pressure dependence of K', but pressure and volume measurements are not yet accurate enough to justify this approach. On the contrary, a second-order expansion often suffices, with which the equation of state reduces to

$$P = 3/2\ K_0\ [(V_0/V)^{7/3} - (V_0/V)^{5/3}]. \quad (5.46d)$$

This expression is the truncated form of (5.46a) for $K_0' = 4$.

Table 5.2. Molar volumes and bulk moduli of some crystals[a]

	V_0 (cm^3/mol)	K_0 (kbar)	K_0'	P_{max} (kbar)
H_2	23.00	3.62 (3)	4.71 (13)	265
Ar	19.57	86.47 (8)	5.21 (10)	800
Ice VII	12.3 (3)	237 (9)	4.15 (7)	1280
C (Graphite)	5.498	338 (30)	8.9 (10)	140
CaO (B1)	16.76	1112 (10)	4.2 (2)	600
CaO (B2)	14.8 (2)	1300 (200)	3.5 (5)	1350
ε-Fe	6.73 (1)	1650 (40)	5.33 (9)	3000
α-Al_2O_3	25.575	2350 (10)	5.0 (4)	500
C (Diamond)	3.417	4360 (60)	3.3 (3)	220

[a] Unless otherwise noted, K_0 and K_0' have been determined with Birch-Murnaghan equations (5.46a) from volume measured at room temperature through X-ray diffraction; P_{max} is the maximum pressure of the measurements. Data of Hemley et al. (1990) for H_2; Richet et al. (1988) for Ar and α-Al_2O_3; Hemley et al. (1987) for ice VII; Hanfland et al. (1989) for graphite; Richet et al. (1988) for CaO; Mao et al. (1990) for ε-Fe; and Gillet et al. (1999) for diamond.

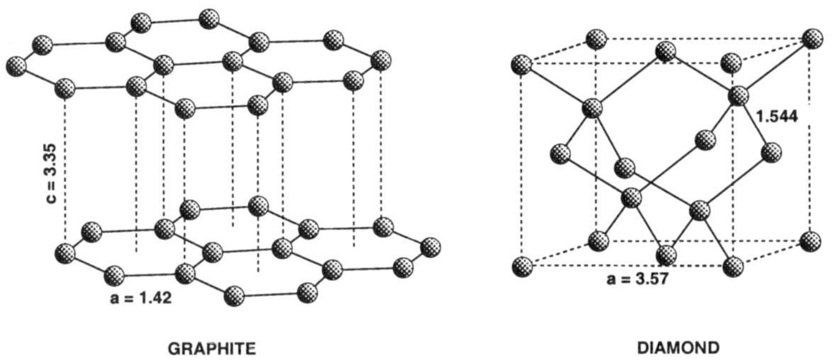

Figure 5.15. Crystal structures of graphite and diamond (distances in Å). The linear compressibility of graphite along the a axis is similar to that of diamond along each of its three axes (K_T= 12500 kbar) and is 35 times lower than along the c axis (K_T = 357 kbar). This illustrates the general feature that short interatomic distances are associated with high bond energies and low compressibilities.

These second- or third-order Eulerian finite-strain equations of state generally give good results up to Mbar pressures. As examples, the bulk moduli of various crystals gathered in Table 5.2 have been determined through fits made with (5.46a) to the measured compression data. Argon and hydrogen are extremely compressible crystals because of the weak van der Waals forces that ensure the crystal cohesion. Thanks to less weak hydrogen bonds, ice VII (the densest polymorph of ice) is less compressible. In metals, such as iron, or ionic crystals, like alumina or lime, a much stronger bonding results in further important compressibility decreases. Finally, its exceptionally strong covalent forces make diamond not only the hardest, but also the known less compressible crystal, in contrast to graphite where a high, but markedly anisotropic compressibility originates in the weak van der Waals bonds between the layers of the structure (Fig. 5.15). In spite of their great K_0 differences, however, most of these crystals have similar K_0' values, ranging from 3.5 to 5.3, which are close to 4, the value for which (5.46a) reduces to (5.46c). Empirically, this constitutes the clearest evidence for the rapid convergence of (5.44) with a Eulerian finite strain equation of state.

Like temperature, pressure induces phase transitions between solids and liquids or between solids. In Table 5.2, there are two sets of data for CaO because the low-pressure phase with an NaCl cubic structure (B1) transforms at 600 kbar (at room temperature) to another cubic phase, having a CsCl structure (B2), with a volume change of -10% (see Fig. 5.14a). The B2 phase is not quenchable to room pressure and reverts instead to the B1 phase on decompression. Likewise, ε-Fe and ice VII are high-pressure phases of iron and ice, which become stable above 150 and 20 kbar, respectively. Their volumes at ambient conditions have to be determined from low-pressure extrapolations of equations of state. This is also the case of argon, which crystallizes at 12 kbar at room temperature.

With any equation of state, the coefficients K_0 and K_0' are adjusted in order to reproduce as well as possible the measured high-pressure volumes. Being determined from an equation relating the pressure to the volume, these moduli are much less precise than the volumes themselves. When available, bulk moduli determined from elastic constant measurements are in principle more accurate; compression measurements can thus be used for determining K_0' more accurately. Note finally that the same values for K_0 and K_0' give different compressions when used in different equations of state. Owing to the specificity of these relations between P, V, K_0 and K_0', it is necessary to state with which equation of state the reported K_0 and K_0' have been determined.

5.5c TEMPERATURE EFFECTS

Until now pressure effects alone have been addressed. How are those of temperature incorporated for condensed phases, whose equations of state are explicitly function of pressure and volume only? Two methods can be used depending whether one takes, along with temperature, pressure or volume as the second independent variable.

Estimations of Thermal Expansion Coefficients

As a state function, the volume can be calculated along any path in the P, T plane. Since thermal expansion coefficients are usually determined at constant pressure, and compressibilities at constant temperature, it is of course preferable to follow either the IAF or the IBF path represented in Fig. 5.16. Along IAF, the volume is

$$V = V(T_0,P_0) + \int_{T_0}^{T} (dV/dT)_{P=P_0} dT + \int_{P_0}^{P} (dV/dP)_T dP. \qquad (5.47)$$

Along IBF, we have

$$V = V(T_0,P_0) + \int_{P_0}^{P} (dV/dP)_{T=T_0} dP + \int_{T_0}^{T} (dV/dT)_P dT. \qquad (5.48)$$

For most contexts in shock-wave and high-pressure physics or geophysics, the variation of temperature with pressure is at most that resulting from an adiabatic compression. Under these conditions, the influence of temperature is relatively minor compared to that of pressure. As an example, the temperature of a substance compressed to 1 Mbar by a shock wave typically increases by less than 1000 K. Assuming K_0= 1000 Kbar, K_0'= 4, and α (thermal expansion coefficient) = 2 10^{-5} K^{-1}, we find a resulting thermal expansion of about 2% whereas the compression is close to -35%. In addition, the pressure dependence of K is generally much better known than the temperature dependence of α. For both of these reasons, it appears preferable to follow

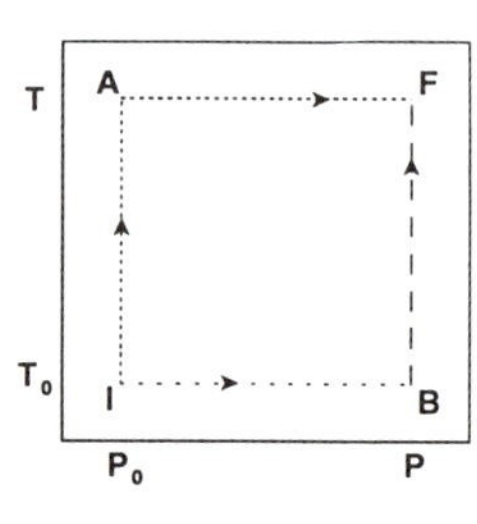

Figure 5.16. Possible paths followed in the P, T plane for calculating a volume at high pressure and temperature.

the path IBF in Fig. 5.16 along which the greatest volume changes can be calculated from directly measured properties.

Passing from this $V(T_0,P)$ obtained from an isothermal equation of state to the desired $V(T,P)$ nonetheless requires knowledge of α at high pressure. One can make use of empirical correlations to obviate the general lack of volume measurements made under simultaneously high pressure and temperature. For instance, recent data suggest that $(\partial \ln\alpha/\partial \ln V)_P$ is approximately constant, with values in the range 5 - 7. Alternatively, one can assume that the product αK_T is almost constant, which gives $\alpha(P) = \alpha(P_0)K_T(P_0)/K_T(P)$ and allows the high-pressure α to be determined from its P_0 (ambient, in general) values and the high-pressure bulk modulus. The same assumption can be made for the thermal Grüneisen parameter ($\gamma = \alpha V K_T/C_v$), or for the ratio γ/V. Owing to the aforementioned relative smallness of dilation corrections and to the fact that α necessarily decreases with increasing pressure, the overall errors are generally acceptable.

Mie-Grüneisen Equation

With the Mie-Grüneisen equation, we take volume as the second independent variable. At constant volume, the variation of P with T is given by (3.38)

$$(\partial P/\partial T)_V = \alpha K_T = \gamma C_v/V. \tag{5.49}$$

As already noted in 3.3c, the pressure generally increases with temperature because this partial derivative has the same sign as α. If the variations of C_v and γ with temperature are small, we readily calculate the pressure change by integrating (5.49)

$$P - P_0 = \frac{\gamma}{V}\, C_v(T - T_0) = \frac{\gamma}{V}\, (U - U_0). \tag{5.50}$$

This approximation works best at elevated temperatures, but a more precise calculation is of course possible if the temperature dependences of C_v and γ are known.

From the thermodynamic equation of state (3.26), we obtain finally the so-called Mie-Grüneisen equation. With 0 K as a reference temperature, it is given by

$$P = -\left(\frac{\partial U}{\partial V}\right)_{T=0} + \frac{\gamma T}{V}\, (U - U_0). \tag{5.51}$$

The Mie-Grüneisen equation of state is commonly used, particularly under conditions justifying the assumption that C_v and γ are actually close to their high-temperature limits. The effects of temperature can be incorporated in the same manner into any other kind of isothermal equation of state for any reference temperature T_0. We would then have

$$P(T,V) = P(T_0,V) + \int_{T_0}^{T} \frac{\gamma}{V}\, C_v dT. \tag{5.52}$$

5.5d CONVECTION AND ADIABATIC GRADIENT

Let us turn to convection to conclude this chapter with another question directly related to equations of state. Consider a temperature difference between the lower and upper parts of a fluid due to heating at the bottom of the column. If the thermal expansion coefficient is negative, the fluid is the densest at the bottom of the column and convection is then impossible. If the thermal expansion coefficient is positive, the heated fluid at the bottom is less dense than the cold fluid at the top so that Archimedean forces could cause the hot region of the fluid to rise and the cold one to sink. This would give rise to a convection cycle whose most familiar example is that of water in a saucepan. But convection is of such general importance that we will assess here the conditions under which it is possible.

When an element of the fluid rises, its pressure decreases because the weight of the fluid column above it decreases. In many industrial or natural processes in which convection takes place, one observes that the fluid velocity is high enough that significant heat exchange takes place only at the upper and lower ends of the column. One can then assume that the temperature decrease caused by the pressure decrease takes place under adiabatic conditions. In a fluid column, these simultaneous pressure and temperature variations are given by the adiabatic gradient (4.16a)

$$\left(\frac{\partial T}{\partial P}\right)_S = -\frac{(\partial S/\partial P)_T}{(\partial S/\partial T)_P} = \frac{\alpha V T}{C_p}. \tag{5.53}$$

For a fluid having a positive α, two cases must be distinguished depending on whether the imposed temperature gradient between the top and bottom of the column is greater or smaller than the value given by (5.53) for the adiabatic decompression of the fluid. In the first case, that of a *super-adiabatic gradient*, the fluid temperature decreases more rapidly than that of its surroundings. The density difference increases, as do the Archimedean forces, and a convective instability develops. In the second case of a *sub-adiabatic* gradient, the temperature of the heated fluid decreases less rapidly than that of its surroundings and convection is impossible. Convection is thus ruled by two factors, the thermal expansion coefficient of the fluid, which determines the magnitude of Archimedean forces, and the temperature gradient, which increases with the temperature difference between the top and the bottom of the fluid column.

Convection is not restricted to fluids. It can manifest itself in solids when considering time scales commensurate with the much higher viscosity of these phases. Although their viscosity is about 10^{20} times higher than that of water, the deep, solid interior of the Earth is subjected to a slow but powerful convection movement having a timescale of the order of one hundred million years. The shallow signatures of this convection are continental drift and sea-floor spreading, at a rate of 1-10 cm/year. The energy needed comes from what remains of the heat accumulated during the Earth formation, and especially from the heat continuously released by long-period radioactive elements (K, U and Th) and their disintegration products. Within the deep Earth, the adiabatic gradient is calculated as follows. As a function of depth, the pressure varies as $dP/dr = -\rho(r)g(r)$, where the density $\rho(r)$ and gravitation constant $g(r)$ at depth r are well known from astronomical and geophysical measurements. From (5.53), the geothermal gradient is calculated with

$$\left(\frac{\partial T}{\partial r}\right)_S = \frac{dP}{dr}\left(\frac{\partial T}{\partial P}\right)_S = -\frac{\alpha g(r)T}{C_p}, \tag{5.54}$$

where C_p is referred to a unit mass of matter. However, the difficulty is that neither C_p nor especially α are known accurately as a function of depth. As a result, depending on which range of values are assumed, calculated temperatures at the center of the Earth vary from 4000 to 7000 K. The more likely estimates of about 6000 K are similar to the temperatures measured for the solar corona. This is, of course, purely coincidental.

Problems

1. Boyle's law can be written as $PV = k_1$, that of Gay-Lussac, $V/T = k_2$. By combining the two laws, one is tempted to write $PT = k_1/k_2$. But this result is clearly incorrect. Why?
2. Show that the work done during the reversible adiabatic expansion of an ideal gas from P_1 to P_2 is $W = C_v T_1 [(P_2/P_1)^{R/C_p} - 1]$.
3. Calculate the Carnot cycle shown in Fig. 5.4 for one mole of water between extreme temperatures and pressures of 373 K and 1 bar, and of 500 K and 10 bar, respectively, taking S(373 K, 1 bar) = 196.4 J/mol K and C_p = 34 J/mol K.
4. An idealized Otto-Beau de Rochas cycle is described by an adiabatic compression of an ideal gas between the points A and B, an isochoric heating (through ignition of the fuel) between B and C, an adiabatic expansion between C and D, and an isochoric cooling betweeen D and A. Represent the P,V, P,T and T,S diagrams of this cycle and show that the maximum efficiency is $\eta = 1 - (V_B/V_A)^{C_p/C_v - 1}$. Regardless of the fact that a real 4-stroke engine does not work reversibly, why the existence of a combustion reaction results in the fact that the actual cycle differs from this idealized one?
5. Same problem for a Diesel cycle, which can be approximated by two adiabats alternating with one isochoric and one isobaric (where the fuel is burned) steps.
6. Among the gases mentioned in Table 5.1 indicate, without doing any calculation, the one for which the ideal gas law is closest to the van der Waals equation of state and the other one for which the difference is largest. In both extreme cases, calculate the "ideal" and van der Waals pressures for a temperature of 600 K and molar volumes ranging from 10^2 to 10^4 cm^3.
7. One mole of a van der Waals gas is expanded adiabatically against a zero pressure from V_1 to V_2. Express the temperature decrease as a function of a, b, C_v and R. Calculate this decrease for helium ($C_v = 3/2\ R$) and steam (C_v = $6R$) by taking the parameters a and b listed in Table 5.1 and V_2 = $3V_1$ = $3\ 10^{-3}$ m^3. Could the difference found between the two gases have been expected? Hint: calculate first $(\partial U/\partial V)_T$, and then $(\partial T/\partial V)_U$.
8. One observes that the enthalpy and internal energy of a substance satisfy the relations $(\partial H/\partial P)_T = 0$ and $(\partial U/\partial V)_T = 0$. Show that the equation of state of this substance is $PV = kT$, where k is a constant.
9. For one mole of gas, the Dieterici equation of state is $P(V - b)\exp(a/RTV) = RT$. As a function of a and b, calculate the critical parameters it yields and express it in a reduced form. Is the calculated critical compressibility factor in better agreement with the experimental data of Table 5.1 than that derived from the van der Waals equation of state? Comment on the difference after having established under what density conditions both equations tend to give the same pressure.

10. It we know as a function of T and P the Joule-Thomson coefficient and the isobaric heat capacity of a gas, can we determine its equation of state? If the answer is negative, what should be known in addition to these quantities?

11. Show that the Joule-Thomson coefficient of a van der Waals gas is $\mu = [2a/RT - b - 3abP/(RT)^2]/C_p$. Does it exist a temperature at which the sign of this coefficient change? From the data of Table 5.1, calculate the temperature change induced by a Joule-Thomson expansion from 10 to 1 bar, at 300 K, for Ar, H_2 and CO_2.

12. From the data of Table 5.2, calculate with Birch-Murnaghan equations of state the work to be done for compressing one mole of H_2, graphite and diamond from 1 bar to 1 Mbar at room temperature. Would the results obtained be very different if one were assuming $K_0' = 4$ in all three cases?

13. Listed below are the cell parameters a and c of solid deuterium (D_2), a crystal with a hexagonal symmetry, which have been measured at 300 K as a function of pressure by Hemley *et al.* (1990). Is the compression isotropic? What would be the 1-bar volume of D_2? Assuming $K_0' = 4$, calculate K_0 for a Birch-Murnaghan equation of state. Would the data be better reproduced with a value of K_0' different from 4 ? Compare the results obtained with those reported in Table 5.2 for H_2, and discuss the differences observed (see also chapter 14 and appendix E).

a (Å)	b (Å)	V (cm^3/mol)	P (kbar)
2.623 (22)	4.252 (35)	7.63 (11)	65 (1)
2.574 (20)	4.201 (33)	7.26 (10)	73 (1)
2.512 (21)	4.108 (35)	6.76 (10)	92 (1)
2.493 (2)	4.050 (5)	6.56 (1)	102 (1)
2.442 (5)	3.963 (7)	6.16 (2)	123 (1)
2.406 (5)	3.894 (8)	5.88 (2)	142 (2)

6
CONFIGURATIONAL CHANGES

Anomalous variations of thermodynamic properties are observed when the configuration of a phase changes. A few examples will first show the diversity of these changes and their influence on the heat capacity. The most dramatic kind of configurational effects, phase transitions, will then be discussed as continuous and first-order transformations. From the critical point of fluids, we will introduce critical exponents and order parameters and illustrate their general usefulness by means of Landau theory for crystal-crystal transitions. We will investigate kinetic transitions, through which systems lose internal equilibrium and acquire a nonzero entropy at 0 K. The most important example will be that of the glass transition, in the light of which a few order-disorder transitions in crystals will be finally discussed.

6.1 CONFIGURATIONS

6.1a TEMPERATURE AND PRESSURE EFFECTS

Contrary to chemical reactions where clearly identifiable species or phases appear and disappear, numerous transformations represent less obvious changes in the structure of a phase induced by slight variations in pressure, temperature or chemical composition. All phases can undergo such *configurational changes* where, as defined in 1.5d, the term configuration designates any microscopic arrangement of matter consistent with a given macroscopic state of the system.

As embodied in Nernst's law, the simplest configuration is achieved by a perfect crystal. Although bond angles and distances can vary due to thermal expansion or compression, only those changes that respect the crystal symmetry can operate. The physical properties vary smoothly with T or P, the heat capacity, for instance, following a typical s-shaped curve like that of Fig. 4.4. If there exist P,T conditions where another crystal structure is more stable, the configuration change is ensured by a phase transition that takes place either abruptly or after anomalous variations of C_p, α and β_T over finite pressure or temperature intervals. In fluids, the lack of long-range order makes available a much wider range of dilation and compression mechanisms than for crystals. This allows the structure of fluids to vary progressively as a function of T and P, for instance through changes in the local environment of atoms. In supercritical fluids, we have already seen in 5.3b how the effects of molecular interactions get stronger when the critical point is approached. Instead of more or less abrupt phase transitions, fluids thus undergo continuous structural changes whose hallmark is the fact that their compressibilities and thermal expansion coefficients are much higher than for crystals.

In all cases, structural changes within a phase are not energetically neutral and they affect not only the heat capacity and entropy, but also volume properties. Conversely, any anomalous variations of these properties signal the existence of configurational changes. In this chapter devoted to these effects, a series of examples will first illustrate the case of temperature-induced transformations. The choice is not arbitrary, since it is for such transformations that available data are the most comprehensive and accurate. These examples will also give a first glimpse at the competition between energy and entropy terms in thermodynamic potentials, in that higher temperatures promote higher entropy phases at the expense of the low-entropy phases stable at lower temperatures.

6.1b A FEW EXAMPLES

Hydrogen Decomposition

As a preamble, let us consider configurational changes due to well-defined reactions with the case of hydrogen heated to very high temperatures. At room pressure, this gas begins to dissociate near 3000 K according to the reaction $H_2 \rightleftharpoons 2\,H$, which is complete above 5000 K. Between 10 000 and 20 000 K atomic hydrogen then ionizes

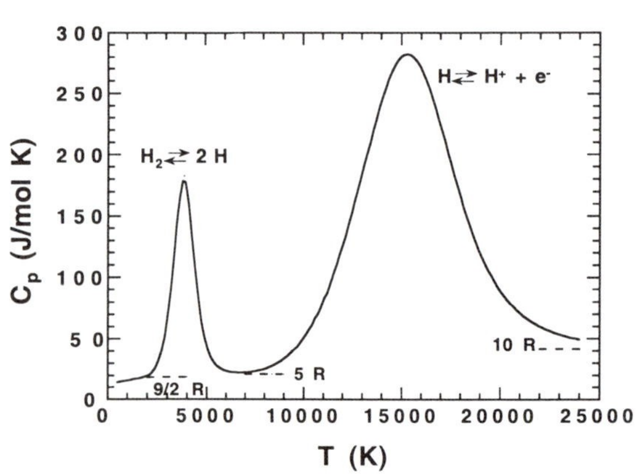

Figure 6.1. Heat capacity of hydrogen (on the basis of two moles of atoms at any temperature). The high-temperature C_p limits for the various species are indicated as dashed lines (see chapter 14), with R = gas constant. Theoretical data from the book by Boulos et al. (1994).

according to the reaction H $\rightleftharpoons$ H^+ + e^-. Both reactions are strongly endothermic. The heat capacity (Fig. 6.1) shows two big peaks representing successively the effects of molecular dissociation and plasma formation. The total enthalpy of each reaction is given by the surface area below the C_p peaks which thus represent a *configurational* heat capacity, i.e., a thermal energy spread over a finite temperature interval that does not result in a temperature increase, but in structural changes whereby configurations of higher energy are reached on heating. Because these reactions are reversible, these enthalpies are progressively released when the plasma is cooled down. The H^+ ions first recombine with the electrons, and atomic hydrogen eventually retransform into molecular hydrogen at low temperature.

Structural Rearrangements in Ice and Water

For ice, the first anomaly to be noted is the departure of the low-temperature heat capacity from an s-shaped curve near 200 K (Fig. 6.2a). More detailed measurements show an abnormal increase near 100 K, which is all the more peculiar as its onset is at different temperatures depending on the thermal history of the sample, as will be described in more detail in 6.4f. Both features are due to changes in the orientation of the hydrogen bonds which tie up together H_2O molecules in the form of a three-dimensional network of H_4O tetrahedra (Fig. 6.2b). Modifying the orientation of these bonds without breaking them is a change that is consistent with the crystal structure and does not require much energy. In a dynamic manner, this bond reorientation begins to operate near 100 K and becomes more extensive as the temperature increases. The heat capacity can then be split in two parts. The energy needed for these changes manifests itself as a configurational heat capacity, whereas the *lattice* C_p is the heat capacity that the crystal would have in the absence of any structural rearrangement.

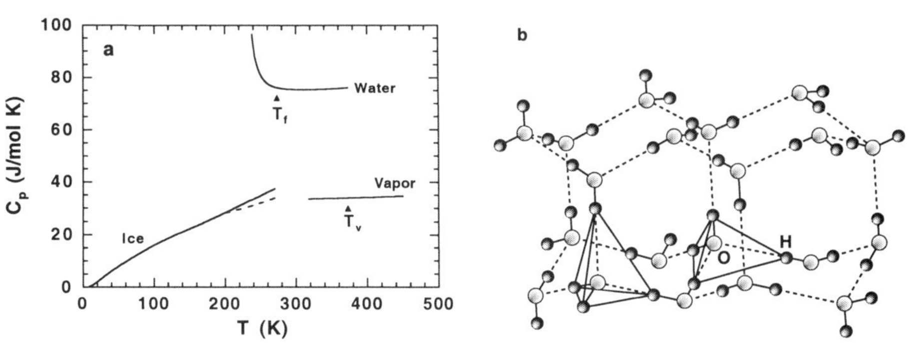

Figure 6.2. (a) Heat capacities of ice, water and water vapor (ideal gas) at 1 bar; lattice heat capacity of ice shown by the dashed curve; T_f = temperature of fusion; T_v = temperature of vaporization; data from Giauque and Stout (1936) and Flubacher et al. (1960) for ice, and from Table 4.1 for water and water vapor (ideal gas). (b) Distorted H_4O tetrahedra and crystal structure of ice: covalent bonds shown as short solid lines, hydrogen bonds as long dashed lines.

When melting takes place at 273 K water molecules become mobile through the severing of about 10 % of the hydrogen bonds (and distortion of the remaining H_4O tetrahedra) associated with an important enthalpy of melting of 6 kJ/mol. Rearrangement of the three-dimensional network of H_4O tetrahedra continues in the liquid state through further hydrogen bond breaking and distortion of tetrahedra. This accounts for the very high heat capacity of water, about half of which is configurational above 0 K. Especially noteworthy is the strong C_p increase observed at lower temperatures for the supercooled liquid (Fig. 6.2a), after a minimum near 310 K which is related to the density maximum observed at 277 K (Fig. 1.2). With decreasing temperature, the variations of density and heat capacity show that the configurational changes are greatest at the highest degrees of supercooling. This indicates that much energy is needed to modify the water structure which, at such low temperatures, tends to prefigure the crystalline molecular arrangement of ice, which should form. At 373 K, in contrast, all remaining hydrogen bonds are severed when water vaporizes and dissociates into independent molecules. The heat capacity of the vapor then displays the smooth increase typical of ideal gases.

Magnetic Transition of Nickel

At low temperature nickel is ferromagnetic. It has a spontaneous magnetization because its external electrons (3d) have their spins preferentially oriented in the same direction (as if the spins of 0.54 electron per Ni atom, on the average, were aligned). With increasing temperature this preferred orientation begins to be lost. Due to strong interactions, the orientation of spins changes collectively, in a *cooperative* manner, through short-lived fluctuations throughout the crystal. These clusters of disoriented spins cover a wide range of sizes, down to the interatomic spacing, and are mutually embedded in regions of preferentially oriented spins. The size of the largest fluctuations away from the ordered background characterizes the *correlation length* of the phenomenon. This length increases with temperature and becomes infinite at the Curie

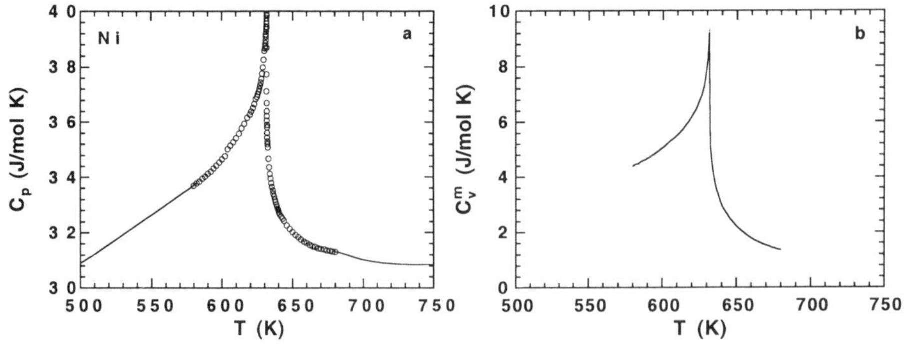

Figure 6.3. Isobaric heat capacity of nickel (a) and derived magnetic contribution to C_v around the Curie temperature (b). Data from Connelly et al. (1971), and baseline C_p from the compilation of Hultgren et al. (1973).

temperature where spin orientation has randomized to the point that nickel has become paramagnetic.

The energy required for these changes manifests itself as a C_p anomaly that begins near 400 K, peaks at the Curie temperature and disappears beyond 700 K (Fig. 6.3a). This rapid but not discontinuous C_p decrease indicates that ordered fluctuations in turn persist above the transition temperature, with a decreasing correlation length. The excess heat capacity observed with respect to the lattice C_v thus represents the energy required for spin randomization at constant volume (Fig. 6.3b), without changes in the crystallographic arrangement of nickel atoms which is left unaffected by the transition.

Order-Disorder in β-Brass

Changes in crystallographic sites are in contrast the essential feature of order-disorder transitions. The standard example is given by β-brass, an alloy made up of about the same numbers of copper and zinc atoms in a body centered cubic lattice (Fig. 6.4a). At low temperatures, the structure is ordered because each atom of one kind is surrounded by eight atoms of the other kind. When the temperature rises, an anomalous C_p increase near 400 K (Fig. 6.4a) indicates that exchange of Cu and Zn atoms has set in between the two different sublattices (Fig. 6.4b), so that the kinds of two crystallographic sites tend to be randomly occupied. Ordered and disordered interpenetrating regions of the crystal coexist, covering again a wide range of scales. In this case, it is for fluctuations of site occupancy that the correlation length becomes infinite at the transition temperature of 741 K. As for the magnetic transition of nickel, this divergence does not imply that ordered domains have disappeared beyond the critical temperature. Although becoming smaller and smaller, they persist because of the existence of residual short-range order. For any atom of a kind, the chance of having as nearest neighbor an atom of the other kind remains in effect higher than 50%. The correlation length thus does not vanish at high temperature for ordered fluctuations, but tends to a value corresponding to a few interatomic distances.

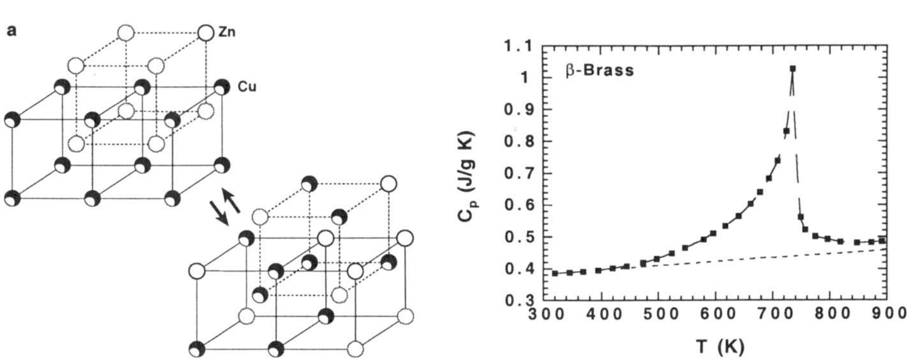

Figure 6.4. (a) Ordered (top left) and disordered (bottom right) distribution of Cu and Zn atoms on crystallographic sites in β-brass; (b) Heat capacity of β-brass (Moser, 1936); lattice heat capacity shown as a dashed line.

6.1c ENERGIES OF TRANSFORMATION

In the preceding examples, "anomalous" variations of the heat capacity have been used as probes of temperature-induced configurational changes. For those cases where the changes actually involve a chemical reaction, we will use the advancement ξ defined in 1.1b to specify the state of the system at any moment. Using ξ in this way, we write for instance the internal energy as

$$\sum(\partial U/\partial n_i)_{V,S,n_j}dn_i = (\partial U/\partial \xi)_{V,S}d\xi. \tag{6.1}$$

As will be seen below, *order parameters* are used instead when the configurational changes lead to a phase transition. These parameters give a direct measure of the structural difference between the phases involved and vary, not as ξ from 0 to 1, but from some finite value to 0 at the temperature of transition. In both cases, however, the relevant parameter is determined unequivocally by only two variables of the system in order to satisfy Duhem's postulate. Although the following discussion will rely on the advancement, identical conclusions would be reached with order parameters.

Let us now examine in more detail the heats of transformation introduced in chapter 2. For a closed system, a heat of transformation is by definition

$$dQ = dU + PdV. \tag{6.2}$$

By inserting in this equation the differential of the internal energy as a function of V, T and ξ, we obtain

$$\begin{aligned} dQ &= (\partial U/\partial T)_{V,\xi}dT + [(\partial U/\partial V)_{T,\xi} + P]dV + (\partial U/\partial \xi)_{V,T}d\xi && (6.3)\\ &= C_v dT + l dV + (\partial U/\partial \xi)_{V,T,\xi}d\xi && (6.4)\end{aligned}$$

where we have used the notation

$$l = (\partial U/\partial V)_{T,\xi} + P. \tag{6.5}$$

With these independent variables, a heat of transformation is the sum of three terms. The first one only has already been mentioned explicitly but, as defined by (6.5), what does the coefficient l represent? Contrary to C_v, this coefficient is not associated with temperature changes. Hence, it is another latent heat, for *volume changes* at constant T and ξ. As to the third one, it is the heat of a chemical reaction.

To obtain a heat of transformation at constant P, let us first write the differential expression of U and V as a function of P, T and ξ, namely,

$$\begin{aligned} dU &= (\partial U/\partial T)_{P,\xi}dT + (\partial U/\partial P)_{T,\xi}dP + (\partial U/\partial \xi)_{P,T}d\xi, && (6.6a)\\ dV &= (\partial V/\partial T)_{P,\xi}dT + (\partial V/\partial P)_{T,\xi}dP + (\partial V/\partial \xi)_{P,T}d\xi. && (6.6b)\end{aligned}$$

In (6.3) we then replace dU and dV by these expressions to obtain

$$\begin{aligned} dQ &= [(\partial U/\partial T)_{P,\xi} + P(\partial V/\partial T)_{P,\xi}]dT + [(\partial U/\partial P)_{T,\xi} + P(\partial V/\partial P)_{T,\xi}]dP \\ &\quad + [(\partial U/\partial \xi)_{P,T} + P(\partial V/\partial \xi)_{P,T}]d\xi \\ &= (\partial H/\partial T)_{P,\xi}dT + [(\partial H/\partial P)_{T,\xi} - V]dP + (\partial H/\partial \xi)_{P,T}d\xi && (6.7a)\\ &= C_{p,\xi}dT + hdP + (\partial H/\partial \xi)_{P,T}d\xi && (6.7b)\end{aligned}$$

with:

$$h = (\partial U/\partial P)_{T,\xi} + P(\partial V/\partial P)_{T,\xi} = (\partial H/\partial P)_{T,\xi} - V. \tag{6.8}$$

That the enthalpy shows up when P is selected in place of V as an independent variable is not a surprise. As a matter of fact, we could have introduced it in this manner. As a function of T, P and ξ, a heat is thus the sum of three terms. These are a heat capacity ($C_{p,\xi}$) at constant pressure and composition, a latent heat h for pressure changes at constant T and ξ, and an enthalpy of chemical reaction at constant T and P. The difference between this heat capacity $C_{p,\xi}$ and C_p as defined in the preceding chapters deserves a few comments, which will be made in the next section. In contrast to heat capacities, the latent heats h and l are not measurable directly. By comparing (6.5) and (6.8) to the expressions (3.26) and (3.29), we conclude that

$$h = -\alpha TV, \qquad l = \alpha T/\beta_T, \tag{6.9a}$$

whence

$$h = l(\partial V/\partial P)_{T,\xi}. \tag{6.9b}$$

6.1d CONFIGURATIONAL HEAT CAPACITY

What does the term of "heat capacity at constant pressure and composition" used for $C_{p,\xi}$ mean? From (6.7a), the heat capacity at constant pressure actually measured is

$$C_p = (dQ/dT)_P = C_{p,\xi} + (\partial H/\partial \xi)_{P,T}(d\xi/dT)_P. \tag{6.10}$$

It is the sum of two terms, the second one being the configurational heat capacity since it describes the effects of temperature on the chemical transformation that takes place in the system

$$C_p = C_{p,\xi} + C_p^{conf}. \tag{6.11}$$

Hence, the first term of (6.10) and (6.11) should rather be called heat capacity at constant pressure and configuration. For crystals, it is the lattice heat capacity.

Like a latent heat, C_p^{conf} is positive because any configurational change must cause an entropy rise when the temperature increases. Since $C_{p,\xi} \geq C_p$, two limiting cases must be distinguished as will be discussed in more detail in 6.3c. If the transformation kinetics are rapid, one measures the sum of the two terms of (6.11) and configurational effects can be determined only if it is possible to define a baseline C_p, i.e., to know the heat capacity that would prevail in the absence of any change. If the reaction kinetics are very slow at the timescale of the measurement, in contrast, equilibrium is not reached and $C_{p,\xi}$ is very close to C_p. In between these two situations, the enthalpy varies with time to reach its equilibrium value: that is to say, the enthalpy is *relaxing*.

Let us finally relate C_p^{conf} to the configurational enthalpy and entropy

$$C_p^{conf} = (\partial H^{conf}/\partial T)_P = T(\partial S^{conf}/\partial T)_P. \tag{6.12a}$$

Between any two temperatures T_0 and T_1, the enthalpy and entropy variations are of course given by

$$H^{conf}(T_1) = H^{conf}(T_0) + \int_{T_0}^{T_1} C_p^{conf}\, dT \tag{6.12b}$$

and

$$S^{conf}(T_1) = S^{conf}(T_0) + \int_{T_0}^{T_1} C_p^{conf}/T\, dT. \tag{6.12c}$$

6.1e CONFIGURATIONAL VS. VIBRATIONAL CONTRIBUTIONS

The negative thermal expansion of water below 277 K has already been alluded to as a configurational effect in 6.1b. For volume properties, configurational parts can also be defined, in particular for the thermal expansion coefficient and the compressibility

$$\alpha^{conf} = 1/V\ (\partial V/\partial \xi)_{P,T}(d\xi/dT)_P. \tag{6.13a}$$

$$\beta_T^{conf} = -1/V\ (\partial V/\partial \xi)_{P,T}(d\xi/dP)_T. \tag{6.13b}$$

As discussed by Prigogine and Defay, the former coefficient is necessarily positive whereas the latter can be positive or negative. In addition, one shows that these two coefficients are related to C_p^{conf} by

$$C_p^{conf} = VT{\alpha^{conf}}^2/\beta_T^{conf}. \tag{6.14}$$

For any system at equilibrium, the so-called Prigogine-Defay ratio Π should thus be unity

$$\Pi = C_p^{conf}\beta_T^{conf}/VT{\alpha^{conf}}^2 = 1. \tag{6.15}$$

6.2 EQUILIBRIUM PHASE TRANSITIONS

6.2a FIRST-ORDER AND CONTINUOUS TRANSITION

When ice melts two phases coexist as long the transformation is not complete. Because water and ice are different phases, going from one to the other phase necessarily entails discontinuities in enthalpy, entropy and other first-order thermodynamic properties. This is why such transitions are termed *first-order*. Probably the most dramatic ones are those between the solid, liquid and gaseous states of pure substances: melting, crystallization, vaporization, sublimation and condensation. Even if only crystal-crystal transitions are considered, the existence of polymorphism makes phase first-order transitions very common; the graphite-diamond or quartz-cristobalite transitions have already been mentioned. As already seen in chapter 5 for liquid-vapor equilibria, a phase is stable only within a finite domain in the P,T plane. Transitions can thus be induced by pressure or temperature changes bringing the phase outside its stability field. As a matter of fact, changes in other intensive variable can trigger a transition at constant P and T. At 95°C and 1 atm, for instance, adding some ethanol to pure water cause partial vaporization of the liquid. These effects of composition variables are important enough that they will be extensively dealt with in chapter 10.

Many transformations such as the ferro-paramagnetic transition of nickel or the order-disorder transition of β-brass take place in such a way that the two phases do not coexist (Figs. 6.3-6.4). There just exists a single point where the change from one phase to the other takes place continuously. There is no discontinuity in first-order properties, but surmised discontinuities in second-order properties led to the denomination of *second-order transitions*. Likewise, n^{th} order transitions would be characterized by discontinuities in n^{th} order properties, lower order properties being continuous. This classification of phase transitions is no longer recommended, however, because extensive measurements made since the late 1960's have shown that the important feature of "second-order" transitions is not discontinuous changes, but divergence of second-order variables at the temperature of transition. The resemblance of the shape of these anomalies with the Greek letter λ (Figs 6.3-6.5) has led to the name of *lambda transitions*. Following an alternative usage that is becoming more common, we will prefer that of *continuous transitions*. The *continuous* qualifier has the additional advantage of embodying the essential fact that there exists an uninterrupted pathway between the two phases involved. This has the important consequence that no phase can persist metastably in the stability field of the other and, also, that these progressive configurational changes can in general be experimentally determined. As will be discussed in the following sections, continuous transitions are not only rather well understood but can be subjected to the same formalism even for quite different kinds of transitions.

Owing to their abrupt occurrence, the configurational changes associated with first-order transitions are in contrast much more elusive. A transformation as important as melting, for example, is still wanting a general explanation. Besides, discontinuous reaction pathways makes metastability possible in first-order transitions, as shown by the cases of supercooled water or superheated water vapor. Actually, another reason why these transitions often do not take place at the equilibrium phase boundary ($\Delta G = 0$, at constant P and T) is that formation of a new phase necessarily involves a surface energy that first opposes the transition. The transition then occurs beyond the nominal boundary, through the nucleation and growth mechanism described in 7.3c, when a significant difference in thermodynamic potential has developed for the reaction kinetics to be non zero. First-order transitions will be treated comprehensively in chapters 9 and 10 from a purely thermodynamic perspective. As will be shown below, from a fundamental standpoint they can be considered in simple instances as continuous transitions aborting when the reaction pathway is abruptly interrupted. That certain relationships do exist between the two kinds of transitions is also demonstrated by the existence of *tricritical points* where transitions change from first-order to continuous as a result of changes in pressure or in an applied field (electric field, for instance, in the case of ferroelectrics).

A point worthy of note is that the great many different properties associated with long-range order give rise to a bewildering diversity of phase transitions in crystals. Inducing changes in the crystal structure, *structural* transitions are probably the most readily detected because they are generally accompanied by marked variations of physical properties. In a general way, it is useful to distinguish *reconstructive* from *distortive* transitions. In the former the two crystals involved have very different structures; like the graphite-diamond transition, they are always first-order. Distortive transitions are in contrast characterized by close structural relationships between the crystals. The two phases being distinguished only by small differences in atomic positions, their common feature is to take place through short atomic displacements. But these are not necessarily continuous as distortive transitions can also be first-order. Depending on whether the configurational entropy varies markedly or not, these transitions are themselves split further into *order-disorder* transitions, where atoms or groups of atoms are distributed among several equivalent positions, and *displacive* transitions.

Changes in thermal expansion had been a serious concern for ceramists long before they could be ascribed to the $\alpha - \beta$ quartz transition. From another practical standpoint, this celebrated transition illustrates the frequent difficulty of identifying the actual order of a transition. Discovered in 1889 by Le Châtelier through dilatometry measurements, it is also marked by the C_p anomaly shown in Fig. 6.5. A configurational heat capacity begins to be observed near 600 K in the α phase and peaks at the extremely high value of 1500 J/mol K at 847 K before vanishing at 860 K for the β phase, less than 15 K above the transition temperature. It has taken nearly 100 years, however, to figure out that this typical lambda transition is in fact made up of a first-order transition between α-quartz and an intermediate (*incommensurate*) phase, stable over less than 2 K, followed by a continuous transition to β-quartz! Regarding first-order variables, another difficulty is that discontinuities can be observed only in an approximate manner. If the enthalpies of water and ice are measured a few degrees apart, melting of ice will appear as a clear discontinuity in the enthalpy-temperature curve. But if the spacing between two measurements were reduced, one would observe instead a continuous, s-shaped variation leading from the enthalpy of ice to that of water because samples are never perfectly pure and temperatures uniform. Since the heat capacity is the slope of this enthalpy curve, it would show a maximum at the inflexion point. This maximum is especially sharp when the resolution of the measurements is high, but never infinite, as illustrated by the remarkable results of Fig. 6.5 for quartz and Fig. 4.6 for selenium melting.

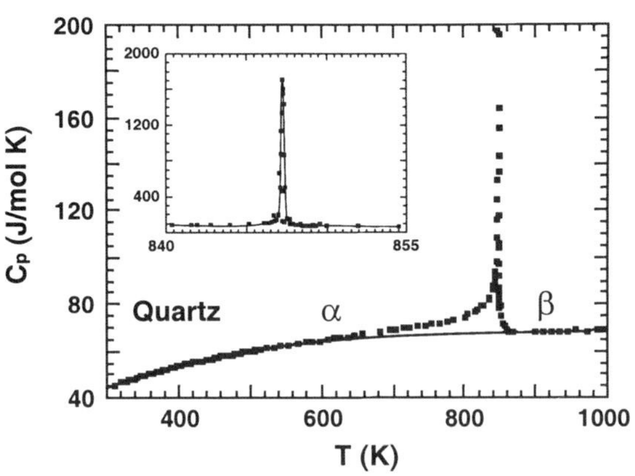

Figure 6.5. Heat capacity of quartz. The α and β forms have nearly the same lattice heat capacity, which is shown as a solid curve. Data of Grønvold et al. (1989).

A few words are now in order about other important kinds of phase transitions in crystals. With the examples of β-brass and nickel we have already mentioned order-disorder and ferro-paramagnetic transitions. Important changes in magnetic properties are also due to *antiferro-* and *ferro-paramagnetic* transitions. The spins are ordered with an antiparallel arrangement in the low-temperature phases involved in the former, whereas the latter affect phases where certain ions have a magnetic moment antiparallel to that of the other ions. On the other hand, useful dielectric properties are observed when the centers of positive and negative charges do not coincide in a crystal. This gives rise to a permanent dipolar moment, even in the absence of any field, which vanishes on heating in *ferro-paraelectric* transitions (at a temperature that strongly depends on the electrical field). We will finally mention the case of ferroelastic crystals, where twinning results from the switch from one to another orientational state induced by a mechanical stress. On heating, this phenomenon disappears at the temperature of a *ferro-paraelastic* transition where both the spontaneous strain and the coercitive stress vanish. In a more general manner, transitions are gathered under the names of *ferroic* and *nonferroic* depending on whether or not they are associated with a change in point group of the crystal symmetry. Ferromagnetics, ferroelectrics and ferroeleastics constitute the primary ferroics. On cooling, they form when a previously unallowed component of magnetization (an axial vector), polarization (a polar vector), or strain (a rank two polar tensor) spontaneously sets in. To complicate the situation further, the onset of second or higher order tensor forms gives rise to secondary or higher order ferroics. This brief summary thus shows that we will just skim over the rich topic of phase transitions in crystals.

6.2b CRITICAL EXPONENTS OF FLUIDS

Continuous transitions are especially interesting because of the *critical phenomena* that take place near the transition temperature. The critical point of fluids is just a special instance of these phenomena that are subjected to similar phenomenological laws. Other examples are homogenization of two phases having initially different chemical compositions (see 7.3d), disordering in an alloy or transitions at Curie and Néel temperatures. Long before being extended to a great many different kinds of transitions, these phenomenological laws have been established for the variations of fluid properties in the neighborhood of the critical point. As a first description of these phenomena, we will thus consider again the critical point of fluids and take CO_2 as an example (Fig. 6.6).

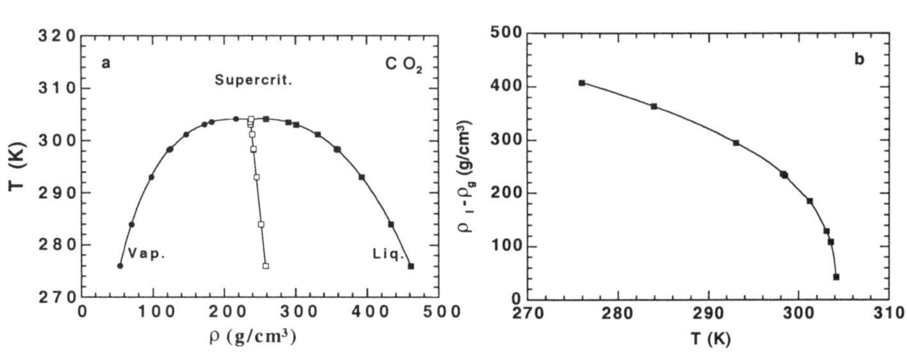

Figure 6.6. Density of CO_2 liquid and gas (a) and order parameter in the neighborhood of the critical point (b). Data of Michels et al. (1937).

A first simple feature is that, along the vapor-liquid coexistence curve, the mean of the liquid (ρ_l) and gas (ρ_g) densities varies linearly with temperature (Fig. 6.6a). Noting that this mean density is equal to the critical density for $T = T_c$, we write this *rectilinear law*

$$1/2\ (\rho_l + \rho_g) = \rho_c[1 + a(1 - T/T_c)] \tag{6.16}$$

where a is a characteristic parameter of the fluid. Turning to the density difference between the liquid and gas (Fig. 6.6b), we then find that it follows a power law

$$(\rho_l - \rho_g)/2\rho_c = B(1 - T/T_c)^{\beta}, \tag{6.17}$$

where B is a constant and β a *critical exponent.* Its value of 0.357, found from the CO_2 data of Fig. 6.6b, happens to be close to 1/3. The β parameter depends little on the nature of molecular interactions, for the parallelism of the lines drawn in a logarithmic plot of $\rho_l - \rho_g$ against $1 - T/T_c$ shows that its value is almost the same for various fluids over very wide intervals of critical temperatures and density differences (Fig. 6.7a).

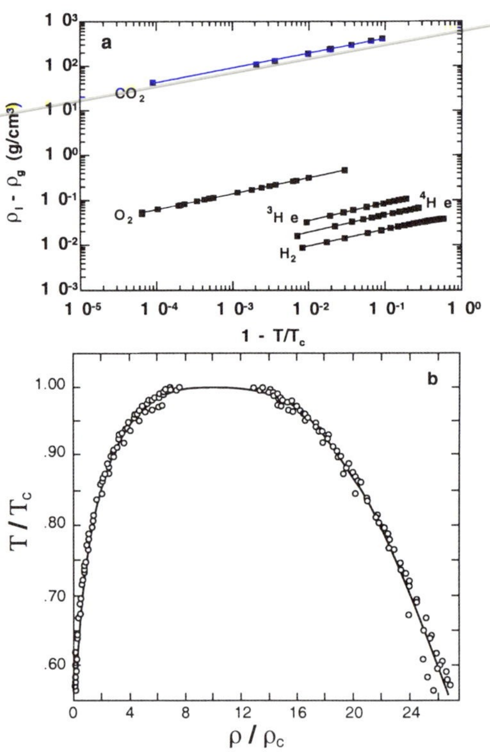

Figure 6.7. Universality of the critical behavior of fluids. (a) Density contrasts between liquid and gas phases, against $1 - T/T_c$, showing the same slopes in the vicinity of T_c and thus the same critical exponent β. Data of Michels et al. (1937) for CO_2, Weber (1970) for O_2, Edward and Woodbury (1963) for ^4He, Sherman (1965) for ^3He and Roder et al. (1962) for p-H_2. (b) Reduced temperature against reduced density for various fluids (Ne, Ar, Kr, Xe, N_2, O_2, CO, CH_4): experimental data and curve calculated from (6.17) and β = 1/3 (from Guggenheim, 1945).

As had been demonstrated by a classic diagram drawn by Guggenheim (Fig. 6.7b), the principle of corresponding states is in fact much better represented by (6.17) than by the van der Waals or Redlich-Kwong equations of state discussed in chapter 5. Owing to the rectilinear law (6.16), it appears that both density differences $(\rho_l - \rho_c)$ and $(\rho_c - \rho_g)$ also follow power laws with the same exponent β. Besides, numerous experiments have shown that the divergences of the isothermal compressibility,

Table 6.1. Critical exponents for limiting power-law variations of fluid properties

Property	Pathway followed ε	$P-P_c$	$\rho-\rho_c$	Power Law	Exponent[a]
$\rho_l-\rho_g$	<0	0	$\neq 0$	$\sim(-\varepsilon)^{\beta}$	$\beta=0.355$ (7)
β_T	>0	0	0	$\sim\varepsilon^{-\gamma}$	$\gamma=1.19$ (3)
	<0	0	$\neq 0$	$\sim(-\varepsilon)^{-\gamma'}$	$\gamma'=\gamma$
C_v	>0	0	0	$\sim\varepsilon^{-\alpha}$	$\alpha=0.10$ (4)
	<0	0	0	$\sim(-\varepsilon)^{-\alpha'}$	$\alpha'=\alpha$
$P-P_c$	0	$\neq 0$	$\neq 0$	$\sim sgn\,\{\rho_l-\rho_g\}\lvert\rho_l-\rho_g\rvert^{\delta}$	$\delta=4.35$ (10)

[a] Critical exponents (and their uncertainties) determined from measurements on ^3He, ^4He, Ar, Kr, Xe, O_2, N_2, CH_4, C_2H_4, p-H_2, CO_2, SF_6, NH_3, H_2O and D_2O (Levelt Sengers et al., 1976).

the isochoric heat capacity and the pressure variation along the critical isotherm obey analogous power laws with, however, different critical exponents. These are summarized in Table 6.1 where we note

$$\varepsilon = 1 - T/T_c. \tag{6.18}$$

There could be as many different critical exponents as different pathways followed for investigating the divergence of a given property. Table 6.1 thus indicates the manner in which the divergences are observed, i.e., either from above ($\varepsilon > 0$) or below ($\varepsilon < 0$) the critical point, or along the coexistence curve ($P - P_c = 0$), the critical isotherm ($\varepsilon = 0$) or the critical isochore ($\rho - \rho_c = 0$). With the assumption that thermodynamic potentials are analytic everywhere except at the critical point, however, one demonstrates that only two such exponents are independent. For the four exponents considered in Table 6.1, one has

$$2 - \alpha = \beta(\delta + 1), \qquad \gamma = \beta(\delta - 1), \tag{6.19}$$

whence, for instance, $\alpha + 2\beta + \gamma = 2$. Since they account for experimental critical data for a variety of fluids, these exponents are termed *universal*.

Calculating universal values for the critical exponents is easy when one assumes that the equation of state of the fluid can be expanded as a Taylor series at the critical point. With the van der Waals equation, for instance, one finds that $\beta = 0.5$, $\gamma = 1$, $\alpha = 0$ and $\delta = 3$. The annoying fact is that such results as obtained from "classic" theory markedly disagree with the experimental values (Table 6.1), especially since no divergence is in fact predicted for β_T or C_v. As a matter of fact, it is the divergence of second-order properties that indicate the nonanalytic nature of thermodynamic potentials at the critical point. In "nonclassic" theories, thermodynamic potentials are thus split into two parts, one (function of critical exponents) with a singularity at T_c, the other without it. Also note that the equations (6.19) have long been established as inequalities. The conjecture that they were actually equalities had to wait renormalization theory (see 6.2d) to be proved. This theory has also demonstrated the equalities $\alpha' = \alpha$, $\gamma' = \gamma$, etc.

6.2c ORDER PARAMETERS

For fluids, $\rho_l - \rho_g$ is a convenient measure of the structural difference between the liquid and gas along the coexistence curve. As such, it plays in (6.17) the role of a macroscopic *order parameter* for describing the extent of configurational differences between the two phases. We have already mentioned in 6.1c that, for a continuous transition, an order parameter η is defined so as to vary in a monotonous fashion in the low-symmetry phase before vanishing at the transition temperature. When applied to the transition between two disordered phases like a liquid and a gas, the term of order parameter may seem ambiguous. This is just a consequence of the fact that η has originally been introduced for describing structural changes within crystalline phases in the neighborood of phase transitions.

If the system is in internal equilibrium, the relevant thermodynamic potential Φ is minimum so that the order parameter is determined by the equation

$$(\partial\Phi/\partial\eta)_{T,P} = 0, \tag{6.20}$$

thanks to which, in accordance with Duhem's postulate, η can be expressed as a function of T and P only. This is obviously the case for the density difference $\rho_l - \rho_g$. When crystalline phases are dealt with, the existence of long-range order allows order parameters to be defined in microscopic terms as will be illustrated by the two examples given below.

To describe the occupation of crystallographic sites by Cu and Zn atoms for the order-disorder transition in β-brass, a long-range order parameter is introduced, $P_0 = 1/N \sum P_R$, where the sum is extended over all N sites R and P_R is +1 or -1 depending on whether the R site is rightly occupied or not. Measurements (by neutron diffraction methods) then show that P_0 is proportional to $(-\varepsilon)^\beta$, with $\beta = 0.305(5)$, a value that is also close to 1/3.

For the magnetic transition of nickel, the relevant order parameter is the magnetization M, which characterizes the manner in which magnetic moments are aligned throughout the crystal. For $0.9 < T/T_c < 0.99$, the zero-field magnetization M_0 is found to be proportional to $(-\varepsilon)^\beta$, with $\beta = 0.375$. Like the pressure (with respect to the volume), the magnetic field H scales with its conjugate variable M as $\mathrm{sgn}\{M\}|M|^\delta$, with $\delta = 4.22(10)$, along the critical magnetic isotherm. Finally the zero-field isothermal susceptibility $\chi_T = \lim_{H\to 0}(\partial M/\partial H)_T$, the analog of the compressibility, varies asymptotically as $(-\varepsilon)^{-\gamma'}$ with $\gamma' = 1.35(2)$ whereas the magnetic contribution to C_v also follows a power law with $\alpha = \alpha' = 0.10$.

When proper order parameters are used, the aforementioned examples thus reveal a striking similarity between the power laws describing phenomena as different as liquid-gas, order-disorder or ferro-paramagnetic transitions near critical points. This similarity contrasts with the great variability of transition temperatures, which strongly depend on the nature of microscopic interactions. Now, a basic aspect of continuous transitions is that the size distribution of the relevant fluctuations widens tremendously when the critical point is approached. Although all length scales then become important, as will be emphasized below, the reason why the variations of physical properties tend to be insensitive to these interactions is that correlations at extremely long distances develop near T_c, involving billions of constituent particles. The correlation length of the relevant fluctuations becomes much longer than the range of the particular forces involved. The similar divergences of correlation lengths, which actually lead to divergences in thermodynamic properties, thus make the critical behavior essentially independent of the specific features of the interactions.

Combining both magnetic and order-disorder aspects, the Ising model has contributed much to shaping these ideas. In a simple 3-dimensional form, it consists of a cubic array of N magnetic dipoles, called spins for short, pointing only either up or down, whose state is described by a number $\sigma = +1$ or -1, respectively. The spins are subjected only to a nearest-neighbor interaction that tends to align them. For parallel and antiparallel pairs of spins, the interaction energy is $-J$ and $+J$, respectively, where J is a positive constant. Summed over all nearest neighbor pairs pq of spins, the total energy is thus $E = -J\ \Sigma\sigma_p\sigma_q$. At 0 K, all spins are either up or down to minimize the energy. When T increases, the increase of the thermal energy leads to partial disordering. Although not amenable to an analytic solution, the statistical mechanical problem can be solved by numerical methods. A lambda anomaly is found for the heat capacity at constant field (and volume) $C_H \sim e^{-\alpha}$, with $\alpha = 0.125$, at a temperature $T_c = 4.510$ J/k where k = Boltzmann constant. At this temperature the order parameter, the magnetization, also diverges with a critical exponent $\beta = 5/16$.

6.2d RENORMALIZATION GROUP THEORY

The universality of critical exponents reviewed in the previous section could be misleading, however, because critical behavior is identical only within well-defined classes of transitions. These are characterized by the dimension of the system and the dimensionality of the order parameter. If the former parameter is in general 3 (but is 2 for the original 2-dimensional Ising model), the latter is more variable because it characterizes the scalar, vector or tensorial nature of the order parameter. It is unity for scalars like the density difference $\rho_l - \rho_g$ or the parameter P_0 describing disordering in β-brass, whereas the order parameter for magnetic transitions in isotropic magnets is a 3-d vector. Critical point of fluids, order-disorder in β-brass and Ising model have similar exponents because they share the same dimensionality for the system and the order parameter. The magnetic transition of nickel belongs to a different class, whose exponents happen not to differ considerably from those just reviewed.

These notions have been established by the renormalization group theory proposed by K.G. Wilson in the early 1970's. The starting point of this involved theory is that a critical phenomenon has the remarkable peculiarity of being determined by events making contributions of equal importance to physical properties in spite of having enormously different lengthscales. Near a critical point, the fluctuation size is limited only by the distance between the particles of the system, on the one hand, and by the size of the system itself on the other hand. Below a critical point, fluctuations of the high-temperature phase are pervasive within the low-temperature phase. The converse holds true above T_c, and it is at this temperature that the correlation lengths of both kinds of fluctuations become equal and diverge. Schematically, Wilson's theory aims at determining how the physical properties vary when the structure is averaged out at progressively longer scales. The critical point can then be identified thanks to the fact that the coexistence of fluctuations having all length scales makes the system invariant with respect to any scale change.

When the order parameter is a scalar, one can characterize any microscopic portion of the system by a single value of this parameter. To reduce the size of the system by a scaling factor b, one groups in clusters of b units all microscopic portions and assign to each cluster the value of the order parameter that predominate in its b portions. The whole process is then redone and each time the order parameter takes on the value representing the state of the majority of clusters that become grouped through the scaling. For $T < T_c$, this procedure is tantamount to lowering continuously the temperature to 0 K because the ordered low-T clusters progressively take over the disordered ones On the other hand, for $T > T_c$ renormalization causes the same effect as increasing the temperature to infinity since it is the disordered high-T clusters that eventually take over. In both cases the evolution is especially rapid when the interaction energy is strong. Besides, the evolution pathway depends on the dimension of the problem since scaling does not operate in the same way in 2 or 3 dimensions or when the order parameter is a scalar or a vector. For the simple case of an Ising model, for instance, the remaining number of spins after n renormalization operations is $1/(2^d)^n$ of the initial value where d is the dimension of space and n the dimensionality of the order parameter. With respect to such scale changes, the thermodynamic potentials must remain unchanged even though all lengths are reduced by a factor b. Thanks to the existence of fluctuations at all length scales, the critical temperature is the only point where the system as a whole is left unchanged by renormalization. The correlation length plays a fundamental role in this respect, being invariant when reduced by a factor b at T_c since it its value is infinity at this point.

In this manner, group renormalization theory allows the critical point of a transition to be identified and the variations of properties to be studied in its vicinity. As noted above, one accounts for a phase transition by including into the expression of thermodynamic potentials terms that are function of critical exponents and present a singularity at the critical temperature. It is by writing that the potentials are invariant at T_c under any scale change that one derives the equalities between exponents used on both sides of T_c as well as the relations (6.19). The exponents can be calculated only if the interaction energy is accurately known. Although they have been made only for simple models, like Ising's, such calculations have nonetheless shown that critical exponents depend only on the dimension of space and symmetry of the order parameter when the interactions are short ranged. This does not apply to situations where long-ranged Coulombic interactions are important.

6.2e LANDAU THEORY

Continuous Transitions

In spite of their diversity, continuous crystal-crystal transitions have the rather general feature that symmetry is generally higher for the high-temperature than for the low-temperature phase. Especially important is the case where the symmetry group of the former is a sub-group of the latter as order parameters were introduced by L. Landau (1908-1968) in 1937 for such transitions. By vanishing at the transition temperature T_c and remaining zero beyond it, the order parameter η characterizes the symmetry breaking that takes place at the transition. At any temperature higher than T_c the Gibbs free energy (or other thermodynamic potential) is thus minimum for $\eta = 0$ (Fig. 6.8). Besides, the stability of the equilibrium is determined by the condition

$$(\partial^2 G/\partial\eta^2)_{T,P} > 0. \tag{6.21}$$

Below T_c the high-symmetry phase is unstable. Internal equilibrium would still be ensured by the minimum condition (6.20) for G, but instability now imposes

$$(\partial^2 G/\partial \eta^2)_{T,P} < 0 \tag{6.22}$$

for positive or negative values close to $\eta = 0$ (Fig. 6.8). As will be shown below, stable equilibrium corresponds to two (or sometimes more) equivalent but non zero values of the order parameter such that both conditions (6.20-6.21) are simultaneously satisfied. We can thus identify T_c with the temperature at which $(\partial^2 G/\partial \eta^2)_{T,P} = 0$.

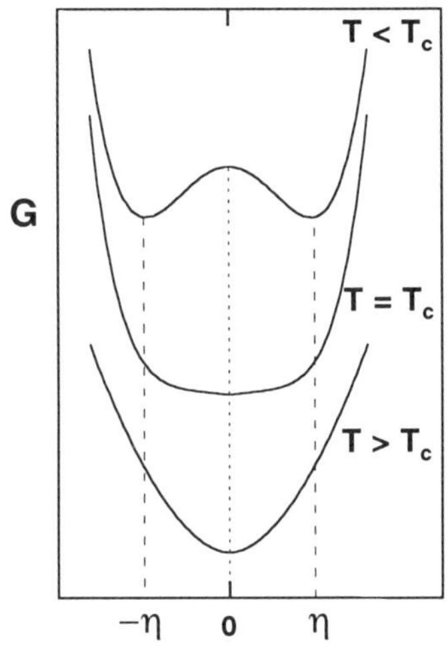

Figure 6.8. Variation of the Gibbs free energy as a function of the order parameter in continuous transitions. Below T_c the low- and high-temperature phases are stable and unstable, respectively. Above T_c only the high-temperature phase is in internal equilibrium.

In the vicinity of T_c, i.e., for small enough values of the order parameter, the basic assumption of Landau theory is that thermodynamic potentials can be expanded as a Taylor series of η

$$G(\eta) = g_0 + g_1\eta + g_2\eta^2 + g_3\eta^3 + g_4\eta^4 + \ldots \tag{6.23}$$

where the g_i coefficients may vary with pressure, temperature and, for solutions, chemical composition. At given T, P and composition, the equilibrium value of the order parameter is determined by the minimum criterion (6.20) as applied to the relevant Gibbs-free energy expansion.

We will consider first a fourth-order expansion of G in η. Because $(\partial G/\partial \eta)_{T,P}$ is 0 above T_c, where $\eta = 0$, the conditions (6.20)-(6.21) imply $g_1 = 0$ and $g_2 > 0$, respectively. Below the transition temperature the instability criterion (6.22) for $\eta = 0$ requires $g_2 < 0$. The transition temperature is thus simply the point where the sign of the g_2 coefficient changes. This suggests that one should expand g_2 in powers of $T - T_c$ and the simplest way of doing it is to write

$$g_2 = a(T - T_c) \tag{6.24}$$

where a is a positive constant. On the other hand, the transition point as a stable equilibrium itself implies $(\partial^2 G/\partial\eta^2)_{T,P} = 0$ at $T = T_c$, which imposes $g_3(T_c) = 0$ and $g_4 > 0$. As it would be fortuitous that both g_2 and g_3 vanish at the transition temperature, one often sets $g_2 = 0$ at all temperatures and assumes in addition that g_4 remains positive at least over some interval in the vicinity of T_c to prevent $G(\eta)$ from vanishing at temperature differing from T_c. With the assumption $g_4 = c$, where c is a positive constant, the simplest fourth-order Gibbs free energy expansion is

$$G(\eta) = g_0(T) + a(T - T_c)\eta^2 + c\eta^4. \tag{6.25}$$

The values of the order parameter minimizing this expansion are the real roots of

$$2a(T - T_c)\eta + 4c\eta^3 = 0. \tag{6.26}$$

As expected, this equation has a single root $\eta = 0$ for $T > T_c$. For $T < T_c$ there are three real roots. The first one, $\eta = 0$, must be rejected because it corresponds to the unstable equilibrium (maximum G) of the high-symmetry phase. The two others are

$$\eta = \pm[a(T_c - T)/2c]^{1/2}. \tag{6.27}$$

Hence the Landau expansion actually describes a continuous transition since the order parameter varies continuously up to T_c as the square root of $T_c - T$. Replacing η by (6.27) in (6.25), we finally calculate the entropy and heat capacity. Starting from

$$S = -(\partial G/\partial T)_{P,\eta} = S_0 - (\partial g_2/\partial T)_{P,\eta}\eta^2, \tag{6.28}$$

where $S_0 = (\partial g_0/\partial T)_{P,\eta}$, we find for the high-symmetry phase at $T > T_c$

$$S = S_0, \qquad C_p = T(\partial S/\partial T)_{P,\eta} = C_{p0}. \tag{6.29a}$$

Below T_c we have for the low-symmetry phase

$$S = S_0 + (T - T_c)a^2/2c, \qquad C_p = C_{p0} + a^2T/2c. \tag{6.29b}$$

The entropy is thus continuous at the transition temperature whereas the heat capacity increases discontinuously by the amount $a^2T_c/2c$ between the low and high-symmetry phases.

In practice, Gibbs free energy expansions more complex than (6.25) are frequently used in applications of the Landau theory. As assumed here, $G(\eta)$ is often independent of the sign of η. In magnetic or order-disorder transitions, for example, it does not matter whether the reference state is taken to be spins up or down, or one crystal sublattice instead of the other. In these cases the symmetry of the order parameter precludes the presence of odd terms in the series expansion. Since $G(\eta)$ is more generally subjected to all the symmetry requirements of the crystal, the actual form of (6.23) has to be determined from group theoretical arguments. (When the Landau theory is used outside of its original context for non-symmetry breaking transitions, these constraints of course no longer apply.) With only even coefficients, a common

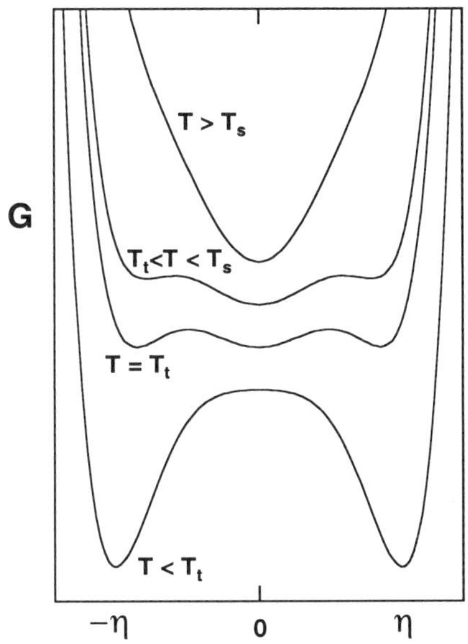

Figure 6.9. Schematic variations of the Gibbs free energy as a function of the order parameter in first-order transitions for various temperatures. Above the temperature T_s an equilibrium state is obtained only for the high-symmetry, high-temperature phase. Between T_s and T_t, the low-temperature phase is metastable with respect to the high-temperature phase. At T_t the two phases have the same Gibbs free energy, and the high-temperature phase becomes metastable below T_t.

form is the so-called "2-4-6" potential

$$G(\eta) = g_0 + g_2\eta^2 + g_4\eta^4 + g_6\eta^6. \tag{6.30}$$

In certain cases of crystals with a high symmetry odd terms are nonetheless allowed. For instance, the also popular "2-3-4" potential is

$$G = g_0 + a(T - T_c)\eta^2 + b\eta^3 + c\eta^4. \tag{6.31}$$

First-Order Transitions

As a matter of fact , the lack of odd terms in $G(\eta)$ expansions is necessary for predicting the existence of a continuous transition. But this condition is not sufficient because negative values of the g_4 coefficient in either (6.25) or (6.30) can lead to different solutions. Writing for instance (6.30) as a function of constant coefficients (i.e., $c < 0$, $e > 0$)

$$G(\eta) = g_0 + a(T - T_c)\eta^2 + c\eta^4 + e\eta^6, \tag{6.32}$$

we can repeat the calculation made above to find $\eta = 0$ for the high-symmetry phase and non zero values

$$\eta = \pm\left[\frac{-c + [c^2 - 3ae(T - T_c)]^{1/2}}{3e}\right]^{1/2} \tag{6.33}$$

for the low-symmetry phase.

The situation is however more complex than for a continuous transition (Fig. 6.9). For temperatures higher than $T_s = T_c + c^2/3ae$, (6.32) has a single root $\eta = 0$. At the transition temperature T_t the value $\eta = 0$ of the order parameter of the high-temperature phase implies that $G = g_0$, and that this value also applies to the other phase. We then find that, as given by

$$T_t = T_c + c^2/34e, \tag{6.34}$$

T_t is higher than the critical temperature T_c. In other words, the order parameter no longer varies continuously before vanishing at $T = T_c$ but undergoes instead a discontinuous change $(-c/6e)^{1/2}$ at the temperature T_t of a first-order transition. At this temperature the entropy change is $\Delta S = -ac/6e$, and there also exists an enthalpy of transition $T_t\Delta S$. We finally note that metastability has become possible because on both sides of T_t the Gibbs free energy minima of the low- and high-symmetry phases correspond to equilibria that are no longer unstable (Fig. 6.9). With the assumption that the Gibbs free energy is still expandable as a series of the order parameter, the Landau theory is thus also capable of accounting for the existence of first-order transitions and for the discontinuous variations of physical properties at the transition temperature.

Limitations of the Theory

From a theoretical standpoint, the Landau theory has the great merit of relating simply first-order and continuous transitions and showing that their main difference concerns the evolution of the order parameter as determined by the thermodynamic potential of the system. In addition, it indicates that a critical point between a crystal and liquid, analogous to that between a liquid and a gas, cannot exist because an order parameter describing the transition between the crystal (an ordered phase) and the liquid (a disordered phase) could never vanish. By definition, the Landau theory is universal since it assumes the same analytical form for the Gibbs free energy regardless of the kind of order parameter. For any expansion used, it gives a variation of η of the form (6.27) for a continuous transition, hence a critical exponent $\beta = 1/2$ for the scaling laws followed by order parameters near transition temperatures. The agreement is therefore poor with the experimentally observed values that are close to 1/3. Likewise, the Landau theory does not predict the divergences of heat capacities and compressibilities. As a result, it cannot work well for magnetic transitions or the critical point of fluids.

Like that of van der Waals, which also yields an exponent $\beta = 1/2$ for the critical point of fluids, Landau theory is a *mean field* theory. All such theories suppose that the order parameter is constant throughout the system. The state of any element is determined by the average properties of the system as a whole, which is tantamount to assuming that interactions are very long ranged at any temperature so that the fluctuating forces exerted on any element can be replaced by a mean field typical of the system. As exemplified by the incorrect value they predict for β, mean field theories necessarily fail near transition temperatures because they average out the local fluctuations and correlations that are at the roots of critical behavior. Hence an intrinsic contradistinction is that, on the one hand, the Landau expansions should converge faster for smaller values of the order parameter, but that, on the other hand, these conditions correspond to the vicinity of the critical temperature where the thermodynamic potentials become nonanalytic.

Although modifications of the Landau theory aim at improving the account taken of fluctuations close to T_c, it is worthy of note that very good results can nevertheless be obtained with the expansions described in this section. This is the case of the $\alpha-\beta$ quartz transition, which is the prototype of displacive phase changes, for which the order parameter describes the rotation of SiO_4 tetrahedra. For such transitions, and especially ferro-paraelastic transitions, Gibbs free energy expansions can even unexpectedly converge over temperature intervals of hundreds of degrees for elevated values of η. The main reason is that in such cases fluctuations become important only very close to the transition temperature.

6.3 KINETIC TRANSITIONS

6.3a NERNST'S LAW AND INTERNAL DISEQUILIBRIUM

Equilibrium phase transitions are strictly thermodynamic in origin. They take place in response to pressure or temperature changes because they confer to the system a lower Gibbs free energy than would have been ensured by the original phases. Kinetic phase transitions are quite different in this respect because they result from the fact that, under various circumstances, the system has not enough time to minimize its Gibbs free energy through adjustments of its configuration. It is then trapped in a state of higher energy than that of the equilibrium configuration. It follows that the system is no longer in internal equilibrium and that its properties depend on its previous history. This point will first be examined with the important case of entropy.

In preamble, we note that Nernst's law is automatically followed if the configurational entropy is zero at any temperature, which is the case of perfect crystals. Let us then consider a phase whose S^{conf} is not zero at a given temperature T. If this phase nonetheless follows Nernst's law, the configurational entropy is obtained from

$$S^{conf}(T) = \int_0^T C_p^{conf}/T\, dT. \tag{6.35}$$

Inasmuch as C_p necessarily tends to 0 when the temperature tends to 0, C_p^{conf} tends to the same zero limit and $S^{conf}(0)$ is itself zero. If, in contrast, this phase were not following Nernst's law, we should write instead

$$S^{conf}(T) = S^{conf}(0) + \int_0^{T_1} C_p^{conf}/T\, dT, \tag{6.36}$$

and would encounter a serious difficulty. Because the configurational entropy has its counterpart in terms of a configurational heat capacity, it should be possible to express $S^{conf}(0)$ as the integral of C_p^{conf}/T between a temperature T_1, to be determined, and 0 K:

$$S^{conf}(0) = \int_{T_1}^0 C_p^{conf}/T\, dT. \tag{6.37}$$

Because $S^{conf}(0)$, C_p^{conf} and T are all positive definite quantities, (6.37) is an absurd equation.

To account for the existence of phases not following Nernst's law, we must therefore assert that these phases are not in internal equilibrium. Their state is not specified by two variables only in addition to composition variables. Their configurational heat capacity has vanished above 0 K, and not at 0 K exactly. At this point, the structure of the phase has been frozen in and remains down to 0 K that of the temperature T_0 at which equilibrium has been lost. The residual entropy at 0 K then represents the configurational entropy frozen in at T_0

$$S^{conf}(0) = S^{conf}(T) - \int_{T_0}^{T_1} C_p^{conf}/T \, dT. \tag{6.38}$$

In this expression, $S^{conf}(0)$ is function not only of temperature (0 K) and pressure (1 bar, for example), but also of the temperature T_0. For a phase that is out of equilibrium, this temperature T_0 cannot be an intrinsic property since its entropy at 0 K would otherwise be determined by T and P only. As will discussed below, this temperature T_0 is kinetic in origin and it should in principle be mentioned whenever a residual entropy is reported for a phase.

6.3b RELAXATION TIMES

Phases not obeying Nernst's law are not in internal equilibrium. How do they lose it? In a simple way, this can be described with the notion of relaxation, which has already been mentioned in 2.4a. As a starting point, we will consider again the rapid expansion of an ideal gas. Macroscopic measurements indicate that, after a short while, the temperature becomes related to the new pressure and volume conditions by $PV = RT$. Of course, this equation characterizes internal equilibrium since it expresses V as a function of only two other variables. It is only with very rapid measurements that one could follow the transitory steps through which the temperature *relaxes* to the new equilibrium value. If we deliver rapidly some heat to a solid substance, through the flash of an infrared laser, the conversion of light into heat (and vibrational energy in the solid), will take place at the 10^{-12} - 10^{-14} s timescale of atomic vibrations. Because of such very short timescales, vibrational relaxation can always be considered as instantaneous with respect to thermodynamic measurements. If temperature is measured at different places, some time will be in contrast needed for the temperature to become uniform throughout the solid. In this case, the (thermal) relaxation is of course governed by the thermal conductivity of the sample.

In contrast to the preceding cases, consider now a different kind of perturbation whereby physical properties do remain constant throughout a phase. As a concrete example, let us take molten boron oxide whose temperature is changed very rapidly from an initial T of 584 K (Fig. 6.10). Measurements show that the new value of the refractive index is reached in less than 1 hour at 573 K and in a few hours at 553 K. It is not attained yet after 50 hours at 532 K and it would not have changed at all at temperatures lower than 520 K. Like any physical property, the refractive index of a phase (with a fixed composition) in internal equilibrium is determined by only two variables, namely, P (1 atm.) and T (the temperature of the experiment). If the refractive index varies with time, so does the state of the melt whose equilibrium configuration is reached only when relaxation is complete and the refractive index has become constant. For molten B_2O_3,*structural* relaxation is thus rapid above 580 K, slow between 580 and 530 K, and almost nonexistent below 520 K where equilibrium can no longer be attained in experimental timescales. The configuration of B_2O_3 liquid has then been frozen in. The material has become a glass.

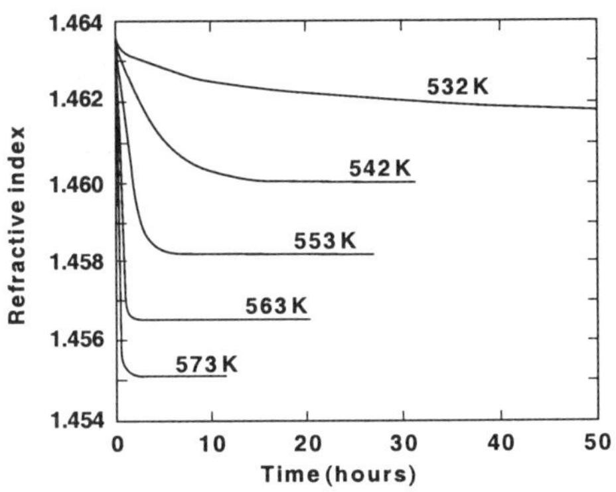

Figure 6.10. Refractive index of B_2O_3 glass against the annealing time of the supercooled liquid at the indicated temperatures. (As accurate measurements are not possible at high temperature, the material is rapidly quenched to a glass whose refractive index differs from that of the liquid by a constant value.) Data of Boesch et al. (1970).

As will be discussed below, relaxation effects can also be important for crystals. To characterize the rate at which a given property Y gets closer to the new equilibrium value Y_e, we define in a general way the relaxation time as

$$\tau_Y = -(Y_t - Y_e)/(\partial Y/\partial t) \tag{6.39a}$$

where Y_t is the value actually measured at instant t. If τ_Y is constant, not depending on the instantaneous value of Y, the relaxation follows an exponential law

$$(Y_t - Y_e) = (Y_0 - Y_e)\exp(-t/\tau_Y) \tag{6.39b}$$

where Y_0 is the initial Y value. After a time τ_Y, the variation of Y then represents a fraction 1/e of the initial difference from the equilibrium value. Relaxation is in fact seldom exponential. For achieving quantitative agreement with experimental data, one must use either a distribution of relaxation times or terms of the form $\exp(-t/\tau_Y)^\beta$ where β is a constant ranging from 0 to 1. As shown in Fig. 6.10 for the refractive index of B_2O_3, the marked increases of relaxation times when the temperature decreases are in any case at the roots of kinetic phase transitions and deviations from Nernst's law.

6.3c VIBRATIONAL VS. CONFIGURATIONAL RELAXATION

In view of the very strong temperature dependence of relaxation times, three cases must be distinguished concerning the variation of a property following a jump in temperature ΔT (Fig. 6.11). Consider for instance the enthalpy (or any first-order property such as the entropy, volume, etc.).

1. At high enough temperatures, τ is short with respect to the timescale of an enthalpy measurement and one observes a concomitant enthalpy jump ΔH (e.g., H_1 in Fig. 6.11a). Since the enthalpy reaches its new equilibrium value instantaneously, the heat capacity as given by ratio of the enthalpy and temperature variations is the sum of the vibrational and configurational parts.

2. At lower temperatures, τ eventually becomes similar to the experimental timescale. A smaller enthalpy jump is first observed, which corresponds to the instantaneous vibrational relaxation, followed by a configurational relaxation which progressively brings the enthalpy to its new equilibrium value. The lower the temperature, the longer is the time required to reach equilibrium (e.g., H_2 and H_3 in Fig. 6.11a).

3. When the temperature has decreased still further, relaxation times have become so long that no configurational relaxation takes place on experimental timescales. The structure of a phase has been frozen in. The configurational heat capacity has vanished and only the small, instantaneous enthalpy decrease associated with the lattice heat capacity is observed (e.g., H_4 in Fig. 6.11a).

In summary, the enthalpy and other first-order properties continuously vary when the temperature decreases, but their derivatives (heat capacity and other second-order properties) decrease markedly in the temperature interval where relaxation times become similar to experimental timescales. In practice, these intervals are often narrow enough that the C_p or α variations can be considered as discontinuous. These rapid variations thus result in kinetic phase transitions, of which a most common example is the glass transition. We have considered here the changes that take place on cooling, but the opposite changes will take place on heating. In view of its nonequilibrium nature, however, the process is characterized by an hysteresis which will be described below in the case of the glass transition.

The distinction between vibrational and configurational properties makes sense only if there is some way to determine their relative importance. Whenever relaxation times are long with respect to the 10^{-12} - 10^{-14} s period of atomic vibrations, vibrations can be considered as taking place instantaneously in a fixed structural environment. As long as this assumption is met, physical properties can be separated into independent vibrational and configurational

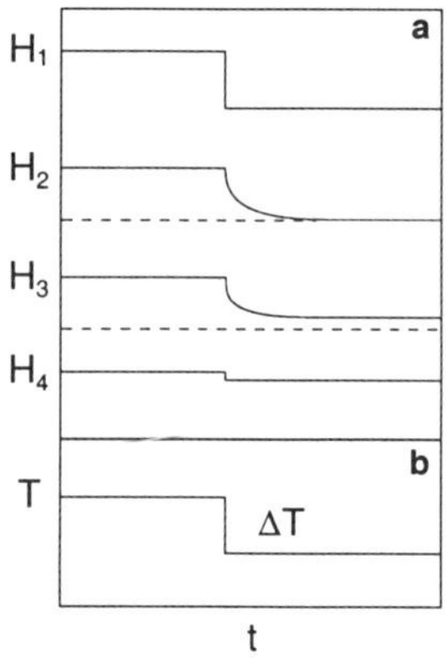

Figure 6.11. Time dependence of the enthalpy (a) for a substance at four temperatures following an abrupt temperature change (b). In the order of decreasing temperatures, H_1 refers to a fully relaxed phase (e.g., a liquid), H_2 and H_3 to a relaxing material, and H_4 to a fully unrelaxed phase (e.g., a glass).

contributions. For liquids, this is actually the case when the viscosity is higher than about 0.1 Pa s for experimental timescales of the order of 1000 s.

6.4 GLASS TRANSITION

6.4a A SCHEMATIC DESCRIPTION

Silicate melts have long been the best known glass-forming liquids, but glasses are also easily made from other oxides (phosphates, borates, germanates...), salts (nitrates, carbonates, chlorides...), some aqueous solutions ($H_2SO_4.3H_2O$...), sugars (glucose, sucrose), alcohols (ethanol, glycerol...) and especially organic polymers (*plastics*). Despite strong differences in chemical composition, all these liquids share a common feature, namely a very high viscosity of about 10^{12} Pa.s when they vitrify with a standard cooling rate of a few degrees per minute (for comparison, the viscosity of water at ambient conditions is about 10^{-3} Pa.s). This universal correlation is a most important feature of the glass transition. It originates in the fact that fluidity and atomic mobility are intimately related. Atomic mobility, which is the hallmark of the molten state, is made evident in daily experience by the ease with which a liquid adjusts to the shape of its container. This is the very reason why liquids flow, even though another well-known feature is that the fluidity diminishes when the temperature decreases. If crystallization does not occur, the viscosity becomes eventually so high that flow can no longer take place on the timescale of an experiment. The resulting material is a *glass*.

A glass is thus a hybrid phase characterized by the fixed atomic arrangement of a crystal and the lack of long-range order of a liquid, whereas a liquid is a phase whose configuration does change rapidly in response to temperature or pressure variations. On two counts the transition between these two phases is a special case of order-disorder reactions in that liquids and glasses are phases where all symmetry elements have been lost, and also in that the rate of the ordering reaction strongly depends on temperature. Anticipating the description of interatomic potentials given in chapter 14, let us examine how the configuration of a glass is determined by local minima of the interatomic potentials in order to give a schematic picture of this transition. Contrary to crystals where these potentials have a long-range symmetry, glasses have only a short-range order because bond angles and distances between next-neighbor atoms are not constant but spread over a range of values (see Fig. 1.7). In a one-dimensional representation (Fig. 6.12), the minima of interatomic potentials are thus separated by barriers with varying heights and shapes.

Now, assume that some heat is delivered instantaneously to a glass. The subsequent temperature rise is solely due to increasing amplitudes of vibration of atoms in their potential energy wells. As for any solid, the heat capacity of the glass is only vibrational in nature. At higher temperatures, thermal energy becomes high enough that atoms can begin to overcome the barriers that lead to the neighboring potential energy wells. This onset of atomic mobility signals that of structural relaxation. For relaxation times longer than observation timescales, however, only the vibrational heat capacity remains measured. If the temperature is increased further or if enough time is left for the new equilibrium configuration to be attained during the measurement, then the configurational heat capacity is also observed. It represents, integrated over

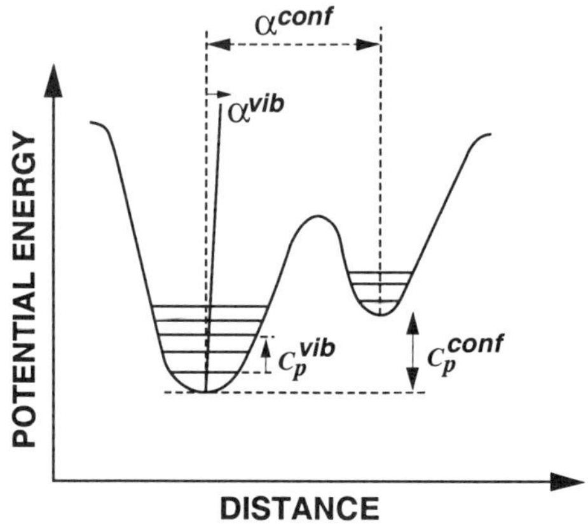

Figure 6.12. Schematic one-dimensional representation of interatomic potentials for an amorphous substance. The configurational heat capacity is determined by the differences between the minima of the various potential energy wells, the relaxation kinetics by the temperature and height of the barriers separating these wells. Likewise, the vibrational and configurational parts of the (linear) thermal expansion coefficient are associated with changes in interatomic distances r.

all atoms, the energy differences between the minima of the potential energy wells that are being explored when the temperature increases.

On heating, the glass transition can thus be viewed as the onset of the exploration by atoms of positions characterized by higher values of interatomic potentials. No matter how complex this dynamical process may be at a microscopic level, this spreading of configurations over states of higher and higher potential energy is the main feature of atomic mobility. This is why configurational heat capacities are positive, whereas relaxation times decrease with rising temperatures because large thermal energies allow the potential energy barriers separating these states to be overcome more easily. We have mentioned in 2.4c that amorphous solids could be prepared by other means than cooling of viscous liquids. At this point it is therefore appropriate to point out that the glass transition has remarkable enough features that one can call *glass* any amorphous solid that undergoes it when heated up regardless of its initial synthesis procedure.

Note that the vibrational and configurational contributions to the thermodynamic properties are not mutually independent. On the whole, configurations of higher energy are associated with lower frequencies of vibrations and thus, as described in 14.3b, with higher vibrational heat capacities and entropies. We also note that a general feature of interatomic potentials is their anharmonic nature, *i.e.*, the forces exerted on the vibrating atoms are not strictly proportional to the displacements from their equilibrium positions. Increasing vibrational amplitudes thus result in increases of interatomic distances (Fig. 6.12), whence the positive thermal expansion coefficients generally observed. Glasses and liquids have such an anharmonic vibrational expansion but the configurations of higher energy that begin to be explored above the glass transition are generally associated with increases in interatomic distances. This is why the thermal expansion coefficient usually increases markedly at the glass transition. (Note, however, that a positive configurational thermal expansion coefficient is not a thermodynamic requirement.) Consider finally the compressibility. For a given configuration, pressure can be viewed as inducing variations in the interatomic potentials characterized by shorter equilibrium distances and steeper slopes around the minima. Since the shape of these potentials determines the vibrational energy levels, this compression will be termed vibrational. For the liquid phase, we must add a configurational contribution to this vibrational one when the pressure changes the potential energy wells themselves and causes significant modifications in short-range order.

6.4b GLASS TRANSITION TEMPERATURE

Instead of measuring the enthalpy as a function of time at constant temperature, let us measure it as a function of temperature at a constant cooling or heating rate. On cooling, relaxation gradually vanishes in response to the continuous increase of relaxation times. The enthalpy then decreases less rapidly because of the important decrease of its temperature derivative, the heat capacity (Fig. 6.13). On heating at the same rate, however, a different pathway is followed, giving rise to the hysteresis shown in Fig. 6.13a. Relaxation does resume at the temperature at which it vanished on cooling, but its first effect is to lower the enthalpy of the glass and bring it closer to the equilibrium values of the supercooled liquid. At higher temperatures, the enthalpy curve of the material has already crossed that of the equilibrium supercooled liquid when relaxation becomes almost complete at the timescale of the experiment. The heat capacity then increases rapidly (Fig. 6.13b), with a bump that depends on thermal history. It is the highest for samples initially cooled down at the slowest rates, or for samples heated up at the highest rates.

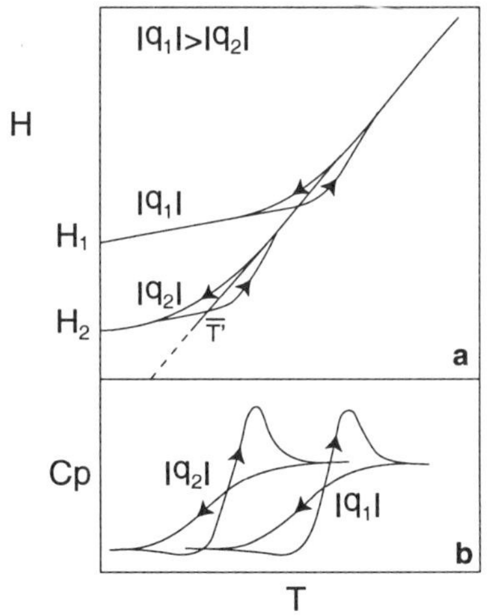

Figure 6.13. Hysteresis of the enthalpy (a) and heat capacity (b) for two different heating or cooling rates q through the glass transition range. The limiting temperature $\overline{T'}$ indicates the end of the transition (see 6.4d). From Moynihan et al. (1974)

The *glass transition range* is the temperature interval where a time dependence of physical properties can be observed at constant T. It has typically a width of a few tens of degrees. For B_2O_3, the measurements of Fig. 6.10 indicate that it is found between about 520 and 580 K. More important is the fact that the contrast between temperature-dependent relaxation times and constant experimental timescales, for a given kind of measurement, causes the glass transition to shift to higher temperatures when the heating or cooling rate is more rapid. This is of course an essential feature of a kinetic transition. For convenience, the glass transition temperature (T_g) is arbitrarily defined as the temperature at which the tangents to the glass and liquid curves of a given property intersect. Contrary to the temperature of an equilibrium transition, T_g is not a fixed point since it varies with the experimental conditions and could even depend on which property is being measured.

Like faster heating or cooling rates, shorter experimental timescales also shift the glass transition to higher temperatures. The aforementioned viscosity of about 10^{12} Pa.s at which a material vitrifies for cooling rates of 1-10 K/min refers to experimental timescales of the order of 10^3 s typical of calorimetry or dilatometry measurements. Now, suppose that a given measurement is made at a timescale of only 10^{-6} s. In practice, this is the case of compressibilities determined from sound velocity measurements typically made at MHz frequencies (see Appendix B). The compressibility change due to the transition from liquid-like to solid-like behavior will then take place at much higher temperatures. Although the T_g shift will vary from one material to another, the viscosity at T_g will be in all cases about 10^3 Pa.s for observational frequencies of ~ 1 MHz. The reason is that relaxation time scales as the viscosity, as can be demonstrated simply for Maxwell's model of *viscoelastic* media.

To account for the (instantaneous) elastic response and the (delayed) viscous response, Maxwell's model views a fluid subjected to a shear stress as a spring and a dash-pot placed in series. Calculation indicates that the system is characterized by a single relaxation time depending on the viscosity η (not to be confused here with an order parameter) as

$$\tau = \eta / G_\infty, \tag{6.40}$$

where G_∞ is the shear modulus at infinite frequency. As to the universal viscosity of 10^{12} Pa.s at T_g, it is due to the fact that G_∞ is of the order of 100 kbar (10 GPa) for most viscous liquids. Although relaxation times in the glass transition range do not depend only on temperature, see below, (6.40) nevertheless gives useful estimates of relaxation times. Consistent with this close link between viscosity and glass transition, it is observed that the variation of T_g with the heating or cooling rate q can be accounted for by the empirical relation

$$\frac{d \ln |q|}{d(1/T_g)} = -\frac{\Delta H_\eta}{R} \tag{6.41}$$

where R is the ideal gas constant and $\Delta H_\eta = R\, \partial \ln \eta / \partial(1/T)$ is the activation energy for viscous flow.

6.4c KAUZMANN PARADOX

Whereas they escape crystallization, why do viscous liquids eventually transform to glasses instead of remaining in the supercooled liquid state? A first answer to this question is purely kinetic and states that it is the increase in relaxation times that causes the glass transition. If experiments could last forever, one could then surmise that any glass would eventually relax to the equilibrium state. The glass transition would then result only from the limited timescale of feasible experiments. In fact, a thermodynamic argument known as Kauzmann's paradox indicates that this answer is not correct.

To expound it, let us first stress once more that the heat capacity of a liquid is higher than that of a crystal of the same composition because of the existence of a C_p^{conf} for the former phase. As a result, the entropy of the liquid decreases much faster than that of a crystal when the temperature diminishes. This difference is shown in Fig. 6.14 for an inorganic and an organic substance, namely, selenium and *o*-Terphenyl ($C_{18}H_{14}$). By extrapolating the entropy of the supercooled liquid to temperatures below the glass transition range, we find that it would become lower than that of the crystal at a temperature T_K that is high enough for making such an extrapolation reasonable. This situation is not thermodynamically forbidden, but it seems unlikely indeed that an amorphous phase, in which matter is spatially distributed, could have a lower entropy than a perfectly crystalline material. We thus conclude that an amorphous phase cannot exist below the temperature T_K, which constitutes the lower bound to the metastability limit of the supercooled liquid. Since internal equilibrium cannot be reached below T_K, the liquid must undergo a phase transition before reaching it. This is of course the glass transition and Kauzmann's paradox suggests that, although it is kinetic in nature, it anticipates a first-order or a continuous transition whose nature is however still debated.

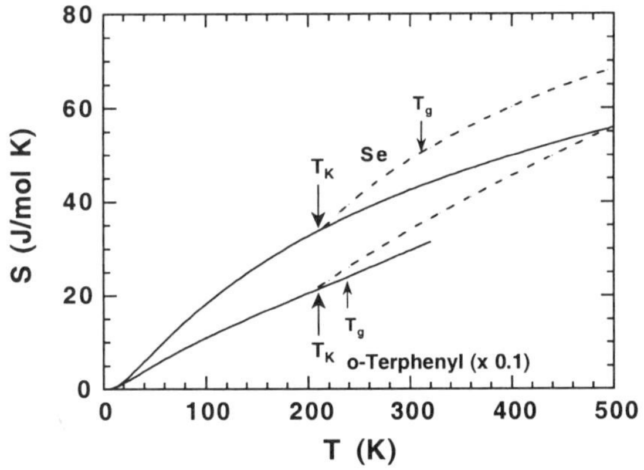

Figure 6.14. Kauzmann's paradox, occurring at temperatures T_K close to 200 K for both *o*-Terphenyl ($C_{18}H_{14}$) and selenium. Calculations made from the entropies of Chang and Bestul (1972, 1974) for the crystalline (solid curves) and liquid (dashed curves) phases. The glass transition temperatures (T_g) also shown indicate the intervals over which C_p extrapolations are made for the supercooled liquids. Data for *o*-Terphenyl divided by 10.

In its original form, which has just been expounded, the Kauzmann paradox implicitly neglects possible differences in vibrational entropy between the amorphous and crystalline phases. This simplification is actually incorrect, but it does not detract anything to the gist of the argument, for taking into account such differences would simply shift slightly T_K. In any case, a more rigorous statement of the paradox states that the catastrophe would take place when the configurational entropy of the supercooled liquid would vanish. Statistical mechanical models of the glass transition attempt to account for this fact. Some predict first-order transitions, others continuous transitions between the liquid and an "ideal" glass phase stable below T_K with zero configurational entropy. The important point to be stressed is that the decrease of configurational entropy plays a major role in the occurrence of the glass transition. Specifically, some models simply relate the increases of relaxation times to decreases of configurational entropy, τ being inversely proportional to the product TS^{conf} so as to become infinite at T_K. In agreement with this view and (6.40) is the fact that the viscosity of glass-forming liquids is represented by empirical laws of the form $\ln\eta = A + B/(T - T_1)$, where A and B are constants and T_1 is the temperature, usually found to be close to T_K, at which the viscosity would indeed diverge. This is consistent with the fact that a liquid with a vanishing configurational entropy would have an infinite viscosity because atoms can move from one place to another only if many configurations are accessible.

6.4d FICTIVE TEMPERATURE

For cooling rates ranging from 0.1 to 100 K/min, the thermal expansion coefficient of vinyl polyacetate decreases by 50% at the glass transition near 300 K. As shown in Fig. 6.15, the rate dependence of T_g has thus the immediate practical consequence that the volume of the glass does not depend only on P and T but also on thermal history. This is of course consistent with the fact that a glass is a phase out of equilibrium. Another observation showing it is the heat capacity hysteresis represented in Fig. 6.13b, which implies that some entropy is created when C_p/T is measured along a heating and cooling cycle through the glass transition range. This is indeed another intrinsic feature of the irreversibility of a nonequilibrium transformation. Since the existence of a reversible pathway between any two temperatures is required to define the entropy of a phase, this would seem to prevent application of equilibrium thermodynamics to glasses. However, the entropy created remains small with respect to the vibrational entropy of the glass as given by integrations of C_p/T from 0 K. In practice, deviations of experimental pathways from a hypothetical reversible one is therefore not a significant problem in calorimetric measurements.

But what meaning must be given to the entropy of a nonequilibrium phase? As long as the glass is kept well below the glass transition range, its configuration and properties do not change with time so that it is tempting to complement the thermodynamic description of the phase by means of order parameters. For a glass, these parameters would characterize the equilibrium configuration that has been frozen in

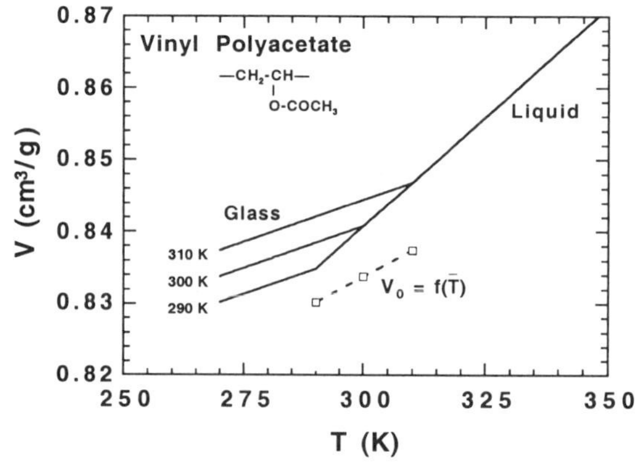

Figure 6.15. Volume of liquid and glassy vinyl polyacetate for three different cooling rates. The dashed line represents the room-temperature density of the glasses against the fictive temperature $\overline{T}$. Note the similar thermal expansion coefficients indicated by the parallelism between the volume curves of the glasses. Drawn from the data of McKinney and Goldstein (1974) and Sasabe and Moynihan (1978).

at the glass transition. Structurally, the difficulty is that the configuration of an amorphous phase cannot be described with a limited number of parameters. Hence we have to content ourselves with a different, empirical approach resting on the fact that, for many kinds of glasses, second-order properties depend little on thermal history so that the enthalpy and volume curves of a given glass formed with different cooling rates are practically parallel. As shown in Fig. 6.15 for the density of vinyl polyacetate, it then suffices to know P, T and the glass transition temperature for determining the properties of the glass. For continuous cooling, this temperature T_g is therefore a third variable describing the state of a glass. It is called the *fictive temperature*, $\overline{T}$.

The fictive temperature plays the role of an order parameter, with the important difference that it does not vanish at the experimentally observed transition temperature. It will be seen below that the fictive temperature varies through annealing of the material in the glass transition range. Hence this temperature is more precisely defined as that at which the glass configuration would be the equilibrium configuration of the supercooled liquid. In any case, the enthalpy and entropy differences between two glasses having fictive temperatures $\overline{T_1}$ and $\overline{T_2}$ are thus

$$\Delta H_a = \int_{\overline{T_1}}^{\overline{T_2}} (C_{pe} - C_{pg}) dT, \qquad \Delta S_a = \int_{\overline{T_1}}^{\overline{T_2}} (C_{pe} - C_{pg})/T \, dT, \tag{6.42}$$

where C_{pl} is the heat capacity of the equilibrium liquid and C_{pg} that of the glass. Analogous equations are written for volume differences in terms of thermal expansion coefficients. Although they are very useful for dealing with first-order thermodynamic properties, we stress that these equations are strictly empirical. Their validity lies on weak sensitivities of second-order properties to thermal history. Because transport properties are more sensitive probes of molecular configurations, use of the fictive temperature is not warranted for characterizing them.

As introduced above, the fictive temperature is well defined only for glasses rapidly cooled below the glass transition range. What happens if the glass is heated up back in the glass transition range, i.e., in a temperature interval where structural relaxation will resume? Like the enthalpy and volume, the fictive temperature must vary to get closer to the equilibrium value, which is the actual temperature of the experiment. Now, experience shows that relaxation times do not depend only on T but also on $\overline{T}$. Inasmuch as the fictive temperature cannot be measured directly, we have to relate it to observable properties. The simplest way of doing it is to define $\overline{T}$ as that part of the given property (H, V, etc.) that relaxes, expressed in temperature units. For enthalpy relaxation, we thus write

$$H(T) = H_e(\overline{T}) - \int_T^{\overline{T}} C_{pg} dT \tag{6.43}$$

where $H_e(\overline{T})$ is the equilibrium enthalpy at temperature $\overline{T}$. Let us differentiate (6.43) with respect to T for obtaining the variation of $\overline{T}$ with temperature under the particular conditions of the experiment

$$(d\overline{T}/dT)|_T = (C_p - C_{pg})|_T / (C_{pe} - C_{pg})|_{\overline{T}}. \tag{6.44}$$

This derivative is the relaxational part of the heat capacity. In the manner of an advancement of reaction, it is zero for the glass and unity for the liquid. At equilibrium, the fictive temperature is equal to the actual temperature $[\overline{T}(T) = T]$, which (6.44) ensures. For continuous cooling, these equations also indicate that $\overline{T}$ tends to a limiting value $\overline{T'}$ which depends on the cooling rate (Fig. 6.13) and is equal to the previously defined glass transition temperature.

6.4e PERMANENT COMPACTION OF GLASSES

The configuration of a glass reflects the conditions under which the liquid structure has been frozen in. In the same way as it is possible to prepare glasses with different fictive temperatures, one can produce dense glasses having kept the configuration of high-pressure liquids. For instance, let us vitrify a liquid, vinyl polyacetate, by cooling it from a high temperature T either at 1 or at 800 bar and measure its volume as a function of temperature (Fig. 6.16). Not only the glass formed at high pressure is denser than its 1-bar counterpart, but it remains so even if the pressure is released to 1 bar. This is shown by the arrow in Fig. 6.16, and is yet another evidence of the fact that glasses are nonequilibrium substances.

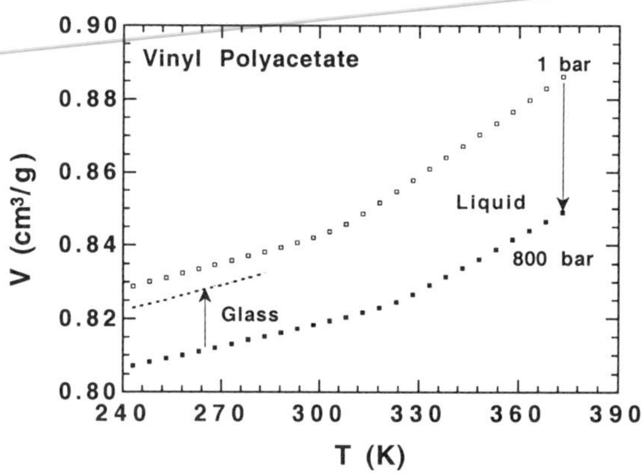

Figure 6.16. Permanent compaction of glasses formed at high pressure at a slow rate of 5 K/h. Open squares: volume of vinyl polyacetate liquid and glass at 1 bar; solid squares: volume at 800 bar; dashed line: 1-bar volumes of permanently compacted glasses formed at 800 bar. Data from McKinney and Goldstein (1974).

This *permanent compaction* is due to the fact that the melt (β_T= 5.9 10^{-5} bar^{-1} at 370 K) is more compressible than the glass (β_T= 2.5 10^{-5} bar^{-1} at 280 K). On decompression to 1 bar, only the purely elastic part of the densification is eventually released when a glass is decompressed, the configurational part being kept. If the pressure dependence of T_g can be neglected, then permanent compaction depends solely on the vitrification pressure and on the compressibility contrast between the liquid and glass. Further heating in the glass transition range is needed for the density of the compacted glass to relax to that of the material formed at 1 bar. The glass transition thus depends not only on temperature and on the frequency of the observation, but also on pressure so that a *fictive pressure* is a second parameter that should be specified for describing the state of a glass.

Glasses compressed at low temperature to pressures of a few tens of kbar also undergo permanent compaction. High pressure can thus induce irreversible configurational changes at temperatures at which the substance is said to be a glass. For given frequencies or experimental timescales, however, the kinetics of pressure- and temperature-induced configurational modifications are markedly different. This dissimilarity originates essentially in that the shape of interatomic potential energy wells vary much less with T than with P. If a high kinetic energy is needed to overcome potential energy barriers at constant pressure, the changes in the shape of these barriers with pressure can lead by themselves to new configurations, at low temperatures, if the pressure is high enough. Without elaborating on this point, we will just point out that processes taking place in glasses at high pressure thus likely mimic to a significant extent those occurring in liquids, whereas this similarity is clearly nonexistent for the effects of temperature.

6.4f GLASS-LIKE TRANSITIONS

This description of the glass transition applies to a variety of kinetically controlled processes in crystals. *Plastic crystals*, characterized by low entropy of fusion and an unusually high plasticity, are good examples of disordered systems with three-dimensional long-range order. When the high-temperature form of cyclohexanol (ChI) crystallizes at 299 K, for instance, the $C_6H_{12}O$ molecules order in a face centered cubic lattice but their regular shape allows them to maintain orientational mobility by rotations around the lattice points. It is through a transition to the low-temperature polymorph (ChII), which is stable below 265 K, that this dynamical behavior vanishes and the orientational disorder disappears. Now, with rapid cooling rates the ChI form can be obtained metastably and kept for long periods of time below 180 K. On further cooling a transition is eventually observed near 160 K (Fig. 6.17). The orientational disorder of $C_6H_{12}O$ molecules is then frozen in within the crystal. In contrast to the ChII form, which obeys Nernst's law, ChI has a nonzero residual entropy of 4.7 (16) J/mol K. Instead of relating a liquid to a crystal, Kauzmann's paradox involves in this case metastable and stable crystals. The similarity with the glass transition phenomenology is such that the name of *glassy crystals* has been proposed for crystals where rotation of molecular groups is freed above a glass-like transition temperature.

Ice is probably the most common glassy crystal. It has long been recognized that it is crystalline only in the position of its molecules but glass-like in their orientation. Whereas the oxygen atoms lie on the sites of an hexagonal lattice (Fig. 6.2b), the hydrogen atoms are arranged in a partially random way so that the orientation of the

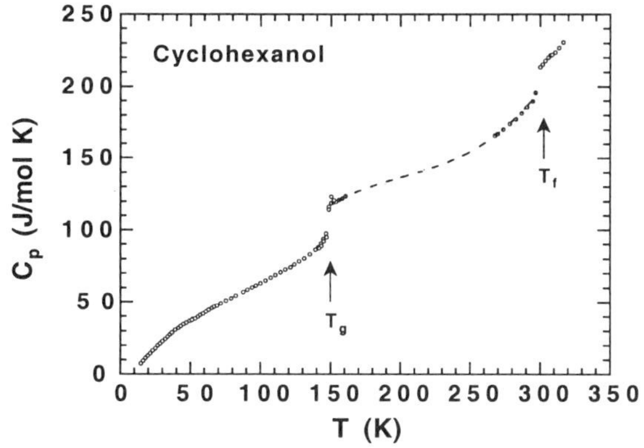

Figure 6.17. "Glass transition" (at T_g) and fusion (at T_f) of the high-temperature polymorph of cyclohexanol. The small C_p bump near T_g is typical of a glass-like transition. Note that calorimetry cannot be made between 160 and 250 K where the stable, ordered polymorph forms readily. Measurements from Adachi et al. (1968) made on heating.

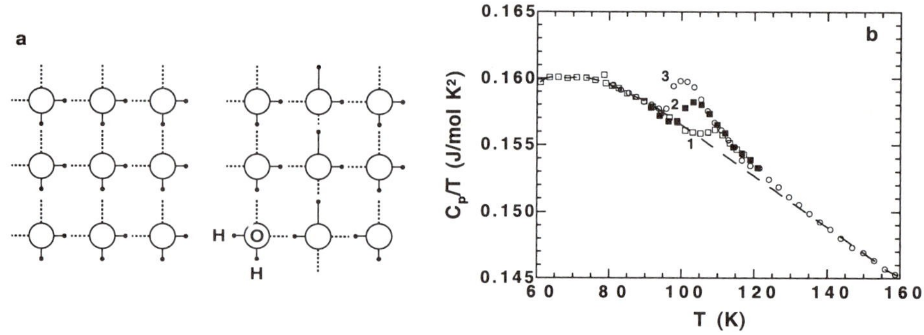

Figure 6.18. (a) Schematic planar representation of ordered (left) and disordered (right) hydrogen bonding between water molecules in ice; (b) C_p increase of ice at the glass-like transition and influence of the thermal history of the sample on this increase. To expand the scale, the measurements have been plotted as C_p/T. Series 1: ice cooled down at a rate of 1 K/min before the measurements made on heating; series 2: ice stabilized at 94.4 K for 71 hours before the measurements; series 3: ice stabilized at 89.4 K for 624 hours. Lattice heat capacity shown as the dashed curve. Note that samples stabilized at the lowest temperatures (i.e., having the lowest fictive temperatures) show the biggest C_p bumps. Data from Haida et al. (1974).

hydrogen bonds linking different H_2O molecules into H_4O tetrahedra is also partially random (Fig. 6.18a). This disordered configuration is frozen in near 100 K. As will be calculated in 9.3f, ice has a residual entropy of 3.4 J/mol K. The relaxation time for hydrogen bond reorientation is 145 hours at 90 K and it decreases to 1 hour at 108 K. The consequence on the heat capacity is shown in Fig. 6.18b. Below 90 K only the vibrational C_p is measured; above 120 K, the configurational C_p is also measured. Between these temperatures C_p^{conf} depends on thermal history. For a sample equilibrated at a temperature higher than that of the experiment, the configurational enthalpy decreases. If the sample is heated again, then recovery of the disordered state requires more heat than needed to heat a non stabilized sample. Hence C_p^{conf} is the highest for samples stabilized at the lowest temperatures.

For most such reactions relaxation times vary so strongly with temperature that transition temperatures depend little on cooling rate. As configurational heat capacities are relatively small, the residual entropies of partially disordered crystals depend little on thermal history. For ice, for instance, S(0) varies by only 0.06 J/mol K between a rapidly quenched sample and another sample equilibrated at 89 K. This insensitivity can lead one to overlook the kinetic nature of the order-disorder reaction and the fact that cation distributions measured at room temperature differ from those prevailing at high temperatures, but represent those that are frozen in at lower temperatures on cooling.

In a last example, we will address the important case of the distribution of ions over crystallographic sites and consider $FeAl_2O_4$, a compound with the so-called AB_2O_4 spinel structure. In a perfectly ordered spinel, the divalent cation A (e.g., Fe) and triva-

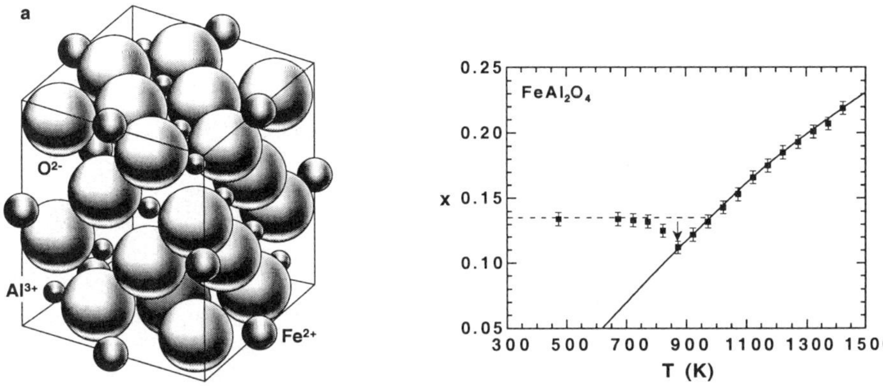

Figure 6.19. (a) Crystal structure (from the book by Kittel, 1996) and (b) degree of inversion against temperature of $FeAl_2O_4$ spinel. Relaxation indicated by the arrow. Neutron powder diffraction data of Harrison et al. (1998).

lent cation B (e.g., Al) would occupy the octahedral and tetrahedral sites, respectively, of an approximately cubic close packed arrangement (Fig. 6.19a). In actual spinels, cations distribute themselves over one tetrahedral site and two octahedral sites per formula unit according to the general scheme $A_{1-x}B_x(A_{x/2}B_{1-x/2})_2O_4$ where parentheses indicate which cations occupy octahedral sites. The *degree of inversion* x then is a convenient variable for investigating the disordering reaction as a function of temperature. It varies from 0 for ordered, "normal" spinels, to 2/3 for a random cation distribution, and even to 1 for "inverse", ordered spinels where all A cations would occupy a tetrahedral site.

For $FeAl_2O_4$, experience shows that x strongly varies with temperature (Fig. 6.19b). The measurements have been made on heating for a sample that was first quenched rapidly down to room temperature. On reheating, x remains constant at about 0.13 because the relaxation kinetics for the ordering reaction is much too slow. It is near 800 K that relaxation sets in, when x begins to decrease and comes closer to the equilibrium value. On the timescale of the experiment, equilibrium is actually reached at about 900 K, above which x increases steadily. Conversely, the data of Fig. 6.19b show that the partially disordered Fe/Al cation distribution over the two kinds of crystallographic sites has been frozen in at about 980 K during the initial quench of the sample. This temperature is the analog of a fictive temperature and the 800-880 K interval is thus the range of a glass-like transition. The existence of this transition results in deviations from Nernst's law. The residual entropy is 7.4 and 6.5 J/mol K for the quenched ($x = 0.134$) and an annealed sample ($x = 0.112$), respectively.

Problems

1. Plot schematically the variations of C_p, H and G on both sides of the transition temperature for first-order, continuous and kinetic transitions.

2. With the van der Waals equation of state, expand $P - P_c$ in powers of density and temperature and demonstrate that the conditions of criticality allow the low-order terms to be written $P - P_c = P_{01}\Delta T + P_{11}\Delta T\ \Delta\rho + P_{30}(\Delta\rho)^3 + ...$ Finally, show that the critical exponents δ, β, γ and α are equal to 3, 0.5, 1 and 0, respectively.
3. Starting from the fact that both S and V are continuous at T_c, show that the pressure dependence of a *second-order* transition is given by $dT/dP = \Delta C_p/TV\Delta\alpha$ and $dT/dP = \Delta\alpha/\Delta\beta_T$, respectively. These expressions are known as Ehrenfest relations. Why are they in fact irrelevant for the *continous* transitions that are actually observed? Would they be applicable to the glass transition? Experience shows that only one of these equations predicts rather well the pressure dependence of T_g as measured under given conditions. Given the existence of Kauzmann's paradox, explain which one it is.
4. The pressure dependence of the transition temperature is given by the Clapeyron equation for first-order transitions. Why would we have two expressions for a second-order transition, and three for a third-order transition? Derive these three equations.
5. *Tricritical* phase transitions can be described by a Landau potential of the form $G(\eta) = a(T - T_c)\eta^2 + d\eta^6$. With regard to continuous and first-order transitions, what does mean the fact that the c coefficient of the potential is zero? What should be the sign of the d coefficient to ensure that this potential has a minimum for the low-temperature phase? How does the order parameter vary as a function of temperature ? Calculate the variations of S, H and C_p as a function of T, and compare the results with those derived in 6.2e for continuous and first-order transitions. How many phases coexist at the critical point? What is the variance of the equilibrium at this point, and how many components are needed for such a transition to be observed?
6. Why is it impossible to imagine an analog of Kauzmann's paradox for the volumes of the crystalline and liquid phases of the same substance?
7. The viscosities of molten boron oxide and silica are given as a function of temperature by $\log\eta$ (Pa.s) $= -1.20 + 1777/(T - 415)$ and $\log\eta$ (Pa.s) $= -8.71 + 2.8610^3/(T - 530)$, respectively. For a cooling rate of 10 K/min, the glass transition temperatures are 1480 and 550 K, respectively. For both substances, calculate the glass transition temperatures for cooling rates of 1 and 1000 K/min. For which substance is the calculation made for a rate of 1000 K/min at best approximate? (N.B. examine for this purpose the variation with T of the activation energy for viscous flow.)
8. The production of flat glass display panels (active matrix liquid crystal display or plasma display panels) requires various treatments in the glass transition range during which the ficitve temperature of the material may vary. A silicate glass used for such applications has a density of 2.76 g/cm^3 at room temperature, a thermal expansion coefficient $\alpha_g = 8.4\ 10^{-6}$ K^{-1} below T_g and $\alpha_l = 14\ 10^{-6}$ K^{-1} above T_g. What is the highest possible variation of the fictive temperature $\overline{T}$ of the glass for the relative change in the room-temperature density to remain lower than the limit of 10^{-5} set by the display panel applications?
9. The bulk modulus of a calcium aluminosilicate is 210 kbar in the molten state and 600 kbar for the glass phase. All other things being equal, what would be the density difference between two samples of this material vitrified at 1 bar and at 30 kbar ? Assume the bulk moduli to be independent of T.
10. It is in fact possible to measure shear wave velocities (c_t) in viscous liquids. Explain why Brillouin spectrocopy (see Appendix B) must be used for this purpose and indicate whether or not such measurements refer to a fixed structure. Indicate how one can then determine the configurational part of the compressibility of the liquid by combining such measurements of c_t and c_l (the longitudinal velocity) with c_l measurements made with ultrasonic methods.

7
CRITERIA FOR CHEMICAL EQUILIBRIUM

This chapter constitutes a brief introduction to chemical equilibrium. After having defined the affinity, we will demonstrate the equivalence of the criteria for chemical equilibrium newly drawn from the affinity with those obtained previously. A discussion of stability of chemical equilibrium with respect to compositional fluctuations will then allow us to introduce the graphical representations of the Gibbs free energy that will be used for discussing phase equilibria in following chapters.

7.1 CHEMICAL REACTIONS

7.1a ADVANCEMENT OF A REACTION

Let us consider chemical reactions in a more specific way by taking again the example of the reaction

$$4\,H_2 + CO_2 \rightleftharpoons CH_4 + 2\,H_2O, \tag{7.1}$$

which will be noted in the general form $\sum \nu_i M_i = 0$, where ν_i is the stoichiometric coefficient of species i. Let us start with a mixture of H_2 and CO_2 to be transformed into CH_4 and H_2O. Each time one mole of carbon dioxide reacts, four moles of hydrogen are consumed whereas one mole of methane and two moles of water are formed. Because the variations of the number of moles are correlated

$$dn_1/\nu_1 = dn_2/\nu_2 = dn_3/\nu_3 = \ldots = d\xi, \tag{7.2}$$

the progress of the reaction can be characterized by a single variable, the advancement ξ. Before the reaction begins to proceed, ξ is zero, and it becomes unity when the transformation is completed. If the numbers of moles of the various components are initially known, then ξ characterizes unambiguously the state of the system. As already noted, the advancement of the reaction is thus a state function and use of ξ incidentally results in somewhat simpler expressions for exact differentials. For example, instead of

$$dU = (\partial U/\partial S)_{V,n_i} dS + (\partial U/\partial V)_{S,n_i} dV + \sum (\partial U/\partial n_i)_{V,S,n_j} dn_i, \tag{7.3}$$

we will write

$$dU = (\partial U/\partial S)_{V,\xi} dS + (\partial U/\partial V)_{S,\xi} dV + (\partial U/\partial \xi)_{V,S} d\xi. \tag{7.4}$$

7.1b AFFINITY

For a reaction proceeding at constant P and T, how does the Gibbs free energy of a system vary? We know that this variation is

$$dG = \sum \mu_i dn_i, \tag{7.5}$$

which (7.2) enables us to rewrite as

$$dG = \sum \mu_i \nu_i d\xi = d\xi \sum \mu_i \nu_i. \tag{7.6}$$

From (2.37), we also know that this Gibbs free energy variation is proportional to the entropy created dS'

$$TdS' = -\sum \mu_i dn_i = -d\xi \sum \mu_i \nu_i = Ad\xi, \tag{7.7}$$

where A is the *affinity* of the reaction. At equilibrium, the entropy created dS' is zero and the equilibrium state is characterized by

$$A = -\sum \nu_i \mu_i = 0. \tag{7.8}$$

In the case dealt with above, this additional state function contributes nothing new with regard to the Gibbs free energy. Hence, why was the affinity introduced, by De Donder (1872-1957)? The major reason is that, as described in 2.4c, the most general evolution criterion is $dS' > 0$.

For *any* chemical transformation, we will thus be able to write

$$A = -\sum \nu_i \mu_i \geq 0, \tag{7.9}$$

or

$$TdS' = -\sum \mu_i dn_i = Ad\xi, \tag{7.10}$$

without limiting ourselves to transformations taking place at constant S and V, S and P, T and V, or T and P. In these particular cases, recall that we have

$$\begin{aligned} A &= -\sum \mu_i \nu_i = -(\partial \Delta U/\partial \xi)_{S,V} = -(\partial \Delta H/\partial \xi)_{S,P} \\ &= -(\partial \Delta F/\partial \xi)_{T,V} = -(\partial \Delta G/\partial \xi)_{T,P} \geq 0. \end{aligned} \tag{7.11}$$

For chemical reactions, the most general evolution criterion is thus (7.9). At equilibrium, the affinity is zero; as long as the affinity is positive, the reaction can proceed spontaneously.

From equations (7.11) and the definition (7.2) of the degree of advancement of a reaction, other expressions are drawn, which could have been used to define the affinity. For example

$$dU = TdS - PdV - Ad\xi. \tag{7.12}$$

In any case, interest in the affinity essentially lies in that it allows one not to consider only the final, equilibrium state, contrary to the situation where one uses the minimum criteria given by U, H, F and G. Throughout any process, (7.10) accounts simply for the creation of entropy, and thus for the irreversible nature of the chemical reaction. With the affinity, it is possible to deal with situations far from equilibrium. This goal of the *thermodynamics of nonequilibrium processes* is however beyond the scope of this book.

Nonetheless, let us have a glance at the rate of a reaction, ν, which is related to ξ by

$$\nu = d\xi/dt, \tag{7.13}$$

where t is time. Because the second principle implies that TdS'/dt is positive, ν and A have necessarily the same sign

$$TdS'/dt = Ad\xi/dt \geq 0. \tag{7.14}$$

This relation is known as the De Donder inequality. While ν must be zero when A is zero, ν can also be zero when A is not. This is the case of a metastable equilibrium: the kinetics of the reaction are zero even though the affinity is not zero.

It is readily observed that ice melts at 0°C. As already mentioned, however, in practice supercooling causes pure water to freeze at lower temperatures, generally between − 5 and − 10°C for large samples, and at at still lower temperatures for small ones. One could bracket more accurately the equilibrium freezing temperature by making longer and longer experiments to get rid of the kinetic phenomenon of supercooling. To prove that equilibrium has actually been reached experimentally, a much better idea would be to observe the freezing rate. It would then suffice to note what happens when small pieces of ice are added at the same temperature as the water. As long as ice grows, the temperature is lower than the freezing temperature because the rate of a reaction has the same sign as the affinity. It is at 0°C exactly that the rate of freezing vanishes, as does the affinity, before becoming negative at higher temperatures. In this simple way, the reversible melting temperature of 0°C would be determined rigorously.

7.2 EQUILIBRIA

7.2a STABILITY OF EQUILIBRIA

Whether an equilibrium is stable or metastable has no consequences for its thermodynamic treatment; the equilibrium criteria are the same in both cases. As an example, we will examine the Gibbs free energy of several forms of silica (SiO_2) plotted in Fig. 7.1 against temperature at room pressure. At low temperature, the phase with the lowest Gibbs free energy is quartz, the commonest crystalline form of SiO_2. At 1100 K, it is another polymorph, cristobalite, that becomes stable before melting at 2000 K. But the transformation of quartz into cristobalite is so sluggish that one can heat quartz above 1100 K and even observe its melting near 1700 K. Like quartz, molten SiO_2 is then metastable, being a supercooled liquid up to the melting point of the stable phase cristobalite at 2000 K. A 1700 K the metastable equilibrium between quartz and liquid is nonetheless characterized by $\Delta G = 0$, like the stable ones between quartz and cristobalite at 1100 K, and between cristobalite and liquid at 2000 K.

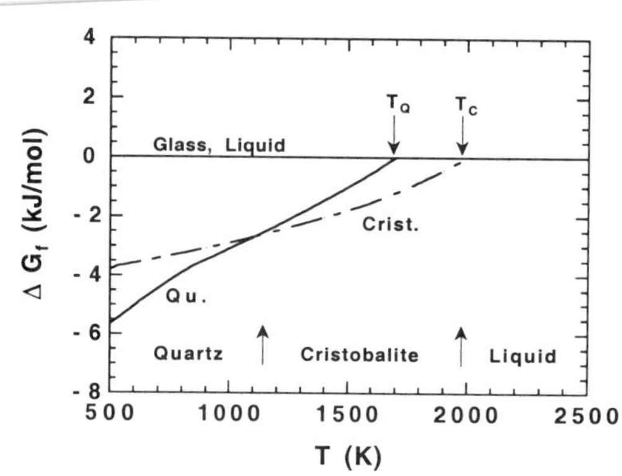

Figure 7.1. Gibbs free energy of formation of quartz and cristobalite from SiO_2 glass (up to 1480 K) and liquid (above 1480 K). The stability limits of the phases are delineated by arrows. These Gibbs-free energies also show the general feature that the melting temperature is necessarily lower for a metastable (quartz, T_Q) than for a stable (cristobalite, T_C) polymorph.

7.2b EQUILIBRIUM CRITERIA

We can now proceed to discussing in more depth the notion of equilibrium in a system. Any equilibrium must be simultaneously mechanical, thermal, and chemical. As implicitly considered until now, the simplest case is that where the influence of gravity can be neglected. Mechanical equilibrium is then attained when the pressure is the same in every phase α, β, γ, etc., of the system

$$P^{\alpha} = P^{\beta} = P^{\gamma} = \ldots = P. \tag{7.15a}$$

For thermal equilibrium, the criterion is that temperature be constant throughout the system

$$T^{\alpha} = T^{\beta} = T^{\gamma} = \ldots = T. \tag{7.15b}$$

Now, what is the criterion for chemical equilibrium? As already noted, the chemical potentials and numbers of moles are the tensions and extensities, respectively, of chemical energy. The equilibrium criterion is thus that the chemical potential of a component should not vary from one phase to another in the system. For a component i partitioned between phases α, β, γ, etc., we have

$$\mu_i^\alpha = \mu_i^\beta = \mu_i^\gamma = \ldots \tag{7.15c}$$

and N equations of the form (7.15c) must be written for an N-component system.

As introduced in this way, the criterion (7.15c) can be accepted as a postulate analogous to (7.15a-b). We hasten to note, however, that (7.15c) is strictly equivalent to the criteria deduced in chapter 2 from the second principle. For transformations at constant P and T, for instance, (7.15c) is equivalent to the criterion of a Gibbs free energy minimum. To prove this simply, let us note that, if (7.15c) were not obeyed at equilibrium, there would exist a phase where the chemical potential of a component would be higher than in other phases. By transferring this component from this phase to another, we could make the Gibbs free energy of the system decrease. Hence, the system would not have been initially in equilibrium. Note finally that we have already demonstrated in 7.1b that a Gibbs free energy minimum is tantamount to a zero affinity: to ensure that further advancement of the reaction does not lower the Gibbs free energy, it is necessary and sufficient that the affinity be zero.

In practice, the choice of a particular equilibrium criterion is essentially a matter of convenience. For transformations at constant P and T, looking for the Gibbs free energy minimum is well suited for equilibrium calculations in complex gaseous mixtures. For heterogeneous equilibrium between solutions, (7.15c) is in general more convenient. In most simple cases, however, the criterion of zero affinity is to be preferred. An important case is that of a system in which several reactions take place simultaneously. The total affinity is then the sum of the affinities of each reaction and equilibrium is reached when the affinities of all reactions have become zero.

The criterion (7.15b) constitutes what is often called the zeroth principle of thermodynamics, namely, that, if two bodies are in thermal equilibrium with a third one, they are themselves in mutual equilibrium. In fact, equality (7.15b) results only from the fact that temperature is a tension. Inasmuch as (7.15a) is not called the zeroth principle of mechanics, the use of a zeroth principle of thermodynamics is misleading in that it would give temperature an unwarranted peculiarity with respect to pressure or chemical potentials.

A more rigorous equilibrium criterion would be that a system has reached equilibrium when no extensity transfer can result from differences in tension factors. In this manner, we would modify (7.15a) in order to include systems in which there is necessarily a pressure gradient at equilibrium. An example is a column of water where the pressure has to be higher at the bottom than at the top, which does not prevent the system from being in equilibrium because no mass transfer originates in this pressure difference. To include the effects of gravitation in the internal energy balance, let M_i be the mass of component i and Γ the gravitational potential at a given height. If the numbers of moles of the components change through some transformation, the internal energy varies by

$$dU = dQ + dW + \sum \mu_i dn_i + \sum \Gamma M_i dn_i = TdS - PdV + \sum(\mu_i + \Gamma M_i)dn_i. \tag{7.16}$$

The differentials of the enthalpy, and Helmholtz and Gibbs free energies as well, are modified in the same way. We would obtain the equilibrium criteria by replacing the usual expressions for chemical potentials by $\mu_i + \Gamma M_i$. Application of these relations to concrete problems is less simple, however, and we refer the reader to the book by Guggenheim, for instance, for a complete treatment.

7.2c EQUIVALENCE OF THE CRITERIA FOR CHEMICAL EQUILIBRIUM

Consider once more the case of ice-water equilibrium at constant P and T. The system is made up of a single component, H_2O, and its Gibbs free energy is minimum at equilibrium. If this Gibbs free energy cannot be lowered, this means that the affinity of the ice-water reaction is zero, and that the chemical potential of H_2O is the same in ice (i) and water (w)

$$\mu^i_{\mathrm{H_2O}} = \mu^w_{\mathrm{H_2O}}. \tag{7.17}$$

We find again the equivalence between the three equilibrium criteria of Gibbs free energy minimum, zero affinity and equality between the chemical potentials of the components.

Let us complicate the situation a little bit by lowering the temperature and adding some NaCl until ice begins to melt. Two phases, the aqueous solution (w) and ice are still present, but there are now two components, NaCl and H_2O. Two reactions have to be considered, which must be simultaneously at equilibrium, namely, water and NaCl exchanges between ice and the solution. The affinity must be zero for both reactions, whence

$$\mu^i_{\mathrm{H_2O}} = \mu^w_{\mathrm{H_2O}}, \qquad \mu^i_{\mathrm{NaCl}} = \mu^w_{\mathrm{NaCl}}. \tag{7.18}$$

This is the same result as given by (7.17), which can be generalized immediately to the case of an N-component system with ϕ phases for which (7.15c) has been given.

We have taken NaCl and H_2O as components. Now, it has long been known that NaCl dissociates in aqueous solutions into Na^+ and Cl^- ions. We also know that water dissociates partially into H^+ and OH^- ions, which coexist with H_2O molecules. Is there, thus, some ambiguity as to the choice of the components? The answer is no, for components can be arbitrarily selected as long as they account for the chemical composition of the system. By taking Na^+, Cl^-, H_2O, H^+, OH^-, etc., we have to consider as many equations (7.15c) as there are components. But all such sets of equations are strictly equivalent because we have made no assumption about the real chemical constitution of the system to arrive at (2.22a) and state that the Gibbs free energy of a system is the sum of the chemical potentials of its components (weighted by the numbers of moles).

In 2.2c, we have already seen that the Gibbs free energy can be expressed in many ways. To give only three limiting cases for an NaCl aqueous solution, we can write

$$G = \sum n_i \mu_i = n_{\mathrm{NaCl}} \mu_{\mathrm{NaCl}} + n_{\mathrm{H_2O}} \mu_{\mathrm{H_2O}} \tag{7.19a}$$

or

$$G = n_{\mathrm{Na^+}} \mu_{\mathrm{Na^+}} + n_{\mathrm{Cl^-}} \mu_{\mathrm{Cl^-}} + n_{\mathrm{H_2O}} \mu_{\mathrm{H_2O}}, \tag{7.19b}$$

or

$$G = n_{Na^+}\mu_{Na^+} + n_{Cl^-}\mu_{Cl^-} + n_{H_2O}\mu_{H_2O} + n_{H^+}\mu_{H^+} + n_{OH^-}\mu_{OH^-}. \tag{7.19c}$$

The only condition to be met is that the total number of moles of components represents correctly the composition of the system. Why should one of these equations be preferred over the others? Once more, the choice is dictated by the question raised and the available information on the phases present in the system. Given the strong influence of ionic dissociation on solution properties, (7.19b) must in fact be preferred over (7.19a).

A last remark. The equilibrium criterion (7.15c) requires that all components be present in all phases. In particular, this implies that there is necessarily some salt in ice in equilibrium with an aqueous solution. Such amounts could be very small, so small as not to be measurable with the most sensitive analytical methods, but there should be enough salt to raise its chemical potential to that of NaCl in the solution where its concentration is much higher. The important conclusion is that, at equilibrium, a phase cannot be really pure in a system with several components and several phases. An example of low mutual miscibility is shown in Fig. 7.2 with the system water-styrene (C_8H_8). In practice, we may nonetheless speak of pure phases whenever impurities have such low concentrations that their contributions to the total Gibbs-free energy of the phase are negligible. In this case, they do not need to be taken into account in equilibrium criteria. This would be the case of styrene in water (Fig. 7.2a), but much less so of water in styrene (Fig. 7.2b).

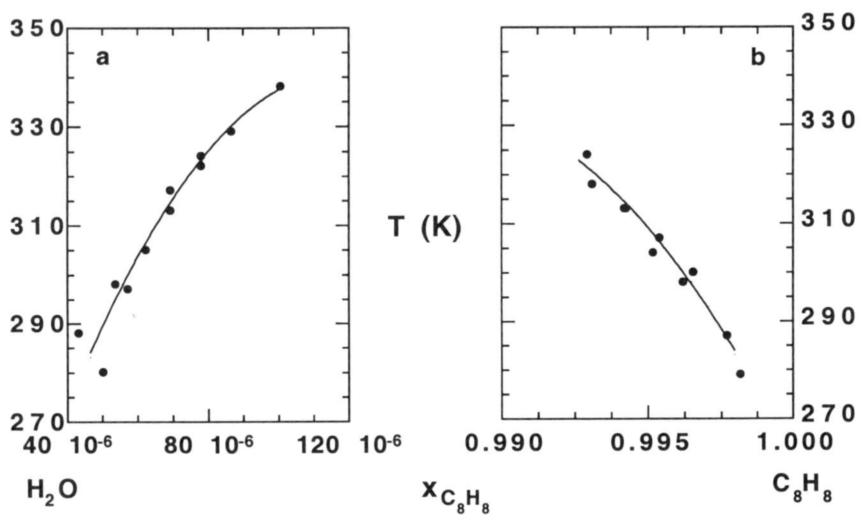

Figure 7.2. Low mutual solubility of styrene (C_8H_8) in water (a) and of water in styrene (b). The two curves in fact represent two opposite branches of a solvus, see Fig. 7.8. Note the difference between the two scales used. Data from Lane (1946).

7.3 PHASE SEPARATION

7.3a GIBBS FREE ENERGY AND UNMIXING

To illustrate differently the mutual relationships between components, phases and equilibrium stabilities, we will discuss in this section the important issue of phase separation and will do it by introducing graphical representations of the Gibbs free energy. For the sake of simplicity we will deal with binary solutions and distinguish several limiting cases for the variation of the Gibbs-free energy with the composition of a homogeneous solution. As a reference state, we will first define a *mechanical mixture* as the coexistence of several non interacting phases, each one keeping its own individuality. Pieces of sugar and salt constitute such a mechanical mixture since these two phases do not react to form a salt-sugar solid solution. Likewise, water and mercury can be said to be immiscible because the extent of their mutual solubility is negligible (solubility of Hg in water at room P and T: 0.2 ppm, or 0.2 mg/liter). As an extensive variable, the Gibbs free energy is proportional to the quantities of the phases present in a mechanical mixture. For a binary mechanical mixture, G thus varies linearly with composition (Fig. 7.3).

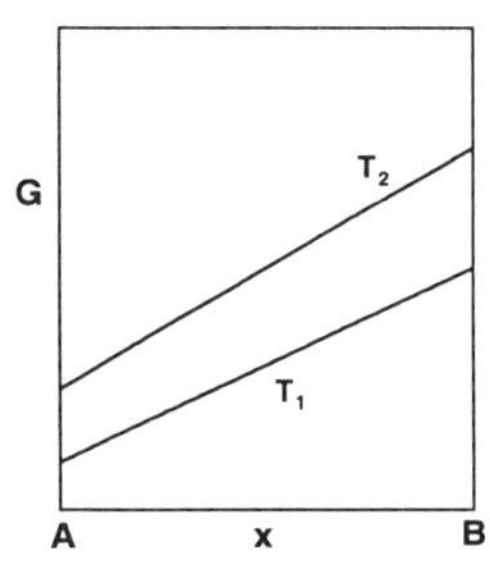

Figure 7.3. Gibbs free energy of a mechanical mixture at two temperatures. Note that the slope of the straight line changes because because the temperature dependence of the Gibbs-free energy should be different for the two endmembers A and B.

Let us now turn to the Gibbs free energy of a real solution. In the following chapters, we will see in various instances how it varies as a function of temperature and composition through a balance between the enthalpy and entropy terms. For illustration purposes, we will just examine two limiting and one intermediate cases for the composition dependence of G at a given temperature. In the first one, the Gibbs free energy varies monotonically with composition as indicated in Fig. 7.4. At any point, G is lower for the solution than for the mechanical mixture of A and B: a single phase is stable, the solution.

Let us then examine the example shown in Fig. 7.5 where in a wide composition interval the stable state corresponds to a separation of the system in two phases. From both endmembers, the Gibbs free energy of the solution first decreases when the other component is added. Close to the points denoted N and N', the Gibbs free energy begins to increase when further addition of the second component is made. Between the points N and N' it shows a local maximum M surrounded by the two inflexion points S and S'. At any point, however, we will assume that the Gibbs free energy of the solution is lower than that of the mechanical mixture of A and B. (From the limiting cases treated in this section, it would be possible to extend this analysis to systems

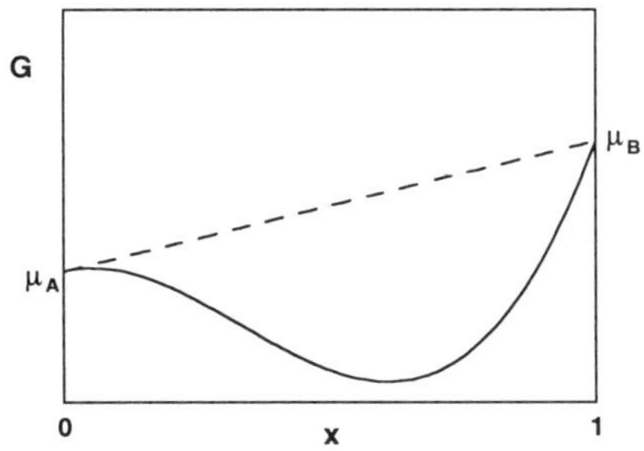

Figure 7.4. Gibbs free energy of a solution stable in the whole composition range of a binary system. Mechanical mixture shown for comparison by the dashed curve.

for which this assumption is not verified in one or several composition domains.) A preliminary division of the system into three regions follows immediately. For compositions between A and N and between N' and B, the solution represents, as in the preceding case, the state with the lowest energy. On the contrary, for any composition M between N and N', the Gibbs free energy of the assemblage of the two phases having the compositions N and N' is lower than that of the initial solution M because it would plot on the line NN'. A *phase separation*, or *unmixing*, must therefore take place.

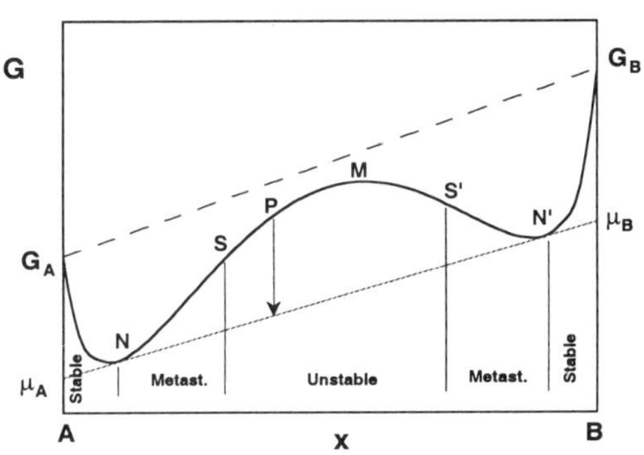

Figure 7.5. Gibbs free energy of a binary solution stable near the endmembers of the solution, unstable at the middle of the composition range, and metastable in between these stable and unstable domains.

How can we determine the compositions x and x' of the *conjugate* phases N and N'? A chemical potential is a partial molar property. The chemical potential of A is represented by the intercept on the A axis of the tangent to the Gibbs free energy curve (see 3.2c). Likewise, the chemical potential of B is represented by the intercept on the B axis of the tangent to G. At equilibrium, the chemical potentials of A and B must be equal in the two phases ($\mu_A = \mu'_A$, $\mu_B = \mu'_B$) whose equilibrium compositions are thus determined by the two points of the G curve that have a common tangent (Fig. 7.5). For this reason, these points N and N' should be distinct from the local minima of G, even though they are usually very close to them.

The next question to be raised is that of the mechanisms through which unmixing takes place. In a fluid, the intrinsic mobility of the components gives rise to short-lived local fluctuations in composition whereby any local enrichment in a component is necessarily accompanied by a concentration decrease in an adjoining region (Fig.

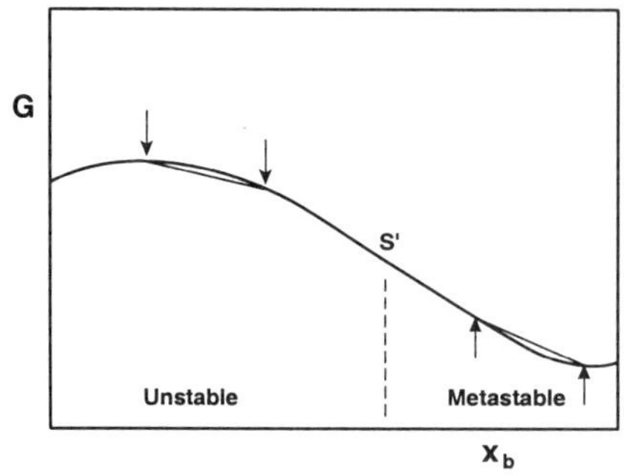

Figure 7.6. Effects of composition fluctuations on the stability of a solution as indicated by a close up of Fig. 7.5 in the vicinity of point S'. Depending on the concavity of the Gibbs-free energy curve, the composition changes induced by the fluctuation(shown as arrows) produced initially either a decrease (left) or an increase (right) of the Gibbs free energy of the solution.

7.6). Analogous composition fluctuations can also develop in a solid at high temperatures, albeit more slowly because diffusion processes are generally much slower in solids than in liquids. In both cases, such fluctuations represent the very first step of an unmixing process, and this leads to distinguishing three different zones in the composition range NN' where unmixing develops. Between S and S', the downward concavity of G has the effect that any composition fluctuation of the initial solution leads to a Gibbs free energy decrease of the system (Fig. 7.6). The initial single phase state is by definition unstable. Between N and S and between S' and N', G has an upward concavity. Any fluctuation then causes an initial Gibbs free energy increase (Fig. 7.6). To produce a phase separation, this Gibbs free energy barrier must be overcome before the line NN' can be reached. The solution is thus in a metastable state.

Consider finally the case represented in Fig. 7.7, which is characterized by a maximum of G between the two endmembers of the system. For any composition M, the Gibbs free energy of the solution would be higher than that of a mechanical mixture of the two components. For any composition, the stable phase assemblage under the given P and T conditions will be a mechanical mixture of the two almost pure endmembers. The qualifier "almost" is necessary here because, according to the discussion of 7.2c, two phases in equilibrium cannot be *absolutely* pure. This implies that the Gibbs free energy of the solution always have local minima near the endmembers, analogous to those of Fig. 7.5. These minima are often so close to the pure endmembers, however, that their existence can be neglected in practice.

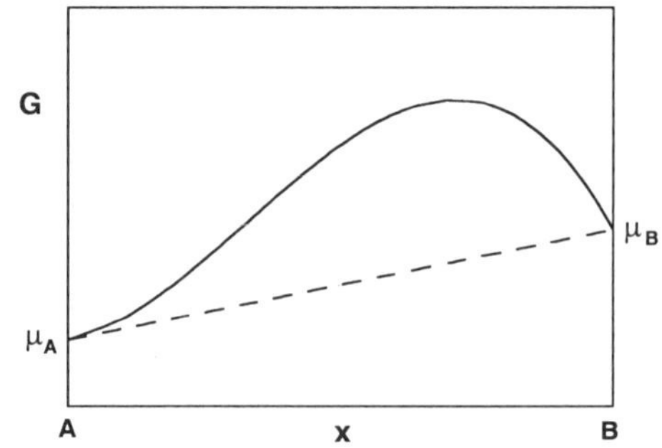

Figure 7.7. Gibbs free energy of a binary solution unstable at any point except in the immediate vicinity of the endmembers where, albeit not apparent on this plot, the situation of Fig. 7.5 must prevail.

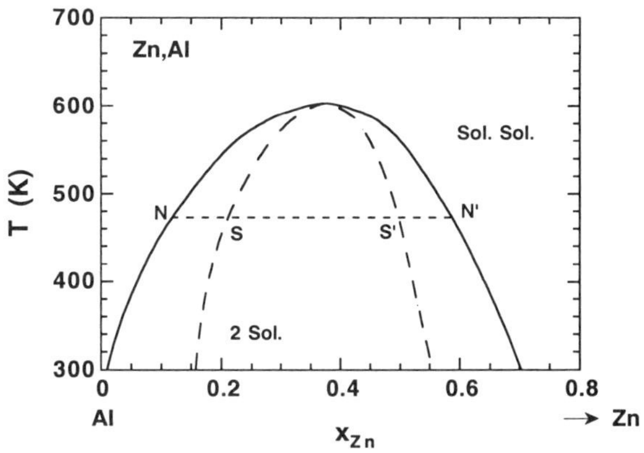

Figure 7.8. Aluminum-zinc phase separation: stability limit between a single (Sol. Sol.) and two solid (2 Sol.) solutions shown by the solvus (solid curve); spinodal represented as the dashed curve. Diagram from Rundman and Hilliard (1967), modified by Cahn (1967).

7.3b SOLVUS AND SPINODAL

The Gibbs free energies of the endmembers and the separated phases change with pressure and temperature so that the boundaries between the stable, metastable and unstable regions shift when T and P vary. What are the mathematical criteria for calculating the locus of these boundaries? We have already remarked that the boundary between the metastable and unstable regions corresponds to the inflexion points of the curves $G = f(x)$. In the pressure-temperature-composition space, the locus of these points S and S' is thus a surface, the *spinodal*. In a binary system at constant pressure, this surface reduces to a curve which is generally bell shaped as shown in Fig. 7.8 for the Al-Zn system. The equation of the curve is

$$(\partial^2 G/\partial x^2)_{P,T} = 0. \tag{7.20}$$

This partial derivative is negative in the unstable region and positive in the metastable region.

The boundary between metastability and stability defines a still more important surface, the *solvus*, sometimes termed the *binodal*, which is tangential to the spinodal at the homogenization temperature (Fig. 7.8). To calculate it, it suffices to make use of the equality between the chemical potentials of the components in the two phases N and N' of compositions x and x', respectively (Fig. 7.5). Between the points N and N', the slope of the common tangent to the Gibbs-free energy curve is given by the variation of G

$$G - G' = (x' - x)(\partial G/\partial x)_{T,P}. \tag{7.21}$$

The point where the solvus is tangential to the spinodal is a critical point since it is the limit beyond which the distinction between the two phases can no longer be made. The *critical temperature* represents the temperature at which the two inflexion points of the Gibbs-free energy curve, which characterize the spinodal, merge. Because the inflexion points are defined by (7.20), the critical point is determined by

$$(\partial^3 G/\partial x^3)_{P,T} = 0. \tag{7.22}$$

Analytically we cannot go beyond this result yet, for the form of a spinodal and a solvus obviously depend on the manner in which G varies with temperature, pressure and especially composition. As will be described in chapter 10, specific models have thus to be established to account for these dependences in order to calculate (7.20) - (7.22).

7.3c PHASE SEPARATION MECHANISMS

The basic differences between phase separation outside and within the spinodal are summarized in Fig. 7.9. Between the spinodal and the solvus, the finite composition difference between the unmixed phases requires to take into account an additional surface energy term in the Gibbs free energy. If a spherical droplet of radius r of phase α appears within phase β, the Gibbs free energy change is

$$\Delta G = 4/3\ \pi r^3(\mu_\alpha - \mu_\beta) + 4\pi\sigma r^2, \tag{7.23}$$

where μ is here the chemical potential per volume unit and σ is the surface tension, which is positive. For small droplets both terms of (7.23) are thus generally positive and the homogeneous phase is metastable because all small changes would raise the Gibbs free energy. Unmixing is thermodynamically possible only for a minimum size such that the first term becomes negative and sufficient to overcome the interfacial Gibbs free energy barrier represented by the second term. This happens through a *nucleation* stage. Only a fraction of the nuclei that form through chance composition fluctuations can reach the minimum size beyond which a second step of growth takes place. This is a *thermally activated* process, for the interfacial barriers are more easily overcome at high temperatures. Nucleation and growth are *heterogeneous* processes because the presence of impurities, walls and other nucleating agents effectively reduce the height of these barriers.

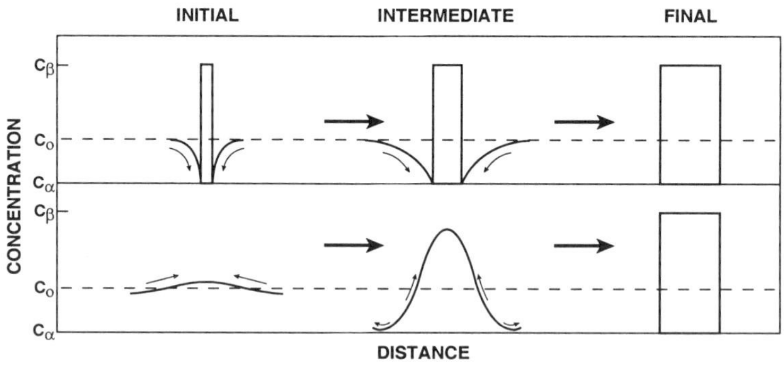

Figure 7.9. Concentration profiles illustrating the differences between nucleation and growth (top) and spinodal decomposition (bottom) mechanisms for phase separation. As indicated by the small arrows, note the opposite directions of the diffusional fluxes with respect to the concentration gradients (redrawn from Cahn, 1968).

The spinodal decomposition is in contrast an *homogeneous* process because it involves only the unmixing phase. The diffusional flux acts in this case against the concentration gradient and the growth of the unmixed phases is limited by the rate of this *uphill* diffusion, which takes place against the concentration gradient. Interfacial energy comes into play only when a significant composition difference has developed, i.e., when the sizes of the unmixed phases are already large enough for this energy to be minor with respect to the chemical potential difference in (7.23). In order to minimize the Gibbs free energy, interfacial energy just promotes the coalescence of each phase. In solids, the continuous nature of spinodal decomposition restricts the process to crystals whose structure is closely related to that of the original phase. The simplest case is of course that of solid solutions whose endmembers have the same crystal structure.

7.3d UNMIXING AS A CRITICAL PHENOMENON

We will conclude this chapter with a closer look at phase separation as a critical phenomenon. The example to be taken to illustrate this aspect is the system water-isobutyric acid ($C_4H_8O_2$), which presents a solvus analogous to that shown in Fig. 7.8. The densities of the coexisting liquids in the vicinity of the homogenization temperature are plotted in Fig. 7.10a. Their variations are similar to those of a liquid and its vapor near the critical point of fluids discussed in 6.2b. The mean density of the two liquids also follows a rectilinear law (Fig. 7.10a) whereas the density difference is again taken as an order parameter in the power law equation

$$\rho_1 - \rho_2 = B(1 - T/T_c)^\beta. \tag{7.24}$$

From the results shown in Fig. 7.10b one finds $B = 0.53$, an exponent $\beta = 0.309$ (4) and a critical unmixing temperature $T_c = 299.144$ K. Although the exponent β differs

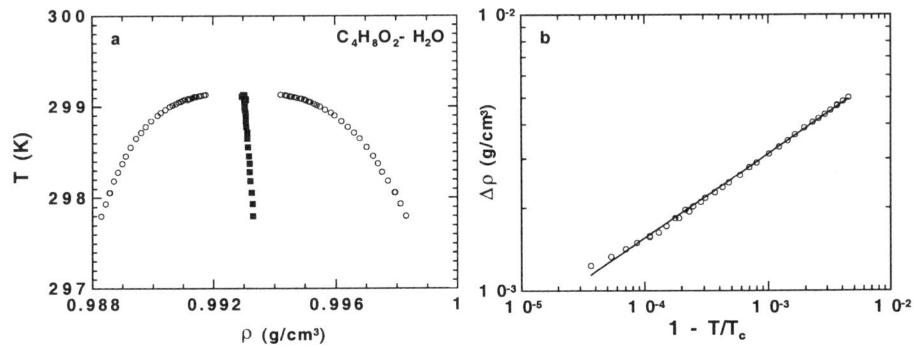

Figure 7.10. (a) Densities of coexisting liquids (circles) and mean density of the liquids (squares) in the system water-isobutyric acid near the critical unmixing temperature; (b) Log-log plot of the density difference between coexisting fluids against $1 - T/T_c$ showing observance of the power law (7.24). Data from Greer (1976).

from the "universal" value 0.355 (7) of fluids, it is nevertheless close to 1/3 like the values found for other transitions for which the space dimension is 3 and the order parameter a scalar, see 6.2c.

Problems

1. The polymorph of germanium dioxide GeO_2 with a tetragonal symmetry is stable up to the temperature T_t = 1319 K, at which the enthalpy of transition to the hexagonal polymorph is ΔH_t= 19.9 kJ/mol. This polymorph melts at T_h = 1389 K with an enthalpy of fusion ΔH_h = 16.8 kJ/mol. What are the temperature and enthalpy of metastable melting of the tetragonal form? (The C_p differences between the polymorphs and the liquid will be neglected.) Check that the metastable polymorph has a lower melting temperature and a higher enthalpy of fusion than the stable form. By what amount would the enthalpy of fusion of the quadratic form change if T_t were changed by 10 K? Does knowledge of a temperature of transition allow one to determine accurately an enthalpy of transition?
2. The CO_2 pressure is about 5 bars in a bottle of sparkling wine. When a bottle is open, why does the liquid not boil, but forms instead lines of bubbles from a few points on the glass surface?
3. Show that, in a binary system, $(\partial\mu_1/\partial x_2)_{T,P} = -x_2(\partial^2 G/\partial x_2^2)_{T,P}$ and that the criteria characterzing a critical mixing temperature can be written $(\partial\mu_1/\partial x_2)_c = -x_2(\partial^2\mu_1/\partial x_2^2)_c = 0$.

8 EQUILIBRIUM AND CHEMICAL POTENTIALS

The ideal gas equation of state gives rise to a very simple expression for the chemical potential as a function of T and P. This expression is so convenient for calculating chemical equilibria that we will define new variables, the fugacity and activity, in order to keep the same analytical form for the chemical potential of real gases and condensed phases. This will be the main theme of this chapter, which will have as a background the importance of created entropy as a driving force in chemical reactions. After having introduced the mass action law for ideal gases, we will see how it can be used for real gases and condensed phases. The notion of ideal solution will also be presented, along with the usefulness of the Gibbs-Duhem equation for calculating activities in the case of binary systems.

8.1 PREAMBLE

8.1a INTEGRATION PATHS OF CHEMICAL POTENTIALS

For the time being, we will leave apart the influence of composition to examine first the variation of the chemical potential of a substance between an initial (P_0, T_0) and a final (P, T) state. In practice, the simplest paths that can be followed are represented as IAF and IBF in Fig. 8.1. Along these paths the variation of the chemical potential is obtained through integration of

$$d\mu = -SdT + VdP. \tag{8.1}$$

Since a chemical potential is a state function these two paths are theoretically equivalent. For experimental reasons, however, the IAF path has to be preferred. To justify this choice without any ambiguity, we will put in parentheses the pressure or temperature at which S or V are integrated as a function of T or P.

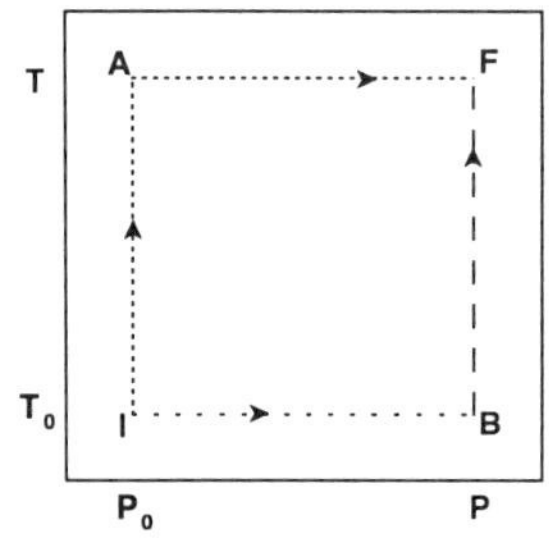

Figure 8.1. *P,T* paths for integrating chemical potentials.

Along the path IAF, we integrate (8.1) first with respect to T, and then to P

$$\mu = \mu(T_0, P_0) - \int_{T_0}^{T} S(P_0)dT + \int_{P_0}^{P} V(T)dP. \tag{8.2}$$

The first integration is straightforward with an adequate analytical expression for S, which depends itself on the coefficients of the empirical equation used for C_p. The second integral, with respect to P, is also readily performed if the equation of state of the phase is known.

Along the other path IBF, we have

$$\mu = \mu(T_0, P_0) + \int_{P_0}^{P} V(T_0)dP - \int_{T_0}^{T} S(P)dT. \tag{8.3}$$

Integration with respect to P is still easier than along the other path, for it suffices to know the room-temperature compressibility but the difficulty raised by the IBF path is that the second integral requires knowledge of C_p at the elevated pressure P. Now, with fluids as possible exceptions, accurate calorimetric measurements cannot be generally performed at high pressures. In most cases, one must calculate the pressure dependence of C_p from the Maxwell relation (3.6d) as follows

$$\left(\frac{\partial C_p}{\partial P}\right)_T = \frac{\partial}{\partial P}\left[T\left(\frac{\partial S}{\partial T}\right)_P\right]_T = T\frac{\partial}{\partial T}\left[\left(\frac{\partial S}{\partial P}\right)_T\right]_P = -T\left(\frac{\partial^2 V}{\partial T^2}\right)_P. \tag{8.4}$$

As the reader may care to check, it is only for ideal gases that C_p does not depend on pressure. In general, the variations of C_p with P are determined from the thermal expansion coefficient and its temperature derivative, which are both derived from the equation of state. Altogether it is simpler to skip this exercise and use directly the equation of state for integrating VdP, thus following the IAF path where only the room-pressure C_p has to be known.

8.1b STANDARD STATE AND CHEMICAL POTENTIAL

As will become apparent below, it is in fact more convenient to take as a reference state the phase of interest at T and P_0 rather than at T_0 et P_0 to calculate chemical potential variations. For a pressure P_0 of 1 bar, the phase is said to be in its *standard state* and its *standard chemical potential* is

$$\mu^\circ = \mu(T_0, P_0) - \int_{T_0}^{T} S dT. \tag{8.5}$$

With this equation, (8.2) is rewritten more simply as

$$\mu = \mu^\circ + \int_{P_0}^{P} V(T) dP. \tag{8.6}$$

The standard potential does not depend on pressure because the *standard pressure* P_0 = 1 bar has been fixed once and for all. In contrast, it depends strongly on temperature. Hence, we stress that μ° is not a constant.

8.1c STANDARD AFFINITY OF REACTION

As defined by (7.8), an affinity of reaction is a difference between the chemical potentials of products and reactants. From (8.6) it is natural to split it in two terms. Here we will be interested only in the first one, the difference between standard potentials, which is termed the *standard affinity* of reaction and is written

$$-A^\circ = \sum \nu_i \mu_i^\circ \tag{8.7}$$

where ν_i is the stoichiometric coefficient of the species or component i participating in the reaction.

To illustrate how a standard affinity is calculated, consider the case of equilibrium melting of a pure substance at constant P. From the definition of the Gibbs free energy, we have

$$\begin{aligned} -A^\circ &= \mu^{ol} - \mu^{oc} \\ &= \Delta H(T_f) + \int_{T_f}^{T} \Delta C_p dT - T[\Delta S(T_f) + \int_{T_f}^{T} \Delta C_p / T \, dT], \end{aligned} \tag{8.8}$$

where the superscripts l and c designate the liquid and crystal phases, respectively, ΔC_p is the heat capacity difference between the liquid and crystal, and $\Delta H(T_f)$ and $\Delta S(T_f)$ are the enthalpy and entropy of fusion at the melting temperature T_f. At the equilibrium temperature T_f, melting is reversible, whence $\Delta H(T_f) = T_f \Delta S(T_f)$. If ΔC_p could be neglected the standard affinity would reduce to

$$-A^\circ = \Delta H(T_f)(1 - T/T_f). \tag{8.9}$$

As shown by this equation, the affinity of the melting reaction is positive for $T > T_f$ and negative for $T < T_f$, consistent with the respective stabilities of the liquid and crystal phases. But, because the heat capacity of liquid is generally greater than that of the crystal, the approximation (8.9) is not valid over wide temperature intervals. A similar conclusion would apply to vaporization and sublimation equilibria.

8.2 IDEAL GAS

8.2a CHEMICAL POTENTIAL OF AN IDEAL GAS

With the ideal gas equation $V = RT/P$, (8.6) is easily integrated

$$\mu(T,P) = \mu^\circ + RT \int_{P_0}^{P} dV/V = \mu^\circ + RT \ln(P/P_0). \tag{8.10a}$$

This equation represents another way of stating $PV = RT$. Alternatively, it could be used to define an ideal gas. With bars as pressure units, the pressure of the standard state is unity and the chemical potential of an ideal gas at the pressure P and temperature T is finally written as

$$\mu = \mu^\circ + RT \ln P. \tag{8.10b}$$

In this case we emphasize that the actual pressure term in (8.10b) is P/P_0, which is a dimensionless number. When using Pascals as pressure units the pressure of the standard state is 10^5 Pa. Only (8.10a) can then be used. This should not be forgotten! In this respect, the use of bars is less risky since both (8.10a) and (8.10b) are correct.

8.2b IDEAL GAS MIXTURES

Let us now turn to ideal gas mixtures and put in contact n_1 and n_2 moles of two gases initially at the same pressures and temperatures in two adiabatic enclosures of volumes V_1 and V_2, respectively. After mixing, we observe that the pressure and temperature have not changed. Hence, the final volume V

$$V = V_1 + V_2 = n_1RT/P + n_2RT/P = nRT/P \tag{8.11}$$

obeys the ideal gas law, with $n = n_1 + n_2$. This is not a surprise, for the equation of state of an ideal gas depends only on the number of moles. Thanks to the lack of any specific parameter, the chemical potential (8.10) is the same for ideal gas mixtures as for pure ideal gases. Likewise, that no heat is exchanged while the gases are mixing directly results from the fact that the internal energy of an ideal gas is function of temperature only.

Nonetheless, we know that the two gases will not spontaneously unmix and return to their initial vessel: mixing is clearly an irreversible transformation through which the Gibbs free energy of the system has decreased. Before looking for a thermodynamic explanation for this irreversibility, we remark that the total pressure of an ideal gas mixture is the sum of the pressures exerted by each gas on the walls of the vessel. If the pressure remains constant when the volume available to the various gases increases, then it follows that the pressure of each gas correlatively decreases. For each gas, this pressure, proportional to the number of moles, is the *partial pressure*

$$p_i = (n_i/\sum n_i)\,P = x_i P, \tag{8.12a}$$

where x_i is the mole fraction of gas i in the mixture. Of course, as stated by the so-called Dalton's (1766-1844) law, the sum of the partial pressures is equal to the total pressure

$$\sum p_i = \sum x_i P = P \sum x_i = P. \tag{8.12b}$$

But why did the Gibbs free energy of the system decrease on mixing? To answer this question, let us examine how a chemical potential varies when an ideal gas is diluted. From (8.10) and (2.22a) we write that, for one mole, the Gibbs free energy of the mixture is

$$\mu = \mu^\circ + RT \ln P = \sum x_i \mu_i \tag{8.13}$$

where μ_i is the chemical potential of gas i in the mixture. For consistency reasons, we will postulate that this potential keeps for the mixture the same form (8.10) as that of the pure gases. We will thus write

$$\mu_i = \mu_i'^\circ + RT \ln f_i. \tag{8.14}$$

Now, the matter is to relate the unknowns $\mu_i'^\circ$ and f_i to measurable properties. For $\mu_i'^\circ$, the answer is obvious. To be useful, the desired chemical potential equation should not change with the mole fraction of gas i, which varies from 0 to 1. The potential $\mu_i'^\circ$ is thus composition independent, which implies to set it equal to μ_i°, the standard potential of the pure gas. On the other hand, f_i must take on values ranging from P, the pressure of the pure gas, to 0, the pressure of the infinitely dilute gas. It happens that these conditions are exactly fulfilled by the partial pressure p_i. Hence, let us check whether the expression

$$\mu_i = \mu_i^\circ + RT \ln p_i \tag{8.15}$$

is consistent or not with the mixing laws of ideal gases. To begin with, we will calculate the partial molar volumes of the gases. Because the standard potential μ_i° does not depend on pressure, we have

$$\overline{V}_i = (\partial \mu_i/\partial P)_{T,n_j} = RT/P. \tag{8.16}$$

Consistent with the ideal gas law, the total volume of the mixture is $V = \sum n_i \overline{V}_i = \sum n_i RT/P$ and the *volume of mixing* is zero.

From the Gibbs-Helholtz equation (3.11c) we find for the partial molar enthalpy of gas i

$$\overline{H}_i = \left[\frac{\partial(\mu_i/T)}{\partial(1/T)}\right]_{P,n_j} = \left[\frac{\partial(\mu_i^\circ/T)}{\partial(1/T)}\right]_{P,n_j} + \left[\frac{\partial(R\ln p_i)}{\partial(1/T)}\right]_{P,n_j}$$

$$= \left[\frac{\partial(\mu_i^\circ/T)}{\partial(1/T)}\right]_{P,n_j} = H_i^*. \tag{8.17}$$

The partial molar enthalpy is the same as the molar enthalpy H_i^* of the pure gas. It follows that the enthalpy of the mixture is

$$H = \sum n_i \overline{H}_i = \sum n_i H_i^*,$$

which is the weighted sum of the pure gas enthalpies and implies that no heat has been exchanged on mixing. The *enthalpy of mixing* is zero. The chemical potential equation (8.15) is thus correct, for it accounts for the two characteristic features of an ideal gas mixture: additivity of the volumes at constant P and T constants, and lack of heat exchanged on mixing.

A famous experiment made by Berthollet (1748-1822) showed that two gas layers made up of gases with different densities (hydrogen, at the top, and carbon dioxide, at the bottom) eventually mix even when there is no convection in the fluid. Mixing proceeds through diffusion, which is a slow process, but is not hampered by the density difference between the two gases. For any component of the system, the existence of diffusion is determined by chemical potential gradients against which gravity cannot oppose through density differences. For the same reason, the constituting gases of air would remain mixed even without any atmospheric circulation.

8.2c ENTROPY OF MIXING

The difference between the final and initial entropies, that is, the entropy created through mixing, can now be calculated directly from the expression (8.15) for the chemical potential. The Gibbs free energy of one mole of a gas mixture is

$$\mu = \sum x_i \mu_i = \sum x_i (\mu_i^\circ + RT \ln p_i) = \sum x_i \mu_i^\circ + RT \sum x_i \ln x_i + RT \ln P. \tag{8.18}$$

By identifying this expression with (8.13), we obtain

$$\mu^\circ = \sum x_i \mu_i^\circ + RT \sum x_i \ln x_i. \tag{8.19}$$

Because $x_i \leq 1$, the term $RT \sum x_i \ln x_i$ is necessarily negative and the standard potential of the mixture is smaller than the weighted sum $\sum x_i \mu_i^\circ$ of the standard potentials of the pure gases. We have already stressed that ideal gases mix without any enthalpy effects. Because $G = H - TS$, the term

$$\Delta S_m = -R \sum x_i \ln x_i \tag{8.20}$$

is the *entropy of mixing* we were looking for. Noting that the entropy is an extensive property, we finally deduce from (8.19) that the *partial molar entropy of mixing* of gas i is

$$\Delta \overline{S}_{mi} = (\partial \Delta S_m / \partial n_i)_{P,T,n_j} = -R \ln x_i, \tag{8.21}$$

and that it is necessarily positive. Regarding chemical equilibrium, the partial molar entropy of mixing plays a fundamental role that will appear more clearly in section 8.3c.

Let us now calculate the entropy of mixing from another perspective, which, in addition, will demonstrate in a quite different manner that the ideal gas law and the expression (8.15) for the chemical potential are equivalent. By definition, there are no interactions between the various ideal gases of a mixture so that we can consider that mixing of gases at constant P and T represents a simultaneous isothermal expansion of all gases from their initial volume V_i to the final volume V of the vessel. As already seen in 5.2c, for such an expansion the molar entropy change is $R\ln(V/V_i)$. For the sake of simplicity, let us again assume that we have a total of one mole of gas in the mixture. With the ideal gas law, the entropy change for the x_i moles of gas i in the mixture is written

$$\Delta S_i = x_i R \ln \frac{V}{V_i} = R\, x_i \ln\left[\frac{RT\ \sum x_i/P}{RT x_i/P}\right] = -R x_i \ln x_i. \tag{8.22}$$

The entropy of mixing is the sum of the entropy increases calculated for all gases in the mixture

$$\Delta S_m = \sum \Delta S_i = -R \sum x_i\ lnx_i.$$

This is the aforementioned expression (8.20), which we had derived in a purely formal way, in the same way as $\Delta S_i = x_i \Delta \overline{S}_{mi}$. By definition x_i is smaller than unity, which implies that ΔS_i and ΔS_m are positive: like for each gas individually, the entropy variation is positive for the whole mixture. This result is consistent with the irreversible nature of the gas expansions, from V_i to V, during which the spatial distribution of molecules has widened. At this point, it is important to note that (8.20) applies only if different gases are mixed. If the gases were identical, the actual spatial distribution of molecules would not vary and this equation would indeed yield a spurious entropy increase. This conclusion is known as Gibbs' paradox. As will be seen in 13.3b, it indicates that distinguishability or indistinguishability of matter is a fundamental issue at a microscopic level. Hence, it is only when mixing distinguishable entities that (8.20) is valid. It is under these conditions that, as aptly summarized in 1803 by W. Henry (1774-1836), of Henry's law fame, "a gas is a vacuum for other gases".

8.3 MASS ACTION LAW

8.3a CHEMICAL EQUILIBRIA BETWEEN IDEAL GASES

Let us have another look at a reaction we have already considered, namely

$$CH_4 + 2H_2O \rightleftharpoons 4H_2 + CO_2, \tag{8.23a}$$

whose affinity is

$$A = -\Delta G = \mu_{CH_4} + 2\mu_{H_2O} - 4\mu_{H_2} - \mu_{CO_2}. \tag{8.23b}$$

As usual, let us denote by ν_i the stoichiometric coefficient of species i. When a gaseous phase is being dealt with, the affinity becomes

$$\begin{aligned} A &= -\sum \nu_i(\mu^\circ + RT\ln p_i) = -\sum \nu_i \mu_i^\circ - \sum \nu_i RT lnp_i \\ &= -\sum \nu_i \mu_i^\circ - RT \sum \ln p_i^{\nu_i}. \end{aligned} \tag{8.24a}$$

At equilibrium, the affinity is zero, which implies that

$$-\sum \nu_i \mu_i^\circ = RT \sum \ln p_i^{\nu_i} = RT \ln \prod p_i^{\nu_i}. \tag{8.24b}$$

Hence, (8.24b) indicates that the product of the partial pressures of the gases is related to the standard affinity (A°) of the reaction by

$$\prod p_i^{\nu_i} = \exp(-\sum \nu_i \mu_i^\circ / RT) = \exp(A^\circ / RT) = k_p. \tag{8.25}$$

For a reaction between ideal gases, the standard affinity A° depends only on temperature. For a given temperature, k_p is the *equilibrium constant*, which is dimensionless because partial pressures actually appear in (8.25) as pressure ratios p_i/P_0. By knowing the standard potential of all the species participating in the reaction, we can calculate the equilibrium partial pressures from the initial amounts of the gases present.

The equation (8.25) is the ideal gas expression of the mass action law discovered by Guldberg (1836-1902) and Waage (1833-1900) from the study of reaction rates. In this respect, we stress that the mass action law represents a dynamic and not a static equilibrium. The reaction is not terminated, but at equilibrium it proceeds at the same rate in the two opposite directions. An important case is that where inert gases are present in the system. If the reaction does not induce a change in the total mole number of gases (i.e., if $\sum \nu_i = 0$), then equilibrium is insensitive to pressure and inert gases play no role at all. In contrast, if $\sum \nu_i \neq 0$, then the presence of inert gases has to be taken into account through their contribution to the total pressure.

Finally, it can be more convenient to express the constant of an ideal gas equilibrium in terms of molarities, $c_i = n_i/V$, rather than of partial pressures. From (8.12a), the partial pressure is related to the molarity by

$$p_i = x_i P = x_i \sum n_i RT/V = c_i RT. \tag{8.26}$$

By denoting $\Delta\nu = \sum \nu_i$, we rewrite (8.25) as

$$k_p = \prod p_i^{\nu_i} = \prod c_i^{\nu_i} (RT)^{\nu_i} = (RT)^{\Delta\nu} \prod c_i^{\nu_i}. \tag{8.27a}$$

Another equilibrium constant can thus be defined,

$$k_c = \prod c_i^{\nu_i} = k_p/(RT)^{\Delta\nu}. \tag{8.27b}$$

8.3b TEMPERATURE DEPENDENCE OF EQUILIBRIUM CONSTANTS

As already noted, k_p does not depend on pressure. From (8.27), it appears that k_c is also pressure independent. In contrast, these two equilibrium constants do depend on temperature. We directly calculate their variations with the Gibbs-Helmholtz equation (3.4c)

$$\left(\frac{\partial \ln k_p}{\partial T}\right)_P = \left[\frac{\partial (A^\circ/RT)}{\partial T}\right]_P = \frac{\Delta H^\circ}{RT^2}, \tag{8.28}$$

an expression known as van't Hoff (1852-1911) equation. Thanks to modern computing means, however, this equation is nowadays less useful than it was at the time of the slide rule for calculating the temperature dependence of the affinity from thermochemical data. But it does remain important in allowing enthalpies of reaction to be calculated accurately from experimentally determined equilibrium constants.

Because $H = U + PV = U + RT$ for one mole of ideal gas, the analog of (8.27) for k_c is readily obtained

$$\begin{aligned}\left(\frac{\partial \ln k_c}{\partial T}\right)_P &= \left(\frac{\partial \ln k_p}{\partial T}\right)_P - \Delta\nu \left(\frac{\partial \ln RT}{\partial T}\right)_P = \frac{\Delta H^\circ}{RT^2} - \frac{\Delta\nu}{T} \\ &= \frac{\Delta H^\circ - RT\Delta\nu}{RT^2} = \frac{\Delta H^\circ - \Delta(PV)}{RT^2} = \frac{\Delta U^\circ}{RT^2}.\end{aligned} \tag{8.29}$$

8.3c THE AFFINITY AS A DRIVING FORCE OF REACTIONS

At a temperature of 400 K, the Gibbs free energy of one mole of methane and of two moles of water is higher than that of a mixture of four moles of hydrogen and one mole of carbon dioxide. The standard affinity of reaction (8.23a) is indeed a sizeable 95.088 kJ. For equilibrium to be reached, calculations nonetheless show that some methane and water are produced at the expense of hydrogen and carbon dioxide, two apparently stable species. What is happening?

In fact, this apparent paradox originates in a possible confusion between the affinity and the standard affinity of reaction that must be avoided. It is the affinity, i.e., a difference between the Gibbs free energies of the gases *in the mixture*, which determines equilibrium, and not the standard affinity, which is the difference between the Gibbs free energies of the *pure* gases. Now, in the affinity of the reaction

$$-A = \sum x_i(\mu^\circ + RT \ln p_i) = -A^\circ + RT\left[\sum x_i(\ln x_i + \ln P)\right], \tag{8.30}$$

the mixing entropy term plays a fundamental role by lowering the chemical potentials of all species with respect to the pure state, especially at the lowest mole fractions.

To illustrate this point in the simplest possible manner, let us consider the transformation M $\rightleftharpoons$ M' between two gaseous species having the same chemical formula. A real case could be that of the isomeric equilibrium between the gaseous forms of *n*-butane and isobutane (C_4H_{10})

$$\underset{\text{n-butane}}{CH_3\text{-}CH_2\text{-}CH_2\text{-}CH_3} \rightleftharpoons \underset{\text{isobutane}}{CH_3\text{-}\overset{\overset{CH_3}{|}}{C}H\text{-}CH_3}$$

Another example could be the equilibrium between ortho- and para-hydrogen which, as a mixture constitutes ordinary H_2 hydrogen. (The spins of the two protons of the molecule are parallel in ortho-hydrogen, and have an opposite sign in para-hydrogen.)

Let us assume that the affinity of such a transformation M $\rightleftharpoons$ M' is negative. The standard potential of M being higher than that of M', we could be tempted to conclude that the species M' cannot transform into M and, conversely, that M could not exist because of its instability with respect to M'. As just stressed, however, the equilibrium criterion is not $A^\circ = 0$ but $A = 0$. If x is the mole fraction of M', that of M is $1 - x$, and we write for the affinity of the reaction

$$-A = -A^\circ + RT[\ln P + x \ln x + (1-x)\ln(1-x)].$$

Let us plot $-A$ against x for a unit pressure, a temperature of 298 K and three different values $-A^\circ = \mu^\circ_M - \mu^\circ_{M'} = 0$, 5 and 10 kJ (Fig. 8.2). Regardless of the initial content of M', we observe that a Gibbs free energy decrease of the system is achieved through production of the apparently unstable species M. The mole fraction of M corresponding to the Gibbs free energy minimum does depend on the standard affinity of the reaction. It is 0.5 when A° is zero, and it decreases, without ever becoming zero, when the absolute value of A° increases. When x_i tends to 0, μ_i indeed tends to $-\infty$ because of the term $x_i \ln x_i$. Before tending to 0 the term $x_i \mu_i$ takes on strongly negative values that favor the formation of a certain quantity, perhaps vanishingly small, but non zero, of every species participating in the reaction.

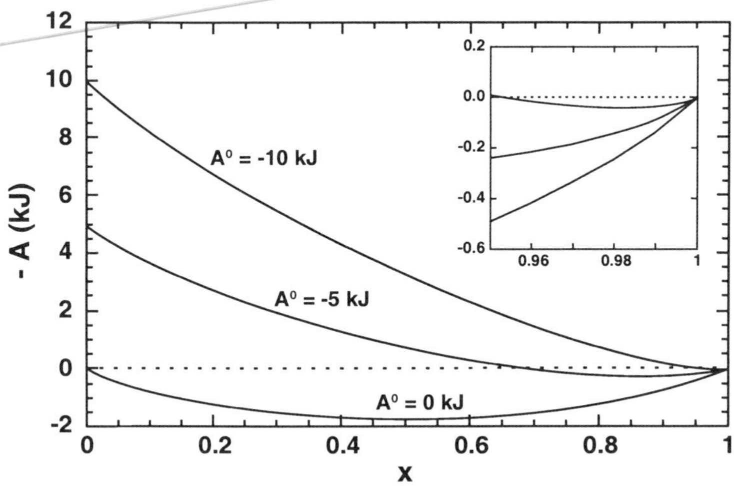

Figure 8.2. Affinity of an isomerization reaction M $\rightleftharpoons$ M' for three different values of the standard affinity of reaction, namely, 0, -5 and -10 kJ. Although the position and magnitude of its minimum strongly depends on the standard affinity of the reaction, $-A$ necessarily becomes negative in the vicinity of the more stable species as shown in the inset.

8.4 REAL GASES

8.4a FUGACITY OF A REAL GAS

The analytical form of the chemical potential of ideal gases leads to such a convenient treatment of chemical equilibria in gaseous phases that Lewis had the very good idea to stick with it for real gases. Hence, we will write for a pure real gas

$$\mu = \mu^\circ + RT \ln f/f_0, \tag{8.31}$$

and call *fugacity* the quantity f introduced so that (8.31) is by definition exact for a real gas. Now, the problem is to relate the arbitrarily defined fugacity to the gas pressure, which is the measurable property. For this purpose, let us keep the pure, ideal gas at T and $P_0 = 1$ bar as the standard state in order that the standard chemical potential be still given by (8.5). In passing, we stress that whether or not the gas is actually ideal under the T, P_0 conditions is here irrelevant. Because chemical potentials have no absolute values, only their variations between two states matter and these can be calculated with respect to any kind of real or hypothetical baseline, i.e., to any standard state, as long as the same baseline is consistently used. By combining (8.6) and (8.31), we find for the fugacity

$$RT \ln f = RT \ln f_0 + \int_{P_0}^{P} V dP = \int_0^P V dP. \tag{8.32}$$

If the fugacity change between any two pressures can be readily evaluated with this equation, the fugacity itself cannot be so because the volume of a gas becomes infinite when the pressure tends to zero. To obviate this difficulty, we introduce a *fugacity coefficient* γ

$$f = \gamma P, \tag{8.33}$$

which should depend on temperature and tends to 1 when P tends to 0. Let us take advantage of the fact that a real gas tends to behave like an ideal gas under low-pressure conditions and replace in (8.32) f by its expression (8.33) before differentiating with respect to P. Remarking that μ° does not depend on P, we find

$$d \ln \gamma = (V/RT - 1/P)dP = (V - V_{IG})/RT \; dP, \tag{8.34}$$

where $V_{IG} = RT/P$ denotes the volume of an ideal gas at the same pressure and temperature as the real gas. The integration problem raised by (8.32) has disappeared

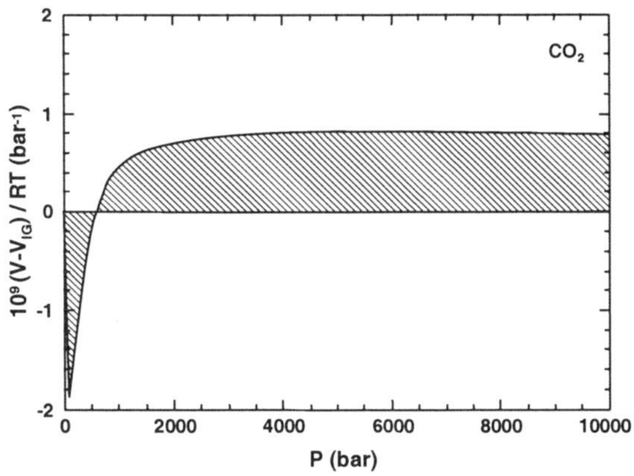

Figure 8.3. Graphical representation of a fugacity coefficient, as the shaded area, for CO_2 at 400 K. Volumes from Bottinga and Richet (1981).

since $V - V_{IG}$ tends to 0 at vanishing pressures. A graphical representation of a fugacity coefficient derived in this way is given in Fig. 8.3.

The fugacity is thus calculated with (8.33) and (8.34) from experimentally determined isothermal volume-pressure relationships. If the equation of state used for a real gas gives an analytical expression for the volume in terms of P and T, (8.34) is integrated directly. Alternatively, (8.34) is integrated by parts

$$\int V dP = PV - \int P dV, \tag{8.35}$$

for van der Waals and other equations lacking such an expression. Denoting by V' the (infinite) volume at the zero pressure P' and by V the volume at P, we have in this case

$$RT \ln\gamma = \left[PV - P'V' - \int_{V'}^{V} P dV\right] - \left[PV_{IG} - P'V'_{IG} - \int_{V'}^{V} P_{IG} dV\right]. \tag{8.36}$$

As an example, we will insert in (8.36) P as given by the van der Waals equation (5.25). Noting that the limit at zero pressure of the $RTV'/(V'-b)$ term is RT, that the divergence of the integrals is avoided thanks to compensation of the terms in $\ln(V'-b)$ and $\ln V'_{IG}$, and, finally, that $PV_{IG} = P'V'_{IG}$ since the temperature is constant, we thus obtain

$$\ln\gamma = V/(V-b) - \ln(V-b)/V - 1 - 2a/RTV. \tag{8.37}$$

8.4b FUGACITIES IN A GAS MIXTURE

Fugacities in real gas mixtures are calculated in the same way as for pure gases. For simplicity reasons, we keep again the same form for the chemical potentials of a real gas whether it be pure or in a mixture

$$\mu_i = \mu_i^\circ + RT \ln f_i = \mu_i^\circ + RT \ln p_i + RT \ln\gamma_i. \tag{8.38}$$

How does vary the fugacity coefficient γ_i with the composition of the mixture? For a pure real gas, we cannot say anything more about the fugacity than about the equation of state parameters, which are determined by specific molecular interactions. Such interactions are more more complex in gas mixtures, and the only rule that can be stated is that γ_i must tend to the fugacity coefficient of the pure gas when p_i tends to P. In short, there will be as many particular cases as different kinds of mixtures.

For a real gas mixture, the pressure is, by definition, the sum of the partial pressures. The simplest case is thus that of a mixture in which the partial pressures would be the pressures that all gases would have if they were pure at the same temperature and volume. This is the so-called Amagat rule. In terms of fugacities, this is tantamount to assuming that the fugacity coefficients are proportional to the mole fractions of the species in the mixture

$$f_i = x_i f_i^* = x_i \gamma_i^* P, \tag{8.39}$$

where f_i^* and γ_i^* are the fugacity and the fugacity coefficient of the pure gas at the same temperature and total pressure. Under this form, this is the Lewis and Randall rule. Let us check that it is consistent with the Gibbs-Duhem equation (2.11). Because the μ_i° terms are composition independent, like the total pressure and the fugacity coefficients, we have indeed

$$\sum x_i d\mu_i = RT\sum x_i d\ln f_i = RT\sum x_i d\ln x_i = RT\sum x_i dx_i/x_i = RT\sum dx_i = 0.$$

As made above for the ideal gas mixture, the reader may check that the partial molar volumes and enthalpies of the gas in the mixture are the volumes and enthalpies of the pure gases at the same pressure and temperature. In other words, both the volume and enthalpy of mixing are zero. The mixture is an ideal one, even though the gases are individually not ideal. This is an ideal mixture of real gases.

We have seen in the preceding section that fugacity coefficients depend on equation of state parameters. As described in 5.4c, these parameters can very often be calculated reliably as a function of composition from simple mixing laws. For such mixtures, the fugacity coefficients can then be calculated and possible deviations from the Lewis and Randall rule ascertained.

8.4c EQUILIBRIUM BETWEEN REAL GASES

Of course, the form (8.38) of chemical potentials has been chosen to arrive directly at the following constant for an equilibrium between real gases

$$k_f = \exp(A^\circ/RT) = \prod f_i^{\nu_i} = \prod p_i^{\nu_i}\gamma_i^{\nu_i} = k_p\prod \gamma_i^{\nu_i}. \tag{8.40}$$

This constant is denoted k_f to distinguish it from k_p, the product of partial pressures.

Not surprisingly, calculating partial pressures from equilibrium constants is more difficult for real than for ideal gases because the various fugacity coefficients should depend not only on pressure and temperature, but also on gas composition. If the relations between composition and fugacity coefficients are known, (8.40) must be solved by an iterative process whereby fugacity coefficients have to be recalculated after each calculation of partial pressures. If the mixture is assumed to obey the Lewis and Randall rule, then (8.40) simplifies to

$$k_f = \prod p_i^{\nu_i}\gamma_i^{*\nu_i}, \tag{8.41}$$

and this equation is solved in the same way as (8.26).

Experimental data that could be used to check the validity of the Lewis and Randall rule at high pressures are not that common. A classical example is the synthesis of ammonia through the reaction 4 $NH_3 \rightleftharpoons 2\ N_2 + 3\ H_2$. With a worldwide ammonia production well over 10^8 tons/year (i.e., about 20 kg/inhabitant per year), few synthesis reactions have indeed so great a practical importance. Industrially, this synthesis represents a compromise between high temperatures, which increase the rate of reaction but lower the theoretical yield, high pressures, which favor the reaction, and the operating cost, which increases faster than the pressure. For T = 723 K we plot in Fig. 8.4 the constants k_p and k_f of the reaction against pressure along with the fugacity

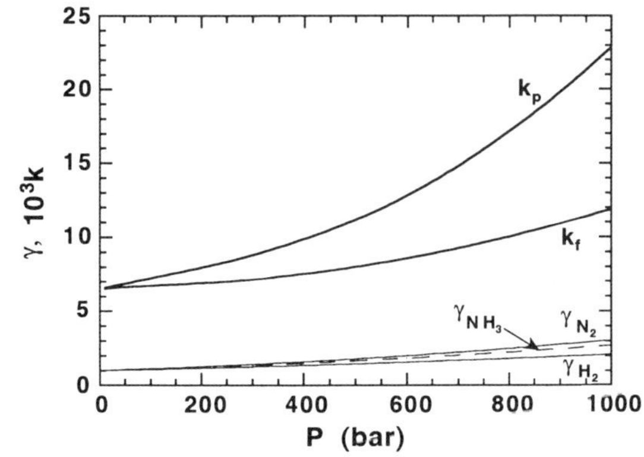

Figure 8.4. Van der Waals fugacity coefficients for H_2, N_2 and NH_3 and equilibrium constants for the synthesis of ammonia, with $k_f = k_p \gamma_{NH_3}/\gamma_{N_2}^{3/2}\gamma_{H_2}^{1/2}$. Data from Dodge (1944) for k_p.

coefficients calculated with (8.37) from the van der Waals parameters of Table 5.1. Whereas k_p begins to increase at the rather low pressure of 10 bars, k_f appears to be almost constant up to 300 bars. This near constancy of k_f indicates that the Lewis and Randall rule constitutes a considerable improvement over the ideal gas mixture model. Its variation with P results instead from the shortcomings of the van der Waals equation, which become manifest at pressures of a few hundred bars.

8.5 CONDENSED PHASES

8.5a ACTIVITY

For real as well as ideal gases, chemical equilibrium is expressed in terms of equilibrium constants because the affinity of the reaction is written as the sum of two terms, one depending on temperature, A°, and the other on composition. Going one step farther with the analytical form of (8.10), Lewis proposed to write in the same way the chemical potential of any phase

$$\mu = \mu^\circ + RT\ln a/a_0 = RT/lna. \tag{8.42}$$

Here, the *activity a* is by definition the variable that allows (8.42) to be exact, and $a_0 = 1$ is the activity of the standard state. (Note that this equation introduces what has been termed the *relative* activity, in contrast to the *absolute* activity defined by $\mu = RT\ln a$, which has been alternatively used by some authors.) Again, we have now to relate the activity to measurable properties. For a gas, the activity is the fugacity ratio f/f_0. But what about a condensed phase? To answer this question, let us once more keep the definition of μ° as the chemical potential of the pure phase at the given temperature T and standard pressure $P_0 = 1$ bar. From (8.6), the activity of the pure phase is simply

$$RT\ln a = \int_{P_0}^{P} V dP. \tag{8.43}$$

For $P = P_0$, $a = 1$ and the activity then increases with pressure.

We have noted that equations of state used for condensed phases rarely lead to direct integrations of equations of the kind (8.43). But, as already made at several times, this does not prevent (8.43) from being integrated by parts. Thanks to the low compressibility of condensed phases, however, the complexity of such calculations can be avoided at not too high pressures with the drastic assumption $a = 1$. For a relatively incompressible phase, the volume would be approximately constant, and the activity given by $RT \ln a = VP$. As an example, let us take a high molar volume V = 50 cm^3 = 5 J/bar. For P = 100 bars, the error made through neglecting the difference between μ and μ° is given by the integral of (8.43) between 1 and 100 bars. It amounts to 500 J only, a figure that is barely greater than typical uncertainties on chemical potentials.

When the pressure is too high to make this approximation acceptable, a change in the definition of the standard state enables the simplification $a = 1$ to be retained. Instead of being the chemical potential of the phase at T and P_0, μ° becomes the potential at T and the high pressure P. With the latter convention, the influence of pressure on the properties of the condensed phase is directly incorporated in the standard affinity of the reaction. This has the great advantage of letting the equilibrium constant of a reaction between condensed phases and gases be expressed solely as a function of partial pressures or fugacities.

Calcite ($CaCO_3$) is the most common form of calcium carbonate. For its decomposition according to the reaction $CaCO_3 \rightleftharpoons CO_2 + CaO$, the equilibrium constant is equal to the fugacity of CO_2

$$k_f = f_{CO_2} = \exp(A^\circ/RT).$$

Knowing A° as a function of temperature, we calculate readily the fugacities of carbon dioxide in equilibrium with calcite plotted in Fig. 8.5. The fugacity coefficient remains close enough to 1 so that the fugacity does not differ appreciably from the gas pressure. If kept in vacuum in a box at temperature T, calcite should thus decompose till the CO_2 pressure reaches the equilibrium value.

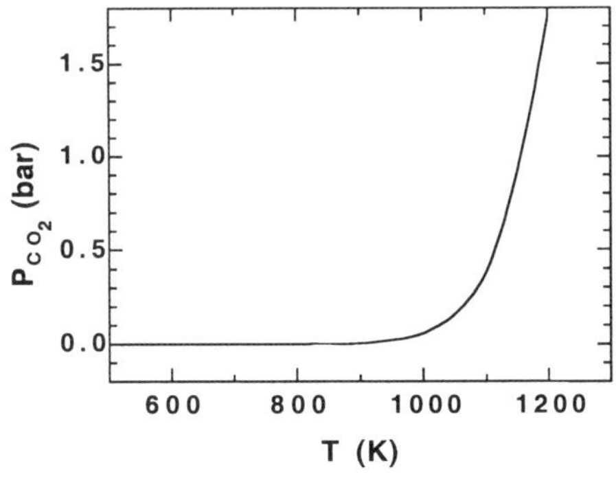

Figure 8.5. Partial pressure of CO_2 in equilibrium with $CaCO_3$. Affinity of reaction from Robie et al. (1979).

8.5b ACTIVITY IN SOLID OR LIQUID SOLUTIONS

For solid or liquid solutions, it will not be a surprise that the chemical potential of component i be defined by

$$\mu_i = \mu_i^\circ + RT \ln a_i. \tag{8.44a}$$

For relating the activity of a component to the composition, it is useful to introduce other parameters, the *activity coefficients*, γ_i

$$a_i = \gamma_i x_i, \tag{8.44b}$$

with which the chemical potentials are written

$$\mu_i = \mu_i^\circ + RT \ln \gamma_i x_i. \tag{8.44c}$$

Owing the extreme specificity of atomic interactions in condensed phases, there are unfortunately no such things as general activity-composition relationships. With the exception of the ideal solution model, which will be presented below, all solution models relate the activity coefficients to the composition through parameters that cannot be calculated from the endmember properties and have instead to be adjusted experimentally. In this respect, the great usefulness of the Gibbs-Duhem equation (2.11) must be emphasized. As a function of the activity coefficient this equation is

$$\sum n_i d\mu_i = \sum n_i d \ln a_i = \sum n_i d \ln \gamma_i x_i = 0 \tag{8.45}$$

at constant temperature and pressure. For a binary system, let us take $n_1 = x_1$ moles and $n_2 = x_2$ moles of the components 1 and 2. Since $dx_1 = -dx_2$, we can write (8.45) in the form $x_1(\partial x_1 \gamma_1/\partial x_1) - x_2(\partial x_2 \gamma_1/\partial x_2) = 0$, and obtain

$$x_1 d \ln \gamma_1/dx_1 - x_2 d \ln \gamma_2/dx_2 = 0. \tag{8.46}$$

This equation therefore indicates how the variation with composition of the activity coefficient of one component determines that of the other.

8.5c ACTIVITY MEASUREMENTS : AN EXAMPLE

Defining activities of components in solutions implies that there exists adequate means to determine them. Although the methods to be used of course depend on the nature of the phases, most of them consists of equilibrating the solution studied at a given temperature and pressure with another phase in which the activity of one or several components is already known. These methods include measurements of osmotic pressures, freezing-point depressions, boiling point elevations, vapor-pressure variations, solubilities or electromotive force (emf) measurements in galvanic cells, whose principle will be discussed in the following chapters. For very fluid liquids, activities can even be obtained from diffusion experiments. As a single example of activity determinations, we will consider here an aqueous solution of NaCl in equilibrium with a vapor phase.

In preamble, let us once more recall that a chemical potential is defined apart from an arbitrary constant, which is however fixed by the convention that the Gibbs free energies of all elements in their reference state at 298 K and 1 atm are zero. Now, experience shows that the vapor in equilibrium with NaCl solutions is essentially made up of H_2O, the vapor pressure of NaCl being negligible (but not strictly zero). If the vapor pressure P is sufficiently low, no significant error is made by assuming that the vapor is an ideal gas. (If this pressure were too high, it would suffice to use below the fugacity in place of P.) From (8.10b), we write that the chemical potential of H_2O in the vapor (g) is

$$\mu^{g}_{H_2O} = \mu^{og}_{H_2O} + RT\ln P. \tag{8.47}$$

For the liquid (l) in equilibrium with the vapor, we have

$$\mu^{l}_{H_2O} = \mu^{ol}_{H_2O} + RT\ln a^{l}_{H_2O} = \mu^{g}_{H_2O}. \tag{8.48}$$

The activity of water in the solution is thus given by

$$\ln a^{l}_{H_2O} = \ln(P/P^*) \tag{8.49a}$$

$$a^{l}_{H_2O} = P\exp(A^\circ/RT), \tag{8.49b}$$

where P^* is the vapor pressure of pure water ($a^{l}_{H_2O} = 1$) at the temperature of the experiment and A° is the standard affinity of vaporization. This affinity can be calculated at any temperature, with (8.8), from the enthalpy of vaporization of water and the heat capacities of water and steam, but it would be here much simpler and more accurate to determine it from the measured P^*= 31.253 10^{-3} bar at 298 K (which gives A° = 8591 J/mol). By means of a vapor pressure measurement (Fig. 8.6), we can therefore determine the activity of water in the solution as a function of temperature and NaCl mole fraction, and then look for a solution model representing correctly the experimental activity-composition relationship.

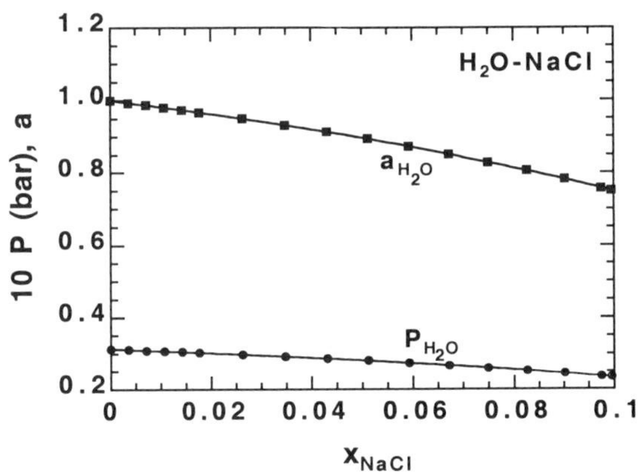

Figure 8.6. Vapor pressure over H_2O-NaCl solutions and derived activity of water at 298 K. Data of Pearce and Nelson (1932).

8.5d INTEGRATION OF THE GIBBS-DUHEM EQUATION

The example of NaCl aqueous solutions illustrates another general feature, namely, that it is often impossible to determine the activity of all the components of a phase with the same method. Because the partial pressure of NaCl of an aqueous solution is much too low to be measured, how can the activity of this component be alternatively determined? Here lies the practical interest of the Gibbs-Duhem equation. In an NaCl-

H_2O solution, the chemical potential of NaCl is

$$\mu_{NaCl} = \mu^{\circ}_{NaCl} + RT \ln a_{NaCl}. \tag{8.50}$$

Now, thanks to (8.45), the chemical potential of NaCl can be obtained from that of H_2O by integration of

$$d \ln a_{NaCl} = -\int x_{H_2O}/x_{NaCl}\, d \ln a_{H_2O}. \tag{8.51}$$

Before doing so, however, we remark that $dx_{H_2O} = -dx_{NaCl}$, whence

$$d \ln x_{H_2O} = -x_{NaCl}/x_{H_2O}\, d \ln x_{NaCl}. \tag{8.52}$$

Through subtraction of (8.52) from (8.51), (8.48) can be rewritten in terms of activity coefficients as

$$d \ln \gamma_{NaCl} = -x_{H_2O}/x_{NaCl}\, d \ln \gamma_{H_2O}, \tag{8.53}$$

and, between any two compositions x'_{H_2O} and x''_{H_2O}, the changes in activity coefficients are given by

$$\Delta \ln \gamma_{NaCl} = -\int_{x'_{H_2O}}^{x''_{H_2O}} x_{H_2O}/x_{NaCl}\, d \ln \gamma_{H_2O}. \tag{8.54}$$

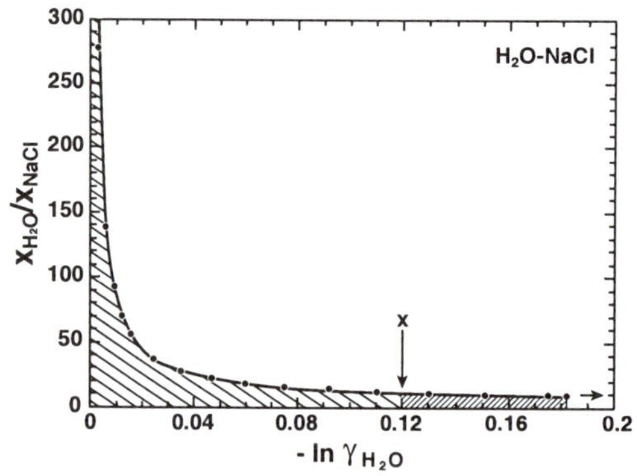

Figure 8.7. Graphical representation of the activity coefficient of NaCl in water at 298 K for the composition x_{NaCl} = 0.079 (for which $\ln \gamma_{H_2O}$ = -0.12). With pure liquid NaCl as a standard state, this coefficient is represented by the the right-hand side shaded area extrapolated to γ_{H_2O} = 1 (pure water); with "dilute" NaCl, it is represented by the left-hand side shaded area. Calculated from the data of Fig. 8.6.

Graphically, the variations of the activity coefficient of NaCl represent an area below the curve x_{H_2O}/x_{NaCl} plotted against $-\ln \gamma_{H_2O}$ (Fig. 8.7). But what about the absolute value of this coefficient? In (8.54), one bound of the integral has to be the mole fraction x_{NaCl} of the composition considered. The other bound depends on the choice of the actual standard potential μ^{ol}_{NaCl}. It would be $x_{NaCl} = 1$ if the standard state were pure liquid NaCl. At this composition γ_{NaCl} would be unity, and integration of (8.54) would be similar to that made for deriving fugacity coefficients of gases. As a matter of fact, such a choice would be unfortunate in two respects. First, the very high melting point of NaCl (1074 K) would require unwarranted extrapolations of the thermochemical data used to calculate the affinity of fusion of NaCl near room temperature. Second, the activity of water itself cannot be determined for NaCl-rich solutions. At 298 K, for instance, saturation is already reached for a low $x_{NaCl} = 0.10$.

In contrast, the activity of water can be measured in the composition interval ranging from pure water to the saturated solution. Integrating (8.51) thus requires to start from $x_{NaCl} = 0$, which imposes to take as a reference the properties of an infinitely dilute NaCl solution. The standard state cannot be an infinitely dilute solution, however, because a standard chemical potential necessarily refers to an ideal solution of unit concentration for which μ_i reduces to μ_i°. We will describe in chapter 11 the various ways in which a standard state is selected depending on which composition variable is used. Here, it will suffice to note that one takes as a standard state a solution of unit concentration in which the activity of the component of interest is obtained from a linear extrapolation of that of an infinitely dilute solution. For brevity, we will call it in the rest of this chapter the "dilute" standard state.

Now, the problem with this choice is that, as apparent in Fig. 8.7, x_{H_2O}/x_{NaCl} then tends to infinity when γ_{NaCl} tends to 1. Although the integral (8.54) converges, it cannot be evaluated accurately from only experimental data because the variations of the water activity cannot be measured accurately at extreme dilution, even with the most sensitive method that could be available. As will be discussed in chapter 11, two cases must be distinguished in this respect. For NaCl and other electrolytes, i.e., substances dissociating into ions in solution, we must integrate analytically (8.54) by relating $\ln\gamma_{H_2O}$ to x_{H_2O}/x_{NaCl} with an expression whose form is indicated by theoretical arguments. For nonelectrolytes, the difficulties are lesser and one generally assumes more simply that an empirical law can be extraplolated down to infinite dilution.

8.6 GENERAL EQUILIBRIUM

8.6a ACTIVITIES AND STANDARD CHEMICAL POTENTIALS

By taking the "dilute" standard state, we have seriously infringed the original definition (8.5) of the standard chemical potential as referring to the pure component at T and $P_0 = 1$ bar. In the preceding sections, some freedom had already been taken with this definition when letting μ° refer to a pressure different from $P = P_0$. As shown by H_2O-NaCl solutions, it is possible to make other choices but a strict prerequisite is that the adopted convention be not only stated explicitly, but also be consistent with the chosen definition of the activity.

To give another example of possible standard states, let us again consider an aqueous solution. With pure water at T and P_0 as a standard state, the chemical potential of water is written

$$\mu = \mu^\circ + RT \ln a, \tag{8.55}$$

and the activity is unity for a unity mole fraction. Suppose that some good reasons lead to taking ice at T and P_0 as a standard state for H_2O. Denoting by $a^\star$ the resulting new activity, we have

$$\mu = \mu^{\circ\star} + RT \ln a^\star. \tag{8.56}$$

The chemical potential does not depend on the particular choice made for the standard state. If μ° and $\mu^{\circ\star}$ are different, then the activities a and $a^\star$ also differ. They are related by

$$a^\star/a = \exp[-(\mu^{\circ\star} - \mu^\circ)/RT], \tag{8.57}$$

which implies an analogous relationship between the associated activity coefficients

$$RT\ln(\gamma^\star/\gamma) = -(\mu^{\circ\star} - \mu^\circ). \tag{8.58}$$

Hence, the activity and activity coefficients depend on the particular standard state. This is why we have denoted the activity either by a, with reference to water, or by $a^\star$ with reference to ice. In addition, we stress that this activity $a^\star$ differs from unity when the mole fraction is unity. In this particular example, the difference $\mu^{\circ\star} - \mu^\circ$ is readily calculated from the enthalpy of fusion of ice and the heat capacities of ice and water, see (8.8), but the choice of the solid phase as a standard state for the liquid is gratuitous. Such a choice could be fruitful in other circumstances, for instance if the solution were not extending to the pure component of interest.

In summary, for any component of a solution the chemical potential may refer to any standard state as long as this state is clearly specified and the activity is defined in a consistent manner with it. To be complete, the description must include the temperature, pressure, chemical composition and structural state of the component in the standard state. It is only for the choice of a pure component as endmember of the solution (at the same P and T) that the activity changes from 0 to 1 when the mole fraction also varies from 0 to 1. This convention is usually followed for continuous solutions. When solubility is only partial, this convention is also generally adopted for the *solvent* (i.e., the most abundant component) whereas the "dilute" standard state is selected for the other components, the *solutes*, whose concentration remain low. Before discussing these aspects in more detail in chapter 11, we emphasize again that the best choice results from either a matter of convenience or the availability of experimental data.

8.6b IDEAL SOLUTIONS

The simplest solution model is that where the activity of a component is equal to its mole fraction

$$a_i = x_i. \tag{8.59a}$$

Like ideal mixtures of gases, these *ideal solutions* have zero enthalpies and volumes of mixing

$$\Delta H_m = \Delta V_m = 0 \tag{8.59b}$$

whereas the entropies and Gibbs free energies of mixing are given by

$$\Delta S_m = -R\sum x_i \ln x_i, \tag{8.59c}$$

$$\Delta G_m = RT\sum x_i \ln x_i. \tag{8.59d}$$

Note that the Gibbs free energy of a solution having nonzero enthalpies of mixing can nonetheless obey (8.59d), to within experimental errors, due to a compensation between the enthalpy and entropy terms. Without an experimental determination of the enthalpy of mixing, it is thus impossible to decide whether a solution is strictly ideal, conforming to (8.59a-d), or behaves only like an ideal solution through following (8.59a) and (8.59d).

8.6c GENERAL EXPRESSION FOR AN EQUILIBRIUM CONSTANT

Consider finally the most general case of reactions involving gaseous, liquid and solid phases. With (8.44), the equilibrium constant may be written

$$k = \prod a_i^{\nu_i} = \exp(A^\circ/RT), \tag{8.60}$$

where, as will be discussed in more detail in 11.2d, activities can be defined with the relevant activity coefficients on any compositions scale (mole fraction, molarity, molality, etc.). The ratio A°/RT is a dimensionless number. It follows that an equilibrium number is also a dimensionless number. As already pointed out for ideal gases in 8.2a, the product of activities present in (8.60) actually involves the ratios a_i/a_i°, which are also dimensionless numbers. This does not prevent units from being often reported for equilibrium constants. For liquids solutions, this practice allows one to know which composition scale is being used.

Unless a contrary convention is adopted, the standard chemical potentials are always temperature dependent. From the Gibbs-Helmholtz equation (3.4c), the variation of k with temperature is thus given by

$$(\partial \ln k/\partial T)_P = \Delta H^\circ/RT^2 = \Delta S^\circ/RT, \tag{8.61}$$

where $\Delta H^\circ = \sum \nu_i H_i^\circ$ and $\Delta S^\circ = \sum \nu_i S_i^\circ$ are the standard enthalpy and entropy of reaction. These are calculated by taking into account *all* the species participating in the reaction, whether gaseous, liquid or solid, and considering the proper standard states (i.e., pure or "dilute" substance, for instance, at P and T of interest).

Contrary to equilibrium constants for gas equilibria, the constant k defined by (8.60) may depend on pressure. This happens when the standard chemical potentials of condensed phases are chosen to be a function of P. The variation of k with P is therefore given by

$$(\partial \ln k/\partial P)_T = \Delta V'/RT, \tag{8.62}$$

with $\Delta V' = \sum \nu_i V_i'$, where the summation is made *only for those species whose chemical potential is pressure dependent.*

8.6d GIBBS FREE ENERGY MINIMIZATIONS

When dealing with the reaction CH_4 + 2 $H_2O \rightleftharpoons$ 4 H_2 + CO_2, we have implicitly assumed that only these four species were coexisting in significant amounts at equilibrium. But how could it be ascertained that the partial pressures of other gases such as CO or O_2 are actually negligible? For examining this issue, let us assume that the only information initially available is that the system kept at some given P and T is made up of carbon, oxygen and hydrogen in some given proportions. Even if we deal with P,T conditions where, to make the problem simpler, a single gaseous phase exists, numerous species such as H_2, O_2, H_2O, CO, CO_2, CH_4, C_2H_2, C_2H_4, C_2H_6, CH_2OH, etc., should be considered.

Various methods have been devised to write complete sets of chemical reactions whose equilibrium constants allow the partial pressures of all reacting gases to be calculated. Such approaches are direct, but cumbersome, and it is nowadays more straightforward to calculate equilibrium compositions by means of Gibbs-free energy minimization methods. The problem is to find the whole set of mole numbers for which the function

$$G = \sum n_i \mu_i = \sum n_i (\mu_i^\circ + RT \ln a_i) \tag{8.63}$$

is minimum. In addition, this minimum is subjected to the constraint that all mole numbers are necessarily positive

$$n_i \geq 0, \tag{8.64}$$

and it must satisfy a set of mass balance equations. Let us denote by N_H, N_C and N_O the total mole numbers of hydrogen, carbon and oxygen of the system. Mass conservation requires that

$$N_H = 2n_{H_2} + 2n_{H_2O} + 4n_{CH_4} + 2n_{C_2H_2} + 4n_{C_2H_4} + \ldots = \sum \nu_{iH} n_i, \tag{8.65a}$$

$$N_C = n_{CO} + n_{CO_2} + n_{CH_4} + 2n_{C_2H_2} + 2n_{C_2H_4} + \ldots = \nu_{iC} n_i, \tag{8.65b}$$

$$N_O = 2n_{O_2} + n_{H_2O} + n_{CO} + 2n_{CO_2} + \ldots = \sum \nu_{iO} n_i. \tag{8.65c}$$

A general method is well suited for finding the extremum of a function whose variables are subjected to such constraints. Let us denote by j the indices H, C and O of the stoichiometric coefficients ν in (8.65), and form a new function L that incorporates the imposed constraints to the function whose minimum is to be determined. This is made with the so-called *Lagrange multipliers* λ_j

$$L = \sum n_i \mu_i - \sum \lambda_j \left(\sum \nu_{ij} n_i - N_j\right). \tag{8.66}$$

We then calculate the mole numbers corresponding to the extremum of G by solving the set of equations

$$\partial L / \partial n_i = 0. \tag{8.67}$$

When replacing L in (8.67) by its expression (8.66), we find that these equations become

$$\mu_i - \sum \lambda_j \nu_{ij} = 0. \tag{8.68}$$

Let us finally denote by N the total number of species dealt with. With 3 Lagrange multipliers, there are $N + 3$ unknowns, satisfying the sign constraints (8.64), to be determined from the 3 linear mass balance equations (8.65) and the N nonlinear equations (8.68). This can be made with available nonlinear optimization codes, some of which do not require very good initial estimates of the various parameters to be computed. The book by van Zeggeren and Storey gives a detailed discussion of more specific methods for solving the problem. To derive (8.68), note finally that no assumption has been made as to the composition dependence of the chemical potentials. These methods are thus also valid for non gaseous substances as well as for nonideal solutions.

Problems

1. At 1 bar, one mole of helium at 400 K is mixed under adiabatic conditions with two moles of argon at 600 K. What are the entropy and Gibbs free energy variations due to this process?

2. Some CoO is placed in a vessel at 1700 K where there is an oxygen pressure of 10^{-6} bar. The Gibbs free energy of formation of CoO at this temperature is -111.23 kJ/mol. Will CoO be reduced to Co for equilibrium to be reached? The enthalpy of formation of CoO is -241.00 kJ/mol at 1700 K and will be assumed to be constant. At what temperature will CoO be in equilibrium with an oxygen pressure of 10^{-6} bar?

3. Pure water is introduced in a vessel. From the Gibbs free energies of formation of water given for three temperatures in the next problem, calculate the partial pressures of hydrogen in equilibrium with steam at these temperatures for total pressures of 1, 100 and 1000 bars. What do mean the very low hydrogen partial pressures calculated at the lowest temperatures? How do these pressures change if half of the total pressure represents the partial pressure of a noble gas like argon?

4. The Gibbs free energies and enthalpies of formation of the gases participating in the reaction (7.1) are listed below. Is it posible to consider that the enthalpy of the reaction is constant? For a total pressure of 2 bars, calculate the equilibrium partial pressures of the gases at temperatures between 400 and 1600 K for an initial mixture of H_2 and CO_2 where both gases have a partial pressure of 1 bar. Same question for pressures ranging from 2 to 300 bars at a constant temperature of 1000 K, for a mixture where H_2 and CO_2 have initially the same number of moles. Redo the latter calculation by considering that the gases obey the van der Waals equation of state with the parameters listed in Table 5.1.

(kJ/mol)	$\Delta G_f^\circ(400\ K)$	$\Delta G_f^\circ(1000\ \mathrm{K})$	$\Delta H_f^\circ(1600\ \mathrm{K})$
CO_2	-394.660	-395.830	-396.250
CH_4	-41.984	19.598	86.240
H_2O	-223.883	-192.557	-158.564

5. The enthalpies of the transition between n-butane and isobutane are -8.87, -8.50 and -7.03 kJ/mol at 298, 900 and 1500 K, respectively. The entropies of this transition are -14.56, -14.18 and -13.72 J/mol K at the same temperatures. Calculate the mole fractions of the two isomers at equilibrium in each case. Compare the effects of a temperature increase, for a given affinity of reaction, to those of a decrease of the affinity at constant temperature.

6. At 298 K, the standard Gibbs free energies of formation of the following species are (in kJ/mol) : CO (gas) : -137.2 ; CO_2 (gas) : -394.4 ; H_2O (steam) : - 228.6 ; H_2O (liq) : -237.1. What are the standard affinity and the equilibrium constant of the reaction CO (g) + H_2O (g) = CO_2 (g) + H_2 (g)? In presence of water, what is the partial pressure of steam at 298 K? If we mix CO, CO_2 and H_2 in such a way that the partial pressure of each gas is 1 bar, and if we then add water in excess to the gas mixture, what will be the final equilibrium partial pressure of each gas?

7. Calculate the fugacity coefficient of gas whose equation of state is $P = RT/(V-b)$. What is the Gibbs free energy variation for the isothermal compression of 1 mol of gas from P_1 at P_2 ? Compare the result with that obtained for an ideal gas. Application: $b = 30$ cm^3/mol, $P_1 = 1$ bar, $P_2 = 50$ bars.

8. Calculate the a and b parameters of the Redlich-Kwong equation of state from the critical data of Table 5.1 for H_2O and CO_2. Show that the fugacity coefficient of a Redlich-Kwong

gas is

$$\ln\gamma = \ln\frac{b}{V-b} + \frac{a}{RT^{1.5}(V+b)} - \ln\frac{V-b}{V} - \frac{a}{bRT^{1.5}}\ln\frac{V+b}{V} - \ln\left[\frac{V}{V-b} - \frac{a}{RT^{1.5}(V+b)}\right]$$

Plot the change in the Gibbs free energies of H_2O and CO_2 between 1 bar and 5 kbar at 1000 K. Compare with the values that one would obtain for an ideal gas.

9. What is the activity of water in an aqueous solution in equilibrium with a vapor phase where the partial pressure of steam is 0.5 bar at 373 K?

9

PHASE RULE AND SIMPLE UNIVARIANT EQUILIBRIA

What is the state of water at 20°C and 1 atm? Liquid, of course, but the answer would be less obvious if the question had been posed for -10°C and 500 bar. To such questions, this chapter will give answers in terms of phase diagrams which represent the stability fields of phases under definite conditions of composition, temperature and pressure. We will first present the phase rule, which fixes the number of intensive variable that can be modified without inducing changes in the number and nature of phases coexisting at equilibrium. We will then review and calculate the phase diagrams of systems made up of phases with constant compositions.

9.1 PHASE RULE

9.1a VARIANCE

Let a multiphase system be in equilibrium at some given pressure and temperature. Let us then change arbitrarily the pressure, temperature and composition, either separately or simultaneously. Will the phases initially at equilibrium remain so? It is to these questions that the phase rule gives definite answers from a comparison between the numbers of thermodynamic variables and equations.

Although the state of a system is characterized by the pressure, temperature and amount and composition of each phase, the criteria for chemical equilibrium concern only the relative stability of the different phases. At 0°C and 1 atm, for instance, a system made up of a droplet of water and one ton of ice is in the same equilibrium as another one made up of an ice fragment floating in a 1-ton tank of water. Because equilibrium criteria do not give any indication on the size of the various phases of the system, the phase rule solely deals with the number of phases, not with their absolute or relative amounts.

Given this aspect, we can describe the composition of a phase in terms of intensive variables such as mole or weight fractions, and not in terms of mole numbers. For c components distributed among ϕ phases, we thus need to know only $\phi(c-1)$ composition variables. Leaving apart phase sizes, we need a total of $\phi(c-1)+2$ variables to specify the state of a system since the pressure and temperature must also be known. Now, if the system is made up of ϕ phases, the equilibrium criterion (7.15c), $\mu_i^\alpha = \mu_i^\beta = \mu_i^\gamma = ...$, represents a total of $\phi - 1$ equations for each independent component. Hence, there are $c(\phi-1)$ equations for determining the $\phi(c-1)+2$ variables needed to specify the state of the system. The difference between these two numbers is the *variance*

$$\nu = \phi(c-1)+2-c(\phi-1) = c+2-\phi. \tag{9.1}$$

The variance is therefore the number of intensive variables that must be fixed before the state of the system can be specified. In turn, it is the number of variables that can be varied without causing a change in the number and nature of the phases that coexist. For this reason the variance is also termed the *number of degrees of freedom*. Systems in thermodynamic equilibrium must have a positive (or zero) variance. If equilibrium is *invariant* ($\nu = 0$), there are just as many equations as variables to be determined; all the variables are unambiguously fixed. If $\nu > 0$, in contrast, the state of the system is determined only after ν intensive variables have been fixed.

Let us take again melting ice as a first example to illustrate the phase rule. With one component and two phases, the system is *univariant* ($\nu = 1$). When the pressure is fixed, ice and water are in equilibrium at a single temperature, 0°C at 1 atm, for instance. Conversely, equilibrium prevails at a single pressure if the temperature is fixed. If both temperature and pressure are fixed, ice and water should no longer be in equilibrium and one of the two phases will disappear. Note, however, that the "freedom" embodied by the phase rule is far from being total. As an example, at no pressure ice and water can be in equilibrium at 1000 K because the relevant intensive variables may be arbitrarily fixed only over restricted intervals where the various phases can actually coexist at equilibrium. This point will be made clearer below with the help of phase diagrams.

9.1b INDEPENDENT COMPONENTS

A subtlety of the phase rule must now be pointed out because in some instances it makes applications less obvious than it could seem at first sight. The number of independent components is the number of components initially selected, minus the number e of equations that mutually relate their mole fractions. We will thus write in a more general way

$$\nu = (c-e)+2-\phi. \tag{9.2}$$

In theory, $c-e$ is determined by the stoichiometry of the various phases of the system. More precisely, its value represents the rank of the matrix formed by the stoichiometric coefficients of elements in the chemical formulae of all substances comprising the system (the so-called Brinkley rule). In practice, this number applies to systems where all equilibria are stable whereas it represents an upper bound to the actual variance when equilibrium is metastable.

Consider for instance a mixture of CO, CO_2 and O_2 at room temperature. The system is made up of a single phase, but there are three components because the quantities of each species can be chosen and varied at will. Its variance is thus 4. When the temperature increases, the kinetics of the reaction CO + 1/2 $O_2 \rightleftharpoons CO_2$ become rapid enough that this reaction actually controls the gas partial pressures. There are two independent components left in the system because the partial pressure of any one species can be calculated from those of the two others and the equilibrium constant of the reaction. When changing from metastable to stable, the equilibrium has become *trivariant* ($\nu = 3$).

This example thus shows that the actual number of relations may depend on the peculiar physical conditions. This number can vary, especially with temperature, but also with pressure or composition in such a progressive way that there are some ranges of condition where, for kinetic reasons, the variance is not well defined operationally. Hence, the number of components cannot be determined unless the stable or metastable nature of the equilibrium is known for the system.

9.1c REPRESENTATIONS OF COMPOSITION

Before determining the variance of a few kinds of equilibria, it might be useful to review briefly the graphical representations of chemical compositions of simple systems. For a binary system A-B, the composition can be represented by the mole fraction of a single component. As done many times in previous chapters, the composition is represented by the point $x_A = 1 - x_B$ as the abscissa on the segment [0 - 1], and any property of the system can then be plotted as the ordinate in a Cartesian diagram (see Fig. 9.2 below).

Within a ternary system A-B-C, the composition can be specified by two mole fractions. Rather than plotting them in a Cartesian diagram, it is preferable to use a triangular representation that respects the symmetrical role of the three components. This representation relies on the fact that a point having the coordinates x_A, x_B, x_C, with $0 < x_i < 1$ and $\sum x_i = 1$, can be represented within an equilateral triangle whose

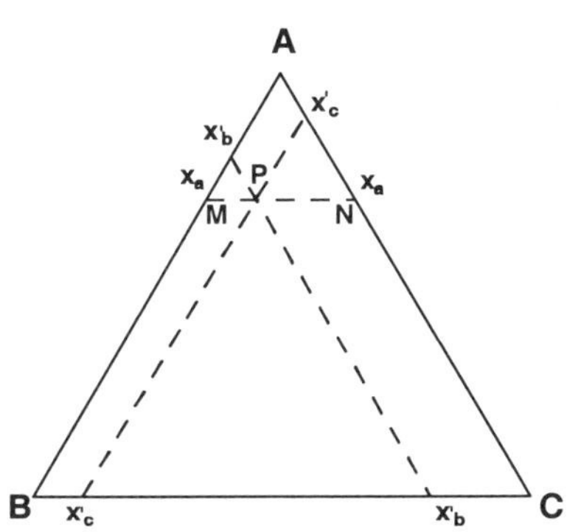

Figure 9.1. Representation of chemical composition in ternary systems.

side length is unity (Fig. 9.1). In this case, the points M and N correspond to the compositions x_A, $x_B = 1 - x_A$, $x_C = 0$ and $x_A, x_B = 0$, $x_C = 1 - x_A$, respectively. Now, the mole fraction x_A remains constant along the line MN whereas the mole fractions of B and C progressively change from 0 to $1 - x_A$, and from $1 - x_A$ to 0, respectively. For any point P between M and N of coordinates x'_A, x'_B and x'_C, we have thus $x_A = x'_A$. Likewise, a line parallel to AB passing through P intersects the sides AC and BC at two points whose positions directly give the mole fraction of C. Finally, the mole fraction of B is given by the intersections of the line parallel to AC passing through P with the sides AB and BC. By a simple geometrical calculation one can check that, for a side of length unity, the three mole fractions sum to unity.

A ternary system is bounded by three limiting *binary joins*. Systems having as endmembers any two compositions within a ternary system, like M and N in Fig. 15.1, represent *pseudobinary joins*. Representing all mixtures that can be formed from the two endmembers, these lines are not to be confused with *tie lines*, which are the lines joining the compositions of the conjugate phases that actually coexist, see Fig. 9.2. Any physical property for substances within a ternary system can be represented as a surface above or as a projection on the composition triangle. The three dimensions of space being used in this way, such representations are impossible for systems having more than three components. For quaternary systems, we can at most represent a composition as a point within a tetrahedra. In fact, the standard practice is to consider *pseudoternary planes*, where the relevant compositions lie on a given plane within the tetrahedra. A similar approach is followed for systems having a higher number of components, whose compositions can be represented graphically only through some projection onto a surface.

9.1d LEVER RULE

In binary and ternary representations, the distance of a given composition to the endmembers or, more generally, to the different phases of the system, is proportional to the mole fraction of the endmember or phase (Fig. 9.2). This is the *lever rule*, which can be used in two different ways for a given composition within the system. If the

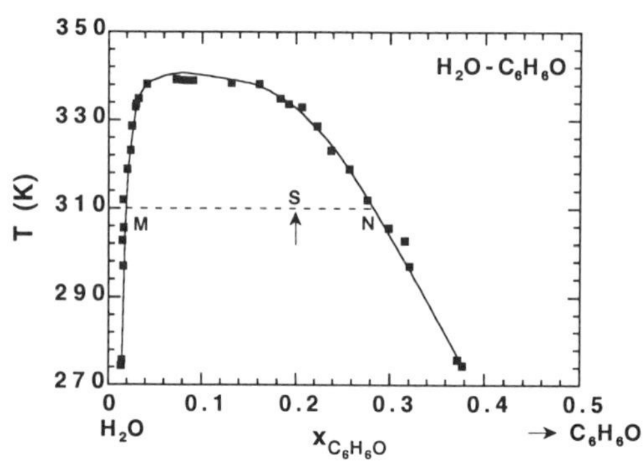

Figure 9.2. Lever rule and solvus of the system phenol-water. At 310 K, a solution S with 20 mol% C_6H_6O separates into two liquids M and N with 2 and 28 mol% C_6H_6O, respectively, whose abundances are in a ratio SN:SM = 8:18 = 1:2.25. The tie line MN joins the compositions of the two phases in equilibrium. Data from the compilation series of Rasmussen et al. (1979).

compositions of all phases are also known, then their relative proportions can be calculated. Conversely, the composition of one phase can be calculated if those of the other phases and their relative proportions are known. The example given in Fig. 9.2 concerns the two immiscible liquids coexisting within the solvus of the binary system phenol-water.

9.2 UNARY SYSTEMS

9.2a EQUILIBRIUM CURVES AND TRIPLE POINTS

A unary system is made up of a single component (this is the case of a *pure substance*, a term which may be unfortunately confusing). When a single phase is present, the equilibrium is *divariant* ($c = 1$, $\phi = 1$, $v = 2$). This is of course the state of internal equilibrium, which prevails whenever two arbitrarily chosen intensive variables characterize the state of the system.

Equilibrium becomes univariant when two phases coexist ($c = 1$, $\phi = 2$, $v = 1$). These phases can be a gas, a liquid or one or several solids. Let us equilibrate water and ice at 0°C and 1 atm before cooling the system down to -1°C. Water freezes, of course, but by raising the pressure to 110 bar we would observe ice melting. While lowering further the temperature, we would reach again equilibrium by raising further the pressure. The same remarks apply to the vaporization of water and the sublimation of ice. All three univariant equilibria plot as curves in the P, T plane and thus delineate the stability fields of the three phases (Fig. 9.3). When two of these curves intersect, the three phases coexist and the system is *invariant* ($c = 1$, $\phi = 3$, $v = 0$). Since equilibrium is achieved at a single pressure and temperature, the third curve has also

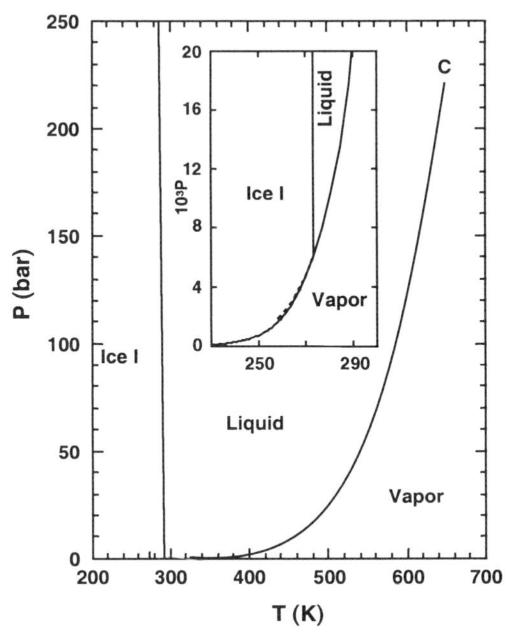

Figure 9.3. Low-pressure phase diagram of H_2O. Note the critical point C and the triple point at 273.16 K and 6.1173 10^{-3} bar. The metastable extension of the equilibrium water vapor pressure curve (dashed curve in the inset), indicates a higher vapor pressure for supercooled liquid H_2O than for ice. Data from Haar et al. (1984) and Wagner et al. (1994).

to pass through this intersection, the *triple point* of a pure substance. Unlike the two other univariant curves, the vaporization curve ends at a point, the critical point. Note finally that each of the three stable equilibria represented as solid lines in Fig. 9.3 become metastable beyond the triple point.

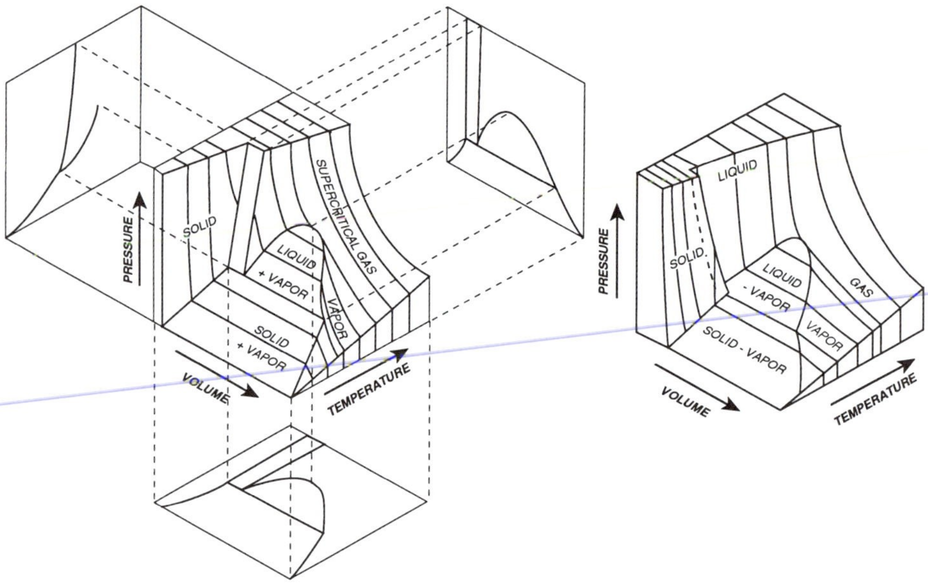

Figure 9.4. Three-dimensional phase diagram of pure substances with positive (left) and negative (right) melting volumes.

These observations represent projections on the P,T plane of the surfaces in the P,V,T space drawn in Fig. 9.4. From one substance to another, liquid-vapor and solid-vapor equilibria keep the same topology because the volumes and enthalpies of transition are always positive. In contrast, as will be described below, the representation of solid-liquid equilibria depends on the sign of the volume of melting. This is the reason why two different topologies have been distinguished in Fig. 9.4.

9.2b POLYMORPHISM

In Fig. 9.3 we show only the stability field of ice I, the H_2O form crystallizing at 0°C and 1 atm. But ice undergoes a series of phase transitions when the pressure is increased and the temperature lowered, giving rise to ice II, III, etc., where the H_4O tetrahedra (shown in Fig. 6.2b for ice I) are packed in a more and more compact manner. Owing to the existence of these polymorphs, the actual stability field of ice I is much narrower than Fig. 9.3 could have suggested. Ice I is not in stable equilibrium with water from 2.1 kbar and 251.2 K (Fig. 9.5a). It is ice III, and then ice IV, VI and VII, whose melting temperatures increase with pressure. Like equilibria between other H_2O phases, the equilibria between two ice polymorphs are univariant. Those involving three phases are invariant and represent as many triple points.

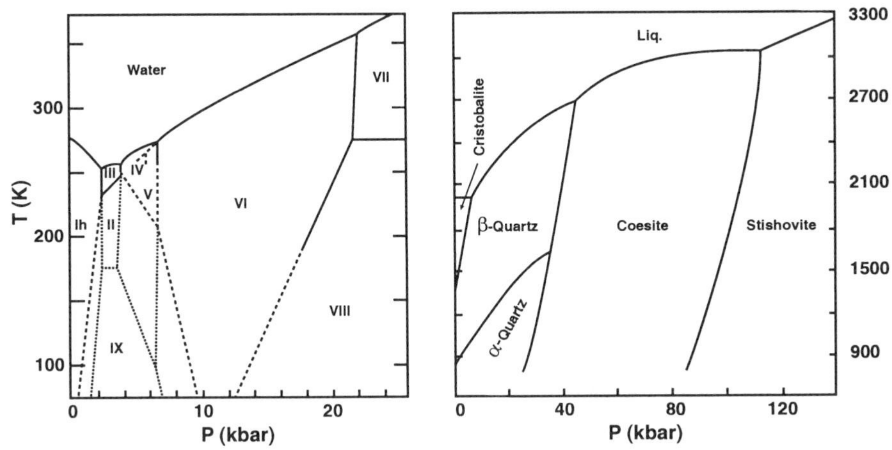

Figure 9.5. High-pressure phase diagrams. (a) H_2O (from Mishima et al., 1984); (b) SiO_2 (from Swamy et al., 1994). These two substances display numerous polymorphs based on H_4O and SiO_4 tetrahedra, respectively, and form three-dimensional open structures (with the exception of stishovite, which is made up of a compact arrangement of SiO_6 octahedra).

Up to this point we have considered only stable equilibria. The metastable extensions of some of these equilibria have also been plotted in Fig. 9.5a. In view of the relatively slight structural and energetic differences between the different ices, the kinetics of their mutual transformations are rapid above 240 K, where phase boundaries are thus well known. The situation is different at lower temperatures where ice polymorphs can be kept metastably. If the possibility of observing the metastable extensions depends on the transformation rates, these extensions are readily calculated as long as extrapolations of thermochemical data for determining the affinity of reaction remain reasonable. For silica (SiO_2), it is above 1200 K that the energy barrier opposing phase transformations can be overcome. At room temperature, most of the various forms indicated in Fig. 9.5b can be kept near room temperature for geologically long periods of time. They are nevertheless in internal equilibrium and equilibria between metastable phases can even be observed.

9.2c VAPOR PRESSURE

Let us put a solid in an enclosure where the pressure is as low as possible. Because of its high cohesion, the solid undergoes a negligible volume change when passing from 1 atm to a vanishing pressure. As given by (3.6d), its entropy increase at constant T, $\Delta S = -\int \alpha V dP$, is thus also negligible. Under, say, adiabatic or isothermal conditions, what can then happen in order to lower the Gibbs-free energy of the system? Sublimation is the answer, for the gas that will form will give rise to a significant en-

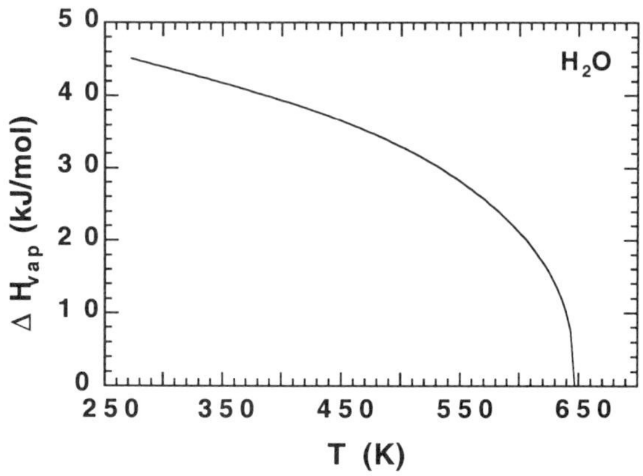

Figure 9.6. Enthalpy of vaporization of water along the liquid-vapor equilibrium curve. Data from Osborne et al. (1939)

tropy increase by occupying the whole available volume. But since some energy needs to be supplied to the solid for severing the bonds that keep within it the would-be gas entities, equilibrium is in fact determined by a compromise between these entropy and enthalpy effects that act in an opposite direction on the Gibbs-free energy of the system. The lower the enthalpy of sublimation, the higher will be the amount of gas produced. At equilibrium, the gas pressure is the saturating vapor pressure.

Vaporization is interpreted in a similar way. Whereas the vapor pressure of the liquid increases with temperature, the enthalpy of vaporization decreases continuously and eventually vanishes at the critical point (Fig. 9.6). Owing to the increasing thermal energy of the liquid and decreasing structural differences between the two kinds of fluids, less heat has to be supplied for vaporization at higher temperatures. All other things being equal, what are the differences between the vapor pressures of the various phases of the same substance? To this question, the topology of the Gibbs free energies plotted in Fig. 9.7 gives an immediate answer. In the stability field of the crystal, one finds that the crystal has a lower vapor pressure than the supercooled liquid because the affinity of sublimation has necessarily a larger magnitude than the affinity of vaporization. Conversely, the vapor pressures of different phases of a single substance are thus a direct measure of their relative stability.

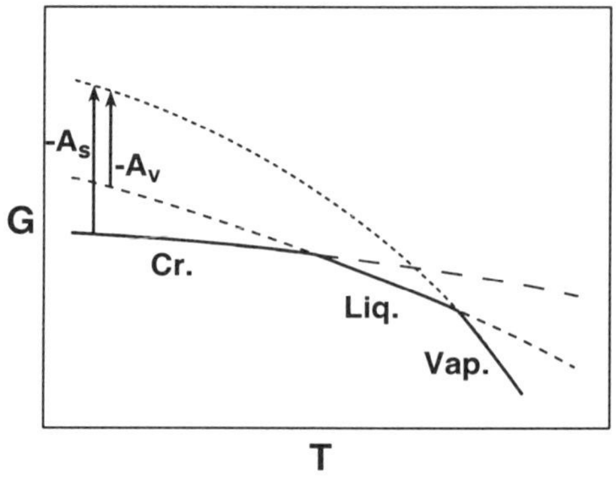

Figure 9.7. Schematic representation of the Gibbs free energies of three phases of the same substance. At a given temperature the stable phase has the lowest Gibbs free energy and, through (9.5), the vapor pressure is a direct measure of the affinities A_v and A_s of vaporization and sublimation, respectively. Solid lines : Gibbs free energies of stable phases; dashed lines: Gibbs-free energies of metastable phases.

9.2d EFFECT OF AN INERT GAS ON VAPOR PRESSURE

Daily experience shows that water vaporizes at ambient temperature and pressure although it constitutes the stable phase of H_2O under these conditions. Why? This apparent paradox simply results from the incomplete nature of the statement. Only water is stable at 25°C and 1 atm if the liquid fills up entirely a container upon which this pressure is exerted. If the container were slightly too large to be filled up by the liquid phase, then the inside pressure could no longer be 1 atm but would be instead the saturating pressure of water at the temperature of the experiment.

But why is air not dry at the surface of a thawing pond since, far away from the P,T conditions of the water triple point, only two phases should coexist? That steam nonetheless coexists with ice and water is due to the fact that the the inert gases of air consitute a third phase of which water must necessarily be a component. We thus increase the number of components to 2 and that of phases to 3 if we enlarge a vessel full of melting ice to introduce an inert gas at a pressure p_A. The equilibrium becomes divariant: two out of the three parameters T, P and p_A enable the third one to be determined. (With N inert gases, the resulting $N-1$ additional degrees of freedom would concern only the $N-1$ additional molar fractions needed for describing the composition of the gaseous phase.) If the temperature is then lowered till ice crystallization, the equilibrium becomes univariant and steam can still coexist with water and ice. Hence, water vaporization when ice is thawing is possible because the volume of the system is not constrained to be that of the condenses phases.

The second effect to be discussed is the dependence of vapor pressures on the pressure of gases that do not react with the condensed phase or dissolve in a significant manner in it. In the presence of inert gases, the enthalpy of vaporization remains unchanged so that the origin of sublimation and vaporization still lies mainly in the entropy increase associated with the expansion of the gas. We have already seen in 8.2c how this entropy change could be determined. As we will see now, the presence of an inert gas slightly displaces equilibrium in favor of the vapor phase whose pressure increases.

For calculating these effects, let us denote by P_A^* the vapor pressure of a species A in equilibrium with the pure condensed phase. Without inert gases, the vapor pressure is determined by the equality between the chemical potentials of A in the vapor (g) and condensed phase (c)

$$\mu_A^g = \mu_A^{\circ g} + RT \ln f_A = \mu_A^c. \tag{9.3}$$

By taking as a standard state the pure condensed phase at the temperature T and pressure P of interest, we have

$$\mu_A^c = \mu_A^{\circ c} \tag{9.4}$$

and obtain immediately

$$f_A = \gamma_A P_A^* = \exp[-(\mu_A^{\circ g} - \mu_A^{\circ c})/RT] = \exp(A^\circ/RT), \tag{9.5}$$

where A_A° is the standard affinity of sublimation or vaporization at temperature T. For the sake of simplicity, let us now assume that the gas is ideal ($\gamma_A = 1$). Equation (9.5) then indicates that the vapor pressure increases very rapidly with temperature, as already shown in Fig. 9.3 by for the case of water. By definition, the *boiling* point is the temperature at which the vapor pressure above the liquid is equal to the atmospheric pressure (which itself depends on elevation).

With inert gases, μ_A^c does not change as long as the solubility of the gases in the condensed phase is negligible. (The case where gas dissolution is significant will be treated in chapter 11.) Replacing the pressure P_A^* by the partial pressure p_A in (9.5), we obtain for the total pressure P

$$p_A = x_A P = \exp(A_A^\circ/RT). \tag{9.6}$$

With the standard states chosen, μ_A^{og} depends only on T whereas μ_A^c depends on P and T. The change in vapor pressure induced by a variation in pressure is thus, see (8.62)

$$(\partial \ln p_A/\partial P)_T = V_A/RT, \tag{9.7}$$

where V_A is the molar volume of the condensed phase A. Since volume is positive, application of an inert gas pressure results in an increase of the vapor pressure. This effect is slight, however, because the molar volumes of condensed phases are generally small. For water at room temperature the vapor pressure increases by only 1‰ between the pure vapor ($P^* = 0.03$ bar) and the vapor under a 1-atm air pressure.

At constant total pressure, note finally that the variation of p_A with temperature is the same with as without inert gases. From (8.61) we have

$$\left(\frac{\partial \ln p_A}{\partial T}\right)_P = \left(\frac{\partial [A_A^\circ/RT]}{\partial T}\right)_P = \frac{\Delta S}{RT} = \frac{\Delta H}{RT^2}, \tag{9.8}$$

where ΔS and ΔH are the entropy and enthalpy of vaporization of A at temperature T.

9.3 SIMPLE UNIVARIANT EQUILIBRIA

The effects of inert gases on vapor pressures represent a brief foray into binary systems to which the next chapter is devoted. Let us now turn to calculations of simple univariant equilibria, i.e., pertaining to phases having a fixed composition whose activity depends only on pressure and temperature. The phase changes of pure substances are special cases of such equilibria. We will examine the P,T stability fields of phase assemblages before illustrating, with temperature-oxygen fugacity phase diagrams, the possibility of other graphical representations of phase equilibria.

9.3a CLAPEYRON EQUATION

Consider several phases in univariant equilibrium at a given pressure P and temperature T. What will be the new equilibrium temperature if the pressure changes from P to $P+dP$, or the new equilibrium pressure if the temperature changes from T to $T+dT$? Because the affinity of the reaction is 0 not only at P and T, but also at $P+dP$ and $T+dT$, the variation of the affinity between the two states is also 0, i.e., $dA = 0$. The chemical composition of all the phases that participate in the reaction is fixed. For all phases, the changes in chemical potential thus reduce to

$$d\mu = VdP - SdT. \tag{9.9}$$

For the reaction, we have

$$\Delta VdP - \Delta SdT = 0. \tag{9.10}$$

At equilibrium, the entropy and enthalpy of the transition are related by $\Delta S = \Delta H/T$. We can therefore rewrite (9.10) in two ways

$$\frac{dT}{dP} = \frac{\Delta V}{\Delta S} = \frac{T\Delta V}{\Delta H}. \tag{9.11}$$

This is the equation originally derived by Clapeyron (1799-1864) from Carnot cycles. As ΔV, ΔH and ΔS are the volume, entropy and enthalpy of the reaction for *the pressure P and temperature T considered*, in principle they cannot be assumed to be constant along the equilibrium curve.

Clapeyron equation is of great historical interest. It revealed the importance of Carnot's ideas, which were then overlooked, and prompted Kelvin and his brother to check, in the first phase equilibrium study, that the melting point of ice decreases when P increases. Usually, a substance is less dense in a liquid than a solid phase. The volume of fusion is then positive. Because ΔS is always positive, the slope of the curve in the P,T plane where the crystal and liquid are in equilibrium (the *melting curve*, for short) is generally positive. The melting curves of ice (I), silicon and gallium are the best known exceptions in this respect. For "normal" melting reactions, one observes that dT/dP generally decreases when the pressure increases. Empirically, this is well accounted for by the Simon equation, $P = A[(T/T_0)^B - 1]$, where T_0 is the 1-bar melting temperature and A and B are adjustable parameters. The causes for these changes in dT/dP slopes are two fold. Enthalpies of fusion increase with temperature because the liquid has a higher heat capacity than the crystal, whereas the volume of fusion decreases with pressure because the liquid is more compressible than the crystal. In both instances, the differences between liquid and crystal are due to the existence of configurational changes in the liquid, see 6.4a. Liquids owe their higher compressibilities to the fact that, lacking long-range order, their structure can sample a whole variety of different atomic arrangements. In contrast, compression mechanisms of crystals are restricted to those changes in bond angles and distances that are compatible with their ordered structure.

9.3b LE CHÂTELIER'S LAWS

By integrating the Clapeyron equation we obtain the univariant equilibrium curves in the P,T plane between two phases or phase assemblages (e.g., Fig. 9.5). Let us now move away from such curves by varying the pressure by an amount dP or the temperature by dT. Which phase assemblage must then disappear? From the partial differentials of the chemical potential $(\partial\mu/\partial T)_{P,n_i} = -S$, and $(\partial\mu/\partial P)_{T,n_i} = V$, we write for the reaction between the two phase assemblages

$$(\partial\Delta G/\partial T)_{P,n_i} = -\Delta S, \qquad (\partial\Delta G/\partial P)_{T,n_i} = \Delta V. \tag{9.12}$$

Let us first consider a temperature rise ($dT > 0$). To satisfy the evolution criterion $A \geq 0$, the entropy of reaction must be positive. A temperature rise will thus make the assemblage whose entropy is the lowest disappear. Conversely, the assemblage with the highest entropy will disappear when the temperature decreases. Likewise, ΔV must be negative for ensuring a positive affinity in response to a pressure increase ($dP > 0$). The densest assemblage is thus stable at high pressure.

These two conditions are known as Le Châtelier's (1850-1936) laws. They constitute special cases of the *moderation theorem*, which states that *any change in one of the variables that determines the state of a system in equilibrium causes a shift in the position of equilibrium in a direction that tends to counteract the change in the variable under consideration.* When high-density phases form through a pressure increase, the effect is actually to reduce the excess pressure exerted. In this respect, the effects of temperature and pressure are thus analogous to those of composition as embodied by the mass action law, which represents another expression of the moderation theorem.

9.3c ENTHALPIES AND ENTROPIES OF TRANSFORMATION

In the Clapeyron equation (9.11), ΔV and ΔH change with pressure and temperature. To calculate their variations, we must not overlook the fact that P and T do not vary independently along the equilibrium curve. For the enthalpy of reaction, the expression (3.30) can be written as

$$d\Delta H = \Delta C_p dT + [\Delta V - T(\partial \Delta V/\partial T)_P]\, dP. \tag{9.13}$$

Making use of the Clapeyron equation to relate the concomitant pressure and temperature variations, we obtain the so-called Planck equation

$$\begin{aligned} d\Delta H/dT &= \Delta C_p + [\Delta V - T(\partial \Delta V/\partial T)_P]\Delta H/T\Delta V \\ &= \Delta C_p + \Delta H/T - \Delta H(\partial \ln \Delta V/\partial T)_P, \end{aligned} \tag{9.14}$$

which gives the variation of the enthalpy of reaction along the equilibrium curve. In this expression, the derivative of the volume of reaction ΔV is obtained from the equations of state of the different phases.

In the case of an equilibrium with a gaseous phase, the volume of the gas is very large with respect to those of the condensed phases as long as the pressure is not of the order of 1 kbar or more. At low pressure the volume of reaction can then be taken as the volume of the single gaseous phase. If, in addition, the gas is assumed to be ideal, then with $\Delta V = RT/P$ we find that $(\partial \ln \Delta V/\partial T)_P = 1/T$, and the Planck equation reduces to

$$d\Delta H/dT = \Delta C_p. \tag{9.15}$$

Let us examine in the same way the entropy variations from the differential of S as a function of P and T :

$$dS = (\partial S/\partial P)_T dP + (\partial S/\partial T)_P dT = -\alpha V dP + C_p/T\, dT. \tag{9.16a}$$

Along the univariant curve, the entropy of transformation thus satisfies the differential equation

$$d\Delta S/dT = -\Delta(\alpha V)dP/dT + \Delta C_p/T = -\Delta(\alpha V)\Delta S/\Delta V + \Delta C_p/T. \qquad (9.16b)$$

For a vaporization or sublimation reaction leading to an ideal gas, we will again assume $\Delta V \approx RT/P$ along with $\Delta(\alpha V) \approx R/P$, whence

$$d\Delta S/dT = (-\Delta S + \Delta C_p)/T, \qquad (9.16c)$$

an equation that could have been directly arrived at from (9.15) and the relation $d\Delta S = d(\Delta H/T)$.

9.3d EQUILIBRIUM BETWEEN SOLIDS

An important case is that of reactions between solids for which one can often neglect above room temperature the C_p differences between two coexisting phase assemblages. Under these conditions, the entropy and enthalpy of reaction are constant and, for not too large pressure ranges, the volume of reaction can also be considered a constant. The slope $\Delta V/\Delta S$ is then constant and the equilibrium curve is a straight line. But we still need to know one point of the equilibrium line for fixing its position in the P,T plane. Calculation of the 1-bar equilibrium temperature is in fact easy because at this pressure the affinity of the reaction is

$$A = A^\circ = -(\Delta H - T\Delta S) = 0. \qquad (9.17)$$

From calorimetric measurements, it thus suffices to solve (9.17) for T

$$T_0 = \Delta H/\Delta S. \qquad (9.18)$$

With these simplifying features, the equation for the equilibrium line between the two phase assemblages is

$$T = \frac{\Delta H}{\Delta S} + \left(\frac{\Delta V}{\Delta S}\right)(P-1) \approx \frac{\Delta H}{\Delta S} + \frac{\Delta V}{\Delta S}P. \qquad (9.19)$$

Note that the temperature T_0 can be negative. This is the case of the ice VI-ice VIII boundary in Fig. 9.5, and just means that one of the two phase assemblages is stable only at high pressure.

9.3e GENERAL CALCULATION OF AN EQUILIBRIUM CURVE

An equilibrium curve is all the more reliable when it has been consistently determined through both experiment and thermodynamic calculation. The most general method for calculating univariant equilibrium curves aims at finding the locus of the points in the P,T plane for which the affinity is zero

$$A(T,P) = 0. \qquad (9.20)$$

Accurate calorimetric measurements are performed at room pressure only. For this reason, let us calculate first the affinity as a function of T from the enthalpy and entropy of the transformation at T_0 and P_0 with

$$-A(T,P_0) = \Delta H(T_0,P_0) + \int_{T_0}^{T} \Delta C_p \, dT - T\left[\Delta S(T_0,P_0) + \int_{T_0}^{T} \Delta C_p/T \, dT\right]. \tag{9.21a}$$

Between (T,P_0) and (T,P), the variation of the affinity is given by

$$-A(T,P) = -A(T,P_0) + \int_{P_0}^{P} \Delta V(T,P) \, dP. \tag{9.21b}$$

With these two equations, the affinity is calculated at any P and T from C_p equations, 1-bar enthalpies and entropies of reaction, and equations of state. Owing to the complicated form of (9.21), (9.20) has no analytical solution for P (or T) once T (or P) has been fixed. Nowadays, however, finding numerical solutions does not raise any serious difficulty.

An alternative procedure consists of integrating the Clapeyron equation from the equilibrium temperature T_0 at $P = 1$ bar, which is calculated through numerical resolution of (9.21a) when ΔC_p cannot be neglected. Whereas the volume of reaction is given by the equations of state, the variations with P and T of the enthalpy (or entropy) of reaction are obtained through numerical integration of (9.14) and (9.16). As an example, let us again take the vaporization or sublimation of a substance to an ideal gas for integrating the Clapeyron equation from the equilibrium conditions T_0 and P_0 (usually 1 bar). To integrate readily (9.11) let us in addition assume that ΔC_p is constant. We then have

$$\frac{dP}{dT} = \frac{\Delta H}{T\Delta V} = P\frac{[\Delta H(T_0) + \Delta C_p(T - T_0)]}{RT^2}.$$

By separating the variables P and T, we finally arrive at

$$\ln\frac{P}{P_0} = \left(\frac{\Delta H(T_0) - T_0\Delta C_p}{R}\right)\left(\frac{1}{T_0} - \frac{1}{T}\right) + \frac{\Delta C_p}{R}\ln\frac{T}{T_0}. \tag{9.22}$$

Equations (9.20)-(9.21) have been derived for calculating equilibrium curves from enthalpies, entropies and volumes of reaction. In turn, it is possible to determine reaction properties from phase equilibria observations. For reactions between solids, volumes are often better known than enthalpies and entropies of reaction. After having determined dP/dT from phase equilibria experiments, one calculates ΔS and ΔH from the experimentally determined Clapeyron slope. For melting reactions, ΔV is beset by large errors whenever it represents a small difference between the great volumes of the liquid and solid phases. As entropies of melting are generally known accurately, then it is preferable to calculate ΔV from the observed Clapeyron slope. This is why phase equilibria data constitute a most useful source of thermodynamic information.

9.3f EXPERIMENTAL TEST OF NERNST'S LAW

Comparisons of experimental and calculated data constitute the only means of checking whether or not Nernst's law is obeyed within a set of phases. As an example, we will take the classic determination of the residual entropy of ice by Giauque and coworkers in 1936-37. This investigation relied on the determination of the standard entropy of water vapor in two entirely different ways. As described in detail below, the first one was obtained through an analysis of the experimental dissociation of magnesium hydroxide at moderate temperature

$$Mg(OH)_2 \rightleftharpoons MgO + H_2O. \tag{9.23}$$

The result found for $T = 298.15$ K was $S_{H_2O} = 188.8$ (1) J/mol K. The second entropy was significantly lower, at 185.2 (2) J/mol K, as determined from the series of transformations

$$\begin{aligned} &\text{ice (0 K)} \rightarrow \text{ice (273 K)} \rightarrow \text{water (273 K)} \\ &\quad \rightarrow \text{water (298 K)} \rightarrow \text{steam (298 K)} \end{aligned} \tag{9.24}$$

through a combination of C_p measurements and determinations of the entropies of melting and vaporization. Because the starting point of this entropy determination was ice at 0 K, the difference of 3.6 J/mol K between the two results was ascribed to the existence of a residual entropy for this phase. Confirmation was given by the fact that statistical mechanical calculations gave a standard entropy of 188.66 J/mol K for steam, in excellent agreement with the former value derived from reaction (9.23).

Table 9.1. Thermodynamic data for reaction (9.23)[a]

T(K)	$S_{Mg(OH)_2}$	S_{MgO}	ΔH_d	P_d (bar)	S_{H_2O}
485.0	105.5	47.98	82 360	$7.157\ 10^{-4}$	205.4
463.1	101.0	45.93	82 760	$2.620\ 10^{-4}$	203.8
298.15	63.1	27.86	84 660	$1.530\ 10^{-7}$	188.8

[a] S in J/mol K, ΔH_d in J/mol.

Under the conditions where the dissociation of magnesium hydroxide takes place at equilibrium, the standard (1 bar) entropy of steam is given by

$$S_{H_2O} = S_{Mg(OH)_2} - S_{MgO} + \Delta H_d/T + R\ln P_d \tag{9.25}$$

where ΔH_d is the dissociation enthalpy and P_d the equilibrium pressure of dissociation. The entropies of $Mg(OH)_2$ and MgO were determined from heat capacity measurements, ΔH_d from C_p values and enthalpy of solution measurements in HF at 298 K. Finally, to ensure that equilibrium was actually reached at the two temperatures investigated, the dissociation

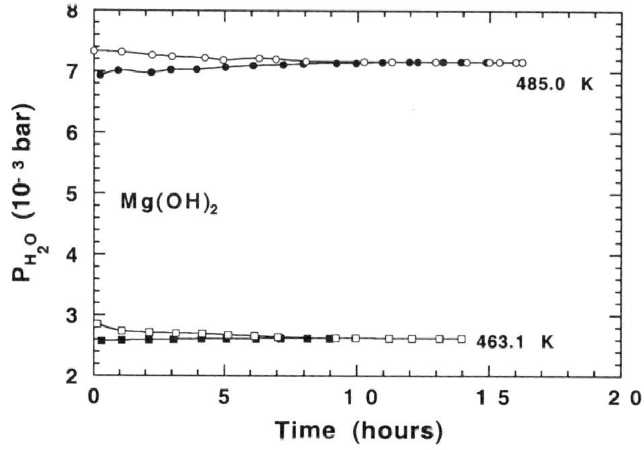

Figure 9.8. Attainment of equilibrium for the dissociation pressure of magnesium hydroxide (after Giauque and Archibald, 1937).

pressure was measured in both cases as a function of time, starting from higher and lower pressures (Fig. 9.8). The calorimetric results are summarized in Table 9.1 (including data thereafter derived for $T = 298$K).

Along the second path, we have (all data in J/mol K) $S_{273} - S_0 = 38.1$ for ice, $\Delta S = 22.0$ for ice melting, $S_{298} - S_{273} = 6.6$ for water, $\Delta S = 147.3$ for the vaporization of water at 298 K, and $\Delta S = -28.8$ J/mol K for the compression of one mole of vapor to 1 bar from 31.26 10^{-3} bar (equilibrium vapor pressure at 298 K). With a negligible correction for the nonideality of water vapor, the calculated standard entropy of water at 298 K is thus 185.2(2) J/mol K.

There are other means of testing the validity of Nernst's law that do not require the existence of a gaseous phase. Let us assume incorrectly that the entropy of a crystal is zero at 0 K. An entropy of reaction as calculated solely from integration of C_p/T will therefore be in error and the calculated Clapeyron slope of a univariant equilibrium involving this phase will not agree with the experimentally determined value. Conversely, it will be possible to determine which value of the configurational entropy of the crystal will yield satisfactory agreement between the calculated and observed slope.

9.3g TEMPERATURE-FUGACITY DIAGRAMS

Phase diagrams most commonly represent the stability fields of phases in the *P*,*T* plane. There are a great many situations where it is useful to consider the influence of other intensive variables on the stability of phase assemblages, especially in the case of open systems. Anticipating somewhat the sections devoted to redox reactions, we will take as an example of such representations a temperature-oxygen fugacity diagram. Let us consider for this purpose the following two reactions

$$Ni + 1/2\ O_2 \rightleftharpoons NiO, \qquad Fe + 1/2\ O_2 \rightleftharpoons \text{"FeO"}.$$

At equilibrium, the oxygen fugacities are given by

$$f_{O_2} = 1/k_p^2 = \exp(A^\circ/RT). \tag{9.26}$$

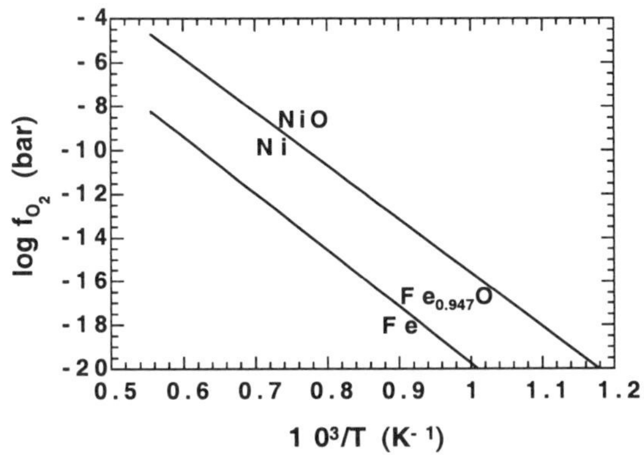

Figure 9.9. Oxygen fugacity for the Ni-NiO and Fe-"FeO" equilibria. Calculations made with the Gibbs free energies of formation from the book by Robie et al. (1979).

They are plotted in Fig. 9.9 against reciprocal temperature. These curves also constitute a phase diagram since they delineate the stability limits of the metals and their oxides in the $T - f_{O_2}$ plane. For open systems, where the oxygen fugacity can be imposed at will, Ni and Fe are stable at very low fugacities, NiO and FeO at very high fugacities. Because of the entropy increase due to the formation of a gas phase, we find the general result that, at constant oxygen fugacity, temperature increases favor the metals (the reduced form) at the expense of the oxides (the oxidized form). At the root of this fact we find again the balance between entropy and enthalpy factors in the Gibbs free energy already described for vapor pressures. In this case the large entropy increase associated with the formation of a gas phase eventually causes the Gibbs free energy to decrease when the temperature is high enough for the $-TS$ term to overwhelm the strong enthalpy increase due to the reduction reaction.

Because iron oxide, "FeO", always includes small amounts of ferric iron, its formula is slightly displaced toward Fe_2O_3, whence the usual notation $Fe_{1-x}O$ where $1-x$ is about 0.95. In Fig. 9.9, equilibrium has been calculated for a $Fe_{0.947}O$ formula, for which $f_{O_2} = 1/k_p^{1.894}$.

Problems

1. The enthalpy of melting of ice is 6.0 kJ/mol at 273.15 K. At this temperature, the densities of ice and water are 0.917 et 0.9998 g/cm^3, respectively. What is the convexity of this curve, and how should it be extrapolated in the domains of stability of the high-pressure ice polymorphs?

2. Calcite and aragonite are the two polymorphs of $CaCO_3$. At 298 K, ΔH = 251 J/mol, ΔS = 4,06 J/mol K and ΔV = 2,78 cm^3/mol for the aragonite-calcite transition. Calculate the stability limits of the two polymorphs by assuming $\Delta C_p = 0$ and that ΔS and ΔV are constant. What is the stable phase stable at 1 bar and 400 K ? What is the equilibrium pressure at 600 K ? From 298 to 600 K, one can in fact consider that ΔC_p = 18.8 J/mole K. What is the entropy change for the aragonite-calcite transformation at 1 bar and 400 K ? What is the difference between this result and the value one would find if the transformation were reversible under these P, t conditions?

3. The volumes (ΔV), entropies (ΔS) and enthalpies (ΔH) of transition between ice polymorphs are given below at the indicated pressures and temperatures (data from the book by Einseberg and Kauzmann, 1969).

Transition	T (K)	P (kbar)	ΔV (cm^3)	ΔS (J/K)	ΔH (J)
I/II	238	2.13	-3.92	-3.18	750
I/III	251	2.08	-3.27	1.67	390
II/III	249	3.44	0.26	5.10	1270
VI/VII	355	22	-1.05	0	0
VII/VIII	278	21	0	-3.89	-1090

Plot the stability limits of the phases as calculated with these data. Compare the results with the phase diagram of Fig. 9.5a. Would the approximations made be valid for calculating the melting curves? From Fig. 9.5b, determine ΔV, ΔS and ΔH for the transition entre β-quartz-coesite and coesite-stishovite transitions. Compare with the values listed above for icc transitions. Why are the melting curves observed at much higher temperatures for silica than for ice polymorphs? What can be said about the density differences between the liquids and crystals?

10
BINARY PHASE DIAGRAMS

Boiling sea water and condensing the vapor was already known in Antiquity as a way of obtaining pure water. It is the composition change due to vaporization that is exploited in this manner. Like vaporization, melting, sublimation and other phase changes are generally incongruent. Why? This question is the main theme of this chapter where we will restrict ourselves to binary systems to unravel the main aspects of phase diagrams that are entangled in multicomponent systems. The correspondence between Gibbs free energies of systems and phase diagrams will again be used to demonstrate the important effects of the entropy of mixing on phase composition. We will also discuss the complexities arising from partial solubility in terms of competing enthalpies and entropies of mixing. Although the focus will be put on temperature-composition diagrams, we will finally present some effects of pressure on phase equilibria.

10.1 SOLUBILITY

10.1a PRELIMINARY REMARKS

In the real world solutions generally do not extend continuously from one endmember to another. The *solubility* is the maximum amount of a component that can be accommodated by a phase under given conditions of temperature, pressure and bulk chemical composition. A continuous solution represents an intermediate case between the two extreme situations represented by separation and association of components in a solution. The former case is unmixing, whose principle has already been described in 7.3. It originates in strong repulsive interactions whereby a component has a chemical potential increasing so rapidly with its concentration that, at a potentially very low level, it induces the decomposition of the solution in two phases to lower the Gibbs free energy of the system. On the other hand, strongly attractive interactions between dissimilar components can in the latter case lead to a Gibbs free energy decrease through formation of single phase. This emphasis put on enthalpic factors should not conceal the fact that entropy of mixing keeps a fundamental importance. One of the main points to be addressed in this chapter is thus the temperature-dependent competition between enthalpy and entropy terms for determining the phases present in a system and the topology of a phase diagram.

The demonstration will rest on the geometrical representations of Gibbs free energies of solutions. Several limiting cases of equilibria will be described. The first, where the two components are never significantly miscible, will be disposed of at once: everything happens as if the system were made up of a mechanical mixture of two pure substances, each undergoing its phase transitions without any interference from the other. Problems begin when the mutual solubility of the two components is total in one phase but zero in the other. This gives rise to eutectic systems, which are especially common for equilibria between two solids and a liquid. Total solubility in both states, vapor and liquid, liquid and solid, or vapor and solid, will then lead to spindle-like diagrams (often termed Roozeboom type I diagrams). Finally, we will review cases where solubility is partial, and others where intermediate compounds form. In a few cases, not necessarily those of pure substances, the phase change takes place without a change in chemical composition. The transformation is then *congruent*. In general, the composition varies and the phase change is said to be *incongruent*.

It is probably useful to add that caution should be exercised before applying the concepts and calculations presented in this chapter to real processes. Especially for equilibria involving solids, the rates of mass and heat transfer exert a considerable influence on actual phase formation and evolution. As will be pointed out for solid-liquid equilibria, qualitative knowledge of these kinetic factors is a prerequisite to any practical application.

10.1b SOLIDS, LIQUIDS AND GASES

By definition, molecular interactions are nonexistent in ideal gases. These gases are mutually miscible in any proportion. Real gases thus also mix completely as long as deviations from ideal gas behavior are not major, but interactions get stronger when the density increases. Beginning often at a few kbar pressure, unmixing in fluids is very common beyond the critical point even for weakly reacting gases. For helium, solubility with ethane (C_2H_6) is not complete at the moderate pressure of 50 bar at 310 K whereas unmixing prevails at 2 kbar and 300 K with xenon and at 10 kbar and 400 K with ammonia.

It will not be a surprise that such fluid-fluid unmixing can take place as an extension of liquid-liquid equilibria found at lower T and P. As a matter of fact, mixing is still less general in liquids than in supercritical fluids. Much better than by the familiar oil-water system, the prevalence of liquid-liquid unmixing is dramatically illustrated by the seven superimposed liquids (heptane, aniline, water, perfluorokerosen, phosphorus, gallium and mercury) shown in the book by Hildebrand and Scott. Thanks

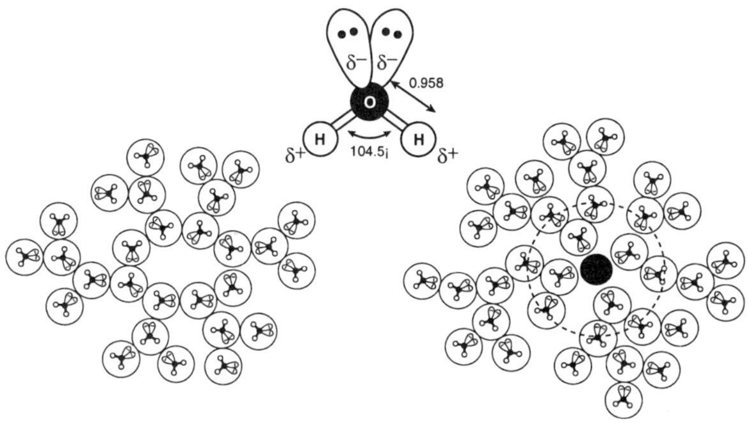

Figure 10.1. Schematic pictures of the tetrahedral structure of the water molecule due to the lone electron pair of the oxygen atom (top, from the book by van Holde et al., 1998), of the random arrangment of H_2O molecules in liquid water (left) and of the local ordering induced by water solvation for an Na^+ ion shown as black circle (right). The excess charge of oxygen and hydrogen atoms are indicated by δ^- and δ^+, respectively.

to the lack of long-range order and the possibility of forming new chemical entities, however, liquid solutions can nonetheless accommodate components having initially different shapes, sizes or types of chemical bonds. The examples of water and ethanol (CH_3CH_2OH) or water and NaCl solutions are well-known cases in point.

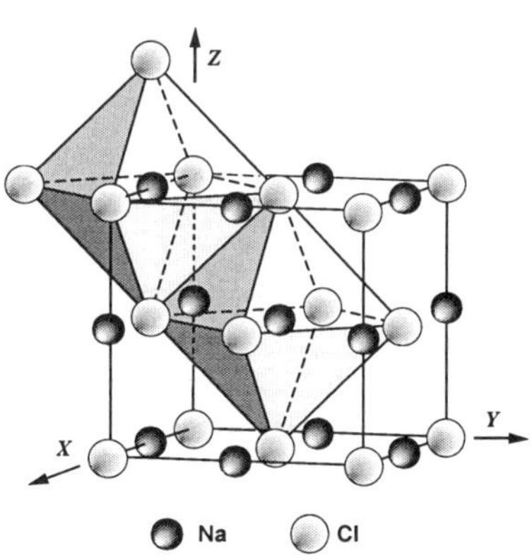

Figure 10.2. Crystal structure of NaCl. The 6 nearest neighbors of an ion Na^+ (or Cl^-) can be represented by a regular octahedron of Cl^- (or Na^+) ions.

When, in addition to high densities, the constraint of long-range order is imposed in crystals, the possibility of mixing two components decreases one step further. Ice, for instance, can host only traces amounts of foreign substances because of the large strain that its crystal lattice would experience through incorporation of entities too different from its H_4O tetrahedra. As exemplified by H_2O and ethanol, or H_2O and NaCl which are not miscible in the solid state, two components do not form a solid solution unless they show a great similarity of structure and properties.

In water, the arrangement of H_2O molecules changes locally around the dissolved Na^+ and Cl^- ions. These ions are *solvated*, i.e., they are surrounded by concentric spheres of water molecules turning the hydrogen atoms toward the Cl^- ions, and the oxygens toward the Na^+ ions (Fig. 10.1). Thanks to the possibility of varying bond angles and distances, the structure changes rather smoothly from the dissolved ions to unperturbed water molecules. The ordering of the ions and water molecules causes the entropy to decrease markedly but the enthalpy is also lowered so that the Gibbs-free energy of solution of NaCl is negative. With respect to water and salt, the solution is stable at 300 K up to 10 mol % NaCl. Beyond this point ionic interactions become very strong. Only two or three water molecules separate the pairs of cations or anions, and the net result of the complex effects that come into play is that a decrease of the Gibbs free energy of the system is ensured by crystallization of the salt.

In the solid state, the NaCl structure is described as a packing of Na^+ and Cl^- ions located at the apex and the face centers of cubes (Fig. 10.2), in such a way that each chlorine ion is surrounded by 6 sodium ions and each Na^+ by 6 Cl^- ions. This face-centered cubic lattice has nothing in common with the ice lattice (Fig. 6.2b). Incorporating significant amounts of ice in NaCl, or of NaCl in ice, would distort the crystal lattices in such a way that long-range order could not be maintained. The mutual solubility of the two solid phases is thus negligible.

10.2 EUTECTIC DIAGRAMS

10.2a GRAPHICAL REPRESENTATION

Let two crystals A and B be in equilibrium with a liquid phase in which the components A and B are mutually miscible. (Although much less frequent for the reasons given above, analogous equilibria between two liquids and a gas phase would be dealt with in the same manner.) At a fixed pressure P, consider the case of a liquid solution whose Gibbs free energy varies with composition as plotted at the top of Fig. 10.3. In contrast, the solids are pure and have a constant composition. Their chemical potentials thus depend only on temperature.

Now, let us fix a temperature T_1 higher than the melting points T_A and T_B of both components. The standard affinities of fusion of the two pure components are positive and the chemical potentials of the solids are higher than those of the liquids (Fig. 10.3). If solid A were coexisting with a liquid of given composition x_B, the Gibbs free energy of the system would plot, against the relative proportion of solid and liquid, along a line like MP_1 in Fig. 10.3. For any composition x_B, we see that the Gibbs free energy of single liquid phase is lower than that of a solid-liquid assemblage. As defined by the minimum of Gibbs free energy, equilibrium is therefore characterized by the absence of any solid phase.

Let us lower the temperature to the melting temperature of the most refractory solid, $T_2 = T_A > T_B$. Again, the Gibbs free energy of a solid-liquid assemblage is higher than that of the liquid alone, with the exception of pure A where $\mu_A^c = \mu_A^l$. Equilibrium between the solid and liquid phases is thus achieved at this single composition.

Let us now fix the temperature between the melting points of the two components: $T_A > T_3 > T_B$. For component B the chemical potential of the solid is higher than that of

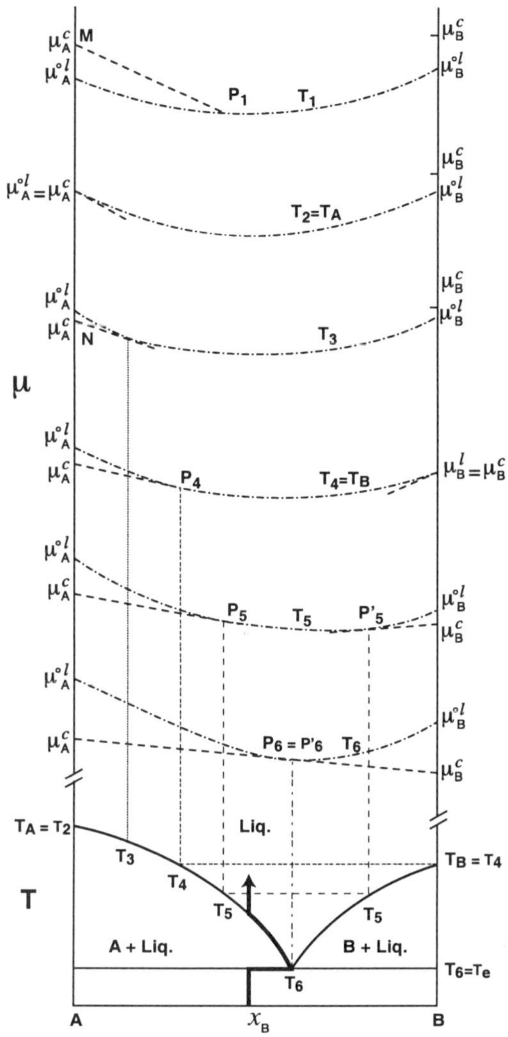

Figure 10.3. Gibbs free energies of mixing and eutectic phase diagram. The thick curve on the phase diagram shows how the composition of the liquid varies when an initially solid assemblage of A and B with composition x_B is heated and melts completely.

the pure liquid, but the converse is true for A. Hence, a liquid plus solid A assemblage has a Gibbs free energy plotting on the line NP_3. For any composition between pure A and the point P_3, this line is below the Gibbs free energy curve of the liquid. For this composition range, the Gibbs free energy of the system thus decreases through partial crystallization of A. What is the amount of solid A crystallizing at temperature T_3 and what is the composition of the liquid in equilibrium with solid A? At equilibrium, the chemical potential of a component must be the same in all phases. In the solution, the chemical potential of A is represented by the intercept on the A axis of the tangent to the Gibbs free energy curve of the mixture for the composition considered. The composition of the liquid in equilibrium with solid A is thus represented by the point P_3 where the line drawn from the chemical potential of solid A is tangential to the Gibbs free energy curve of the liquid. From the temperature T_A we can begin to plot the phase diagram shown at the bottom of Fig. 10.3. As to the relative proportions of conjugate liquid and solid at equilibrium, they can be determined with the lever rule because we know the composition of both phases as well as the bulk composition of the system. Note finally that the only liquids to crystallize partially at T_3 are those whose composition is more A-rich than the P_3 composition.

Let us keep cooling. The reasoning is the same as in the previous case with the difference that the standard affinity of fusion of A continues to increase. The touching point P of the tangent thus keeps shifting away from A and crystallization takes place for liquids continuously more B-rich. In Fig. 10.3, this effect is shown by the point P_4 for the temperature T_4 which has been taken as the melting point of pure B, T_B. Hence, the Gibbs free energy of the system can now also decrease through crystallization of B, in the same way as described above for A.

Let us move to a temperature T_5 lower than T_B. For A-rich compositions, crystallization of A and enrichment in B of the residual liquid continue whereas an analogous evolution is taking place for B-rich compositions. The liquid is not yet crystallizing only if the composition of the system plots between the points P_5 and P'_5. The chemical potential differences between the liquid and pure solid phases vary with temperature independently of each other for A and B. Hence, there is no reason why the tangents to the Gibbs free energy curves of the liquid starting from A and B coincide. A liquid is thus in equilibrium either with A or with B, but not simultaneously with the two solids.

There is an important exception in this respect, however. Because the liquid composition in equilibrium with A or B shifts toward B or A when the temperature decreases, it happens necessarily that the points of tangency P_6 and P'_6 eventually coincide at a given point, the *eutectic* (= "which melts easily"). In Fig. 10.3, this is shown at temperature T_6 where the two solids are in equilibrium with the same liquid. Both of them crystallize simultaneously in the proportions of the eutectic composition, and the temperature will not decrease until crystallization is complete. Crystallization ends at this point regardless of the initial composition of the liquid. The last droplet of liquid thus disappears at the eutectic temperature.

The phase diagram shown at the bottom of Fig. 10.3 is readily drawn from these observations. It represents the composition of the liquids in equilibrium with A or B at temperatures that would have been continuously lowered. Two curves are obtained in this way, the two *liquidus* branches, which meet at the eutectic. At a given composition the liquidus indicates the temperature at which the first crystal forms, whereas the temperature at which the last droplet of liquid crystallizes is given by the *solidus*. Crystallization ending at the eutectic temperature, the solidus is the horizontal line passing through the eutectic point.

What happens when an assemblage of solids A and B of composition x_B melts? The same paths are followed backward (Fig. 10.3). The composition of the first droplet of liquid formed is that of the eutectic. This composition as well as the temperature remain constant as long as enough heat has not been supplied to melt entirely either A or B, depending on the composition of the system. When one of the two solids has been totally melted, the temperature increases and the proportion of molten A or B increases. The liquid then follows the liquidus of A or B until its composition matches that of the initial solid assemblage. At this point, fusion is of course complete.

Let us finally review the variance of the different equilibria. At the eutectic, three phases are present for two components. Equilibrium is univariant. If the pressure is fixed, as is usually the case, the temperature and composition of the liquid are determined. Likewise, a single eutectic pressure and composition are found for an arbitrarily chosen temperature. A eutectic thus behaves as a pure, congruent substance. Along the liquidus, two phases only coexist and equilibrium is divariant. At a given pressure, we can fix either the composition of the liquid, or the temperature. Equilibrium is divariant below the solidus as well as above the liquidus. We could of course fix either the temperature and pressure, or pressure and composition, or composition and temperature.

In the middle ages, how were sherbets prepared in Summer? The only available cold source, snow kept in ice houses, was not cold enough for this purpose. Solution: pour some salt on the ice. On the water side, the liquidus of the system H_2O-NaCl is of the eutectic type (Fig. 10.4). As everybody knows, adding a little bit of salt causes ice to melt, but the trick is that this requires a considerable enthalpy of 320 J/g, which the system must provide by lowering its temperature (a good thermal insulation helping in this respect). With 10% of salt, by weight, the temperature can be lowered in this way to 266.5 K, and even the 255 K of modern freezers are reached with 21 wt% of salt. The effect is analogous to the cold sensation felt when alcohol or ether evaporate on the skin. In a cursory way, one could say that enthalpy is very efficiently extracted through entropy production. The evolution of a system of course depends on the factors that are controlled, temperature in the latter examples, or heat exchanged in the case of sherbets.

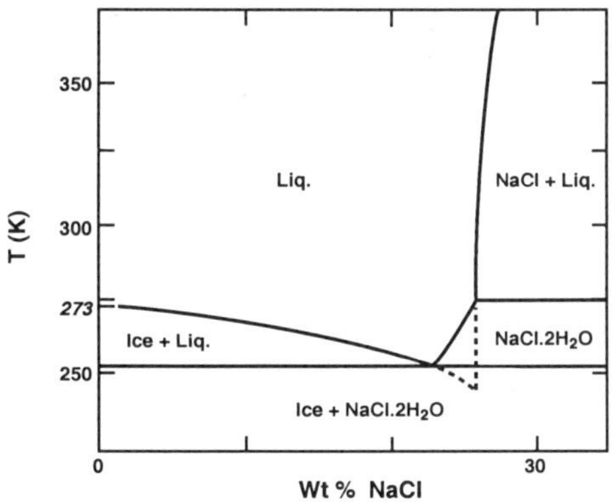

Figure 10.4. NaCl-H_2O phase diagram. On the NaCl side of the diagram, note the complication arising from the existence of the compound NaCl.$2H_2O$ which melts incongruently (see 10.6b). The dashed curves are the metastable extensions of the two liquidus branches.

10.2b CALCULATION

Examination of Fig. 10.3 shows that two factors play a major role in eutectic phase equilibria. For each endmember of the system, the temperature at which the Gibbs free energy of the liquid becomes higher than that of the solid is determined by the enthalpy of fusion and the heat capacity difference between the liquid and solid. (These C_p differences are due to the configurational changes that take place in the liquids, see

6.4a: the smaller these are, the less rapidly a liquid loses its excess enthalpy and entropy with respect to the crystal, and the lower is the endmember melting point). For the whole system, then, it is the entropy of mixing that gives rise to the curvature of the Gibbs free energy of the liquid phase and to the ensuing chemical composition difference between the solid and liquid phases. These two conclusions will now be set into an analytical form.

Along both liquidus branches equilibrium is attained between a pure solid and the solution. For component A, the equilibrium criterion is

$$\mu_A^c = \mu_A^l. \tag{10.1}$$

Let us take as a standard state the pure phase at the temperature and pressure of interest. In the solution the chemical potential of A is

$$\mu_A^l = \mu_A^{\circ l} + RT\ln x_A^l + RT\ln\gamma_A^l, \tag{10.2}$$

whereas that of pure solid A is of course

$$\mu_A^c = \mu_A^{\circ c}. \tag{10.3}$$

For an ideal solution ($\gamma_A^l = \gamma_B^l = 1$), the chemical potential equality (10.1) becomes

$$\mu_A^{\circ c} = \mu_A^{\circ l} + RT\ln x_A^l, \tag{10.4}$$

and the equation of the liquidus branch of A is

$$x_A^l = \exp(A_A^\circ/RT), \tag{10.5}$$

where the standard affinity of fusion is given by (8.8). With the approximation (8.9) for the standard affinity, we obtain the so-called Schröder-van Laar equation

$$\ln x_A^l = -\frac{\Delta H(T_A)}{R(1/T - 1/T_A)}, \tag{10.6}$$

which is function of the enthalpy and temperature of fusion of pure A.

The other liquidus is calculated in the same way. With its constant temperature, the solidus is then determined by the intersection of the two liquidus branches. For a real solution, the calculation is the same. It suffices to replace in (10.2) the activity coefficient by its temperature and composition dependent expression and to solve, in general numerically, the analog of (10.4) obtained in this way. If the approximation $\Delta C_p = 0$ is basically unsatisfactory for the calculation of standard affinities, it is, in contrast, valid for calculating the slope of the liquidus at the endmember compositions. Denoting the entropy of fusion of A by ΔS_A, we çalculate from (10.6)

$$(dT/dx_A)_l = RT_A/\Delta S_A. \tag{10.7}$$

In a general way, the melting point is lowered less when the entropy of fusion is high. This conclusion expresses the fact that the entropy of mixing contributes least to the total entropy of the liquid when the entropies of melting of the components are the highest.

10.3 SPINDLE-LIKE DIAGRAMS

10.3a GRAPHICAL REPRESENTATION

We will then consider the other limiting case where miscibility is total between the two components in the two phases in equilibrium. As previously made for eutectic systems, we will examine how chemical potentials vary when the temperature of the solution is lowered progressively.

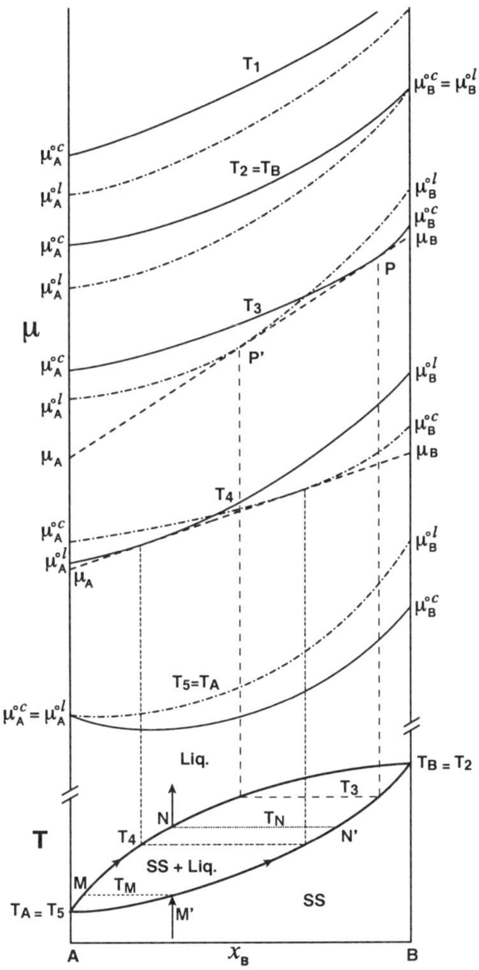

Figure 10.5. Gibbs free energies of mixing and spindle-like phase diagram. SS = solid solution. The arcs M'N' and MN represent the composition changes of the crystal and liquid, respectively, when a solid solution is melted.

Of course, the liquid is the stable phase if we start from any temperature T higher than the temperature of fusion of the most refractory component ($T > T_B$). The Gibbs free energy of the solid is higher than that of the liquid (Fig. 10.5). Crystallization begins at $T_2 = T_B$ ($> T_A$), solely for B. At temperatures between T_A and T_B, the standard affinity of fusion is negative for B and positive for A. The Gibbs free energy curves of the solid and liquid solutions thus intersect, as indicated in Fig. 10.5. Between the compositions P and P' the Gibbs free energy is lower for a liquid plus crystal assemblage than for one of the two phases only. What are the compositions of the two phases? As usual, the chemical potentials of A and B must be equal in the liquid and solid phases. At equilibrium, the compositions of the phases are thus determined by the points where a single line is tangential to both Gibbs free energy curves.

When the temperature decreases the two compositions shift toward the less refractory component A whose (still positive) standard affinity of fusion keeps decreasing. When the affinity vanishes at $T = T_A$, the compositions of the solidus and the liquidus coincide at pure A. Point by point, we can plot a diagram in the temperature-composition plane which delineates three domains (Fig. 10.5, bottom). Above the liquidus, the system is trivariant ($\nu = 3$) with a single (liquid) phase and two components. Three intensive variables, composition, temperature and pressure, must be specified to characterize the state of the system. The variance reduces to 2 between the solidus and liquidus where two phases are coexisting. If the pressure and temperature are fixed, the compositions of the solid and of liquid in equilibrium are determined. Likewise, one can fix the pressure and composition of the liquidus, and the temperature and composition of the solidus will be determined. Finally, equilibrium is again trivariant below the solidus.

Let us summarize these conclusions by examining the melting of a solid with a composition 30 mol % B (Fig. 10.5). When the solidus is reached at the point M' the first droplet of liquid forms. As given by the intersection of the liquidus with the isotherm T_M, the liquid composition M shifts toward the less refractory A component. Correlatively, the composition of the solid solution slightly shifts toward B. At this point, as well as at the following steps, the relative proportions of conjugate solid and liquid phases can be calculated with the lever rule. When the temperature increases further, the amount of liquid formed increases and its composition shifts toward A. By a mass balance effect, that of the solid continues to shift toward B. Melting is of course complete, at point N, when the liquid composition reaches that of the starting solid; the composition of the residual solid is then given by the intersection N' of the solidus with the T_N isotherm. Conversely, the same path is followed in reverse when a liquid is cooled. The first solid crystallizes when the liquidus is reached for the composition considered. The solid composition then follows the arc N'M', while that of the liquid follows the arc NM, and crystallization ends when the crystal composition becomes the same as that of the starting liquid.

10.3b CALCULATION

For calculating this melting equilibrium the starting point is the equality of the chemical potentials of A and B between the two phases

$$\mu_A^l = \mu_A^c, \qquad \mu_B^l = \mu_B^c. \tag{10.8}$$

For the sake of simplicity, let us again assume ideal solutions for both phases while noting that the calculation would be conducted in a similar way for real solutions provided that the adequate expressions for the activity coefficients are used. We then write

$$\mu_A^l = \mu_A^{\circ l} + RT \ln x_A^l, \qquad \mu_B^l = \mu_B^{\circ l} + RT \ln x_B^l, \tag{10.9}$$

$$\mu_A^c = \mu_A^{\circ c} + RT \ln x_A^c, \qquad \mu_B^c = \mu_B^{\circ c} + RT \ln x_B^c. \tag{10.10}$$

With these expressions the equilibrium criteria are

$$x_A^l = x_A^c \exp(A_A^\circ/RT), \qquad x_B^l = x_B^c \exp(A_B^\circ/RT). \tag{10.11}$$

These equations depend on the mole fractions of a given component in two different phases. For both the solidus and liquidus, they have to be rewritten so that they will depend on the mole fractions of the two components in the same phase. For this purpose, let us make use of the mass balance equations for the mole fractions

$$x_A^l + x_B^l = 1, \qquad x_A^c + x_B^c = 1. \tag{10.12}$$

By summing the two equations (10.11) and eliminating one of the two mole fractions with $x_B^c = 1 - x_A^c$, we obtain along the solidus

$$x_A^c \exp(A_A^\circ/RT) + (1 - x_A^c) \exp(A_B^\circ/RT) = 1, \tag{10.13}$$

and the equation of the solidus is finally

$$x_A^c = \frac{[1 - \exp(A_B^\circ/RT)]}{[\exp(A_A^\circ/RT) - \exp(A_B^\circ/RT)]}. \tag{10.14}$$

By rewriting (10.11) in the form $x_A^c = x_A^l \exp(-A_A^\circ/RT)$, we derive in the same manner the equation of the liquidus

$$x_A^l = \frac{[1 - \exp(-A_B^\circ/RT)]}{[\exp(-A_A^\circ/RT) - \exp(-A_B^\circ/RT)]}. \tag{10.15}$$

10.3c LIQUID-VAPOR EQUILIBRIA

Like Gibbs free energies, phase diagrams change with the total pressure and complete representations should thus be made in P,T,X space (X = composition). Because the properties of condensed phases vary little with pressure, the T,X sections described above for liquid-solid equilibria often do not change appreciably up to pressures of a few kbar. Of course, this does not hold true for equilibria involving vapor phases, which can thus be represented as sections in either the P,X or T,X planes. As an example of such representations, the isothermal O_2-N_2 and isobaric CH_4-N_2 phase diagrams are reproduced in Fig. 10.6. Comparing them, we see that the opposite effects of temperature and pressure on the stability of the vapor phase manifest themselves by the inverted positions of the vaporization and condensation curves in the two representations. The former curves (the *bubble* lines) indicate when the first bubble of gas evaporates, the latter the *dew* lines) when the first droplet of liquid condenses. We will not delve into calculations of vaporization and sublimation equilibria because these are performed along the lines already described for solid-liquid equilibria, but we do emphasize the strong shift of both curves to higher temperatures when the pressure is increased.

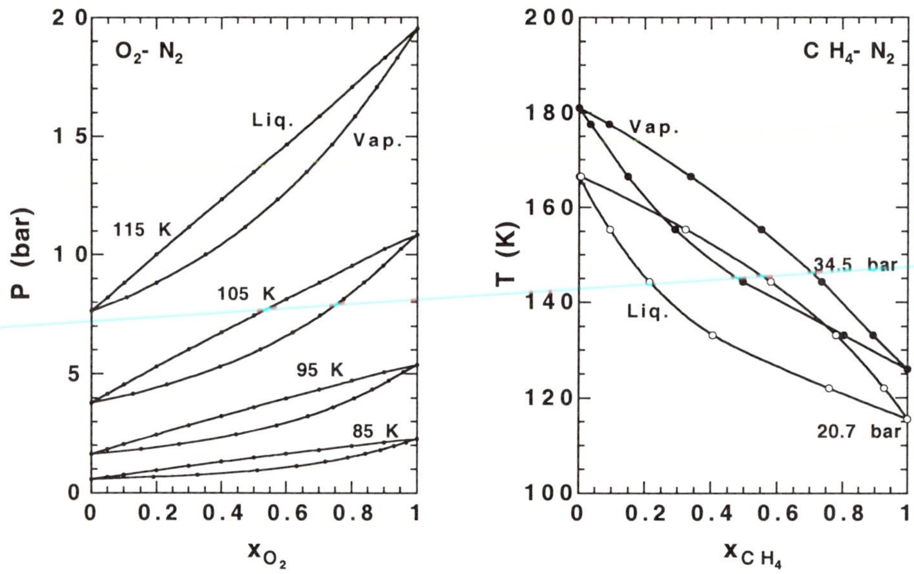

Figure 10.6. Liquid-vapor phase diagrams. Left: P,X diagrams of N_2-O_2 at the temperatures indicated (Dodge and Dunbar, 1927); right: T,X diagrams CH_4-N_2 at the pressures indicated (Bloomer and Parent, 1953).

At this point a few words about the manner in which phase diagrams are experimentally determined might be useful. For solid-liquid diagrams, two cases must be distinguished. When melting is sharp, the solidus and liquidus temperatures can be obtained from measurements of physical properties that depend sensitively on the solid or liquid nature of a sample. As a function of temperature, the electrical conductivity or the heat capacity thus show marked kinks when fusion begins or ends. For glass-forming substances, melting is usually too sluggish in this respect. The *quenching method* is then the standard procedure. After rapid cooling to ambient conditions, one examines the state of the substance heated for long periods of time at given temperatures. The solidus is the lowest temperature at which the first piece of glass is observed, the liquidus, the lowest temperature at which no crystal is observed. For liquid-vapor equilibria, the simplest measurements are those of the partial pressures of both components as a function of composition. At constant temperature, the sum of the partial pressures gives the vaporization curve, when plotted against the composition of the liquid, and the condensation curve, when plotted against the vapor composition (which is itself determined from the partial pressures of the components). Note finally that such measurements yield not only the phase diagram, but also, as described in 11.3, the activities of both components through plots of the partial pressure against the concentration.

10.4 PARTIAL SOLUBILITIES

10.4a GIBBS FREE ENERGY OF MIXING

In the previous cases, the composition difference between the solid and liquid in equilibrium originated in the existence of a solution for at least one phase. The possibility to mix components thus resulted in an entropy production, and in a concomitant decrease of the Gibbs free energy of the system. But what about real solutions where enthalpies of mixing are non zero? Announcing the main theme of the next chapter, we will take here a single example for illustrating the general effects of the competition between enthalpy and entropy factors in a real binary solution. Not to complicate the demonstration, we will deal with temperatures at which no other phase change takes place. In the following, unmixing will therefore constitute the only kind of transformation.

In real solutions an important fact is that entropies of mixing cannot be negative even though they differ from $-R\sum x_i \ln x_i$. They are positive and smallest when intermolecular interactions are the strongest. Schematically, enthalpies of mixing are positive or negative depending on the overall repulsive or attractive nature of these interactions. When negative, these enthalpies contribute to stabilization of the solution. For understanding the origins of unmixing, the relevant case is thus that of positive enthalpies of mixing which oppose the negative entropy term ($-TS_m$) in the expression of the Gibbs free energy.

Let us assume that the enthalpies of mixing are symmetric with respect to composition and can be represented by the equation

$$\Delta H_m = x(1-x)W, \tag{10.16}$$

where W is an *interaction parameter* to be adjusted in order to reproduce the experimental data. As the most favorable case for miscibility, let us finally assume that the entropy of mixing is that of an ideal solution. The Gibbs free energy of mixing then is

$$\Delta G_m = x(1-x)W + RT\sum x_i \ln x_i. \tag{10.17}$$

As a matter of fact, this expression defines a *regular* solution. Now, let us examine how the variation of ΔG_m with temperature influences the Gibbs free energy of the mixture as given by $G = \sum x_i \mu_i^\circ + \Delta G_m$. For this purpose, we will take W = 20 kJ, a typical value for high-temperature systems, and plot ΔG_m at different temperatures (Fig. 10.7). At high temperatures, the entropy term predominates in (10.17) because of the T factor in the $-TS$ term. As considered in previous sections, the Gibbs free energy of the mixture has a regular curvature and the solution is stable as long as melting, crystallization or other phase changes do not take place.

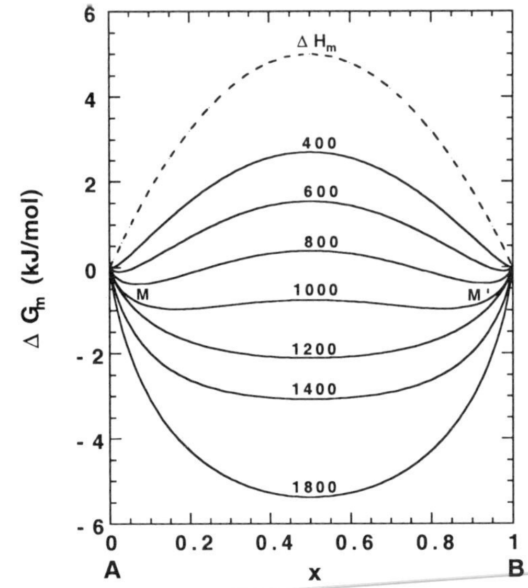

Figure 10.7. Gibbs free energies of mixing for a binary regular solution at the temperatures indicated (W = 20 kJ). Note how the enthalpy of mixing ΔH_m (which is included as a dashed curve and is assumed to be constant), is progressively overwhelmed by the temperature dependent entropy of mixing term.

When the temperature decreases, the relative contribution of ΔH_m to ΔG_m increases. Progressively two bumps develop in the Gibbs free energy curve of the mixture, which, as already described in detail in 7.3, cause two phases to separate. The stable state is still the homogeneous solution for compositions between A and M or between M' and B. In between these two regions, the solution separates into the conjugate phases of compositions M and M' which are determined by the points where the two bumps share a common tangent. Their proportions are calculated with the lever rule from the initial composition of the homogeneous solution.

The progressive growth of the bumps at lower temperature has the consequence that the points M and M' move away from each other and shift toward A and B, respectively. In a temperature-composition diagram, equilibrium between the two separated phases is represented by the bell-shaped solvus shown in Fig. 10.8. By giving at any temperature the composition of the conjugate phases, the solvus separates the trivariant domain of the homogeneous phase from that, divariant, of the equilibrium between the two separated phases.

10.4b SPINODAL, CRITICAL POINT AND SOLVUS CALCULATIONS

The spinodal is the locus of the inflexion points of the Gibbs free energy curve for the mixture (see 7.3). As seen in 7.3, it is characterized by

$$(\partial^2 G/\partial x^2)_{P,T} = 0. \tag{10.18}$$

At the critical point, the Gibbs free energy of the mixture presents a single extremum where the two spinodal branches merge. The critical point is thus defined by

$$(\partial^3 G/\partial x^3)_{P,T} = 0. \tag{10.19}$$

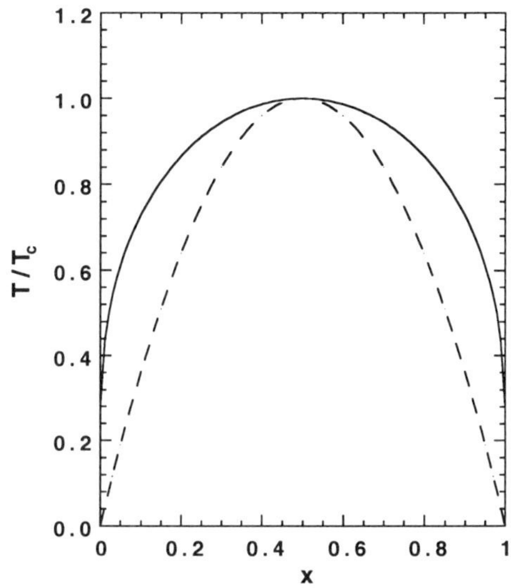

Figure 10.8. Solvus (solid curve) and spinodal (dashed curve) for a binary regular solution drawn as a function of the reduced temperature T/T_c

Since $(\partial G/\partial x)_{P,T} = (\partial \Delta G_m/\partial x)_{P,T}$, we have

$$(\partial G/\partial x)_{P,T} = W(1-2x) + RT[\ln x - \ln(1-x)] = 0. \tag{10.20}$$

The equation of the spinodal is thus

$$(\partial^2 G/\partial x^2)_{P,T} = -2W + RT[1/x + 1/(1-x)] = 0, \tag{10.21}$$

from which we find that

$$T = 2Wx(1-x)/R. \tag{10.22}$$

From (10.19), we could then determine the critical composition. In the present instance, however, it is simpler instead to make use of the symmetrical nature of a regular solution model, which imposes $x = 0.5$ for the critical composition. From (10.22), we find directly in this way

$$T_c = W/2R, \tag{10.23}$$

i.e., a value of 1203 K in the special case of Fig. 10.7.

Finally, for calculating the solvus itself we need to write once more that the chemical potentials of the two components are equal in the two phases

$$\mu_A = \mu_A' \qquad \mu_B = \mu_B'. \tag{10.24}$$

Let x_A and x_A' be the mole fractions of component A in the two conjugate phases. Those of component B are

$$x_B = 1 - x_A, \qquad x_B' = 1 - x_A'. \tag{10.25}$$

Because the solution is not ideal we must write for each component

$$\mu^\circ + RT\ln x + RT\ln\gamma = \mu'^\circ + RT\ln x' + RT\ln\gamma'. \tag{10.26}$$

The standard potentials μ° and μ'° are identical. As will be seen in 11.4b, the activity coefficient for a phase with composition x is given by

$$RT\ln\gamma = W(1-x)^2. \tag{10.27}$$

This allows (10.26) to be rewritten as

$$RT\ln(x_A/x'_A) = W[(1-x'_A)^2 - (1-x_A)^2], \tag{10.28a}$$
$$RT\ln(x_B/x'_B) = W[(1-x'_B)^2 - (1-x_B)^2]. \tag{10.28b}$$

For a given temperature, equations (10.25) and (10.28) constitute a system of four equations with four unknowns. In a general case, these would have to be solved numerically. For a regular solution, the symmetry with respect to two components imposes that $x_A = 1 - x'_A$. The equations (10.28a-b) are therefore the same and give for the solvus

$$T = \frac{W}{R}\frac{1-2x_A}{\ln[(1-x_A)/x_A]} = 2T_c\frac{1-2x_A}{\ln[(1-x_A)/x_A]}. \tag{10.29}$$

A principle of corresponding states results from the last equation: any solvus obtained with a regular solution model plots against T/T_c on the same master curve. Likewise, for the spinodal we have

$$T/T_c = 4x(1-x). \tag{10.30}$$

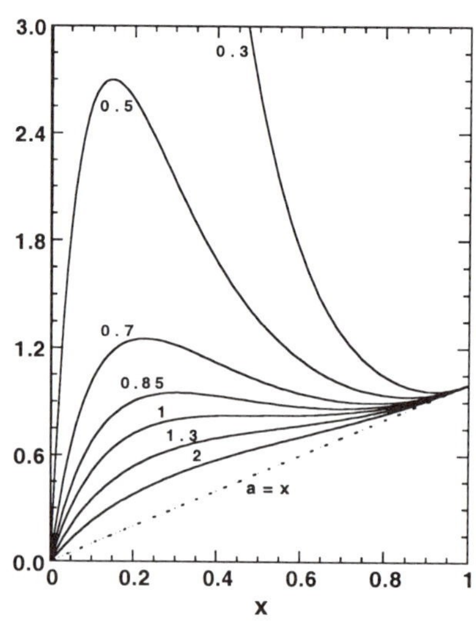

Figure 10.9. Reduced activity for a component in a binary regular solution as given by (10.3) for the reduced temperatures indicated. The critical point of the solvus is characterized by the horizontal slope of the activity curve for x=0.5 at T/T_c=1.

From (10.23) and (10.27) the activity can also be written in a reduced form

$$a = x\exp[2x(1-x)^2\, T/T_c]. \tag{10.31}$$

The activities calculated from this equation are compared in Fig. 10.9 to the ideal values $a = x$ for a series of reduced temperatures. As shown by Fig. 10.9, the critical point and incipient stability indeed correspond to the temperature at which the activity has an horizontal slope for the composition $x = 0.5$. We thus have

$$\left(\frac{\partial \ln a}{\partial x}\right)_{P,T} = \left(\frac{\partial^2 \ln a}{\partial x^2}\right)_{P,T} = 0 \tag{10.32}$$

10.4c CRITICAL MIXING TEMPERATURES

From the above discussion it should not be concluded that all solvus are of the kind shown in Fig. 10.8 although these are indeed the most common ones with an *upper critical mixing temperature*. Other kinds of systems display a *lower* critical mixing temperature (Fig. 10.10a) or even a closed solubility loop (Fig. 10.10b). In particular, aliphatic amines and alcoholic ethers in aqueous solutions form an homogeneous solution over limited temperature and composition domains. All these kinds of solvus can in principle be stable or metastable. Since the kinetics of unmixing get slower when the temperature decreases, the lower critical temperature of a metastable solvus is unlikely to be observed. For a liquid, crystallization should take place before it could be reached. In contrast, a metastable solvus with an upper critical temperature can be observed, for example in supercooled liquids when high viscosities impede crystallization but not liquid-liquid unmixing.

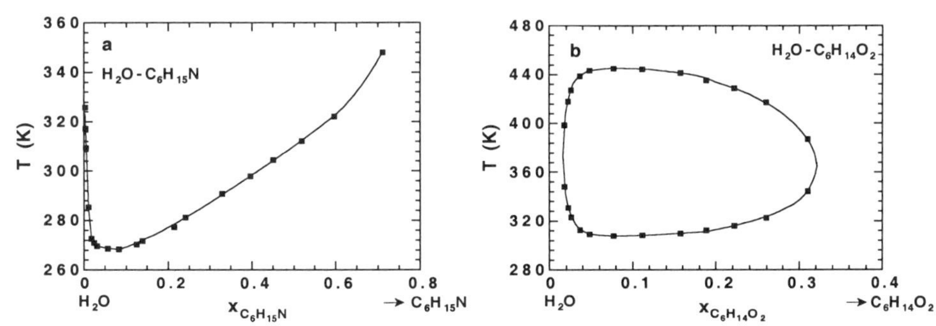

Figure 10.10. (a) Solubility of dipropylamine ($C_6H_{15}N$) in water (Hobson et al., 1941) ; (b) Solubility of 1-propoxypropan-2-ol ($C_6H_{14}O_2$) in water (Cox et al., 1927). Note that systems showing a lower critical temperature would actually be of the loop type if decomposition or vaporization were not taking place before the loop could be closed at higher temperatures.

In the particular case of the regular solution model, we have seen that a positive enthalpy of mixing was at the origin of phase separation and the existence of an upper critical temperature. In a more general way, it can be shown that the kind of solvus is determined by the sign of $(\partial^2 \Delta H_m / \partial x^2)$. If this partial derivative is negative, the enthalpy of mixing is positive and the system has an upper critical temperature. If positive, there is a lower critical temperature, implying that the solubility decreases with increasing temperatures. In this case, the entropy of mixing is clearly nonideal. Indeed, it must be smaller than for an ideal solution and be associated with a relatively small negative enthalpy of mixing and a moderate, positive heat capacity of mixing. For the diagrams showing a closed solubility loop, the enthalpies and entropies of mixing must in addition change markedly with temperature, which implies that the heat capacity of the solution varies in a strongly nonlinear fashion with the composition of the solution. The relative scarcity of such diagrams indicates that this delicate balance of thermodynamic functions seldom prevails.

10.5 AZEOTROPES, FRACTIONATION

10.5a AZEOTROPES

Noting that solubility is highest in the high-temperature phases, we will take advantage of liquid-vapor equilibria to introduce *azeotropes* (= "not to boil with change") with the example of the water-formic acid and bezene-ethanol systems (Fig. 10.11). These diagrams are characterized by the existence of a point, the *azeotropic point*, where the boiling and condensation curves are mutually tangential. These diagrams are very common and they actually occur in two kinds, positive or negative azetropes, depending on whether the pressure extremum is a maximum or a minimum (and, correlatively, whether the temperature extremum is a minimum or a maximum). The reader may care to examine by himself that this difference translates into opposite trends regarding the variation of the composition of a vapor in equilibrium with a liquid. As will be discussed below, azeotropic diagrams are in fact intermediate between

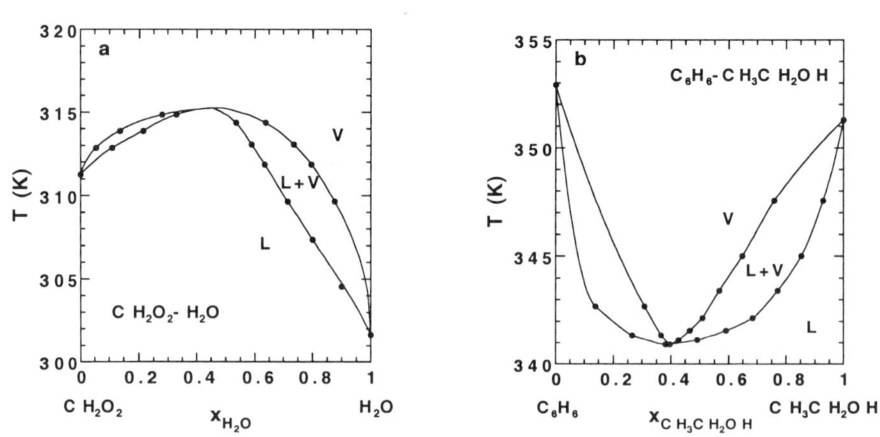

Figure 10.11. Azeotropic phase diagrams. (a) formic-acid-water at 0.066 bar (Kushner et al., 1966); (b) benzene-ethanol at 0.99 bar (Tyler, 1912).

eutectic and spindle-like diagrams. Like in the latter, mutual solubility is total in both phases but, like in the former, there exists an intermediate composition, the azeotropic point, where the solution behaves as a pure, congruently vaporizing substance, acting as a composition barrier dividing the phase diagram in two parts. This is why an azeotropic transformation is said to occur when, in a closed system, the mass of one of the phases increases at the expense of the other, without changing the composition of either phase.

For a positive azeotrope, the corresponding Gibbs free energies are plotted at the top of Fig. 10.12. Since they are similar to those described in detail for spindle-like systems, their high-temperature variations need not be reviewed. Rather, we will directly set the temperature to a value intermediate between the melting or vaporization temperatures of the solution endmembers and see that the basic feature of these diagrams is the little pronounced curvature of the Gibbs free energy in the low-temperature

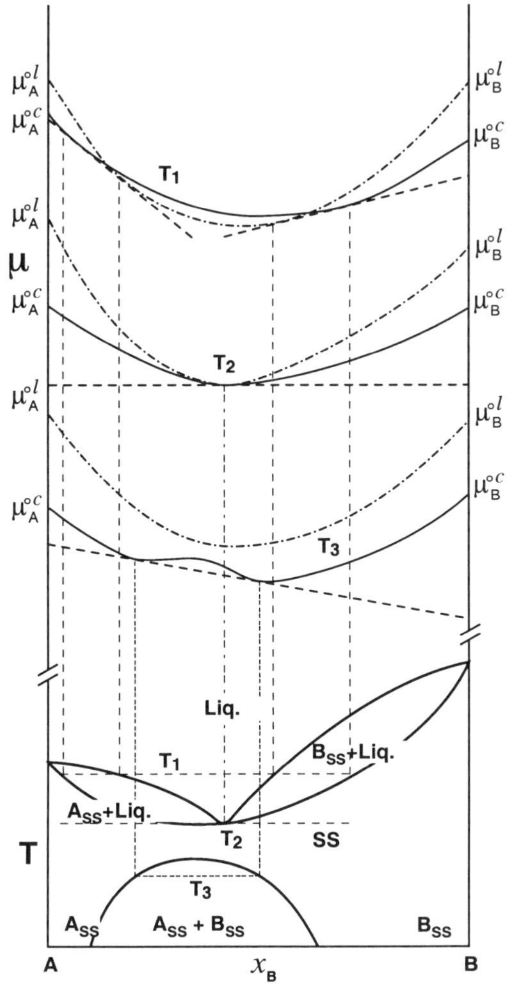

Figure 10.12. Gibbs free energies of mixing and positive azeotrope for liquid-solid equilibrium. SS = solid solution. Note the solvus shown below the azeotropic point.

(solid or liquid) compared to the high-temperature (liquid or vapor) phase. This can be due to a lower entropy of mixing in the former than in the latter and causes these two curves to intersect at two places. On both sides of the diagram, equilibrium is thus achieved by a two-phase assemblage (solid + liquid or liquid + vapor), whose compositions are as usual obtained from the common tangent method. At the center of the diagram, in contrast, a single phase is stable (liquid or vapor) as long as both Gibbs free energy curves have not eventually become tangential at the azeotropic point.

Because of this curvature difference, ideality is necessarily a poorer assumption in the low- than in the high-temperature phase. The existence of an azeotropic point thus indicates definite nonideality, and unmixing should generally be observed at lower temperatures, were it not for kinetic reasons, when the positive enthalpy of mixing predominates in ΔG_m (Fig. 10.12).

10.5b PARTITION COEFFICIENTS, CHEMICAL FRACTIONATION

In the case of equilibrium melting described by a spindle-like diagram, the component having the lowest melting point is enriched in the liquid phase. Likewise, the gas with the lowest vaporization temperature is enriched in the vapor, at constant P, as is, at constant T the gas having the highest vapor pressure. These features constitute the general basis of *chemical fractionation*, which is used to separate substances by partial melting, fractional crystallization or distillation. At equilibrium, any such chemical composition difference between phases can be characterized by a *partition coefficient* K. For a melting equilibrium in a eutectic or spindle-like system, this coefficient is defined by

$$K_A = \frac{[\mathrm{A}]_{\mathrm{liq}}/([\mathrm{A}]+[\mathrm{B}])_{\mathrm{liq}}}{[\mathrm{A}]_{\mathrm{crist}}/([\mathrm{A}]+[\mathrm{B}])_{\mathrm{crist}}}, \tag{10.33}$$

where [A] designates the concentration of A in a phase. For instance, let us use mole fractions as compositional variables. For a eutectic system, the partition coefficient K_A then reduces to x_A^l whereas it is x_A^l/x_A^c for a spindle-like system. In the latter case, the partition coefficient is readily obtained from (10.33) and (10.14)-(10.15), or from the analogous expressions for real solutions, if the standard affinities of fusion are known.

In this whole discussion, we have assumed that the two coexisting phases are in chemical equilibrium at all times. This assumption is generally valid for liquid-vapor transformations thanks to the high fluidity of the phases which makes continuous homogenization of their compositions possible. For reactions involving solid phases, this assumption is clearly wrong because the rate of homogenization relies on diffusion which is a very slow process in solids, even in the vicinity of the melting point. In practice, considerable time would be needed for solid-liquid and solid-vapor equilibria to take place as described above. Hence, we must consider another limiting case whereby melting and crystallization are too rapid for the crystal to reequilibrate its bulk composition continuously with the liquid. Fusion taking place at the surface, everything happens as if the solid were melting in successive, thin layers of constant composition. This purely kinetic effect would then erase all traces of chemical fractionation because the liquid would have the same composition as the solid. In contrast, in the same limiting case the laws of equilibrium govern the reverse crystallization process locally. The composition of the first layer of crystalline solid is determined by that of the liquid (Fig. 10.5). Because the composition of the liquid changes, so does that of the second solid layer. Without any reequilibration between its successive layers, the solid shows, from the outside to the inside, a continuous variation of chemical composition that parallels that of the liquid.

High chemical fractionation is ensured through incomplete reactions or, still better, through separation of the new phase and repetition of the process. Let us take the example of the water-ethanol system, which is azeotropic (Fig. 10.13). If 10% of a 90% water-10% ethanol mixture with 10 vol% ethanol (i.e., 3.3 mol% ethanol) is vaporized at room pressure, the vapor formed will contain about 30 mol% ethanol (Fig. 10.12). Let us isolate and completely condense this vapor, then vaporize only 10 % of this new liquid. The new vapor represents only 1 % of the initial amount of liquid, but its composition is almost 60 mol % ethanol (83 vol%). After repeated distillation, pure water can be approached for the liquid but the existence of an azeotropic point limits the highest ethanol content to about 90 mol%. To obtain "absolute" ethanol, distillation must be made after mixing with a third, entraining component. Although the ethanol-benzene diagram is also azeotropic (Fig. 11.11b), benzene is well suited in this respect because the azeotropic point is not a barrier at high ethanol concentration.

Rather than proceeding in a step-wise fashion, one industrially fractionates in a continuous way to reduce production costs and to process huge amounts of chemicals. In such processes, the liquid is alternately vaporized and condensed in superimposed *plates* in a distillation column, the number of *theoretical plates* representing the number of vaporization and condensation steps needed to prepare the desired distillate from a liquid with a given composition. For solids, an important process is *zone refinement*. The material to be purified is progressively melted, and then recrystallized, from one of its ends to the other. Because the impurities generally partition in favor of the liquid phase, they are found with a much lower concentration in the material that recrystallizes behind the hot front.

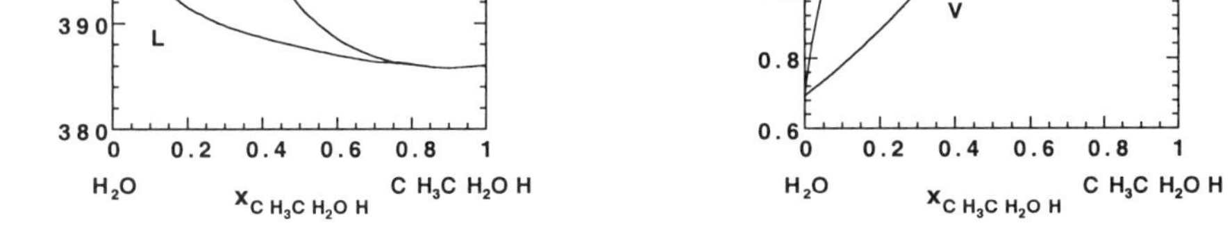

Figure 10.13. Fluid-phase equilibria in the system ethanol-water for a constant pressure of 1 bar (a) and for a constant temperature of 363 K (b). Note the wide azeotropic region close to the ethanol endmember. Diagrams from van Zandijcke and Verhoye (1974) and Pemberton and Mash (1978), respectively.

10.5c NERNST PARTITION LAW

An interesting, closely related question is partitioning of a solute coexisting between two phases. These can be a solid and a liquid, a liquid and a vapor, or two immiscible fluids. Regardless of the fact that ideality could be quite an inappropriate assumption for the main components of the host phases, there are many instances where solute partitioning can be dealt with by assuming that the solute is ideally diluted in both of the host phases. If this is the case, then equilibrium partitioning of solute i between phases α and β is determined by

$$\mu_i^{\alpha} = \mu_i^{\beta}. \tag{10.34}$$

From this condition rewritten as

$$\mu_i^{\circ\alpha} + RT\ln x_i^{\alpha} = \mu_i^{\circ\beta} + RT\ln x_i^{\beta}, \tag{10.35}$$

we obtain directly

$$K_i = \ln(x_i^{\alpha}/x_i^{\beta}) = -(\mu_i^{\circ\alpha} - \mu_i^{\circ\beta})/RT, \tag{10.36}$$

and thus find that the partition coefficient is independent of the actual compositions of either host phase. This result is known as Nernst partitioning law. If not observed at high dilution, the reason is most often that the solute reacts with one or both solvents. The equilibrium constants of such reactions have then to be taken into account in the calculation of the partition coefficient.

10.6 VARIATIONS ON BINARY THEMES

10.6a FROM EUTECTICS TO SPINDLE-LIKE DIAGRAMS, THE "RABBIT EARS"

This subtitle summarizes the changes induced by a progressive increase of the mutual solubility of two phases. In a eutectic diagram, the profound lack of miscibility in the low-temperature phases is in fact tantamount to a solvus with an infinite critical temperature. Hence, the theme to be discussed here is how a increasing mutual solubility affects the topology of stability fields and results in a progressive change from eutectic to spindle-like diagrams through azeotropes. Taking melting equilibria again as examples, we will examine how these combine with unmixing, which has been hitherto dealt with separately.

When a solvus is present below the liquidus in a eutectic diagram, the Gibbs free energies of the liquid and the solid phases resemble those previously plotted for azeotropic systems (Fig. 10.14). Owing to greater, positive enthalpies of mixing, the double minimum in the G curve of the solid phase appears in a higher temperature interval, when the G curves of the liquid and solid phases are still intersecting at two places. With decreasing temperature, the liquidus and solidus lines are both heading for the solvus curve. They meet it when, at the temperature T_2 of Fig. 10.14, the tangent to these two minima M and M' is also tangential to the G curve of the liquid at the point L. There, the Gibbs free energy of any phase assemblage plots on the same line MM' regardless of the proportions of liquid and solid.

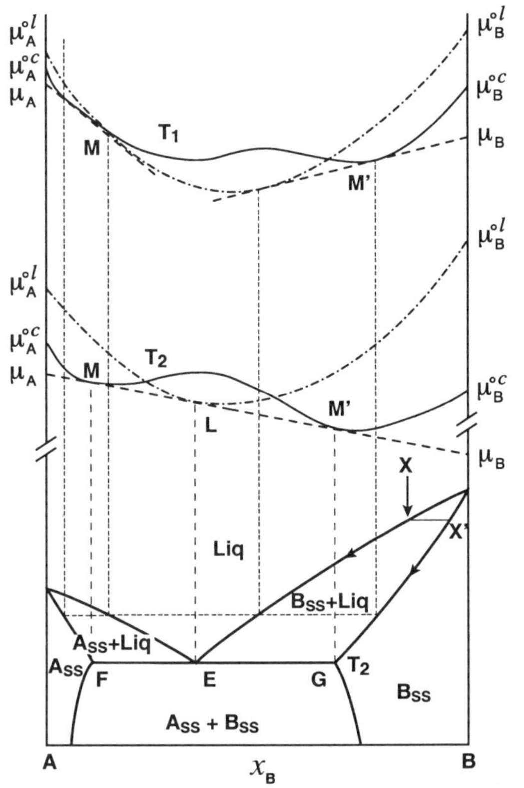

Figure 10.14. Gibbs free energies of mixing and eutectoid diagram. SS = solid solution.

To summarize the evolution followed in this diagram, let us cool a liquid of composition X (Fig. 10.14). The evolution followed at first is analogous to that already described in spindle-like diagrams, with the liquid and solid compositions moving along the arcs XE and X'G, respectively. Two cases have, however, to be distinguished. If composition X is more B rich than G (or more A rich than F), crystallization ends when the solid composition reaches the initial composition of the melt. If X is between G and F, the solvus is reached before crystallization is complete. The solid can no longer get more B- or A-rich. Instead, it phase separates into two other solids with the compositions F and G. Three phases coexist at this point in a two-component system: equilibrium is univariant. Because the pressure is already fixed, the composition of the residual liquid as well as the equilibrium temperature are thus determined. At this temperature, only solids with composition G or F can coexist with a liquid of composition E. Therefore, both solidus branches must intersect the solidus at the same temperature where, in becoming horizontal lines, they experience an abrupt change in slope. This temperature is also that at which both liquidus branches actually meet.

The point where the liquidus branches meet resembles a eutectic in that it also behaves as a pure substance. The liquid has the same composition as the solid made up of two phases with the compositions F and G whose proportions do not change. The "rabbit ears" are the two univariant equilibrium domains between the liquidus and solidus. The same topology can be observed when a continuous solid solution decomposes and yields two phases that have limited mutual miscibility. These analogs of eutectics are termed *eutectoids*. They are easily recognized in many systems because they give rise to characteristic lamellar textures showing alternate bands of each phase.

Finally, we emphasize that a strict eutectic does not exist. Any component has a finite solubility in a phase. As discussed in 7.2, it happens very often that this solubility is in practice negligible. In this respect, note that progress in microanalytical methods has shown the frequent occurrence of limited solid solutions at a level of a few % in systems that were previously thought to be eutectic.

A series of binary phase diagrams having a metallurgical, ceramic or mineralogical importance will illustrate the progressive change from eutectic to spindle-like diagrams. These diagrams have in common magnesia, MgO. Because chemical bonds are essentially ionic, the crystal structure of MgO is to a first approximation determined by the radius and the electrical charges of the Mg^{2+} et O^{2-} ions. The MgO structure is the same as that of NaCl (see Fig. 10.2), with both ions located at the apices and at the center of the faces of cubes. In agreement with this predominantly ionic character, the MgO bond length (2.106 Å) is close to the sum of the ionic radii of Mg^{2+} (0.72 Å) and O^{2-} (1.40 Å) in 6-fold coordination.

Let us start with melting in the MgO-BeO system. Although Be and Mg belong to the same column of the periodic table, Be^{2+} has so small an ionic radius (0.45 Å) when surrounded by six O^{2-} ions that it cannot be incorporated within the MgO lattice. Conversely, the BeO structure cannot accommodate Mg^{2+}. Melts are less demanding. If the liquid solution is probably far being from ideal, the mutual solubility of BeO and MgO is nonetheless total. Hence, the MgO-BeO phase diagram is eutectic (10.15a).

Let us then turn to the MgO-CaO diagram (Fig. 10.15b). In lime (CaO), the calcium ion Ca^{2+} is also divalent but its ionic radius (1.00 Å) is much larger than that of Mg^{2+}. Although both cations can mutually substitute, at contents higher than about 10% the positive strain energy of the lattice due to size mismatch becomes too high to be compensated by the entropy of mixing term in the Gibbs-free energy of the solution. The stable phases are lime with a small amount of MgO (CaO_{ss}) and MgO with a small amount of lime (MgO_{ss}), where ss designates a limited solid solution. This phase separation gives rise to the "rabbit ears" of Fig. 10.15b where the metastable extension of the solvus is included as a dashed curve.

Qualitatively, the MgO-ZnO diagram (Fig. 10.15c) is similar to the MgO-CaO diagram. The Zn^{2+} ion is also divalent, but its ionic radius (0.74 Å) is close to that of Mg^{2+}. The critical temperature of the solvus is not much higher than the eutectoid temperature. The solubility of MgO in ZnO remains limited whereas MgO can accommodate up to 50% ZnO. As a matter of fact, this strong asymmetry does not point to some crystallographic peculiarities, but results mainly from the much lower melting temperature of ZnO with respect to MgO. All other things being equal, see eqn (10.5), the crystal with the highest melting point has indeed the longest liquidus branch in a eutectic diagram.

Although unfortunately lacking, the next step would be the azeotrope shown at the bottom of Fig. 10.12, where the solvus would have contracted and passed below the solidus. (When the critical temperature of a solvus becomes, in this way, very low with respect to the solidus, the kinetics of unmixing tend to be much too slow for phase separation to take place even though it would ensure a stable equilibrium. The homogeneous phase then represents a metastable state.)

The final step of the evolution is shown by the spindle of the MgO-NiO system (Fig. 10.15d). Nickel can substitute in all proportions for magnesium on the sites of the MgO lattice. Even if not ideal, the solid solution is continuous because the ionic radius of Ni^{2+}

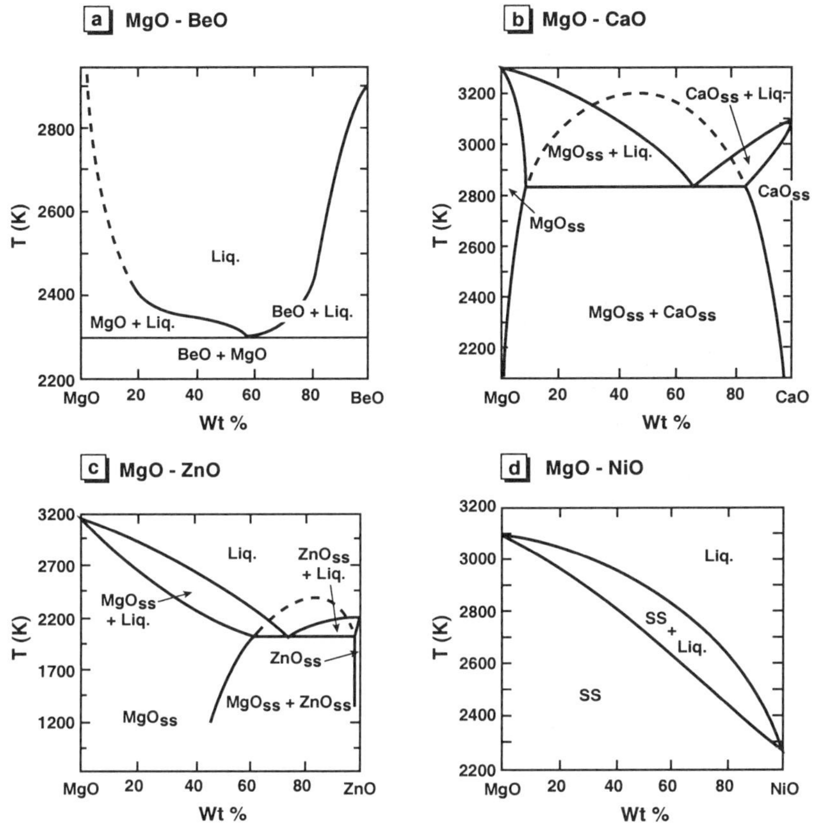

Figure 10.15. Binary phase diagrams between MgO and oxides of other divalent elements (Levin et al., 1964). Estimated stable or metastable phase boundaries drawn as dashed curves.

(0.69 Å) is very close to that of Mg^{2+}. In fact, the difference is the same (but of opposite sign) as between Zn^{2+} and Mg^{2+}. The differences between the NiO-MgO and ZnO-MgO diagrams thus show that, albeit an important factor, the ionic radius is not the sole parameter determining solubility in ionic crystals.

10.6b FORMATION OF INTERMEDIATE COMPOUNDS

The last important case to be discussed is that where attractive interactions make it possible for two widely dissimilar components to associate and form a new phase. The "NaCl" structure of MgO has nothing in common with the structure of quartz (or other SiO_2 polymorphs) which is related to that of ice with a three-dimensional open network of SiO_4 tetrahedra joined by their apices. As shown by the MgO-SiO_2 phase diagram (Fig. 10.16), the Mg^{2+} and Si^{4+} cations can nevertheless satisfy their bonding requirements better by forming intermediate compounds in which they share O^{2-} anions. In Mg_2SiO_4, this is achieved in the form of alternating isolated MgO_6

octahedra and SiO_4 tetrahedra whereas chains of octaedra fit together with chains of tetraedra in $MgSiO_3$. Both compounds have strongly negative Gibbs free energies of formation from the oxides (-57 and -35 kJ/mol, respectively, at 298 K), significantly negative volumes of formation (8 and 3%, respectively) and they separate the stability field of MgO from that of SiO_2 in the phase diagram so that the two endmembers cannot coexist (Fig. 10.16).

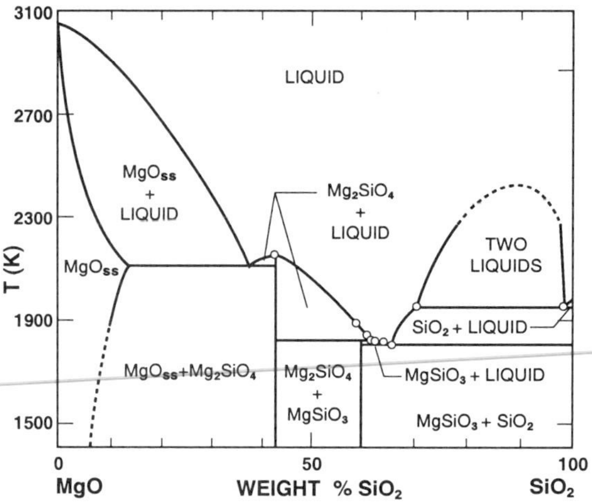

Figure 10.16. MgO-SiO_2 phase diagram (drawn from the observations of Bowen and Andersen, 1914, and Greig, 1927). Extrapolated phase boundaries drawn as dashed curves.

On the SiO_2-rich side of this diagram, another interesting feature is seen with stable liquid unmixing that develops between a few % and 30 wt% MgO. In the liquid state also, MgO and SiO_2 associate only if their relative amounts are adequate to build a common structure. Even close to 2000 K this requires a molar MgO:SiO_2 ratio of at least 1:1 in the liquid phase. At lower MgO contents, the melt splits into an almost pure SiO_2 liquid and an MgO-rich phase. The solvus is thus stable, and its metastable extension below the solidus is shown as dashed lines. In passing, note that Fig. 10.16 is of great interest in geophysics. It illustrates the fact that the Earth is not made up of oxide minerals because all major oxides form a variety of intermediate compounds with the most abundant one, namely, SiO_2. Especially important in this respect are Mg_2SiO_4 and $MgSiO_3$ (with, in both cases, $\approx$ 10% Fe substituted for Mg) which are the main phases present below the thin Earth's crust down to a depth of about 700 km.

10.6c CONGRUENCY, INCONGRUENCY

The various diagrams reviewed in this chapter show that phase changes are in general incongruent, in contrast to the melting, freezing or condensation of pure substances such as water which are congruent transformations. However, the congruent or incongruent nature of a transformation is not directly related to the apparent number of components of the system. In complex systems, congruency is observed at eutectic or azeotropic points where the new and original phases share the same composition.

As indicated by yet another feature of the MgO-SiO_2 system, however (Fig. 10.17), congruency is not necessarily the rule for "pure" substances. Instead of melting directly, the compound $MgSiO_3$ first transforms at 1830 K and 1 bar to a SiO_2-rich liquid and crystalline Mg_2SiO_4. Equilibrium has thus to be discussed within a binary system (whose components could be SiO_2 and MgO, or SiO_2 and Mg_2SiO_4). When three phases coexist in this way, the system is univariant ($c = 2$, $\phi = 3$) and defines a *peritectic* point (P in Fig. 10.17): since the pressure is fixed, the temperature cannot be raised before $MgSiO_3$ has entirely decomposed. When the system has become divariant again ($c = 2$, $\phi = 2$), the temperature can be increased and Mg_2SiO_4 begins to melt, inducing a progressive enrichment of the liquid in MgO. Melting is eventually complete at 1850 K when the composition of the liquid matches the original $MgSiO_3$ stoichiometry.

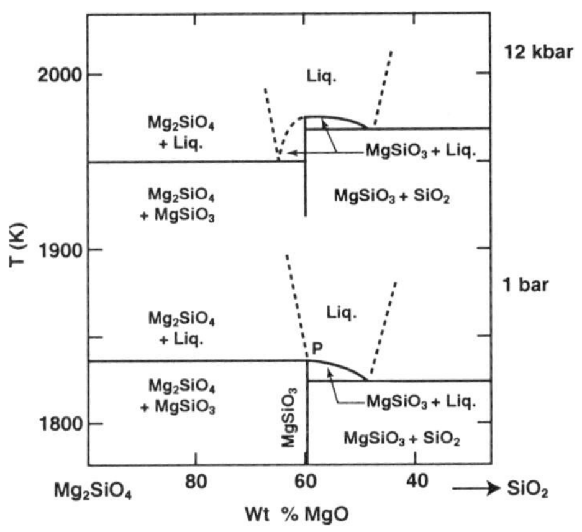

Figure 10.17. Incongruent melting of $MgSiO_3$ at 1 bar and congruent melting at 12 kbar (Chen and Presnall, 1975).

Such incongruent phase changes are common for pure substances, but $MgSiO_3$ has the particularity that the temperature interval of its incongruent melting contracts at higher pressures. Above 2 kbar, melting has become congruent such that $MgSiO_3$ is then a well-behaved pure substance with a univariant melting curve. This example thus shows, not only that it is through experimental evidence that the actual number of components can be ascertained for a system, but also that this number as well as the congruent or incongruent nature of a phase transformation can vary with pressure. This is why the notion of pure substance may be ambiguous.

10.6d MULTICOMPONENT SYSTEMS

Although this chapter is devoted to binary systems a few remarks on multicomponent systems are in order because in the next chapter we will focus on more theoretical aspects about them. We will first note that, if "simple" systems can have very complicated phase relationships, multicomponent systems do not necessarily have complex phase diagrams. Ternary, quaternary and higher order systems are composed of the same basic ingredients found in binary systems, namely, liquidus, solidus and solvus surfaces and eutectic, peritectic and azeotropic points which then involve more than two components. Joining the eutectic (or peritectic) points, one also finds the *cotec-*

tie lines, which can be viewed as the locus of the eutectics of pseudobinary joins, of which an example will be given in Fig. 11.5.

Independently of the fact that these diagrams should be represented graphically as two- or three dimensional projections of hypersurfaces, one should not be surprised that the amount of work required to determine them increases more rapidly than the number of components. Compared to the wealth of available binary diagrams, diagrams for multicomponent systems are thus scarce and their description, even cursory, is beyond the scope of this work. We just refer the interested reader to specialized books or available compilations of phase diagrams and point out that currently more effort is made to model them thermodynamically, taking advantage of binary or ternary information, than to determine them experimentally. We will sketch in the next chapter how such models can be set up.

10.7 PRESSURE EFFECTS

10.7a COMPOSITION AND TEMPERATURE OF CRITICAL MIXING

Equilibrium is univariant at the critical point of a binary solvus. The critical point thus defines a curve in the pressure, temperature, composition space along which the criteria (10.18-19) must be satisfied. In a differential form, these criteria can be written

$$d\left(\frac{\partial^2 G}{\partial x^2}\right) = \left[\frac{\partial}{\partial P}\left(\frac{\partial^2 G}{\partial x^2}\right)\right]_c dP + \left[\frac{\partial}{\partial T}\left(\frac{\partial^2 G}{\partial x^2}\right)\right]_c dT = 0, \tag{10.37}$$

$$d\left(\frac{\partial^3 G}{\partial x^3}\right) = \left[\frac{\partial}{\partial P}\left(\frac{\partial^3 G}{\partial x^3}\right)\right]_c dP + \left[\frac{\partial}{\partial T}\left(\frac{\partial^3 G}{\partial x^3}\right)\right]_c dT + \left[\frac{\partial}{\partial x}\left(\frac{\partial^3 G}{\partial x^3}\right)\right]_c dx = 0. \tag{10.38}$$

where x stands for x_1 or x_2. The analog of the Clapeyron equation is drawn from (10.37). Since a partial derivative does not depend on the order of the differentiation, thanks to (3.3c-d) this equation can be rewritten as

$$\frac{dT_c}{dP} = -\frac{(\partial^3 G/\partial x^2 \partial P)_c}{(\partial^3 G/\partial x^2 \partial T)_c} = \frac{(\partial^2 V/\partial x^2)_c}{(\partial^2 S/\partial x^2)_c}. \tag{10.39}$$

Making use of this result to evaluate dT_c/dP in (10.38), we find in the same manner that the variation of the critical composition with P is

$$\frac{dx_c}{dP} = \frac{(\partial^3 S/\partial x^3)_c(\partial^2 V/\partial x^2)_c - (\partial^3 V/\partial x^3)_c(\partial^2 S/\partial x^2)_c}{(\partial^2 S/\partial x^2)_c(\partial^4 G/\partial x^4)_c}. \tag{10.40}$$

It is not a surprise that the volume determines the variation with pressure of the critical temperature. For an upper critical temperature, $(\partial^2 S/\partial x^2)_c$ is negative and the sign of (10.39) thus differs from that of $(\partial^2 V/\partial x^2)_c$. For a lower critical temperature, dT_c/dP has the same sign as $(\partial^2 V/\partial x^2)_c$. Following Prigogine and Defay, we note that (10.40) has the special interest of being one of the rare equations relating fourth derivatives of the Gibbs free energy to direct observations.

10.7b EUTECTIC EQUILIBRIUM

For eutectic systems, an analog of the Clapeyron relation demonstrated in the previous chapter can apply only to the univariant equilibria, i.e., to the eutectic point. The situation is less simple than in chapter 9, however, because the composition of both phases is no longer constant. For each component, we must take into account the changes in chemical potentials when writing that the variation of the affinity for the transfer between the solid and liquid phases is zero. Taking the same standard state for component i in both phases, the μ_i° terms cancel out and we thus have

$$-\Delta S_i dT + \Delta V_i dP + RT\Delta \ln a_i = 0. \tag{10.41}$$

We will then again assume an ideal solution, i.e., $a_i = x_i$, before recalling that the Gibbs-Duhem equation (8.46) at constant T and P states that

$$x_1(\partial \ln x_1\gamma_1/\partial x_2) + x_2(\partial \ln x_2\gamma_2/\partial x_1) = 0. \tag{10.42}$$

It remains now to eliminate first ∂x_2 and then ∂T in (10.41) to obtain the variations with pressure of the temperature and composition of the eutectic point

$$\left(\frac{dT}{dP}\right)_e = \frac{-V_l + \Sigma x_{jl}V_{jc}}{-S_l + \Sigma x_{jl}S_{jc}} = \frac{\Sigma x_{jl}(\overline{V}_{jl} - V_{jc})}{\Sigma x_{jl}(S_{jl} - S_{jc})}. \tag{10.43a}$$

$$\left(\frac{dx_2}{dP}\right)_e = \frac{x_1\Delta S_1\Delta S_2}{RT(\partial \ln x_2\gamma_2/\partial x_2)(x_1\Delta S_1 + x_2\Delta S_2)}\left(\frac{\Delta V_1}{\Delta S_1} - \frac{\Delta V_2}{\Delta S_2}\right), \tag{10.43b}$$

where $\overline{V}_{jl}$ is the partial molar volume of component j in the liquid, the subscripts c and l designating the crystal and liquid phases, respectively, $\Delta S_1 = \overline{S}_{1l} - S_{1c}$, and similarly for ΔV_1.

Finally let us consider a divariant equilibrium along a liquidus branch. We then have one solid in equilibrium with the two-component solution. At constant temperature, we rewrite (10.42) as

$$\left(\frac{dx_2}{dP}\right)_T = -\frac{\Delta V_2}{RT(\partial \ln x_2\gamma_2/\partial x_2)_{T,P}}. \tag{10.44}$$

Because of the scarcity of experimental solid-liquid equilibria data at high pressure, these equations are not extensively used and there is no real need to treat more complex cases. For liquid-vapor equilibria, in contrast, a wealth of data has been gathered up to a few kbar pressure. Different approaches are then often used, in that one calculates P,X or T,X phase equilibria directly at the temperature or pressure of interest with models in which the pressure dependence of thermodynamic properties is directly incorporated.

Problems

1. The compounds $NaAlSi_3O_8$ (albite, Ab) and $CaAl_2Si_2O_8$ (anorthite, An) form a solid solution, the feldspars, which are one of the main constituent of igneous rocks. The solidus (T_{sol}) and liquidus (T_{liq}) temperatures are listed below as a function of the mol fraction of Ab. Plot the solid-liquid phase diagram and indicate the variance of the various equilibria. At equilibrium, what would be the temperature and the solid composition at which melting of a crystal with $x_{Ab} = 0.6$ would be complete? Same question if the crystal is in complete disequilibrium with the melt. With the assumption that the solid and liquid solutions are ideal, calculate this phase diagram by neglecting the C_p difference between the liquid

(l) and crystalline (c) phases. The enthalpies of fusion are 64.3 kJ/mol at 1391 K for Ab, and 133.0 kJ/mol at 1830 K for An. In the melting interval, one can in fact represent the heat capacities by the following equations (all data in J/mol K): $C^c_{pAb} = 283.01 + 0.031\ T$, $C^l_{pAb} = 261.8 + 61.87\ 10^{-3}\ T$, $C^c_{pAn} = 234.7 + 0.081\ T$, $C^l_{pAn} = 400.8 + 20.24\ 10^{-3}\ T$. Is better agreement achieved with the experimental diagram if one takes into account the C_p differences between the liquid and solid phases (for both of which it will be assumed that C_p varies linearly with composition).

x_{Ab}	0	0.17	0.33	0.50	0.67	0.75	0.80	0.87	1
T_{sol}	1823	1738	1645	1560	1478	1448			1391
T_{liq}	1823	1794	1763	1723	1667	1635	1607	1538	1391

2. The boiling temperatures of formic acid (CH_2O_2) and di-n-butyl ether ($C_8H_{18}O$) are 413,3 and 373,8 K, respectively. The temperatures at which the liquid and vapor phases of their mixtures coexist at 1 atm are listed below. Plot the liquid-vapor phase diagram and indicate the variance of the various equilibria. The mole fractions of formic acid in the liquid and vapor phases are denoted by x^l_f and x^v_f, respectively.

T (K)	393.3	387.2	382.7	379.3	376.8	375.0	373.8	373.2	373.1
x^l_f	0.2	0.3	0.4	0.5	0.60	0.7	0.8	0.9	0.95
x^v_f	0.548	0.667	0.747	0.803	0.844	0.875	0.901	0.926	0.946

3. Compare the two kinds of azeotropes and their different implications on chemical fractionation. Qualitatively, determine whether they correspond to positive or negative deviations of activities from a Raoult's law.

4. Sketch the Gibbs free energy-composition diagram for binary systems showing a solvus with a lower critical temperature.

5. Sketch the Gibbs free energy-composition diagram for binary systems where an intermediate compound forms eutectics with both endmembers.

6. Show that we have $(\partial^2 H \partial x^2)_c = T(\partial^2 S \partial x^2)_c$ at a critical point of unmixing. If the entropy of mixing of the solution is ideal, is a lower critical unmixing temperature possible? Sketch how the molar volume of the solution must vary with composition for dT_c/dP to be positive for upper and lower critical unmixing temperatures.

11
SOLUTIONS AND SOLUTION MODELS

We describe in this chapter various models commonly used for relating the activity to the composition of condensed phases for binary and multicomponent systems. Because the ideal solution model provides a reference frame against which real solutions are assessed, we will begin with another discussion of ideal entropy of mixing. We will also review excess properties and the activity coefficients pertaining to the various composition scales before discussing solvent properties, such as osmotic pressure or changes in partial pressure and freezing point, which mainly depend on solute concentration. A few nonideal solution models of general interest for multicomponent systems, like the regular and subregular models, will then be examined before discussing specific cases like multisite and reciprocal solutions. We will finish with a description of the specific problems raised in electrolytes by ionic dissociation and Coulombic interactions and show how they are accounted for by the Debye-Hückel and Pitzer models.

11.1 ENTROPY IN IDEAL SOLUTIONS

11.1a CONFIGURATIONAL ENTROPY IN A CRYSTAL

For ideal gases, the creation of entropy in a mixture can be interpreted as resulting from an irreversible expansion of the individual gases into the whole volume of the system. As derived in 8.2b, the entropy of mixing is

$$\Delta S_m = -R \sum x_i \ln x_i \tag{11.1}$$

where x_i is the mole fraction of gas i. In a more general way, we have defined as ideal any solution where the enthalpy of mixing is zero and (11.1) also applies. In the previous chapter, this entropy equation has been used to account for phase equilibria. We even used it to calculate a solvus, which seems strange since the assumption of ideal mixing is inconsistent with the existence of the repulsive interactions that eventually cause phase separation. This naturally raises the question of the meaning of (11.1) for condensed phases. Hence the first aim of this chapter will be to sketch the nature of the configurational changes embodied by this equation.

For this purpose, we will select a system of ice crystallites and examine how the configurational entropy increases with the number of atoms. This ice has the peculiarity of being made up of a mixture of $H_2^{16}O$ and $H_2^{18}O$ where the ^{16}O and ^{18}O isotopes are in a 3:1 ratio. In ice I, whose crystal structure is depicted in Fig. 6.2b, the unit-cell is made up of two H_2O molecules and has two distinct crystallographic sites where oxygen atoms are bonded to four hydrogen atoms to form a distorted tetrahedron. Because the following calculation would be less illustrative with only two such sites, we will take as an example another ice polymorph where there would exist four distinct oxygen sites. The distribution of the two kinds of oxygen isotopes over the four crystallographic sites will characterize particular configurations of the system which are amenable to macroscopic measurements. We can then imagine two limiting distributions of isotopes on the sites whereby either ^{18}O occupies only one site and leave the three others to ^{16}O, or both kinds of isotopes distribute randomly over the four sites. Because there are a great many different ways of achieving a given (macroscopic) configuration, we will denote by *complexion* each of these atomic arrangements which would be distinguished from each other only through individual observation of all atoms of the system.

We can now proceed and note that the ordered isotope configuration would be observed at very low temperature where the small energy difference between the four sites is non negligible. This results in the perfectly ordered distribution of isotopes that minimizes the enthalpy, which is the dominant term in the Gibbs free energy when T is very low. With only one way of distributing the two kinds of isotopes on the four sites, there would be only one complexion and the configurational entropy would thus be zero.

The second case of a disordered isotope configuration would prevail in the high-temperature limit. To calculate the associated entropy, let us first consider a fictive "crystallite" which would be made up of only four water molecules at rest in a given orientation. There are four different ways of distributing the ^{18}O isotope on these four distinct sites. Since the ^{16}O isotopes are indistinguishable, their 3! (= 3 x 2 x 1) = 6 arrangements are the same and only the four possible complexions defined by the single ^{18}O isotope have to be counted.

Denoting by Ω this number of complexions, we remark that the total number of complexions is $\Omega_1 = 4N_A$ (N_A= Avogadro number) for a set of $4N_A$ water molecules making up one mole of this peculiar crystallite. Let us then imagine different crystallites made up of 8 H_2O molecules but still having four kinds of tetrahedral sites. Among the eight oxygen sites, there are 8 and 7 ways of distributing the first and second ^{18}O isotopes, respectively. Because two ^{18}O isotopes are indistinguishable, however, there are only 8 x 7/2 = 28 complexions. For a set of $4N_A$ water molecules and $N_A/2$ cristallites, there would be $\Omega_2 = 28N_A/2 = 14N_A$ complexions. If the $4N_A$ water molecules were forming crystallites made up of 12 H_2O molecules, we would find 12 x 11 x 10 / 3! = 220 complexions for each crystallite and a total number of $\Omega_3 = 220N_A/3$ complexions. In the more general case of crystallites each having N unit cells, we count

$$\Omega_N = \frac{4N(4N-1)(4N-2)...(3N+1)}{N!} = \frac{(4N)!}{(3N)!\,N!} \tag{11.2}$$

complexions for each crystallite, whereas

$$\Omega_{N_A} = \frac{(4N_A)!}{(3N_A)!\,N_A!} \tag{11.3}$$

complexions characterize the configurational state of the whole system of $4N_A$ water molecules.

The important fact is that Ω increases tremendously with N. First of all, let us note that

$$\ln N! = \sum \ln N \sim \int_1^N \ln N\, dN = N \ln N - N, \tag{11.4}$$

where in this formula, known as Stirling's approximation, the sum has been replaced by an integral thanks to the great magnitude of N in macroscopic systems. Calculations show that the error made with the latter approximation is already smaller than 10^{-6} when N = 5 10^6. For any real system, where N is at least 10 orders of magnitude higher, we can thus safely rewrite (11.3) as

$$\ln \Omega_{N_A} = 4N_A \ln(4N_A) - 3N_A \ln(3N_A) - N_A \ln N_A = -4N_A(3/4\ \ln 3 - \ln 4). \tag{11.5}$$

As a matter of fact, this equation is the same as the entropy of mixing expression (11.1). With $x_1 = 3/4$ and $x_2 = 1/4$ we have in this case

$$\Delta S_m = -4R(3/4\ \ln 3 - \ln 4). \tag{11.6}$$

From this example, we are thus tempted to conclude

$$\Delta S_m = R/N_A\ \ln \Omega = k \ln \Omega, \tag{11.7}$$

where $k = R/N_A$ is the Boltzmann constant. Since the entropy of mixing is purely configurational in origin, we can write

$$S = k \ln \Omega. \tag{11.8}$$

As will be shown in 13.4, this relation between the configurational entropy and the number of *energetically equivalent complexions* in a system is exact.

To have a closer look at the meaning of this formula derived in another way by Boltzmann (1844-1906), let us examine how the *average* number of complexions per

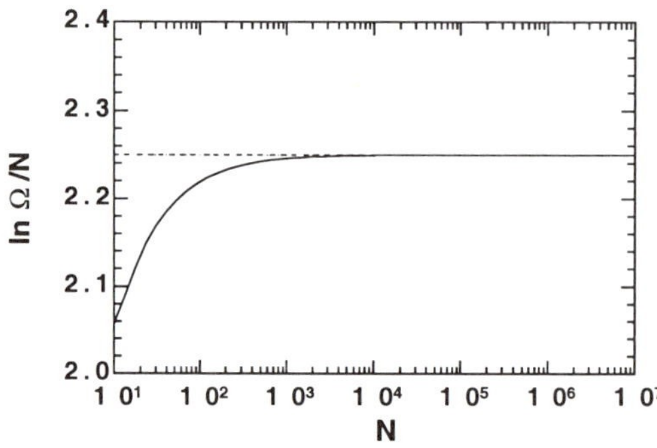

Figure 11.1. Variation of $\ln \Omega/N$ as given by (11.2) with the number N of crystallites for an isotopic mixture of ^{16}O and ^{18}O in a 3:1 ratio. Exact values: solid line; values calculated with Stirling's approximation: dashed line. Per oxygen atom, the 2.25 limit of $\ln \Omega/N$ indicates that the average number of complexions is 1.75.

crystallite $\ln \Omega/N$ increases with the number N of crystallites (Fig. 11.1). We observe that, after an initial increase, the entropy reaches a plateau for values of N of the order of 10^4 although the total number total of complexions actually keeps increasing. This demonstrates that what matters is not the total number of complexions but the number of most probable complexions. Hence, we could imagine a complexion where all ^{18}O isotopes would concentrate in a part of the ice crystallite and ^{16}O in another. Even though there is a huge number of such complexions, this number is actually negligible with respect to the number of complexions where the distribution is more homogeneous on a small scale. This is why the "average" distribution already prevails for a relatively modest number of crystallites. Mathematically, this happens as soon as Stirling's approximation becomes adequate (Fig. 11.1). In (11.5), $\ln \Omega$ then becomes an extensive variable by becoming proportional to N.

In ordinary ice, the $^{18}O/^{16}O$ ratio (= 0.002) is small enough to make isotope mixing negligible from an entropy standpoint. As pointed out by L. Pauling in 1935, however, the residual entropy of ice can be explained along similar lines. In a mole of ice, there are 2^{2N_A} complexions for the hydrogen bonds because each hydrogen can adopt two positions in between two adjacent oxygen atoms (Fig. 6.18a). Not all of them are possible, however, because they would give rise to unstable structural entities. For a given oxygen atom, indeed, 10 out of the 16 conceivable arrangements for the four hydrogens can be ruled out: one of them yields H_4O^{2+}, 4 give OH^-, 4 other H_3O^+ and a last one O^{2-}. Only 6/16 of the hydrogen complexions are thus allowed. They yield a configurational entropy $S = k \ln[2^{2N_A}(3/8)^{N_A}]$ = 3.37 J/mol K, which differs by less than 10% from the value 3.64 J/mol K derived from calorimetric measurements (see 9.3f).

11.1b CONFIGURATIONAL ENTROPY IN GASES AND LIQUIDS

The point to be examined now is the significance of (11.7) for fluid or solid solutions in which well-defined sites cannot be identified. For instance, how could we count the number of complexions in a fluid where particles are continuously moving in an apparently random way? Keeping with the simple approach that we have been following, let us consider gases and define an average volume per particle, be it an atom or a molecule. Without trying to give at this moment a more precise meaning to the concept of complexion for a fluid, we will consider this volume to be the analog of a crystallographic site in a solid. Let then V_1 and V_2, and Ω_1 and Ω_2 be the average volumes and the numbers of complexions for two different states of a gas made up of N particles, and consider constant temperature conditions to make sure that the entropy associated with the energy of the individual gas molecules is constant. When going from state 1 to 2, the change in the number of complexions would be given by

$$\Omega_2/\Omega_1 = (V_2/V_1)^N, \tag{11.9}$$

$$\ln \Omega_2/\Omega_1 = N \ln(V_2/V_1). \tag{11.10}$$

By inserting (11.9) in (11.8), we find

$$S_2 - S_1 = Nk \ln(V_2/V_1), \tag{11.11}$$

which is the result found in 1.4d for the entropy change due to isothermal expansion of a gas. With this crude argument we have thus extended the formal treatment of ion mixing in a crystal lattice, which owes its very existence to strong interatomic interactions, to isothermal expansion of an ideal gas where the configuration described by average volumes are not affected at all by molecular interactions.

As described in 13.3a, there exists an important difference between the counting made for a gas and a crystal. Whereas the molecules of a gas are indistinguishable, the atoms of a crystal could be indentified from their position in the lattice. For N gaseous molecules, there are thus $N!$ indistinguishable complexions, but this factor depends only on N so that it cancels out when a ratio of numbers of complexion is calculated. This is why (11.11) yields the correct entropy change due to expansion.

Now, what about liquids? These phases are intermediate between gases and crystals because a lack of long-range order coexists with definite short-range order and high atomic mobility. In a general way, however, there is not yet any simple approach with which use of (11.1) can be justified formally. Despite the wealth of information drawn from statistical mechanics, complexion counting cannot be made realistically for real liquids. Hence, use of (11.1) remains mainly justified by practical reasons. Experimentally, indeed, the thermodynamic properties of a variety of solutions are correctly accounted for by an ideal solution model. As is actually observed for crystals, it is not fortuitous that ideal mixing is most closely approximated in sytems where the components are the most similar. To give well-known examples, benzene (C_6H_6) and toluene (C_7H_8) solutions are nearly ideal, in contrast to water and acetone (CH_3COCH_3) solutions. As in the former case, however, one remarks that the apparent ideality often results from compensation between interactions originating in size differences and molecular forces.

11.2 MIXING AND EXCESS PROPERTIES

11.2a MIXING PROPERTIES

Mixing properties have already been introduced (see 8.2c and 8.6b). These are the differences between the solution properties and those of a mechanical mixture of the endmembers having the same bulk composition. Taking as usual the symbol Δ and the subscript m for designating a mixing property, we write similarly for any extensive property Y

$$\Delta Y_m = Y - \sum x_i Y_i^\circ, \tag{11.12}$$

where x_i and Y_i° are the mol fraction and property of endmember i. In the case of an ideal solution, we have also noted in 8.6b that the enthalpy and volume of mixing are zero whereas the Gibbs free energy of mixing is

$$\Delta G_m = RT \sum x_i \ln x_i. \tag{11.13}$$

Mixing properties can also be defined for intensive properties, such as thermal expansion coefficients and compressibilities. Obvious exceptions are temperature, pressure and chemical potentials, which cannot vary with composition in any system at equilibrium. Like for partial molar properties, many of the relationships established in the preceding chapters can be transposed to mixing properties. To give only one example, we have

$$\Delta H_m(T) = \Delta H_m(T_o) + \int_{T_0}^{T} \Delta C_{pm} dT, \tag{11.14}$$

and this equation shows that a solution has a constant enthalpy of mixing only if the heat capacity of mixing is zero at any temperature.

11.2b PARTIAL MOLAR MIXING PROPERTIES

It is also useful to define partial molar properties for mixing parameters. As a partial molar property, a chemical potential can for instance be expressed in terms of partial molar enthalpies and entropies

$$\mu_i = \overline{H}_i - T\overline{S}_i. \tag{11.15}$$

Denoting by H_i° and S_i° the enthalpy and entropy of the endmember i of the solution, and by $\Delta\overline{H}_{mi}$ and $\Delta\overline{S}_{mi}$ the partial molar mixing properties of component i in the solution, we write

$$\overline{H}_i = H_i^\circ + \Delta\overline{H}_{mi}, \tag{11.16a}$$
$$\overline{S}_i = S_i^\circ + \Delta\overline{S}_{mi}. \tag{11.16b}$$

Since $\mu_i^\circ = H_i^\circ - TS_i^\circ$, the chemical potential has the form

$$\mu_i = \mu_i^\circ + \Delta\overline{H}_{mi} - T\Delta\overline{S}_{mi} = \mu_i^\circ + RT \ln a_i. \tag{11.17}$$

This expression is especially useful in thermodynamic modeling of solutions as it relates the activity of a component to its partial molar enthalpies and entropies. With (11.1), we obtain for an ideal solution

$$\mu_i = \mu_i^\circ + RT \ln x_i, \tag{11.18}$$

but we will remark that a condition has been implicitly imposed to derive (11.18), namely, that the standard state is the pure component at P and T of interest.

11.2c EXCESS PROPERTIES

Because of the importance of ideal solutions as reference models, it is useful to define yet another category of variables, the excess properties. They represent differences between the properties of a given solution (Y) and those (Y^{id}) that the solution would display if it were ideal

$$Y^{xs} = Y - Y^{id}. \tag{11.19}$$

Excess properties therefore allow any problem dominated by mixing processes to be treated more simply. In view of its direct relationship with activity coefficients, the excess Gibbs free energy is a case in point. It is given by

$$G^{xs} = G - [\sum n_i\mu_i^\circ + RT\sum x_i \ln x_i]. \tag{11.20a}$$

Because the Gibbs free energy is

$$G = \sum n_i(\mu_i^\circ + RT\ln a_i) = \sum n_i(\mu_i^\circ + RT\ln x_i + RT\ln\gamma_i), \tag{11.20b}$$

we can rewrite (11.20a) as a function of the activity coefficients

$$G^{xs} = RT\sum n_i \ln\gamma_i. \tag{11.21}$$

With this equation the nonideality of the solution as a whole is thus simply related to that of the components individually. With respect to volume, enthalpy or heat capacities, mixing and excess properties are in contrast identical. To ensure consistency, it is nevertheless preferable not to mix them especially when writing Maxwell or other relations between mixing or excess properties such as

$$(\partial P/\partial T)_{V,n_i} = (\partial S^{xs}/\partial V^{xs})_{T,n_i}. \tag{11.22}$$

Since the Gibbs free energy of mixing can be determined from activity measurements, it is in principle possible to derive excess enthalpies and entropies from

$$H^{xs} = \left[\frac{\partial(G^{xs}/T)}{\partial(1/T)}\right]_P, \qquad S^{xs} = -\left(\frac{\partial G^{xs}}{\partial T}\right)_P, \tag{11.23}$$

and even excess heat capacities in two ways

$$C_p^{xs} = (\partial H^{xs}/\partial T)_P, \tag{11.24a}$$

$$C_p^{xs} = T(\partial S^{xs}/\partial T)_P. \tag{11.24b}$$

Unfortunately, available measurements rarely lend themselves to such determinations in view of the considerable work that would be needed to know G^{xs} over wide enough composition and temperature intervals. In addition, because of experimental uncertainties, the accuracy of the calculated data decreases considerably each time a differentiation is performed. Hence, it is preferable to take the opposite approach and measure instead excess heat capacities to determine the temperature dependences of H^{xs}, S^{xs} and G^{xs}. If the excess Gibbs free energy is known, then (11.21) indicates that activity coefficients can in turn be determined as a function of composition with

$$RT\ln\gamma_i = (\partial G^{xs}/\partial n_i)_{P,T,n_j}. \tag{11.25}$$

It follows from the Gibbs-Helmholtz equation (3.4) and the relations (11.22)-(11.23) that the temperature and pressure dependences of activity coefficients are given by

$$\left(\frac{\partial \ln\gamma_i}{\partial T}\right)_{P,n_j} = -\frac{1}{RT^2}\left(\frac{\partial H^{xs}}{\partial n_i}\right)_{P,n_j}, \tag{11.26a}$$

$$\left(\frac{\partial \ln\gamma_i}{\partial P}\right)_{T,n_j} = \frac{1}{RT}\left(\frac{\partial V^{xs}}{\partial n_i}\right)_{T,n_j}. \tag{11.26b}$$

11.2d A FAMILY OF ACTIVITY COEFFICIENTS

Up to this point we have been using only mole fractions as composition variables. The activity coefficients referring to this particular composition scale are termed *rational* to emphasize the fact, demonstrated below, that they have the value unity in any ideal solution (as long as the standard state is the pure component at the T and P of interest). For electrolytes, however, molarities (c_i) or, especially, molalities (m_i) are commonly used instead of mole fractions. This requires the definition of new activity coefficients with regard to these other composition scales, viz.

$$\mu_i = \mu_i^{\circ c} + RT\ln a_i^c = \mu_i^{\circ c} + RT\ln c_i + RT\ln\gamma_i^c, \tag{11.27}$$

$$\mu_i = \mu_i^{\circ m} + RT\ln a_i^m = \mu_i^{\circ m} + RT\ln m_i + RT\ln\gamma_i^m, \tag{11.28}$$

where γ_i^c and γ_i^m are the *molar* and *molal* activity coefficients, respectively, the latter being also termed *practical* activity coefficient. Rational activity coefficients are dimensionless. The other coefficients γ_i^c and γ_i^m are also dimensionless because the $RT\ln c_i$ and $RT\ln m_i$ terms of (11.27) and (11.28) are actually of the form $RT\ln(c_i/c_i^{\circ c})$ and $RT\ln(m_i/m_i^{\circ m})$, for the activity, and consequently the activity coefficient, are both unity for the standard state.

The various activities and activity coefficients necessarily differ from each other. To calculate for instance the difference between the molar and molal coefficients, let us subtract (11.28) from (11.27)

$$\ln\gamma_i^c = \ln\gamma_i^m + \ln m_i/c_i + (\mu_i^{\circ m} - \mu_i^{\circ c})/RT. \tag{11.29}$$

For solute i at infinite dilution, the ratio c_i/m_i tends to the density ρ_s of the pure solvent (see Table 4.1) and both activity coefficients become unity so that (11.29) reduces to

$$-\ln\rho_s + (\mu_i^{\circ m} - \mu_i^{\circ c})/RT = 0. \tag{11.30}$$

Because $\mu_i^{\circ m} - \mu_i^{\circ c}$ does not depend on concentration, we can make use of this equation to rewrite (11.29) for any concentration as

$$\ln\gamma_i^c = \ln\gamma_i^m + \ln(m_i/c_i) + \ln\rho_s, \tag{11.31a}$$

or

$$\gamma_i^c = \gamma_i^m \rho_s m_i/c_i, \tag{11.31b}$$

where the ratio m_i/c_i can be calculated from the conversion formula of Table 4.1.

To relate the rational to the molar and molal activity coefficients, one shows in a similar way that

$$\gamma_i = \gamma_i^m(1 + 0.001M_s \sum m_j), \tag{11.32}$$

$$\gamma_i = \gamma_i^c[\rho + 0.001(M_s \sum c_j - \sum c_j M_j)]/\rho_s, \tag{11.33}$$

where M_s is the molecular weight of the solvent, ρ the density of the solution and the summations are made over all the solute species. For an ideal solution ($\gamma_i = 1$), it results from these formulae that only the rational activity coefficients are *consistently unity*. Except at very high dilution, the molar and molal activity coefficients do not give such a real measure of nonideal behavior. In passing, note that (11.32)-(11.33) are also valid for mixed solvents. One has to use as an effective molecular weight $w_s/M_s + (1 - w_s)/M_s'$, where w_s and $(1 - w_s)$ denote the weight fractions of solvent components having molecular weights M_s and M_s', respectively.

11.2e STANDARD STATES: SOLUTES VS. SOLVENT

The existence of different composition variables has long been a source of confusion. It is often stated that molarities should be used only at constant temperature and pressure because they vary with these parameters even when the composition does not change. For this reason, the formulae (11.26) for the derivatives of activity coefficients do not apply to the molar coefficients whereas they are valid for the molal ones. The situation is in fact a little more complicated than it might seem, for molalities have their own drawbacks. For instance, molal standard quantities diverge when the solvent concentration tends to 0, which is formally awkward if not practically worrisome. But discussing the pros and cons of each concentration scale and associated standard state is clearly beyond the scope of this book. We prefer instead to emphasize that what matters above all is consistency. As already discussed for gases in *8.4a*, the standard state is just a fixed baseline from which the change in chemical potential due to any process can be calculated.

An essential fact is therefore that the activity coefficients of the components and the affinity of the reaction must be calculated consistently with the concentration scale. For the solvent, there is little ambiguity since the pure substance at the temperature and pressure of interest is always selected as the standard state. For solutes, the situation is more subtle when the same choice cannot be made. As briefly discussed in 8.5d, one must then make use of properties of an infinitely dilute solution. In this case, Fig. 11.2 summarizes the difference between the standard states selected for the solvent and the solute in a binary system. The two components no longer play a symmetrical

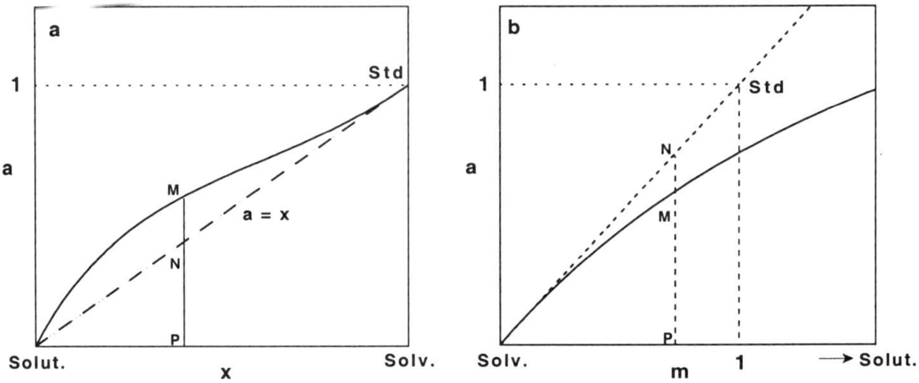

Figure 11.2. Comparison between the activity coefficients of the solvent and a solute. (a) For the solvent, γ = MP/NP directly measures the extent to which the activity of the solution departs from that of an ideal solution (dashed line). (b) For a solute, the activity-composition relationship is known over a limited composition interval so that γ = MP/NP measures the deviation of the activity from the tangent drawn at inifinite dilution (dotted line). As will be seen in 11.4a, Henry's law is valid in the composition range where the activity curve and this tangent coincide. In both cases the standard state is denoted by Std.

role, since for both of them γ_i tends to 1 when the concentration of the solvent tends to 1. For a solute, the standard state is selected so that the activity coefficient thus tends to 1 when the concentration tends to 0 at any temperature (and, generally, also at any pressure). In the molality scale, for instance, we write for the standard chemical potential of a solute

$$\mu_i^\circ = \lim_{\Sigma m_i \to 0} (\mu_i - RT \ln \gamma_i) \tag{11.34}$$

in the general case of a multicomponent solution. Similar equations will be written for other composition scales.

It remains to specify more precisely the standard state of the solute. It must correspond to a unit concentration (in the molality or molarity scale) to ensure that μ_i is then equal to μ_i°. This state is usually termed *hypothetical*, however, for its properties do not correspond to those of any real solution. As shown in Fig. 11.2b, its activity is determined through extrapolation of the activity of an infinitely dilute solution. Likewise, the partial molar enthalpy, volume or heat capacity are equal to the properties that an infinitely dilute solution would have. In contrast, the entropy and chemical potential diverge at infinite dilution because of the presence of a term in the logarithm of concentration in their expressions. Whereas we can write $\overline{H_i^\infty} = H_i^\circ$, we have $\overline{S_i^\infty} \neq S_i^\circ$, where the superscript ∞ indicates infinite dilution and $^\circ$ a standard quantity. As defined by (11.34), the standard chemical potential therefore differs from the potential of an infinitely dilute solution.

In this respect, let us stress again that the standard potentials μ_i°, $\mu_i^{\circ c}$ and $\mu_i^{\circ m}$ need to be distinguished although they all refer to the properties of one mole of substance in solutions of unit concentration. The various activity coefficients do not tend to 1 in the same way when the concentration of a solute (molar fraction, molarity or molality) tends to 0. With (11.30), we have for instance seen in the previous section that that $\mu_i^{\circ m} - \mu_i^{\circ c} = RT\ln\rho_s$, where ρ_s is the density of the solvent. We also note that the variations with temperature and pressure of the activity and activity coefficients depend on the standard state, and that the calculations made below for concentrations expressed in mole fractions are also valid for the molal scale.

Writing a chemical potential in the form $\mu_i^\circ/T = \mu_i/T - R\ln x_i - R\ln\gamma_i$, we first obtain from the Gibbs-Helmholtz equation

$$\left[\frac{\partial(\mu_i^\circ/T)}{\partial T}\right]_{P,n_j} = \left[\frac{\partial(\mu_i/T)}{\partial T}\right]_{P,n_j} - R\left[\frac{\partial\ln\gamma_i}{\partial T}\right]_{P,n_j}$$

$$= -\left[\frac{\overline{H}_i}{T^2}\right]_{P,n_j} - R\left[\frac{\partial\ln\gamma_i}{\partial T}\right]_{P,n_j}. \tag{11.35a}$$

Consider the limiting case where x_i tends to 1. If γ_i also tends to 1, the last term of this equation is zero and we have

$$\left[\frac{\partial(\mu_i^\circ/T)}{\partial T}\right]_{P,n_j} = -\frac{H_i^*}{T^2}, \tag{11.35b}$$

where H_i^* is the molar enthalpy of the pure solvent. Since the derivative of the standard chemical potential does not depend on the concentration of the solution, this formula is valid for any value of x_i. We can therefore write for the activity coefficient

$$\left(\frac{\partial\ln\gamma_i}{\partial T}\right)_{P,n_j} = \frac{H_i^* - \overline{H}_i}{RT^2}. \tag{11.36}$$

If γ_i tends to 1 when x_i tends to 0 (as for a solute), (11.35a) is written

$$\left[\frac{\partial(\mu_i^\circ/T)}{\partial T}\right]_{P,n_j} = -\frac{\overline{H_i^\infty}}{T^2}, \tag{11.37}$$

where $\overline{H_i^\infty}$ is the partial molar enthalpy of the solute i at infinite dilution. We then obtain the derivative of the activity coefficient

$$\left(\frac{\partial\ln\gamma_i}{\partial T}\right)_{P,n_j} = \frac{\overline{H_i^\infty} - \overline{H}_i}{RT^2}. \tag{11.38}$$

In addition to the choice of the composition scale, we recall that there exist two alternative conventions regarding the T,P conditions of the standard state, see 8.5a. The original "TP_0" convention selected for gases refers the standard state to 1 bar pressure. The affinity of a reaction is then independent of P. Since equilibrium conditions (and equilibrium constants) cannot depend on the choice of a standard state, it follows that activities vary with pressure. The price to be paid is that the activity coefficient of the solute in its standard state would generally differ from unity except at 1 bar. This is why the alternate "TP" convention is often preferred, whereby the pressure of interest is selected instead. In this case the affinity of a reaction depends on both pressure and temperature whereas activities depend only on temperature.

With the "TP_0" convention, the variation with pressure of the activity coefficient of the solvent is then given by

$$\left(\frac{\partial \ln \gamma_i}{\partial P}\right)_{T,n_j} = \frac{\overline{V}_i - V_i^*}{RT}, \tag{11.39}$$

where V_i^* is the volume of the pure solvent and $\overline{V}_i$ is its partial molar volume in the solution. For the solutes, we have alternatively

$$\left(\frac{\partial \ln \gamma_i}{\partial P}\right)_{T,n_j} = \frac{\overline{V}_i - \overline{V_i^\infty}}{RT}, \tag{11.40}$$

where $\overline{V_i^\infty}$ is the partial molar volume of i at infinite dilution.

11.2f A NEW LOOK AT EQUILIBRIUM CONSTANTS

We have seen in 8.6c that the most general expression for the equilibrium constant of reaction is given by eqn (8.60)

$$k = \prod a_i^{\nu_i} = \exp(A^\circ/RT), \tag{11.41}$$

where ν_i is the stoichiometric coefficient of species i and A° the affinity of the reaction. As mentioned in previous sections, the standard chemical potentials depend on the concentration scales used for species in solutions. The aforementionned difference in standard potentials thus translates into differences in equilibrium constant. This dependence of equilibrium constants on the concentration scale is analogous to that described in 8.3a for ideal gases.

We will denote the equilibrium constants by k, k_c and k_m for the mole fraction, molarity and molality scales, respectively. From (11.41), these constants appear to be different because the affinity of the same reaction depends on which scale is used. For instance, the relation (11.30) between $\mu_i^{\circ m}$ et $\mu_i^{\circ c}$ indicates that

$$k_c = k_m \rho_s^{n_0}, \tag{11.42}$$

where n_0 is the the number of chemical potentials considered when calculating the affinity of the reaction. For electrolytes, this number is not determined by the number of ionic species participating in the reaction, but by that of the electrically neutral components whose activities and standard potentials are usually considered (see 11.6c). Likewise, we find from the other relationship derived in a similar manner between $\mu_i^{\circ m}$ and μ_i°

$$k = k_m (M_s/1000)^{n_0}. \tag{11.43}$$

The difference between the molal and molar equilibrium constants therefore depends on the solvent density. It is minor for water near room pressure and temperature, but can be important under other conditions or for other solvents.

11.3 COLLIGATIVE PROPERTIES

Under the qualifier of *colligative* one gathers thermodynamic properties, like the volume for ideal gases, that are the same for specific classes of substances regardless of their chemical composition. For liquid solutions they include osmotic pressure, freezing point depression and boiling point elevation because, at low enough dilution, these properties depend only on the total amount and not on the nature of solutes. Historically, they were very important in the early days of chemical thermodynamics as they brought much information about species in solutions at a time, the turn of 20th century, where little was known about atoms and molecules. They are still of great interest although, as will be seen in the next chapter, for electrolytes interest has somewhat shifted toward electrochemical measurements. In both cases, however, dilute solutions are still important in that they represent boundary conditions for real solutions. Usually emphasis is put on liquid-vapor equilibria. Because phase equilibria have already been dealt with in the previous chapter, we will focus instead on activities. Incidentally, this will have the advantage of making applications of Raoult's and Henry's laws immediate for all phases, including solids, and not only to liquids.

11.3a OSMOTIC PRESSURE

When used for separating solutions with different concentrations of solutes, membranes porous to only some solutes allow important *osmotic pressures* to develop. To give a classic example of the phenomenon, let us consider two compartments containing either pure water or a dilute aqueous solution of sucrose ($C_{12}H_{22}O_{11}$), separated by a membrane porous to the small water molecules but impermeable to the big sucrose molecules. Hence, there are large chemical potential differences across the membrane for both water and saccharose. Because sucrose cannot diffuse into pure water, the only way to achieve chemical equilibrium is to let the water diffuse from pure water to the aqueous solution. To suppress such a water intake, one should keep the volume of the solute compartment constant, by exerting a certain pressure Π on it. This pressure must be tuned rather finely. If slightly too low for a given solution, water keeps flowing in. If slightly too high, water flows out. Alternatively, it is possible to perform the experiment in a quite different way with an *osmometer*. With this apparatus shown in Fig. 11.3, solvent diffusion ends as soon as the height difference between the columns of solvent and solution has reached a value corresponding to the aforementioned equilibrium pressure Π.

That equilibrium is reached means that the chemical potential of the solvent in the solution compartment (μ_s^l) has been raised, by the action of pressure, to the chemical potential of pure solvent (μ_s°). Now, this pressure dependence of the chemical potential is related to the partial molar volume of the solvent by

$$(\partial \mu_s^l / \partial P)_T = \overline{V}_s. \tag{11.44}$$

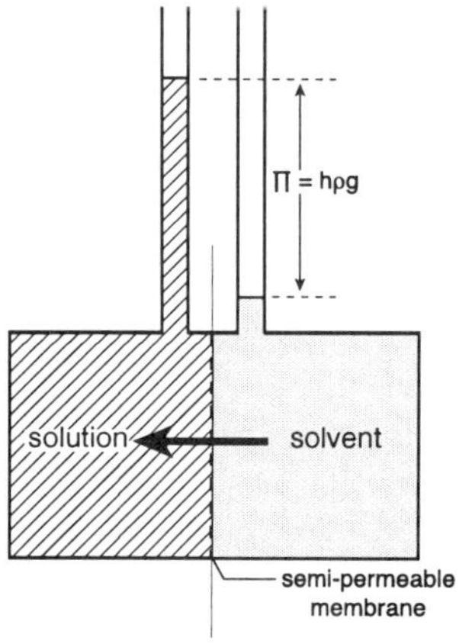

Figure 11.3. Osmotic pressure : schematic representation of an osmometer.

At equilibrium, we have $\mu_s^\circ = \mu_s^l$, where μ_s° is the chemical potential of pure solvent and μ_s^l that of the solvent in the liquid solution. This implies that

$$\mu_s^\circ = \mu_s^{\circ l} + RT\ln a_s + \int_{P_0}^{P_0+\Pi} \overline{V}_s\, dP. \tag{11.45}$$

Taking the pure solvent at P_0 and the temperature of the experiment as the standard state, we have $\mu_s^{\circ l} = \mu_s^\circ$, whence

$$RT\ln a_s = -\int_{P_0}^{P_0+\Pi} \overline{V}_s\, dP. \tag{11.46}$$

Because Π is typically of the order of 10 bars, one can generally neglect the variation of $\overline{V}_i$ with pressure. In this case we have

$$RT\ln a_s = -\overline{V}_s\Pi. \tag{11.47}$$

If the partial molar volume of the solvent is known independently, the activity of the solvent can thus be determined with (11.47) from osmotic measurements. But it is only at low dilution, i.e., when the solution can be assumed ideal, that this equation simplifies to

$$RT\ln x_s = -\overline{V}_s\Pi, \tag{11.48}$$

and that osmotic pressure is actually a colligative property. Under these conditions, the molecular weight of the solute can be obtained from the given mass of solute added and the value of $1 - x_s$ as determined from the measured osmotic pressure.

With biological cell walls, semi-permeable membranes have been working for eons. Owing to large total solute molarities of about 0.5, the osmotic pressure in animal cells is of the order of 10 bars, whence the ease with which a cell blows up when immersed in water or dilute aqueous solutions. Made in the 19th century, the very first measurements indicated that the osmotic pressure of a 1% sucrose solution is 53.5 mm Hg (0.07 bar). As illustrated as early as 1927 with hemoglobin (M = 67 000), these experiments have long been the main means of determining molecular weights of macromolecules. (Whereas the concentration of these big molecules in solution is too low to make freezing point depression measurements accurate, membranes work best when there is a large size difference between solvent and solute molecules). For a very dilute solution, note finally that the sum $\sum x_j$ of the mole fractions of the solutes can be approximated by $\sum n_j/n_s$, where n_s is the mole number of solvent. This allows (11.48) to be rewritten as $\Pi = \sum n_j RT/V = RT\sum c_j$, where c_j denotes the molarity of solute j. This formula derived by van't Hoff led to the conclusion that the osmotic pressure would be the pressure exerted by the solute if, in the gaseous state, it were occupying the same volume as the solution. This formal analogy with the ideal gas equation of state can be misleading, however. The entropy-driven tendency of occupying all available space actually is a common point between the gas molecules and the components of a solution, but the comparison is valid only between an ideal gas, on the one hand, and a solution that is not only ideal, but also highly dilute on the other hand.

11.3b RAOULT'S LAWS

The laws determined by Raoult (1830-1901) describe several manifestations of ideal solution behavior. This is why ideal and Raoultian are often taken as synonymous terms. According to Raoult's first law, the partial pressure of a component is proportional to its mole fraction in the solution

$$P_i = x_i P_i^*, \tag{11.49}$$

where P_i^* is the vapor pressure of pure i at the same temperature. In section 8.5c, we found that the vapor pressure ratio P_i/P_i^* is equal to the activity of component i (standard state: pure component at P and T of interest). For an ideal solution, in which activities are equal to mole fractions, Raoult's law is thus accounted for at once.

Raoult's second law states that equimolar solutions of the same solvent have the same freezing point. More precisely, the difference between the freezing temperature of a pure solvent and its liquidus temperature in a solution (in short, the freezing point depression) is proportional to the total solute mole fraction

$$\Delta T = (1 - x_s)K_s, \tag{11.50}$$

where x_s is the solvent mole fraction and K_s a characteristic parameter of the solvent. Because $\ln x_s \sim 1 - x_s$ for a dilute solution, this law is a much simplified form of the Schröder-van Laar equation (10.6)

$$\ln x_s = -\frac{\Delta H(T_s)}{R}\left(\frac{1}{T} - \frac{1}{T_s}\right), \tag{11.51}$$

where K_s is related to the temperature and enthalpy of fusion of the pure solvent, T_s and $\Delta H(T_s)$, by

$$K_s = \frac{RTT_s}{\Delta H(T_s)} \sim \frac{RT_s^2}{\Delta H(T_s)}. \tag{11.52}$$

The freezing point depression in dilute solutions thus depends only on the mole fractions of the dissolved species, not on their nature.

With regard to solid-liquid equilibria, the solvent is stabilized in the solution because its chemical potential and activity are lowered through the entropy produced on mixing with the solute. For liquid-vapor equilibria, miscibility is in contrast total in both phases and the latter effect is not relevant. As described by Raoult's first law, the partial pressures are then proportional to the mole fractions. When nonvolatile solutes are added to a liquid solution, the entropy-of-mixing contrast is again effective. It gives rise to Raoult's third law, which states that the elevation of the boiling point of the solvent is proportional to the total mole fraction of nonvolatile dissolved species, $\Delta T = (1 - x_s)K'_s$, where the constant K'_s is given in terms of the enthalpy of vaporization by an expression similar to (11.52). If either K_s or K'_s are known, then ΔT measurements in a binary system allow the mole fraction of the solute to be determined. From the mass of dissolved solute, the molecular weight of the solute is then obtained from its mole fraction.

11.4 GENERAL SOLUTION MODELS

Given the important experimental effort needed to measure thermodynamic properties and phase equilibria, modeling rightly appears as an effective and time saving way of understanding and predicting the properties of multicomponent systems. For reasons that are not all purely thermodynamic, however, specific models have been devised for specific kinds of substances. Petroleum engineers, organic chemists, metallurgists, ceramists or geochemists all have their favorite models which form the basis of long computer codes fitted to their particular needs. Reviewing in this section even a fraction of available solution models would thus be impossible, especially since this diversity of models is not going to disappear, especially for nonelectrolytes. Hence, we will present instead a few basic models of general interest. As such, they cannot address particular features of forces in solutions, but they will allow us to emphasize the need for internal consistency between activity coefficients when multicomponent systems are dealt with.

11.4a HENRY'S LAW

While measuring gas solubility in water, W. Henry found in 1803 that the amount of dissolved gas is proportional to the gas pressure. For a gas j, we thus have

$$P_j = k_j c_j \tag{11.53}$$

where k_j is the *Henry's law constant* and c_j the molarity of dissolved j. In fact, this law is a special instance of the more general case where the activity coefficient of a solute is constant at low concentration. Because all different ways of expressing concentrations become proportional to mole fractions in the case of dilute solutions, see 4.1, Henry's law can express this more general statement with any compositional variable. For mole fractions we thus write

$$a_j = k_j^x x_j. \tag{11.54a}$$

A similar equation is written for molalities

$$a_j = k_j^m m_j, \tag{11.54b}$$

and for weight fractions, which are often convenient when dealing with minute solubilities in solid solutions

$$a_j = k_j^w w_j. \tag{11.54c}$$

An important implication of Henry's law is drawn from the Gibbs-Duhem equation (8.45). For a binary system at constant temperature and total pressure, we have

$$x_1 d\ln a_1 + x_s d\ln a_s = 0 \tag{11.55}$$

where the subscripts 1 and s denote the solute and the solvent, respectively. Because $dx_1 = -dx_s$ and $d\ln\gamma_1 = 0$ for the solute, we rewrite (11.55) as

$$dx_1 + x_1 d\ln\gamma_1 + dx_s + x_s d\ln\gamma_s = x_s d\ln\gamma_s = 0, \tag{11.56}$$

and conclude that the activity coefficient of the solvent is also constant, i.e., that $a_s = k_s x_s$. If the pure solvent is taken as a standard state we must have $k_s = 1$ in order that a_s tends to x_s when x_s tends to 1. Hence, the solvent is Raoultian when the solute follows Henry's law, a conclusion also arrived at for any number of solutes as long as these are all Henrian. As apparent from (11.49), Raoult's first law is a special case of Henry's law with $K = 1$. (If written as $P_s = k_s x_s$ for expressing the partial pressure of the solvent above a liquid solution, Raoult's law thus implies $k_s = P_s^*$).

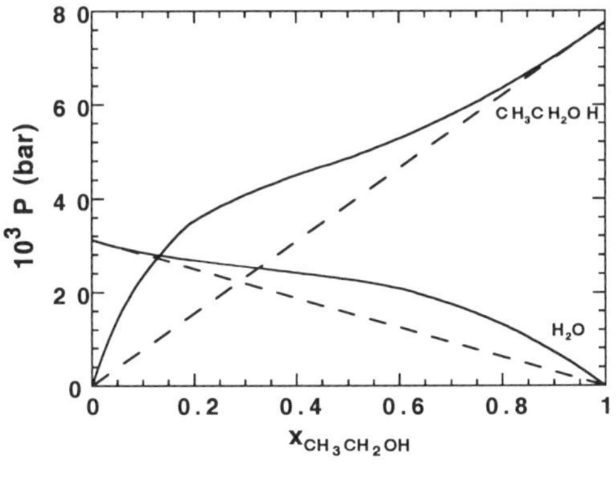

Figure 11.4. Positive deviations from Raoult's law: partial pressures of water and ethanol at 298 K.

The validity of these laws for describing the limiting behavior of real solutions is illustrated in Fig. 11.4 for water-ethanol solutions. For both components, Raoult's law is followed at high contents where the partial pressure curves become tangential to the (dashed) $P_i = x_i P_i^*$ Raoult lines. At high dilution, in contrast, observance of Henry's law is clearly indicated by the differences between Raoult's lines and the initial slopes of the vapor pressure curves. For ethanol, the partial pressure is given by $P_e = 0.315\ x_e$ bar when x_e tends to 0. At any mole fraction, we observe that the vapor pressures of both ethanol and water are higher than predicted by Raoult's law. Their deviations from this law are thus positive, indicating that the vapor phase is stabilized with respect to the liquid because of predominantly repulsive interactions in the latter. If these interactions were essentially attractive, then the vapor pressures would be lower than for an ideal solution and the deviations from Raoult's law would be said to be negative.

With respect to Henry's law, recall that activities in continuous solutions can in principle be referred to either the pure component or the hypothetical solution of unit concentration (at the same P and T) as a standard state. With pure ethanol as a standard state, for $x_e = 0.3$ we have $\gamma_e = P_e/x_eP_e^* = 44.1\ 10^{-3}/(0.3 \text{ x } 77.6\ 10^{-3}) = 1.89$; alternatively, $\gamma_e = P_e/x_ek_e = 44.1\ 10^{-3}/(0.3 \text{ x } 0.315) = 0.467$ with the hypothetical molar solution as a standard state. We again recall that solute activities cannot be measured at vanishingly small concentrations, see 8.5d, but have to be determined instead from extrapolations of available data with appropriate solution models. A few examples will be described below for electrolytes. For dilute solutions of nonelectrolytes, the importance of Henry's law stems from the fact that it is held to be generally valid down to infinite dilution, which is tantamount to assuming that da_j/dx_j remains finite when x_j tends to 0. In some cases, it is at concentrations much lower than 1% that Henry's law is eventually obeyed. Especially when dealing with solids, one should be cautious with Henry's law by keeping in mind that several linear activity-composition regimes could exist at successively lower dilutions whereby different "Henry's law" constants would be observed. In other words, extrapolating actual measurements at finite concentration down to infinite dilution is not necessarily warranted.

11.4b REGULAR SOLUTIONS

Perhaps the regular solution constitutes the most widely used kind of nonideal model for nonelectrolytes. The calculation of a solvus made in 10.4a has already shown its attractiveness. From the expression (10.17) of the Gibbs free energy of mixing for a binary system, we find that

$$G^{xs} = W_{12}x_1x_2. \tag{11.57}$$

The W_{12} interaction parameter must often be considered as temperature or even pressure dependent whereas its sign is the same as that of the (positive or negative) deviations from Raoult's law. For both components the standard state is taken as the pure substance at P and T of interest. With (11.25) activity coefficients are then readily calculated. For component 1, we have

$$RT\ln\gamma_1 = W_{12}(1-x_1)^2 = W_{12}x_2^2. \tag{11.58}$$

With a regular solution one assumes an ideal entropy of mixing and thus ascribes the excess Gibbs free energy to a sole enthalpy of mixing term

$$H^{xs} = \Delta H_m = W_{12}x_1x_2. \tag{11.59}$$

When measured for solutions conforming to (11.58), however, enthalpies of mixing are often inconsistent with (11.59). As far as phase equilibrium is concerned, this is not necessarily a real problem because what matters practically is the activity. Splitting the excess Gibbs free energy into enthalpy and entropy terms can thus be arbitrary to some extent. When interaction parameters are temperature dependent, one clearly highlights this intermingling of enthalpy and entropy terms by using the relations (11.23) to differentiate (11.57)

$$H^{xs} = \left[\frac{\partial(G^{xs}/T)}{\partial(1/T)}\right]_P = x_1 x_2 \left[W_{12} - T\left(\frac{\partial W_{12}}{\partial T}\right)_P\right], \tag{11.60a}$$

$$S^{xs} = -\left(\frac{\partial G^{xs}}{\partial T}\right)_P = -x_1 x_2 \left(\frac{\partial W_{12}}{\partial T}\right)_P, \tag{11.60b}$$

$$C_p^{xs} = -x_1 x_2 T \left(\frac{\partial^2 W_{12}}{\partial T^2}\right)_P. \tag{11.60c}$$

In addition, excess volumes become relevant when interaction parameters also depend on pressure

$$V^{xs} = \left(\frac{\partial G^{xs}}{\partial P}\right)_T = x_1 x_2 \left(\frac{\partial W_{12}}{\partial P}\right)_T. \tag{11.61}$$

11.4c SERIES EXPANSIONS

Expressing thermodynamic variables as series expansions was originally proposed in 1895 by Margules (1856-1920), a meteorologist, for vapor pressures in binary systems, viz.

$$\ln \frac{P_1}{P_1^* \, x} = a_1(1-x) + 1/2\, b_1(1-x)^2 + 1/3\, c_1(1-x)^3 + \ldots \tag{11.62}$$

$$\ln \frac{P_2}{P_2^* \, (1-x)} = a_2 x + 1/2\, b_2 x^2 + 1/3\, c_2 x^3 + \ldots \tag{11.63}$$

where x is the mole fraction of component 1. These expressions reduce to Raoult's law when the excess parameters a, b, etc., vanish, and they tend asymptotically to Henry's law when either mole fraction tends to 0. For instance, the Henry constant of component 2 is equal to $\exp(b_2/2 + c_2/3)$. The number of independent parameters can be reduced through differentiation of these expressions and application of the Gibbs-Duhem equation. In this way we find that

$$a_2 = a_1 = 0, \qquad b_2 = b_1 + c_1, \qquad c_2 = -c_1. \tag{11.64}$$

Since vapor pressures are but one particular kind of variable, nowadays a more general practice consists of expanding the Gibbs free energy. For a binary system, we write

$$G^{xs} = x_1 x_2 \left[A_0 + A_1(x_1 - x_2) + A_2(x_1 - x_2)^2 + \ldots\right]. \tag{11.65}$$

Partial differentiations with respect to the number of moles yield the activity coefficients of both components

$$RT\ln\gamma_1 = x_2^2\,[A_0 + A_1(3x_1 - x_2) + A_2(x_1 - x_2)(5x_1 - x_2) + ...], \tag{11.66a}$$
$$RT\ln\gamma_2 = x_1^2\,[A_0 - A_1(3x_2 - x_1) + A_2(x_2 - x_1)(5x_2 - x_1) + ...], \tag{11.66b}$$

which automatically satisfy the Gibbs-Duhem equation. These expressions are known as Redlich-Kister expansions and their adjustable parameters may also be assumed to depend on T and P. They reduce to those of ideal solutions when all the parameters are zero whereas we find the case of regular solutions with $A_0 \neq 0$ and $A_1 = A_2 = ... = 0$. Setting to zero all odd parameters defines a symmetric model with respect to the two components. With $A_0 \neq 0$ and $A_1 \neq 0$, the model reduces to the asymmetric model of *subregular* solutions, which is also widely used and has as a main feature the fact that interaction parameters vary linearly with composition. Again, if needed to account for experimental data, all these parameters can be assumed to be temperature as well as pressure dependent.

There is usually no need for terms of higher order than 2 in the series expansion of G. Without any special interactions, which would make these expansions basically inappropriate, failure to converge may suggest that the selected components are not the correct ones. As will be seen below in the case of electrolytes, changing the manner in which mole fractions are calculated induces marked changes in activity-composition relationships because the transformation of mole numbers into mole fractions is strongly nonlinear. On the other hand, we stress that, with every model, one tries to find convenient equations accounting for complex interactions. Hence, one should not be too demanding regarding the atomistic significance of parameters in excess Gibbs free energy expansions. After all, it is already surprising that the ideal solution model has more than a reference value. When summed up, complex interatomic interactions can thus manifest themselves in a very simple form, and it is mainly for modeling them better that it is useful to know the individual enthalpy and entropy contributions to G^{xs}.

11.4d EXTENSION TO MULTICOMPONENT SYSTEMS

When dealing with multicomponent systems the simplest assumption that can be made is that all interactions are binary in nature. At the roots of this assumption is the primary importance of interactions between nearest atoms, which also supports the pairwise nature of the interatomic potentials commonly used in theoretical calculations (see 14.4c). Even when higher-order interactions are significant, neglecting them does not necessarily give poor results because various kinds of effects often average out in complex systems. As a result, modeling phase equilibria can be more difficult for binary than for multicomponent systems!

For complex systems, some schemes embody binary interactions by assuming excess Gibbs free energies to be the appropriate sum of those derived for limiting binary systems (G_{ij}^{xs}). As proposed by Kohler (1960), for example, these are weighted as follows according to the relative mole fractions of the components considered

$$G^{xs} = \sum_{i \neq j} (x_i + x_j)^2 G_{ij}^{xs}, \tag{11.67}$$

where one recalculates the excess free energies of the binary systems by normalizing to 1 the mole fractions of the i and j components.

The simplicity of such weighting schemes makes them attractive, but their consistency with the particular kind of model used for the binaries has to be checked because it is not automatically ensured. For instance, (11.67) is inconsistent with the subregular model because, as the reader may care to check, the activity coefficients yielded by this equation for a ternary system do not tend exactly to the coefficients of the limiting binary systems. Following a procedure that ensures consistency with respect to the Gibbs-Duhem equation has thus many advantages. As imposed by the assumption of binary interactions and the search for a minimum number of adjustable parameters, many models represent variations on similar themes. For the regular solution model, extension to multicomponent systems is straightforward through

$$G^{xs} = \sum_{i<j} W_{ij} x_i x_j \tag{11.68}$$

where $W_{ij} = W_{ji}$ and $W_{ii} = 0$. From this expression we obtain (see Appendix D)

$$RT \ln \gamma_i = \sum_{i<j} W_{ij} x_j - G^{xs}. \tag{11.69}$$

In the same way, the subregular model can be extended to n-component systems, and the existence of ternary interactions can be accounted for through additional parameters W_{ijk}. For the excess Gibbs free energy of mixing, the generalized expression of the subregular model is

$$G^{xs} = \sum_{i<j} x_i x_j (W_{ij} x_j + W_{ji} x_i) + \sum_{i<j<k} W_{ijk} x_i x_j x_k, \tag{11.70}$$

where $W_{ij} \neq W_{ji}$. After tedious calculations (see Appendix D), one eventually finds for the activity coefficients

$$RT \ln \gamma_i = 2 \sum_j x_i x_j W_{ji} + \sum_j x_j^2 W_{ij} + \sum_{j<k} x_j x_k W_{ijk} - 2G^{xs}. \tag{11.71}$$

In contrast to what we have discussed above for (11.67), one can check that this expression reduces to the correct one for limiting binary systems. (Note that a factor 1/2 is often introduced in the G^{xs} term of the regular solution model, simply because the summation is then made over all values of i and j.)

11.5 OTHER SOLUTION MODELS

To complement the preceding discussion of thermodynamic models we will present a few other important solution models in this section. Particular attention will be paid to solutions where structurally equivalent atoms mix. A very common case, already mentioned several times, will be that of solid solutions where different elements mix over different crystallographic sites.

11.5a TEMKIN MODEL

Let us first turn to solutions where, for obvious electrostatic reasons, an anion cannot mix with a cation and vice versa. With the model proposed by Temkin in 1945 for molten salts, one assumes that the solution can be divided into two independent subsolutions in which the anions and cations substitute independently for each other. As will be assumed below for aqueous solutions of electrolytes, the activity of a salt A_mX_p is then supposed to be equal to the *product* of the activities of its constituting ions A and X raised to the appropriate power

$$a_{A_mX_p} = a_A^m\, a_X^p. \tag{11.72}$$

Taking molten NaCl as an example, we write

$$a_{\mathrm{NaCl}} = a_{\mathrm{Na^+}}\, a_{\mathrm{Cl^-}}, \tag{11.73a}$$

which is consistent with the additivity of chemical potentials

$$\begin{aligned}\mu_{\mathrm{NaCl}} &= \mu_{\mathrm{Na^+}} + \mu_{\mathrm{Cl^-}} = \mu^\circ_{\mathrm{Na^+}} + RT\ln a_{\mathrm{Na^+}} + \mu^\circ_{\mathrm{Cl^-}} + RT\ln a_{\mathrm{Cl^-}}\\ &= \mu^\circ_{\mathrm{NaCl}} + RT\ln(a_{\mathrm{Na^+}}\, a_{\mathrm{Cl^-}}) = \mu^\circ_{\mathrm{NaCl}} + RT\ln a_{\mathrm{NaCl}}.\end{aligned} \tag{11.73b}$$

In the case of an ideal solution, we would have

$$a_{\mathrm{NaCl}} = x_{\mathrm{Na^+}}\, x_{\mathrm{Cl^-}}, \tag{11.74}$$

where the mole fractions are those of the ions in their respective anion and cation solutions. For a mixture of molten NaCl and KBr, where Na^+ and Cl^- have the same mole fraction, the activity of NaCl would be

$$a_{\mathrm{NaCl}} = x^2_{\mathrm{NaCl}}. \tag{11.75}$$

As will also be described below for aqueous solutions, the activity of NaCl would differ considerably from the case where there would not be any ionic dissociation ($a_{\mathrm{NaCl}} = x_{\mathrm{NaCl}}$). The existence of ionic dissociation could thus be ascertained readily from comparisons between calculated and measured activities.

11.5b RECIPROCAL AND MULTISITE SOLUTIONS

The Temkin model postulates only the existence of two (or more) subsolutions, it deals with a special case of *substitutional* solutions, i.e., solutions where specific sets of ions or atoms mutually substitute as if they were forming by themselves one or several independent solutions. The simplest substitutional solutions are so common as not warranting specific comments, like the solid and liquid solutions MgO-NiO, described in 10.6a, or NaCl-NaBr (where miscibility is complete in the liquid state only, not in the solid state owing to too large a difference in the ionic radii of Cl^- and Br^-).

The situation becomes more complex for *reciprocal* solutions, i.e., when two substitutions take place simultaneously. For instance, consider the double substitution

Na/Cs, Cl/I occurring in the quaternary system NaCl-NaI-CsCl-CsI. The solid-liquid equilibria are of the eutectic type as represented in Fig. 11.5. This diagram does not favor any of the four components since any solution within the system is represented in an unambiguous manner by the intersection of the two lines defining constant Na/Cs and Cl/I ratios. However, a peculiar feature is that only three out of the four components are actually independent because these are related by the reaction NaCl + CsI $\rightleftharpoons$ NaI + CsCl. It is thus possible to chose the three independent components in four different ways. This variety of choice is a characteristic feature of reciprocal solutions. From a practical standpoint, it strongly constrains thermodynamic modeling as the properties of the solution must of course be independent of the choice of components. In the special case of *multisite* solutions, we will now examine this point by determining under which conditions such a solution is actually ideal before calculating the activity of its components.

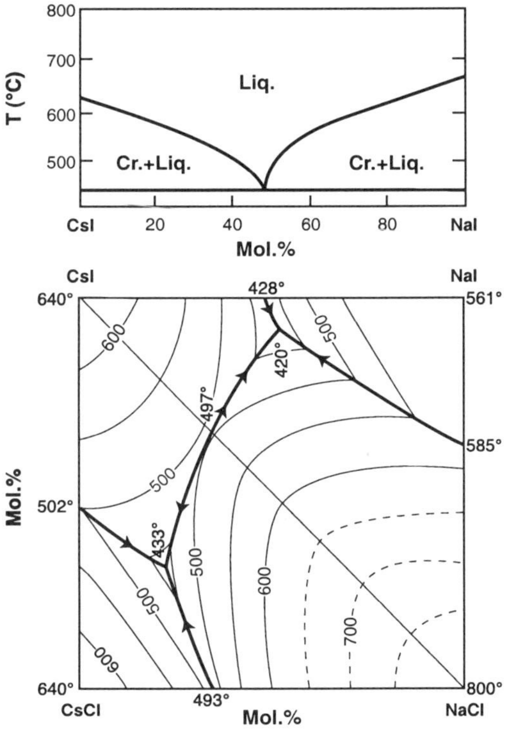

Figure 11.5. Reciprocal solution : eutectic melting relationships in the NaCl-NaI-CsCl-CsI system. The binary eutectics are linked to the two ternary eutectics at 420 and 433°C by cotectic lines whose slope is indicated by the arrows. To complement the isotherms shown as projections on this diagram, the composition-temperature phase diagram of one of the limiting binary systems, CsI-NaCl, has been included on the top of the figure. From the compilation of Levin et al. (1969).

A multisite solution is a solid solution which involves mixing of two pairs of ions (or atoms) on two crystallographically, and thus energetically, distinct sites. As an example, we will deal with a formula unit $(A,B)_m(X,Y)_pZ$, where m and p denote the total numbers of A and B, and X and Y substituting atoms, respectively, and by Z the rest of the formula unit. Without ion exchange between the two sites, this reciprocal solution has four possible components, namely, A_mX_pZ, B_mX_pZ, A_mY_pZ and B_mY_pZ, of which three can be selected as independent components. The Gibbs free energy can be written

$$G_{ss} = \sum_i x_i \mu_i^\circ + G^{xs} - T\Delta S_m, \tag{11.76}$$

where ΔS_m is the ideal entropy of mixing and the summation $\sum_i x_i \mu_i^\circ$ is made over the three independent components can be performed in four different ways. Let us take, for example, the three components A_mX_pZ, B_mX_pZ and A_mY_pZ, and define by $x_A = n_A/(n_A + n_B)$ the fractional occupancy of the first site by A, by x_X that of the second site by X, etc. The summation $\sum_i x_i \mu_i^\circ$ is then

$$\sum_i x_i \mu_i^\circ = x_B \mu^\circ_{B_mX_pZ} + x_Y \mu^\circ_{A_mY_pZ} + (x_X - x_B)\mu^\circ_{A_mX_pZ}. \tag{11.77}$$

Alternatively, we can consider all four components, which would no longer be independent, whose standard chemical potential should then be weighted by the relevant mole fraction for each component. If the solution is ideal, there is no interaction between the atoms which distribute themselves randomly between the crystallographic sites. For each component the mole fraction is then proportional to the probablity of occurrence. We have thus in this case

$$\sum_i x_i \mu_i^\circ = x_A x_X \mu^\circ_{A_mX_pZ} + x_A x_Y \mu^\circ_{A_mY_pZ} + x_B x_X \mu^\circ_{B_mX_pZ} + x_B x_Y \mu^\circ_{B_mY_pZ}. \tag{11.78}$$

Obviously, G_{ss} cannot depend on the choice of components, and the same applies to the chemical potentials of A_mX_pZ, B_mX_pZ, etc. Furthermore, if internal equilibrium is reached the additivity of chemical potentials implies that

$$\mu_{A_mX_pZ} + \mu_{B_mY_pZ} = \mu_{B_mX_pZ} + \mu_{A_mY_pZ}. \tag{11.79}$$

By comparing the expressions (11.77)-(11.79) written for the two kinds of components, we conclude that

$$G^{xs} = x_B x_Y (\mu^\circ_{A_mX_pZ} + \mu^\circ_{B_mX_pZ} + \mu^\circ_{A_mY_pZ} + \mu^\circ_{B_mY_pZ}) \tag{11.80}$$

where the standard states are the pure components at P and T of interest. If the solution is ideal, then G^{xs} is zero. It follows that the solution can be so only if the relation between the standard chemical potentials

$$\mu^\circ_{A_mX_pZ} + \mu^\circ_{B_mY_pZ} - \mu^\circ_{A_mY_pZ} - \mu^\circ_{B_mX_pZ} = 0 \tag{11.81}$$

is obeyed. The same result would have been obtained if any other three possible set of components had been chosen. By looking only at standard Gibbs free energies, it is thus possible to ascertain whether a reciprocal solution can be ideal or not.

11.5c ACTIVITY IN A MULTISITE SOLUTION

With a new approach let us then calculate the activity of components in a reciprocal solution, which will furthermore allow us to generalize the Temkin model. From

(11.17) the activity of any component i is

$$\mu_i = \mu_i^\circ + RT \ln a_i = \mu_i^\circ + \Delta\overline{H}_{mi} - T\Delta\overline{S}_{mi}. \tag{11.82}$$

For an ideal solution the partial molar enthalpy of mixing $\Delta\overline{H}_{mi}$ is of course 0. Hence we just need to know the partial molar entropy of mixing $\Delta\overline{S}_{mi}$, which will be derived here from the equation (11.8) that relates the configurational entropy of an ideal solution to the number of complexions Ω by $S = k \ln \Omega$.

From the definition of a partial molar property we directly write

$$\Delta\overline{S}_{mi} = R \ln(\Omega_1/\Omega) \tag{11.83}$$

where Ω_1 is the new number of complexions when the size of the crystal is increased by just one formula unit of component i. Combining (11.82) and (11.83), we obtain

$$RT \ln a_i = -T\Delta\overline{S}_{mi} = RT \ln(\Omega/\Omega_1). \tag{11.84}$$

This expression can be rewritten as

$$a_i = \ln(\Omega/\Omega_1)\ . \tag{11.85}$$

For a crystal made up of a total of n_A atoms A, n_B atoms B, n_X atoms X and n_Y atoms Y, respectively, the reasoning held in 11.1a indicates that the number of complexions is

$$\Omega = \frac{(n_A + n_B)!}{n_A!\, n_B!} \frac{(n_X + n_Y)!}{n_X!\, n_Y!}. \tag{11.86}$$

If one formula unit of the A_mX_pZ component is added to the crystal, this number becomes

$$\Omega_1 = \frac{(n_A + n_B + m)!}{(n_A + m)!\, n_B!} \frac{(n_X + n_Y + p)!}{(n_X + p)!\, n_Y!}. \tag{11.87}$$

Because m and p are negligible compared to the very large n_A and n_B, we make no error by rewriting (11.87) as

$$\Omega_1 = \frac{(n_A + n_B)^m\, (n_A + n_B)!}{(n_A)^m\, n_A!\, n_B!} \frac{(n_X + n_Y)^p\, (n_X + n_Y)!}{(n_X)^p\, n_X!\, n_Y!}. \tag{11.88}$$

Taking the ratio of these two expressions we find readily that Ω/Ω_1 is equal to $x_A^p x_X^m$. As a function of fractional occupancies, we write from (11.85) that the activity of A_mX_pZ in the solid solution $(A,B)_m(X,Y)_pZ$ is

$$a_{B_mX_pZ} = x_A^m\, x_X^p, \tag{11.89}$$

and we would obtain analogous expressions for the other components. This result is similar to the activity (11.72) postulated in Temkin's model. Besides, the relation $S = k \ln \Omega$ and Stirling's approximation as applied to (11.86) yield for one mole of the solid solution $(\mathrm{A,B})_m(\mathrm{X,Y})_p\mathrm{Z}$

$$S = \Delta S_m = -R[m(x_A \ln x_A + x_B \ln x_B) + p(x_X \ln x_X + x_Y \ln x_Y)]. \tag{11.90}$$

This result is generalizable in a similar way if exchange takes place on more than two sites or if more than two elements distribute over a given site. A last remark. The ideal solution model presented here is just but a starting point. Enthalpy of mixing terms can be included in the chemical potentials. Regular or subregular models are especially convenient in this respect as they consider the configurational entropy of the solution to be ideal, which allows entropy equations of the form (11.90) to be kept for the substitutions considered. We will finally see in 11.6 that solutions of electrolytes sharing common ions constitute very important instances of reciprocal solutions. These are strongly nonideal and we will demonstrate for them that there exists a simple relationship between the activity coefficients of nonindependent components.

It is with (11.90) that the ideal configurational entropy of $FeAl_2O_4$ spinel, mentioned in 6.4f, has been calculated. More generally, (11.90) also applies to the more general case of $(Fe,Mg^{2+})(Fe,Al,Cr^{3+})_2(O^{2-})_4$ spinels in their whole composition domain.

11.5d FLORY-HUGGINS MODEL

Closely related to the regular solution model is the Flory-Huggins model, which has been extensively used for polymers. These macromolecules have very long lengths but lateral dimensions comparable to those of "usual" molecules. Their solutions deviate largely from ideality at finite concentrations and their two-phase equilibria are markedly asymmetric under conditions of limited miscibility. In view of their very large size, the mole fraction is no longer an adequate compositional parameter. It is replaced by the volume fraction $\phi_i = n_i V_i^* / \sum n_i V_i^*$, where n_i and V_i^* are the number of moles of component i and its partial molar volume, which is taken as the volume of the pure component. For a binary system, we then have

$$RT \ln \gamma_i = W_{12} \phi^2, \tag{11.91}$$

with

$$\Delta H_m = W_{12} \phi_1 \phi_2, \qquad \Delta S_m = -R \sum n_i \ln \phi_i. \tag{11.92}$$

For these equations to apply, the mixed molecules can have different sizes but must have similar cross sections. The entropy of mixing results from the increase in volume over which molecules distribute themselves in the mixture as compared to the volume that was initially available to the pure components. The thermodynamic properties are then virtually independent of the structure of the macromolecules and especially of their flexibility.

11.5e ATHERMAL SOLUTIONS

To conclude this section we will briefly mention athermal solutions which are characterized by zero enthalpies of mixing but nonideal entropies of mixing. We have thus

$$\Delta H_m = H^{xs} = 0, \qquad G^{xs} = -TS^{xs} = \sum n_i RT \ln \gamma_i. \tag{11.93}$$

These solutions constitute a rare case where entropies of mixing could be determined directly from activity measurements. Classical examples are solutions of rubber or polystyrene with toluene or benzene. We will make few comments about this class of solutions. We will note that one can account formally for their peculiar behavior by expressing the Gibbs free energy of mixing in terms of volume fractions, and that they reduce to ideal solutions when the mixed molecules have equivalent shapes and sizes. Indeed, the lack of atomic interactions, associated with nonideal entropies of mixing, is most likely to be found when the components mixed have a similar structure but widely different sizes. In this case the inadequacy of the standard ideal model is revealed by the fact that the actual entropies of mixing can be higher than the ideal values, showing that mixing must then be considered at the scale of units smaller than the molecule. This is the basis of *lattice* models with which one evaluates, in terms of site occupancies, the number of configurations available when the components are mixed. The procedure is similar to that described above for multisite solutions.

11.6 ELECTROLYTES

11.6a IONIC DISSOCIATION AND SPECIATION

Electrolytes are substances that conduct electrical current by the movement of the positive and negative ions into which they dissociate. Although solid electrolytes exist, most of them are liquid but not necessarily aqueous solutions. Molten cryolite (Na_3AlF_6) plays for instance a fundamental role in aluminum production through high-temperature reduction of aluminum hydroxides. In view of their general importance, however, we will deal only with aqueous solutions of electrolytes.

When freezing point depressions and lowering of vapor pressures by solutes began to be systematically measured at the end of the 19th century, it became clear that aqueous solutions of salts were not following the usual trends, even at high dilution. The main argument advanced by Arrhenius (1859-1927) to support the concept of electrolyte dissociation into ions in aqueous solutions was that, for the same solute concentration, the freezing point depression of water was about twice as great for alkali halides than for sucrose. This indicated that NaCl dissociates into Na^+ and Cl^- ions. Thermochemically, an important consequence of complete ionic dissociation of NaCl is that the mole fraction of water, for example, should be calculated with

$$x_{H_2O} = \frac{n_{H_2O}}{n_{H_2O} + n_{Na^+} + n_{Cl^-}} = \frac{n_{H_2O}}{n_{H_2O} + 2n_{NaCl}}, \tag{11.94}$$

and not, as would be done without dissociation, as

$$x_{H_2O} = \frac{n_{H_2O}}{n_{H_2O} + n_{NaCl}}. \tag{11.95}$$

Although the activity of water is of course independent of the way in which its mole fraction is defined, the activity-composition relationship $a = \gamma x$ differs markedly

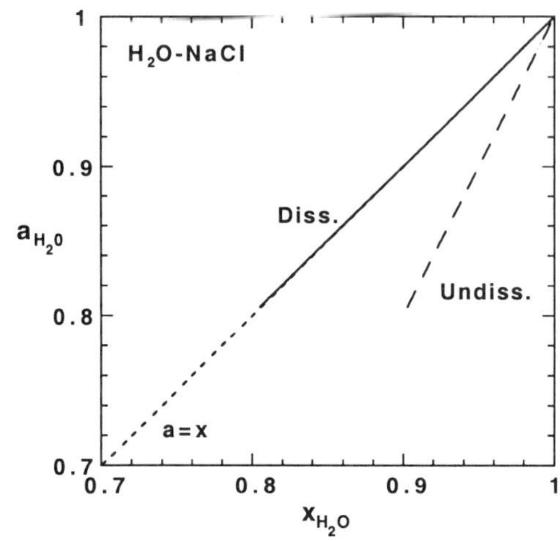

Figure 11.6. Activity of water at 298 K in H_2O-NaCl solutions against H_2O mole fractions, calculated with (solid line) and without (dashed line) ionic dissociation for NaCl. For comparison the values $a = x$ are shown as the dotted line. Note that saturation is achieved, for an activity of about 0.8, at different water concentrations for the two different mole fractions. Data from Hamer and Wu (1972).

depending on whether (11.94) or (11.95) is used to calculate the mole fraction x. Experience shows that the activity coefficient of water remains close to 1 up to much higher NaCl concentration with the former than with the latter definition (Fig. 11.6). With the latter, deviations from ideality are strong and even appear at high dilution where Raoult's law is not followed.

Accounting for ionic dissociation does not necessarily ensure ideal solution behavior, however, as shown in Fig. 11.7 by the activity of water in HCl solutions. If the solution appears less nonideal when HCl dissociation into H^+ and Cl^- ions is assumed, the improvement is very far from yielding good agreement with the experimental data. This discrepancy originates in long-range Coulombic attraction, between ions of unlike charge, in a solvent having a high dielectric constant. This causes strong ionic association and has major consequences even at very low dilution. No matter how mole fractions are defined, this is why electrolyte solutions are strongly nonideal and have to be dealt with separately.

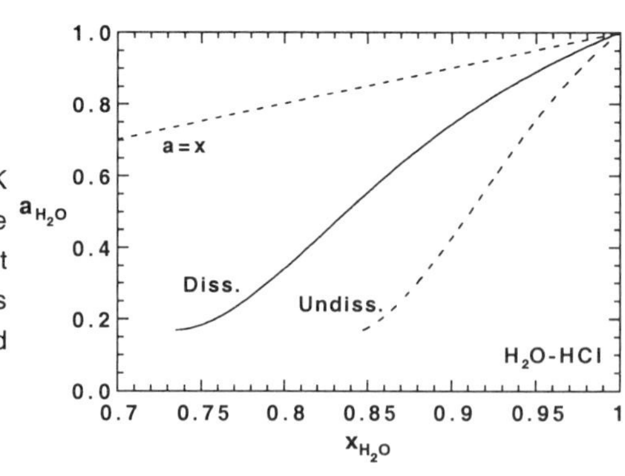

Figure 11.7. Activity of water at 298 K in H_2O-HCl solutions against H_2O mole fractions, calculated with and without ionic dissociation for HCl. Same curves as in Fig. 11.6. Data from Hamer and Wu (1972).

11.6b IONIC ACTIVITIES

Let us consider the dissociation of an electrolyte e in an aqueous solution. By itself, dissociation has no special consequences on the manner in which the activity of the electrolyte is either defined or measured. In any case, it can be expressed as

$$\mu_e = \mu^\circ + RT\ln a_e, \tag{11.96}$$

once a standard state appropriate to the problem of interest has been selected. In contrast, ionic activities cannot be measured because electroneutrality prevents the cation and anion mole numbers from being varied independently. Now, the criteria for chemical equilibrium do not depend on the nature of the components in the solution. The mass action law, in particular, applies equally to neutral and ionic species. Furthermore, the proven existence of ions in solution, for instance by electrical conductivity measurements, makes it mandatory to define activities and activity coefficients for ions. To give a single example, calculations of solubility products would be impossible without defining these properties (see chapter 12).

For the total or partial dissociation of electrolyte e into ν_+ cations and ν_- anions, the additivity of chemical potentials allows us to write

$$\mu_e = \nu_+\mu_+ + \nu_-\mu_-, \tag{11.97}$$

whence

$$\mu_e^\circ + RT\ln a_e = \nu_+[\mu_+^\circ + RT\ln a_+] + \nu_-[\mu_-^\circ + RT\ln a_-], \tag{11.98}$$

where μ_+ and μ_- are the chemical potentials of the cations and anions, respectively. (The electrolyte is said to be uni-univalent (1-1) if $\nu_+ = \nu_- = 1$; di-univalent (2-1) if $\nu_+ = 2$ and $\nu_- = 1$, etc.) Not knowing the actual degree of dissociation of the electrolyte in solution and not looking to determine the individual concentrations of undissociated and ionized entities, we can arbitrarily select for the undissociated electrolyte a possibly hypothetical standard state such that

$$\mu_e^\circ = \nu_+\mu_+^\circ + \nu_-\mu_-^\circ. \tag{11.99}$$

For the ions, in contrast, μ° refers as usual to the hypothetical solution of unit concentration at P and T of interest. With this convention, the equilibrium constant of the dissociation reaction becomes unity. In other words, we then have

$$a_e = (a_+)^{\nu_+}(a_-)^{\nu_-}. \tag{11.100}$$

Contrary to high dilution where ion activities are simply proportional to molalities, cations and anions may have different activities at higher concentration. It is thus convenient to introduce a *mean chemical potential* and a *mean activity* for the electrolyte, viz.

$$\mu_\pm = 1/\nu\ (\nu_+\mu_+ + \nu_-\mu_-) = 1/\nu\ \mu_e, \tag{11.101a}$$

$$a_\pm = [(a_+)^{\nu_+}(a_-)^{\nu_-}]^{1/\nu} = a_e^{1/\nu}, \tag{11.101b}$$

where $\nu = \nu_+ + \nu_-$ is the total number of ions. A number of useful definitions will follow from this geometric mean.

11.6c IONIC ACTIVITY COEFFICIENTS

Although these and the following definitions hold for any concentration variables, molality is the scale generally used. *Individual ionic activity coefficients* are then defined by

$$\gamma_+ = a_+/m_+, \qquad \gamma_- = a_-/m_-. \tag{11.102}$$

Since the standard state is the hypothetical solution of unit molality, both coefficients tend to 1 when the molality tends to 0. If dissociation is complete, we can introduce the *mean ionic molality* $m_\pm$ from the electrolyte molality m_e

$$m_\pm = [(m_+)^{\nu_+}(m_-)^{\nu_-}]^{1/\nu} = m_e^{1/\nu}. \tag{11.103}$$

This allows the mean activity to be related to the experimentally determined activity coefficient γ_e of electrolyte e by the *mean ionic activity coefficient* $\gamma_\pm$

$$\gamma_\pm = a_\pm/m_\pm = [(\gamma_+)^{\nu^+}(\gamma_-)^{\nu^-}]^{1/\nu} = \gamma_e^{1/\nu}. \tag{11.104}$$

In passing, note that if several electrolytes sharing common ions are dissolved in the same solution, all the resulting activities are not mutually independent. Let us take the example of NaCl, NaBr, KCl and KBr in aqueous solution. Like that of the other components of this reciprocal solution, the activity of NaCl has the form $a_{\text{NaCl}} = a_{\text{Na}^+}\, a_{\text{Cl}^-}$, which implies that

$$\frac{a_{\text{NaCl}}}{a_{\text{KCl}}} = \frac{a_{\text{Na}^+}\, a_{\text{Cl}^-}}{a_{\text{K}^+}\, a_{\text{Cl}^-}} = \frac{a_{\text{Na}^+}\, a_{\text{Br}^-}}{a_{\text{Na}^+}\, a_{\text{Br}^-}} = \frac{a_{\text{NaBr}}}{a_{\text{KBr}}}. \tag{11.105}$$

Consistent with the discussion of reciprocal solutions made in 11.5, we find that the activities of only three of the four components are independent.

With this formalism, ions take part in chemical equilibria calculations in a similar way to neutral components. Consider for instance the reaction

$$\text{Fe}^{2+} + \text{Ce}^{4+} \rightleftharpoons \text{Fe}^{3+} + \text{Ce}^{3+}, \tag{11.106}$$

which is commonly used to titrate ferrous iron in aqueous solutions. At equilibrium, we have

$$K_r = \frac{a_{\text{Fe}^{3+}} a_{\text{Ce}^{3+}}}{a_{\text{Fe}^{2+}} a_{\text{Ce}^{4+}}} = \frac{m_{\text{Fe}^{3+}} m_{\text{Ce}^{3+}}}{m_{\text{Fe}^{2+}} m_{\text{Ce}^{4+}}} \frac{\gamma_{\text{Fe}^{3+}} \gamma_{\text{Ce}^{3+}}}{\gamma_{\text{Fe}^{2+}} \gamma_{\text{Ce}^{4+}}}. \tag{11.107}$$

Although individual ionic activity coefficients are not measurable, their ratios have the interesting feature of being unambiguously evaluated. For simplicity, let us now assume that Cl^- is the only anion in the solution. Thanks to relations like $\gamma_{\text{Ce}^{3+}}\, \gamma_{\text{Cl}^-}^3 =$

γ_{CeCl_3}, we write

$$\frac{\gamma_{Fe^{3+}}\gamma_{Ce^{3+}}}{\gamma_{Fe^{2+}}\gamma_{Ce^{4+}}} = \frac{\gamma_{Fe^{3+}}\gamma_{Cl^-}^3\gamma_{Ce^{3+}}\gamma_{Cl^-}^3}{\gamma_{Fe^{2+}}\gamma_{Cl^-}^2\gamma_{Ce^{4+}}\gamma_{Cl^-}^4} = \frac{\gamma_{FeCl_3}\gamma_{CeCl_3}}{\gamma_{FeCl_2}\gamma_{CeCl_4}}, \tag{11.108}$$

and thus express the activity and activity coefficients of ions in terms of those of neutral components, which are determined experimentally.

Two problems remain to be solved, however, before applying these concepts to practical problems. The first one concerns determinations of activity coefficients of ionic solutes at very high dilution where, as we have seen in 8.5d, the integral of the Gibbs-Duhem equation diverges. The second difficulty stems from the very strong variations of the activity coefficients of electrolytes with the composition of the solution. Fortunately, most ions in aqueous solutions are relatively simple entities amenable to theoretical calculations that would be impossible to perform in a general manner for nonelectrolytes. Not only these calculations yield important expressions which may be fitted to experimental data, but they also provide the right limiting laws for determining activity coefficients at high dilution.

11.6d OSMOTIC COEFFICIENT

Among common experimental techniques, electrochemical and vapor pressure measurements are of particular interest for dilute solutions. As shown in Fig. 11.7 for NaCl-H_2O solutions, however, measuring the activity of the solvent at high solute dilution brings little information because of the small magnitude of the deviations from ideality. In any case, the activity of the solute at very high dilution is generally calculated through integration of the Gibbs-Duhem equation with the *osmotic coefficient* (ϕ_m). Originally defined as a direct measure of departure from ideality, ϕ_m was simply related to the activity and mole fraction of the solvent by

$$\ln a_s = \phi_m \ln x_s. \tag{11.109}$$

The coefficient owes its name to the fact that, with (11.48), it is directly related to the osmotic pressure by

$$\Pi = -\phi_m RT \ln x_s/\overline{V}_s. \tag{11.110}$$

For a 0.01 molal aqueous solution of uni-univalent electrolyte, the mean activity of the solute and the osmotic coefficient are about 0.90 and 0.97, respectively, whereas the awkward value 1.00001 of the activity coefficient of the solvent might incorrectly suggest that the solution is almost ideal. Hence it is for numerical reasons that ϕ_m is a convenient variable to deal with very dilute solutions. For dilute binary solutions, x_s depends on the molality of a solute j that dissociates into ν_j ions as (see Table 4.1)

$$\ln x_s = \ln(1-x_j) = \ln\left(1-\frac{\nu_j m_j}{\nu_j m_j + 1000/M_s}\right) \sim \frac{-\nu_j m_j M_s}{1000}. \tag{11.111}$$

From the definition (11.109) of ϕ_m it follows that

$$\ln a_s = -\phi_m \nu_j m_j M_s / 1000, \tag{11.112}$$

and, conversely, that

$$\phi_m = -1000 \ln a_s / \nu_j m_j M_s. \tag{11.113}$$

Although this equation has been derived for dilute solutions, the osmotic coefficient has proven convenient enough that it is nowadays defined by (11.113) and is generally used for reporting the activity of the solvent up to the highest solute molalities. From (11.113), we can thus write for a change in chemical potential of the solvent

$$\mu_s - \mu_s^\circ = RT \ln a_s = -\phi_m RT \nu_j m_j M_s / 1000. \tag{11.114}$$

In the more general case of a multicomponent solution, we define ϕ_m in the same manner

$$\phi_m = -1000 \ln a_s / \sum_j \nu_j m_j M_s, \tag{11.115}$$

where m_j now designates the molality of any ionic and neutral species as well (the latter being considered as ions with zero electrical charge). A change in chemical potential of the solvent is then

$$\mu_s - \mu_s^\circ = RT \ln a_s = -\phi_m RT \sum \nu_j m_i M_s / 1000. \tag{11.116}$$

As obtained from the Gibbs-Duhem equation, the relation (8.45) between the activities of the solvent and the solute in a binary solution allows us to write

$$d \ln a_j = -\frac{x_s}{x_j} d \ln a_s. \tag{11.117}$$

Making use of the relation $a_j = (\gamma_j m_j)^{\nu_j}$ for the solute activity, we then write the activity coefficient of the solvent

$$d \ln \gamma_s = d\phi_m + (\phi_m - 1)/m_j \, dm_j. \tag{11.118}$$

Remarking that ϕ_m is unity when the solute molality is zero, we obtain on integration

$$\ln \gamma_s = (\phi_m - 1) + \int_0^{m_j} (\phi_m - 1)/m_j \, dm_j. \tag{11.119}$$

As already noted, an essential difference between electrolyte and nonelectrolyte liquid solutions is that $d(1 - \phi_m)/dm$ diverges at infinite dilution for the former whereas it tends to a finite limit for the latter. For electrolytes, integrating (11.119) thus requires the availability of an analytical expression for relating the osmotic coefficient to the molality of the solute at high dilution. Empirically, there was early evidence that $1 - \phi_m$ was tending to $m^{1/2}$. Because of the fundamental importance of Coulombic interactions in electrolyte solutions, Lewis and Randall indeed pointed out that the

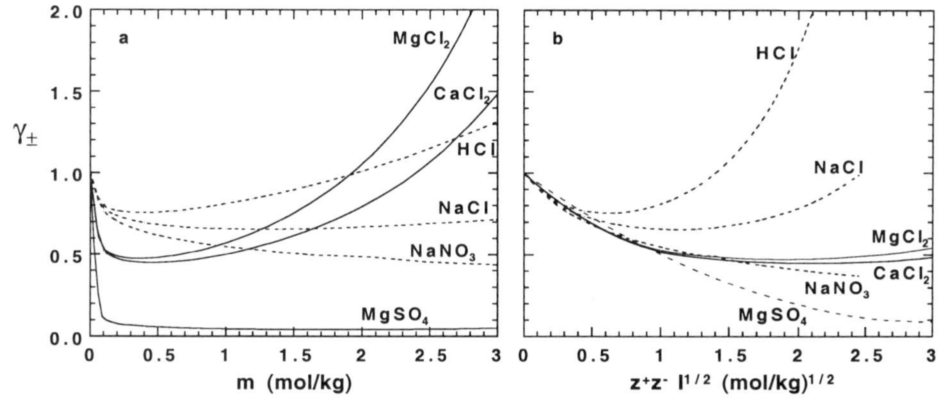

Figure 11.8. (a) Mean ionic activity coefficients of various electrolytes against the molality for uni-univalent (HCl, NaCl, HNO_3), di-univalent ($CaCl_2$, $MgCl_2$) and di-divalent (Mg_2SO_4) electrolytes; (b) Same coefficients against the square of the ionic strength times the product of the charges, see eqn (11.123). The first plot shows that the activities of electrolytes of the same kind tend to the same limit at infinite dilution. In the second, the activity variations are not only simpler but the chosen representation allows all the data to collapse on the same curve at infinite dilution. Note that the strong initial decrease of the activity coefficient is a specific feature of electrolytes. Data from Hamer and Wu (1972) for (1-1) electrolytes, and from the book by Robinson and Stokes (1959) for the others.

basic parameter was the *ionic strength*, I, which they defined as

$$I = 1/2 \sum m_i z_i^2 \tag{11.120}$$

where z_i and m_i and are the ionic valence and molality of ion i. They even claimed in 1921 that "in dilute solution the activity coefficient of any ion depends solely upon the total ionic strength of the solution." Since then this statement has always been experimentally verified (Fig. 11.8). It is to demonstrate it that Debye (1884-1966) and Hückel theoretically calculated in 1923 the proportionality factor between the ionic strength and the activity coefficient at high dilution.

11.6e DEBYE-HÜCKEL MODEL

The Debye-Hückel model rests on the following assumptions: (i) strong electrolytes are completely dissociated, without any formation of pairs by the ions that are formed; (ii) ions are undistorted point charges creating a symmetrical Coulombic field; (iii) all ion interactions result from Coulombic forces; (iv) the mutual electrostatic energy of two ions at their closest distance of approach is small compared to their mean thermal energy of translation; (v) the presence of ions has no effect on the dielectric constant of the solvent. On this basis Debye and Hückel found that the activity coefficient of an ion i at high dilution is

$$\ln \gamma_i = -A_m z_i^2 I^{1/2}. \tag{11.121}$$

In this equation A_m is a characteristic constant of the solvent, namely

$$A_m = \left(\frac{2\pi N_A \rho_s}{1000}\right)^{1/2} \frac{e^3}{(4\pi\varepsilon_0)^{3/2}(\varepsilon kT)^{3/2}}, \tag{11.122}$$

where N_A is Avogadro constant, e the electron charge, ε_0 the permittivity of free space, k Boltzmann constant, ρ_s the density of the solvent and ε its dielectric constant. When concentrations are expressed as molalities, one has $A_m = 1.1762$ in SI units for water at 298 K. On the molarity scale, $A_c = A_m/\rho_s^{1/2}$.

As defined by (11.104), the mean ionic activity coefficient of Debye-Hückel's model is given by

$$\ln\gamma_\pm = -A_m|z_+z_-|I^{1/2}. \tag{11.123}$$

With this expression, $\gamma_\pm$ conveniently appears as a product of two terms, one (A_m) chararacteristic of the solvent and the other (I) of the concentration of the solution. Hence, the activity coefficient of the solute depends only on ionic strength and tends to unity when the molality tends to 0.

The Debye-Hückel model has the great merit of making calculations of the activity coefficients of solutes possible thanks to the correct limiting law it provides for integrating the Gibbs-Duhem equation at very high dilution. In addition, it is applicable without any modification to electrolyte mixtures since the only relevant parameter is the known ion charge. Furthermore, through the excess Gibbs free energy of mixing, it allows one to extrapolate correctly other thermodynamic properties such as enthalpies, heat capacites and volumes to zero concentration. Because it does not discriminate between different ions having the same charge and denies any form of ion association, however, its validity breaks down at rather low ionic strengths. These are typically of the order of 0.001 mol/kg (and much less in some instances), which corresponds, for univalent ions, to still very large inter-ionic distances of the order of 100 Å or more. In chemical equilibria calculations, the practical usefulness of Debye-Hückel's original model is thus limited to highly dilute solutions.

For higher molalities, Debye and Hückel later improved their model by taking into account the fact that electrostatic potentials do not derive from point charges, but from ions having a finite radius. By considering ions as non polarizable rigid spheres, they obtained for solutions of a single electrolyte

$$\ln\gamma_\pm = \frac{-A_m|z_+z_-|I^{1/2}}{1 + B_m s_i I^{1/2}}, \tag{11.124}$$

where s_i is the size of the hydrated ion and B_m a parameter given by

$$B_m = \left(\frac{8\pi N_A}{1000}\right)^{1/2} \frac{e\rho_s^{1/2}}{(4\pi\varepsilon_0)^{1/2}(kT\varepsilon)^{1/2}}, \tag{11.125}$$

whose value, on a molality scale, is $0.3287\ 10^{10}$ in SI units for water at 298 K (i.e., using meters for s_i). On the molarity scale, $B_c = B_m/\rho_s^{1/2}$. The more flexible equation (11.124) tends to (11.121) at high dilution. It works well for single electrolytes up to an ionic strength of 1 mol/kg where the average distance between univalent ions is of the order of 10 Å. It can be also used for mixed electrolytes, but with an average ionic radius (often taken as 1 Å) which has the consequence that the model has a more resticted concentration range of applicability.

The starting point of the modified Debye-Hückel model is a given ion i around which one calculates the distribution of the other ions. For an ion j at distance r from i, the interaction energy is given by the product of the charge $z_j e$ and the average electrostatic potential $\psi_i(r)$ around i. One assumes that the interaction energy follows a Boltzmann distribution (see 13.4), so that the molarity of ion j around i is $c_j = c_m \exp[-z_j e \psi_i(r)/kT]$ where c_m is the mean molarity. This gives the charge density $\rho_i(r) = \Sigma z_j e c_j \exp[-z_j e \psi_i(r)/kT]$. Using the approximation $\exp(-x) \sim (1-x)$, one solves the Poisson equation $\nabla^2 \psi(r) = -\rho(r)/\varepsilon$ for determining the electrostatic potential associated with the ion distribution. To do it, one rewrites more simply the Poisson equation as $\nabla^2 \psi = K^2 \psi_i(r)$ where the parameters that do not depend on r are gathered in the Debye length $1/K$ defined as $K^2 = (4\pi e^2/\varepsilon kT)\Sigma c_i z_i^2$. Assuming for simplicity that the mass of 1000 cm^3 of dilute solution is 1 kg, we have $N_i = 1000 N_A m_i$ (N_A = Avogadro number) and then find that $K^2 = 2\ 10^3 N_A I/\varepsilon kT$, where I is the ionic strength. In this way integration of the Poisson equation yields

$$\psi = \frac{z_j e}{4\pi\varepsilon(1+Ka)} \frac{\exp(Ka)}{r\exp(Kr)}$$

where a is the distance at which the ion j is in contact with the reference ion i. Writing that $\psi_i(r)$ is the sum of the potential due to the reference ion i alone ($z_i e/4\pi\varepsilon r$), and that due to all other ions $[\psi_a(r)]$, and noting that the potential energy V of ion i is the product of its charge $z_i e$ and the potential $\psi_a(r)$ due to the ionic atmosphere, one eventually finds that $V = -z_i^2 e^2 K/4\pi\varepsilon(1+Ka)$. Counting the separation of each pair of ions only once (i.e., introducing a factor 1/2), one calculates that the work done to transfer N ions from the real solution to an ideal solution, where they would be infinitely separated, is $N z_i^2 e^2 K/8\pi\varepsilon(1+Ka)$. This work is equal the Gibbs free energy variation. Since $\Delta G = -RT \ln\gamma_i$, the activity coefficient (11.124) is derived after replacement of K by the expression defined above.

11.6f PITZER MODEL

At molalities higher than unity, few water molecules remain separated from an ion by more than a few other molecules. It is then necessary to deal with the variation with concentration of the dielectric constant of the solution and with the existence of ion association, which reduces the effective number of ions and thus the ionic strength. If the equilibrium constants of these complexes are known (see 12.3c), one can calculate the concentration of free ions by an iterative method. One first assumes all ions to be free in order to calculate the ionic strength, the ionic activity coefficients, the resulting concentrations of the complexes and free ions. The calculation is then repeated until convergence is achieved. It is however more convenient to use activity coefficients which account directly for the various factors operating in concentrated solutions. This has been made in various ways, for instance by adding extra terms in the expression of the activity coefficent. Such equations are useful because they are valid up to unit molalities, but they cannot be used for solutions of several electrolytes where different ions have of course different sizes. Because adding empirical coefficients obviously does not ensure rapid convergence, attempts have been made to arrive at more precise expressions on a sounder physical basis.

We will thus conclude this discussion of aqueous electrolyte solutions with the semiempirical model of Pitzer. This model is widely used because it works very well up to ionic strengths higher than 10 mol/kg even though it depends on a small number of fitted parameters. Its starting point is an expression proposed by Guggenheim,

$$\ln\gamma_\pm = \frac{-A_m|z_+z_-|I^{1/2}}{1+I^{1/2}} + 2\beta m, \tag{11.126}$$

where specific properties of each solute are dealt with by the parameter β. In addition to the long-range electrostatic interactions, treated in a Debye-Hückel manner, it is assumed in the Pitzer model that the properties of electrolyte solutions result from short-range interactions between pairs and triplets of ions. Although complex, these various interactions can be addressed in a rather simple manner since the β parameter is a function of the ionic strength through $\beta = \beta^0 + \beta^1_\pm \exp(-\alpha I^{1/2})$ where α is a constant. The mean ionic activity coefficient is then expressed as the sum of three terms

$$\ln\gamma_\pm = \ln\gamma_1 + \ln\gamma_2 + \ln\gamma_3, \tag{11.127a}$$

where

$$\ln\gamma_1 = -|z_+z_-|\frac{A_m}{3}\left[\frac{I^{1/2}}{1+bI^{1/2}} + \frac{2}{b}\ln(1+bI^{1/2})\right], \tag{11.127b}$$

$$\ln\gamma_2 = m\frac{2\nu_+\nu_-}{\nu}\left[2\beta^0_\pm + \frac{2\beta^1_\pm}{\alpha^2 I}[1-\exp(-\alpha I^{1/2})(1+\alpha I^{1/2}-\frac{1}{2}\alpha^2 I)\right], \tag{11.127c}$$

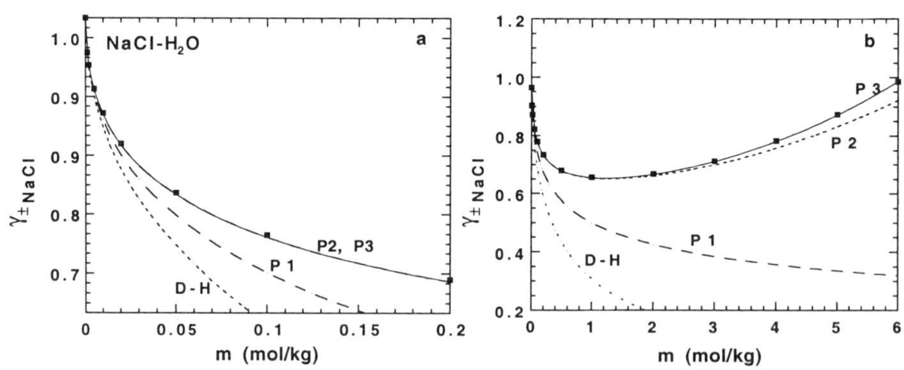

Figure 11.9. Activity of NaCl at 298 K in aqueous solutions at high dilution (a) and at high concentration (b). Experimental data: squares (Hamer and Wu, 1972), and values given by the models of Debye-Hückel (D-H) and Pitzer with the first (P1), first and second (P2) and all three (P3) terms of (11.127).

Table 11.1. Pitzer parameters for various (1-1) and (2-1) electrolytes[a]

	$(2\nu_+\nu_-/\nu)\beta^0_\pm$	$(2\nu_+\nu_-/\nu)\beta^1_\pm$	$2(\nu_+\nu_-)^{3/2}/\nu\, C_\pm$
HCl	0.1775	0.2945	0.00080
HNO_3	0.1119	0.3206	0.00100
LiCl	0.1494	0.3074	0.00359
NaCl	0.0765	0.2664	0.00127
NaOH	0.0864	0.2530	0.00440
$NaNO_3$	0.0068	0.1783	-0.00720
KCl	0.0483	0.2122	-0.00084
KOH	0.1298	0.320	0.00410
NH_4Cl	0.0522	0.1918	-0.00301
$MgCl_2$	0.4698	2.242	0.00979
$CaCl_2$	0.4212	2.152	-0.00064
$Ba(OH)_2$	0.2290	1.600	
$FeCl_2$	0.4479	2.043	-0.01623
$ZnCl_2$	0.3469	2.190	-0.16590
Li_2SO_4	0.1817	1.694	-0.00753
Na_2SO_4	0.0261	1.484	0.00938
K_2SO_4	0.0666	1.039	

[a]Data from Pitzer and Mayorga (1973)

$$\ln\gamma_3 = \frac{2(\nu_+\nu_-)^{3/2}}{\nu} m^2 C_\pm. \tag{11.127d}$$

These expressions are less complicated than they seem because the same parameters $b = 1.2$ and $\alpha = 2.0$ are adequate for pure as well as for mixed electrolytes. In (11.127b), A_m is the Debye-Hückel constant so that the equations (11.127) depend on only three specific parameters for each electrolyte, namely, $\beta^0_\pm$, $\beta^1_\pm$ and $C_\pm$. The important short-range interactions between ion pairs are accounted for by the $\beta^0_\pm$ and $\beta^1_\pm$ terms. As to the simpler $C_\pm$ term, it represents interactions between triplets of ions whose contribution to $\ln\gamma_\pm$ is usually small, which allows its dependence on ionic strength to be neglected. The relative importance of the three terms is shown in Fig. 11.9 for NaCl-H_2O solutions. Although the complete expression (11.127a) is needed for matching the experimental data up to the saturation concentration, good agreement is obtained with the first two terms and the first one already predicts better activity coefficients in dilute solutions than the Debye-Hückel expression.

In summary, the Pitzer model deals with ion interactions in a virial-like manner in terms of parameters which have a physical meaning and depend in a simple way on the ionic strength of the solution. The expressions (11.127) work well from simple (1-1) to more complex (5-1) electrolytes with small, multiply charged "structure making"

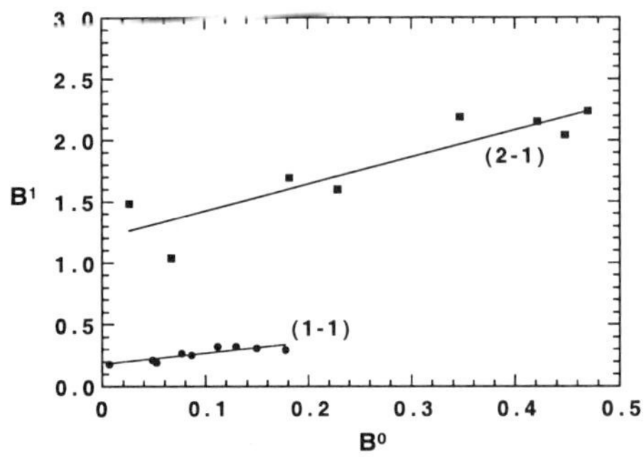

Figure 11.10. Relationship between the $B^0 = (2\nu_+\nu_-/\nu)\beta^0_\pm$ and $B^1 = (2\nu_+\nu_-/\nu)\beta^1_\pm$ coefficients of the Pitzer equation for (1-1) and (2-1) electrolytes

ions like Mg^{2+} and Ca^{2+} as well as large, single charged "structure-breaking" ions such as Rb^+ and ClO_4^-. Pitzer parameters are listed in Table 11.1 for various (1-1) and (2-1) electrolytes. Another interesting feature is the strong correlation they show for a given type of electrolyte (Fig. 11.10). This is especially noteworthy for (1-1) electrolytes for which the parameters of substances as diverse as acids, bases and salts plot nicely on the same line. Thanks to such correlations, a fairly good representation of electrolyte properties can thus be obtained with the single parameter $\beta^0_\pm$.

We also emphasize that the Pitzer model has been designed to be applicable to mixed electrolytes. We refer the reader to the extensive literature on this topic, however, because the relevant expressions are somewhat cumbersome, depend on whether or not the mixed electrolytes have a common ion, and are a function of specific interaction coefficients. We will finally note that the excess Gibbs-free energy calculated from the Pitzer model is, for n_w kg of water

$$\frac{G^{xs}}{n_e RT} = -A_m \frac{I^{1/2}}{(1+bI^{1/2})+m^2(2\nu_+\nu_-)}\left\{2\beta^0_\pm + \frac{2\beta^1_\pm}{\alpha^2 I}\right\} [1-\exp(-\alpha I^{1/2})(1+\alpha I^{1/2})] + m^3[2(\nu_+\nu_-)^{3/2}]C_\pm. \qquad (11.128)$$

Since volume is the partial derivative of G with respect to P, at constant T, it is possible to extend the Pitzer model to the various volume properties (partial molar volumes, etc.) and to the properties of solutions at high pressure.

Problems

1. In the crystalline state CO has a residual entropy of 4.6 J/mol K. Propose a very simple statistical model that would account for this result. For NO, the residual entropy is only 3.1 J/mol K. Noting that the size difference between N and O atoms is larger than between C and O, explain the difference found between the residual entropies of the two crystals.

2. To prevent a sparlking wine from losing its CO_2 once the bottle has been opened, a practice consists of introducing a spoon in the neck of the bottle. Is this procedure thermodynamically grounded? Can it work?

3. With (11.52), one relates the freezing point depression of a solvent to its enthalpy of crystallization, which is assumed constant, and the mole fraction of the solute. With the data given in a problem of chapter 3, calculate how the enthalpy of fusion of ice would depend

on temperature, and then derive a modified equation for the freezing point depression of water. By inspection of Fig. 6.2a, indicate in which temperature range this equation will be valid.

4. The activity coefficient of component 1 of a binary system is represented by $RT\ln\gamma_1 = W_{12}x_2^2$. Show that that of component 2 is given by $RT\ln\gamma_2 = W_{12}x_1^2$.

5. Show that the partial molar enthalpy and entropy of mixing of component i for a multicomponent solution following the Flory-Huggins model are given by

$$\Delta\overline{H}_{mi} = -V_i\sum_j\sum_k(A_{ij} - A_{jk}/2)\phi_j\phi_k, \quad \Delta\overline{S}_{mi} = -R[\ln\phi_i + \sum_j\phi_j(1 - V_i/V_j)]$$

where V_i is the molar volume of component i, which is assumed to be independent of the composition of the solution.

6. Plot and comment on the fugacity-composition diagrams of H_2O-CO_2 mixtures calculated at 1000 K from Redlich-Kwong equations of state for total pressures of 1, 2 and 10 kbar. The parameters a and b of H_2O and CO_2 will be determined from the critical data of table 5.1, those of mixed fluids from the mixing rules described in 5.4c. Note that the expression of the fugacity coefficient of a Redlich-Kwong gas is given in a problem of chapter 8.

7. In the solution model set by Van Laar, the excess Gibbs free energy of a binary system is $G^{xs} = Ax(1-x)/[(xA/B) + (1-x)]$, where A and B are interaction parameters. Show that the activity coefficient of the component of molar fraction x is given by $RT\ln\gamma = A\{1 + Ax/[B(1-x)]\}^{-2}$.

8. The vapor pressures of pure ethanol and benzene at 318 K are 0.294 and 0.227 bar, respectively. The vapor pressures of their solutions at this temperature are listed below as a function of the ethanol mole fraction in the liquid (x_l^e) and vapor (x_l^e) phases.

P (bar)	0.391	0.403	0.407	0.407	0.407	0.405	0.386	0.360	0.312
x_e^l	0.1	0.2	0.3	0.35	0.4	0.5	0.7	0.8	0.9
x_e^v	0.293	0.333	0.360	0.371	0.381	0.401	0.470	0.536	0.666

Plot the phase diagram and calculate the activity coefficient of both components with regular, subregular and Van Laar (see the previous problem) solution models from only the data for the azeotropic point. Compare the activities given by the three models for the whole composition range to the experimental data. Which one works best?

9. The solubility product (see 12.3d) of cesium perchlorate ($CsClO_4$) in aqueous solution is 3.95 10^{-3} at 298 K. Calculate with the original and modified Debye-Hückel models the molalities of the two ions at saturation.

10. Calculate the activity coefficients given by the Debye-Hückel and Pitzer models for a few (1-1) and (2-1) electrolytes listed in Table 11.1, and ascertain for which electrolytes the results of both models agree up to the highest molalities.

12
EQUILIBRIA IN ELECTROLYTE SOLUTIONS

Especially in aqueous solutions, chemical equilibrium is often established through acid-base and redox reactions, i.e., through exchange of protons and electrons, respectively. For calculating acid-base equilibria we will make extensive use of Sillén diagrams with which simplifications are readily made in pH calculations. We will then illustrate how standard properties of individual ions are determined and describe complexation and precipitation reactions. With regard to redox reactions, emphasis will be put on emf measurements and their use as a means for determining thermochemical properties. Finally, the connection between acid-base and redox reactions will be made through potential-pH diagrams.

12.1 ACIDS AND BASES

12.1a CONJUGATE SPECIES

With *oil of vitriol* (H_2SO_4) or *aqua regia* (a mix of HCl and HNO_3), acids are as old as chemistry and alchemy. This is probably why their long history has been characterized by confusion. In the 17th century, the special taste and corrosiveness of acids were for instance attributed to the action of small, sharp pointed corpuscules. One century later Lavoisier thought himself that the element he named *oxygen* ("acid producer") was required to form acids from sulfur or phosphorus. It is much later, when ionic dissociation was recognized, that the modern concepts of acid-base reactions could be established. According to ideas proposed by the Danish chemist J.N. Brønsted

(1879 1947) in the early 1920's, an acid AH is a substance donating a proton

$$AH \rightleftharpoons A^- + H^+, \tag{12.1}$$

whereas a base, B, is a substance receiving a proton

$$B + H^+ \rightleftharpoons BH^+. \tag{12.2}$$

As indicated by these reactions, the novelty introduced by Brønsted was that any acid is *conjugated* to a base (AH to A^-), and any base is conjugated to an acid (B to BH^+).

The acid-base properties exhibited by solutions where protons are absent later lead Lewis to the broad generalization that an acid is a substance capable of accepting a pair of electrons, and a base a substance capable of donating a pair of electrons. The *neutralization* of an acid by a base then corresponds to the formation of a covalent bond in which both electrons of the shared pair are provided by the base. As an example, one of the three free electronic doublets of the OH^- ion is used to bond with a proton H^+ and form a water molecule. With Lewis' concept of acids a new connection between acid-base and redox reactions was therefore established. A given species can in fact play a dual role. The sulfur ion S^{2-} acts as a base toward water ($S^{2-} + 2\,H_2O \rightleftharpoons H_2S + 2\,OH^-$), and a as reductant toward ferric iron ($S^{2-} + 2\,Fe^{3+} \rightleftharpoons S + 2\,Fe^{2+}$), whereas acid and oxidizing functions can be exerted simultaneously by an acid like $HClO_4$. In spite of numerous analogies, concerning for example the notion of conjugate species, the differences between both kinds of reactions are such that these will be treated separately in this chapter.

12.1b HYDROGEN POTENTIAL (PH)

To begin with we will restrict ourselves to probably the most important case of acid-base reactions in aqueous solutions. Because they exert a strong control on chemical reactivity, acid-base properties play a considerable role not only in industrial processes, but also in natural waters or biological media as well. Proton exchange lasting typically less than a millisecond at room temperature, these reactions are characterized by their extremely rapid rates so that we will consider that equilibrium is always reached. But all acids and bases do not have the same propensity for donating or accepting protons. To characterize the acid-base state of a solution, we thus have to specify the concentration of the exchanged species, the proton. These concentrations are most often very small, expressed by negative powers of 10, but vary by a great many orders of magnitude. This makes it convenient to define the pH (*hydrogen potential*) by

$$pH = -\log[H^+] \tag{12.3}$$

where $[H^+]$ denotes the concentration of H^+ ions. We will express this concentration, like those of the other species, on the molality scale (number of moles per kg of solvent) as molalities do not vary with pressure and temperature.

The definition of pH (originally as "power hydrogen", denoted by p_H) owes much to beer in that it was given in 1909 by the Danish chemist Sørensen (1868-1939), working at the Carlsberg laboratories, who also set up electrometric and colorimetric methods for pH measurements. For instance, the concept of pH allowed Sørensen to account for the fact that the activity of invertase, an enzyme that hydrolyzes sucrose into glucose and levulose, is determined by the acidity of the solution. This biochemical activity is in fact optimal for a limited range of hydrogen ion concentrations (pH 4-7). Depending on the concentration scale used, one distinguishes two kinds of pH, $pm_{\rm H}$ and $pc_{\rm H}$ for the molality and molarity scales, respectively. It would be more rigourous to define pH = -log a_{H^+}, but this definition suffers from the problems raised by the determination of individual ionic activity coefficients at high ionic strength. These aspects and the various methods of pH measurement, with the glass electrode in particular, are discussed in the book by Bates (1964).

12.1c WATER IONIZATION

As indicated by electrical conductivity measurements, the purest water is partially ionized even though it is electrically neutral. If protons exist in an aqueous solution, hydroxyl ions (OH^-) are also present and it follows that the partial dissociation of water can be described by the reaction

$$H_2O \rightleftharpoons H^+ + OH^-. \tag{12.4}$$

To take into account proton solvation by water molecules, we can further write

$$H_2O + H^+ \rightleftharpoons H_3O^+, \tag{12.5}$$

where H_3O^+ designates the *hydronium ion*. As expressed by (12.4), a proton is donated only if there is another species in the solution ready to bond with it. By combining (12.4) and (12.5), we describe the *ionization* (or *autoprotolytic*) equilibrium of water in a second manner

$$2H_2O \rightleftharpoons H_3O^+ + OH^-. \tag{12.6}$$

To the reaction (12.4) is associated the *ionization constant* of water or, more simply, its *ion product*

$$k_w = a_{\rm H^+}\, a_{\rm OH^-}/a_{\rm H_2O}, \tag{12.7}$$

which is a function of the activities of the various species. For reaction (12.6), we write in the same way

$$k'_w = a_{\rm H_3O^+}\, a_{\rm OH^-}/a^2_{\rm H_2O}. \tag{12.8}$$

According to the discussion of 2.2c, the equality $\mu_{\rm H_3O^+} = \mu_{\rm H_2O} + \mu_{\rm H^+}$ is satisfied as long as the system is in internal equilibrium. Now, this equality not only accounts for the equilibrium of reaction (12.5), but it dictates that the affinities of reactions (12.4) and (12.6) are equal, and thus that $k'_w = k_w$. As a matter of fact, this conclusion represents another way of stating that the chemical potential of a species does not vary within any system at equilibrium. This potential is constant for H_2O molecules whether these are in close contact or not with the protons. Regardless of the number of H_2O molecules constituting the solvation sphere of a proton, the constant k_w unambiguously characterizes the ionization equilibrium of water. To describe the acid-base state of a solution, the choice of the species H^+, H_3O^+ (or even $H_5O_2^+$, etc.) is thus thermodynamically arbitrary. Although the notation H^+ is the simplest, we will use here the formula H_3O^+ in order to emphasize the importance of proton solvation in

Table 12.1. Ion product of water (in the molality scale): pk_w at various pressures (bar) and temperatures (from Marshall and Franck, 1981).

T (K)	1	100	250	50	1000
298	13.995	13.96	13.90	13.81	13.65
350	12.66	12.63	12.58	12.51	12.37
400	11.89	11.86	11.80	11.73	11.56
450	11.44	11.39	11.32	11.21	11.01
500	11.18	11.13	11.04	10.87	10.64

acid-base equilibrium, and we will also keep in mind that hydroxyl ions are not free but are also solvated.

For sufficiently dilute solutions, the activities of H_3O^+ (or H^+) and OH^- ions can be replaced by their molalities and that of H_2O can be taken as unity

$$k_w = [H^+][OH^-] = [H_3O^+][OH^-]. \tag{12.9}$$

Various means (thermochemistry or electrical conductance measurements) indicate that $\log k_w$ is -14.0 at 298 K and 1 bar (Table 1). By using the notation

$$pk = -\log k \tag{12.10}$$

and defining in the same way as the pH

$$p\text{OH} = -\log[\text{OH}^-], \tag{12.11}$$

we conclude that

$$p\text{H} + p\text{OH} = pk_w, \tag{12.12}$$

independently of any reaction that could take place in a sufficiently dilute aqueous solution. In the case of pure water, electroneutrality of the solution imposes $[H^+] = [OH^-]$, whence the neutral value $p\text{H} = 7$ at room pressure and temperature.

12.1d ACTIVE AND INACTIVE SPECIES

Owing to its ionization equilibrium, water has the very important particularity of being an *amphoteric* solvent, that is, of behaving as a base with respect to an acid

$$\text{AH} + \text{H}_2\text{O} \rightleftharpoons \text{A}^- + \text{H}_3\text{O}^+, \tag{12.13}$$

and as an acid with respect to a base

$$\text{B} + \text{H}_2\text{O} \rightleftharpoons \text{BH}^+ + \text{OH}^-. \tag{12.14}$$

At equilibrium, the equilibrium constants are related to the activities (a) of the species by

$$k_a = \frac{a_{H_3O^+}\, a_{A^-}}{a_{AH}\, a_{H_2O}}, \tag{12.15}$$

$$k_b = \frac{a_{BH^+}\, a_{OH^-}}{a_B\, a_{H_2O}}. \tag{12.16}$$

Depending on the values of k_a and k_b two kinds of acids and bases must be distinguished. For *strong* acids and bases, dissociation can be considered as complete; the conjugate bases and acids are then *inactive*, like Cl^- for HCl, or Na^+ for NaOH. In the case of *weak* acids and bases, such as H_3PO_4 and NH_3, respectively, dissociation is in contrast partial, at least in certain composition intervals. Arbitrarily, zero values of pk_a and pk_b can be chosen as a limit between strong acids and bases and weak acids and bases.

Alternatively, we could write the equilibria (12.13) and (12.14) in the form

$$A^- + H_2O \rightleftharpoons AH + OH^-,$$
$$BH^+ + H_2O \rightleftharpoons B + H_3O^+.$$

One can check that the product of the constants of *conjugate equilibria* is equal to k_w, the ion product of water. The two alternatives of course lead to the same result. It is to conform to common usage that these reactions are written in the more familiar forms (12.13) and (12.14).

12.1e EQUILIBRIUM CONSTANTS

We will express the equilibrium constants of reactions in terms of molalities

$$k_a = \Pi a_i^{\nu_i} = \Pi (m_i \gamma_i)^{\nu_i}, \tag{12.17}$$

where ν_i is the stoichiometric coefficient of species i and the product is taken over all ionic or neutral species, γ denoting the practical activity coefficient. To convert activities into concentrations the simplest situation is of course that of dilute solutions. The activity of water is then unity and the activities of the other species are equal to the molalities

$$k_a = [H_3O^+][A^-]/[AH], \tag{12.18}$$

$$k_b = [BH^+][OH^-]/[B]. \tag{12.19}$$

If needed, the relevant activity coefficients as derived from Debye-Hückel (see 11.6e) or other models can be used for calculation of the equilibrium constant. In this regard, an important feature of these solutions is that acid-base species have often low molalities with respect to inactive ions. The ionic strength of the solution is then essentially pH independent, which makes calculations of the equilibrium concentrations easier because the activity coefficients of the active species can be assumed to be constant. As will be seen below the principle of pH calculations remains the same whether or not activity coefficients are taken into account. For the sake of simplicity, we will thus consider dilute solutions.

12.2 PH CALCULATIONS

12.2a PRINCIPLE

All pH calculations rely on three different kinds of equations. First, we have an equilibrium constant for every acid-base pair present in the solution. Second, we write mass conservation equations; for an acid or a base of molalities m_a and m_b, respectively, these are

$$[\mathrm{A}^-]+[\mathrm{AH}]=m_a, \tag{12.20}$$

$$[\mathrm{B}^+]+[\mathrm{BOH}]=m_b. \tag{12.21}$$

Third, electrical neutrality must also be taken into account. In the case where only the acid AH or the base B are present in the solution, we have

$$[\mathrm{H_3O^+}]=[\mathrm{A}^-]+[\mathrm{OH}^-], \tag{12.22}$$

$$[\mathrm{B}^+]+[\mathrm{H_3O^+}]=[\mathrm{OH}^-]. \tag{12.23}$$

As will be shown in the next section, one finds readily the solutions to these equations in simple cases by using the equilibrium constants and mass balance relations to express the concentrations of the various species in terms of $[\mathrm{H_3O^+}]$ (or $[\mathrm{OH^-}]$) in the electroneutrality equation. With the pH (or pOH) found by solving this equation, one can then calculate the concentrations of the other species.

Calculations are made in a similar manner in complex solutions, but one is then faced with practical difficulties when the number of species becomes large. For n acid-base pairs (including water) present in a solution at given P and T, $2n-1$ concentrations must be determined (not $2n$ because we are not interested in minor changes in the concentration of the H_2O molecules). There are $n-1$ equilibrium constants, one electroneutrality constraint and $n-1$ mass balance equations, which gives a total of $2n-1$ equations. When n increases, however, the electroneutrality and mass balance equations become complicated, in a gratuitous way in fact, because most terms are actually negligible. This is why one must discard in these equations all species whose concentration is insignificant. But making the right approximation can be a tricky exercise in view of the considerable variations of the concentrations of the relevant species with pH, which one actually trie to determine. Fortunately, the diagrams popularized by the Swedish chemist L.G. Sillén (1916-1970) give a simple and clear answer to this problem.

12.2b SILLÉN DIAGRAMS

Let us begin with strong acids. To calculate their pH from a Sillén diagram, we reason the other way and represent the concentrations of species as a function of pH that would be fixed (Fig. 12.1). Due to almost total dissociation, the AH concentration is much too low to be plotted. Hence, the approximation $[\mathrm{A}^-]=m_a$ is obvious. Equilibrium being characterized by electroneutrality, the pH is determined by the point N where the lines $[\mathrm{A}^-]$ and $[\mathrm{H_3O^+}]$ meet because the concentration of OH^- is also

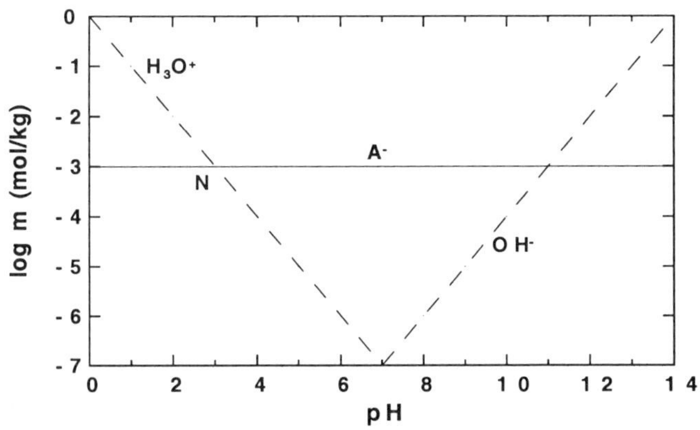

Figure 12.1. Sillén diagram for a strong acid with a molality of 10^{-3} mol/kg.

negligible. Analytically, we find from (12.15) and (12.20)

$$pH = -\log m_a. \tag{12.24}$$

For a strong base, we would find in the same manner pOH = - log m_b, i.e., pH = 14 + log m_b.

With regard to strong acids, weak acids differ by the fact that the concentration of their conjugate base strongly depends on pH (Fig. 12.2). From (12.15) and (12.20), we have indeed

$$[A^-] = \frac{k_a}{k_a + [H_3O^+]}, \tag{12.25}$$

$$[AH] = \frac{m_a\,[H_3O^+]}{k_a + [H_3O^+]}. \tag{12.26}$$

We then readily plot the Sillén diagram (Fig. 12.2). Regardless of the acid concentration, this plot shows that $[OH^-]$ would be comparable to $[A^-]$ only at a very high pH. When electroneutrality is satisfied, at the point N, the only anion having a

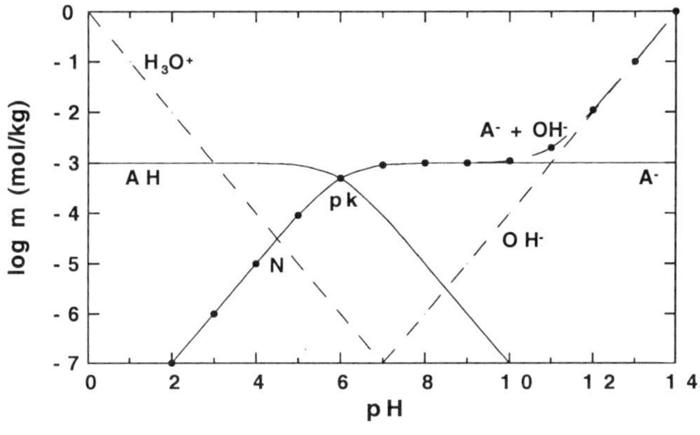

Figure 12.2. Sillén diagram for a weak acid with $pk = 6$ and a molality of 10^{-3}. The sum of the negative charges is represented by the curve with the solid circles.

significant concentration is A^- but its concentration is low compared to that of the acid AH. The two approximations $[A^-] = [H_3O^+]$ and $[AH] = m_a$ thus follow from this observation. By combining them to (12.15), we obtain

$$pH = 1/2\ (pk_a - \log m_a). \tag{12.27}$$

For a weak base, we would have in the same way $p\text{OH} = 1/2\ (pk_b - \log m_b)$, i.e., $p\text{H} = 14 - 1/2\ (pk_b - \log m_b)$. In passing, note that $[A^-]$ is necessarily equal to $[AH]$ for $p\text{H} = pk_a$. For $p\text{H} = pk_a + 1$, $[AH]$ represents only 10% of $[A^-]$, and the converse holds true for $p\text{H} = pk_a - 1$. It is this very strong change in the concentration of acid-base species that must be exploited when making any approximation.

If we lower the acid concentration, the curve $[A^-]$ in Fig. 12.1 and the curves $[AH]$ and $[A^-]$ in Fig. 12.2 are displaced downward. As a function of concentration, this allows acid-base pairs to be placed in distinct domains in a $\log m, pk$ diagram (Fig. 12.3) which indicate how strongly they affect the pH of the solution. Especially noteworthy is the fact that, at low enough concentrations, strong and weak acids (and bases) have eventually no longer any influence on pH. Regardless of their pk, these *minor* pairs have a speciation that is determined instead by the pH of the solution. For illustration purposes, the diagram of Fig. 12.3 has actually been designed to describe the main acid-base pairs of shallow sea waters. Under these conditions of ionic strength, the pk of water is not 14 but 13.7 and the only active pairs are those of carbonic acid, H_2CO_3/HCO_3^- and HCO_3^-/CO_3^{2-}, and the first pair of boric acid, $H_3BO_3/H_2BO_3^-$. For such complex solutions, pH calculations can be considerably simplified when all inactive species are grouped into an *alkaline reserve*, which is by definition insensitive to pH, temperature and pressure. In a Sillén diagram, the pH is then given by the point where the horizontal line of the alkaline reserve intersects the curve representing the sum of the electrical charges of active species. (Owing to the propagation of analytical errors, the alkaline reserve cannot be calculated precisely as the sum of the

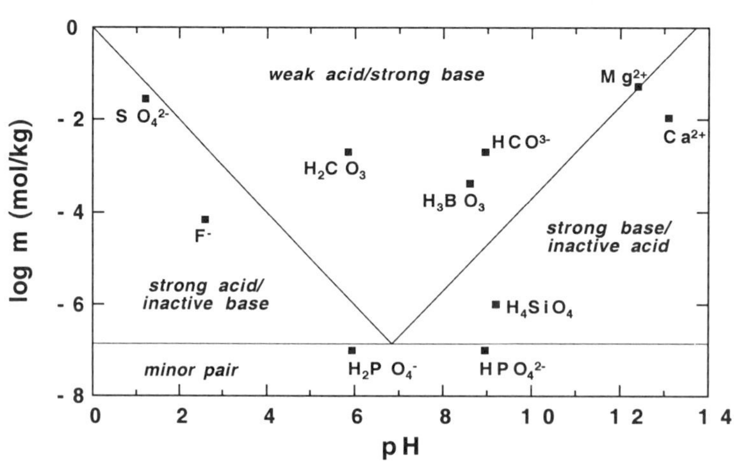

Figure 12.3. Characterization of acid-base pairs by their pk and concentrations. The position of the main pairs present in superficial marine waters are included (from the book by Michard, 2001).

concentrations measured for the various ions. By titration of the solution, however, it is possible to measure it directly with good accuracy.)

12.2c MULTIPLE ACID-BASE FUNCTIONS

Many substances have several acid-base reactions each having a specific *pk*. Well-known examples are carbonic acid, with the two H_2CO_3/HCO_3^- and HCO_3^-/CO_3^{2-} pairs already mentioned, and phosphoric acid, with the three pairs $H_3PO_4/H_2PO_4^-$, $H_2PO_4^-/HPO_4^{2-}$ and HPO_4^{2-}/PO_4^{3-}. For such systems Sillén diagrams are especially useful for calculating the *p*H either graphically or analytically, through the right approximations they allow one to make at once.

Let us consider a substance with three acid-base functions. We will denote as AH_3, AH_2^-, AH^{2-} and A^{3-} its different species and k_1, k_2 and k_3 the constants of the relevant equilibria

$$k_1 = [H_3O^+][AH_2^-]/[AH_3], \tag{12.28a}$$
$$k_2 = [H_3O^+][AH^{2-}]/[AH_2^-], \tag{12.28b}$$
$$k_3 = [H_3O^+][A^{3-}]/[AH^{2-}]. \tag{12.28c}$$

Mass conservation is written as

$$m = [AH_3] + [AH^{2-}] + [AH_2^-] + [A^{3-}], \tag{12.29}$$

and, noting

$$F = 1 + k_1/[H_3O^+] + k_1k_2/[H_3O^+]^2 + k_1k_2k_3/[H_3O^+]^3, \tag{12.30}$$

we can express the species concentrations as a function of *p*H by

$$[AH_3] = m/F, \tag{12.31a}$$
$$[AH_2^-] = \frac{mk_1}{F[H_3O^+]}, \tag{12.31b}$$
$$[AH^{2-}] = \frac{mk_1k_2}{F[H_3O^+]^2}, \tag{12.31c}$$
$$[A^{3-}] = \frac{mk_1k_2k_3}{F[H_3O^+]^3}. \tag{12.31d}$$

It just remains to plot the concentrations of the various species as a function of concentration to make a *p*H calculation.

Let us use these equations for calculating the *p*H of a solution of lysine (Fig. 12.4), one of the *amino acids* that make up proteins, which are thus important polyacids. Differing in that matter from the aforementioned inorganic substances, their acid functions can be due to various molecular groups (COOH, NH^+ and NH_3^+). For lysine, which is electrically neutral as a pure substance, the first acid function is that of a COOH group whereas the two others are due to NH_3^+ groups that are formed by the action of water on NH_2 groups. We will denote as $LysH_3^{2+}/LysH_2^+$, $LysH_2^+/LysH$ and

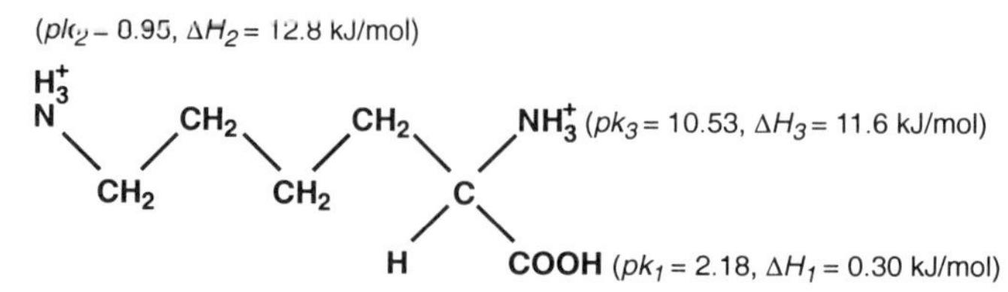

Figure 12.4. Structure of lysine, and pk and enthalpies of its acid-base reactions.

LysH/Lys$^-$ these three acid-base pairs and plot in Fig. 12.5 their Sillén diagram as calculated from eqns (12.31) for a molality of 0.02. The sum of positive and negative charges is then easily calculated as a function of pH. As shown in Fig. 12.5, it is at point N, for $p\text{H} = 9.75$, that electroneutrality is satisfied. Note in passing the important biological fact that the pH of protein solutions depends strongly on temperature because the acid-base reactions of amino-acids have generally high enthalpies.

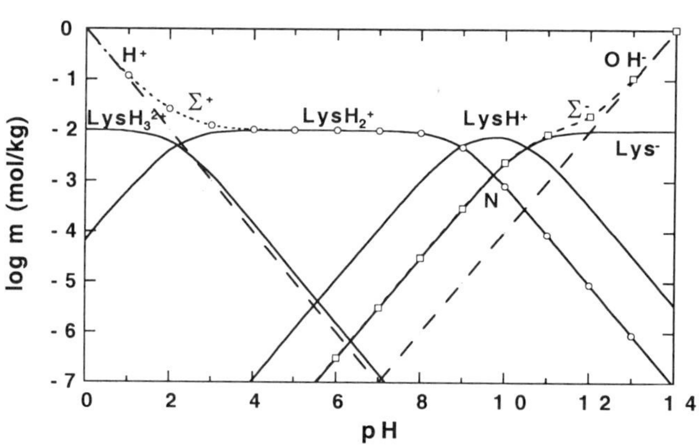

Figure 12.5. Sillén diagram of lysine for a molality of 0.02 mol/kg. The sums of positive and negative charges are represented by the curves with the open circles and squares, respectively.

12.3 IONS: STANDARD PROPERTIES AND INTERACTIONS IN SOLUTION

12.3a STANDARD STATE OF IONS

For any ion, we recall that the standard state usually chosen is a hypothetical ideal solution of unit molality, i.e., a solution whose activity is obtained by extrapolation up to a unit molality, with Henry's law, of the activity of an infinitely dilute solution, see 11.2e. But we have already mentioned in 11.6b that it is impossible to let the concentration of an anion vary independently of those of the cations, or vice versa, because of the condition of electrical neutrality. The stoichiometry of all dissociation reactions imposes well-defined relationships between the numbers of moles of anions and cations so that partial molar properties of individual ions, among which the most important one is the chemical potential, cannot be determined.

It is nonetheless possible to solve formally this problem through yet another convention that is akin to that adopted in 4.5b for defining formation properties from the elements. With it, at any temperature and pressure one sets to zero the Gibbs free energy of formation of the hydrogen ion, H^+, in an aqueous solution where its molality would be unity. In order that ΔG_f° remains zero at any P and T, the enthalpy of formation, entropy, heat capacity and volume of the proton must also be set to zero at any temperature and pressure. Against this standard state, the properties of all other ions can be determined from equilibrium constants of ionic reactions. Those of H_3O^+ (= H^+ + H_2O), for instance, are then identical to those of water. Likewise, by knowing the properties of AH and B, we can determine those of A^- and BH^+ from the equilibria (12.1) and (12.2). Progressively, the properties of any other anion or cation might also be determined.

12.3b STANDARD PROPERTIES OF IONS : THE EXAMPLE OF THE HYDROXYL ION

The important example of the hydroxyl ion will now be taken to describe how the standard properties of ions can be calculated once the properties of the proton have been arbitrarily set to zero. Here, the requisite information will come from the ion product of water as listed in Table 12.1 over wide temperature and pressure ranges. The results obtained are summarized in Table 12.2 along with the relevant properties of water. We note in passing that they will illustrate how the properties of OH^- in an ideal, hypothetical solution of unit molality are determined from experiments made at very high dilution, where the molality of OH^- is as low as 10^{-7}.

Table 12.2. Water ionization and thermodynamic properties of H^+ and OH^- ions in their standard state, at 298 K and 1 bar, as determined from the ion product of water.

Unit	Ionization reaction		Water	H^+	OH^-
kJ/mol	$\Delta G_I = 79.85$	ΔG_f°	-237.14	0	-157.3
kJ/mol	$\Delta H_I = 54.93$	ΔH_f°	- 285.83	0	230.9
J/mol K	ΔS_I = - 88.5	S°	69.95	0	-18.6
J/mol K	ΔC_{pi} = - 135	C_p°	75.19	0	- 59.8
cm^3/mol	ΔV_I = -21.6	V°	18.07	0	-3.6

At the different pressures and temperatures listed in Table 12.1, the Gibbs free energy of the ionization reaction of water is directly given by $\Delta G_I = -2.303\ RT \ln k_w$, where the constant 2.303 represents ln 10. The Gibbs free energy of formation of water being known, that of the OH^- ion is given by

$$\Delta G^\circ_{fOH^-} = \Delta G_I + \Delta G^\circ_{fH_2O} - \Delta G^\circ_{fH^+} = \Delta G_I + \Delta G^\circ_{fH_2O}.$$

To derive the enthalpy of formation in the same way from the enthalpy of ionization, ΔH_I, let us first remark that, at 1 bar, the equilibrium constants of Table 12.1 are reproduced by the empirical equation

$$\ln k_w = 35.777 - 13.98410^3/T + 5.417610^6/T^2 - 0.466610^9/T^3.$$

It then suffices to differentiate this expression with respect to $1/T$ to calculate ΔH_I with the Gibbs-Helmholtz equation (3.4),

$$\Delta H_I = -R\left[\frac{\partial \ln k}{\partial(1/T)}\right]_P,$$

and find the value of ΔH°_f given in Table 12.2. Differentiating with respect to T the ΔH_I obtained, we determine ΔC_p for the ionization reaction and the heat capacity of OH^-. From the other empirical expressions obtained from the data of Table 12.1

$$\Delta G_I = 73,581 - 46.993T + 0.2273T^2 \text{ (J) for } P = 1\text{bar},$$

$$\Delta G_I = 79,871 - 2.1616P + 0.210^{-3}P^2 \text{ (J) for } T = 298 \text{ K},$$

we finally determine the entropy of the hydroxyl ion with $\Delta S_I = -(\partial\Delta G_I/\partial T)_P$ and its volume with $\Delta V_I = (\partial\Delta G_I/\partial P)_T$.

The negative values derived for S°, C°_p and V° are nothing to be worried about: we have actually calculated the molar properties of OH^- in aqueous solutions at very high dilution, which are assumed to be constant up to the unit molality of the hypothetical standard state. Because all properties of H^+ have been arbitrarily set to zero, only the sum of the properties of H^+ and OH^- ions have a physical significance. (With $C^\circ_{p_{OH^-}}$ = -60 J/mol K for instance, 55% of water molecules should be dissociated for the heat capacity of water to vanish!) We nonetheless find that the sum of S°, C°_p and V° for H^+ + OH^- are also negative. This is a consequence of ion solvation which, as already described in 10.1b, gives rise to a more compact and ordered arrangement of the solution around the ions. Remark finally that the derived C°_p for OH^- is subject to a large error because it is given by a second derivative of $\ln k_w$. This is why one obtains a better constrained result of about -117 J/mol K by simultaneously exploiting a number of such ionic equilibria involving the OH^- ion.

12.3c COMPLEXATION REACTIONS

Undissociated acids and bases illustrate the fact that there exist soluble ionic or neutral species that are made up of other species present in the solution. To limit confusion between acid-base and complexation reactions, however, ones denotes by *complex* only species in which there is no proton and hydroxyl ion. A complex is often made up of a big species that bear free electronic doublets (generally an anion), called a *ligand*, to which a cation bonds reversibly. Complexation reactions are in general rapid so that equilibrium prevails. The complexation reaction can be written

$$\text{M (single ion)} + \text{L (ligand)} \rightleftharpoons \text{ML (complex)}, \tag{12.32}$$

and its equilibrium constant expressed as a function of the activities of the species and the standard affinity of reaction

$$k = a_{ML}/a_M a_L = \exp(A^\circ/RT). \tag{12.33}$$

The main action of complexes is to reduce greatly the effective concentration of the species involved. In turn this allow the concentration of the metal to be buffered in a solution. The redox potential of a metal is modified by complexing because the stability constant of a given ligand depends on the valence of the metal.

Complexes in which the cation has several bonds with the ligand are called *chelates*. Their stability constants are generally very high. The metal can often bond with more than one molecule of ligand

$$\mathrm{ML} + \mathrm{L} \rightleftharpoons \mathrm{ML}_2, \tag{12.34}$$

in which case the *cumulative stability constant* $\beta_n = k_1 k_2 ... k_n$ is the product of the various equilibrium constants. The metal can also bond to several different ligands. An interesting example is provided by the mercury ion Hg^{2+} which forms complexes with chloride ions Cl^- ($HgCl^-$, $HgCl_2^\circ$, $HgCl_3^-$ and $HgCl_4^{2-}$) and hydroxyl ions as well ($Hg(OH)_2^\circ$ and $HgOH^+$). The question is then to know how the concentration of the various complexes vary with that of the ligands Cl^- and OH^-. Sillén diagrams, which are drawn in the same way as described above for *p*H calculations, are very useful in this respect. A convenient way to summarize the whole issue is to represent in a log $[Cl^-]$-*p*H diagram the conditions where the two most abundant species have the same concentration (Fig. 12.6). Such a plot is not a phase diagram, but a representation of the domains where the various species present in a single phase, the solution, predominate at a given temperature and pressure. It happens that a given species is never the most abundant, which is the case of $Hg(OH)^+$ in Fig. 12.6, but we emphasize that the concentration of every species remains nonzero, although it can be very small.

Complexing play a very important role in natural, biological and industrial processes as many organic compounds (including amino acids) are very effective ligands. This holds especially true for heavy metals. Although these are trapped in the soil through complexation with organic matter, complexation with stronger ligands can conversely be used for remediation of polluted soils and waters. In the same way, Pb, As and Hg poisoning

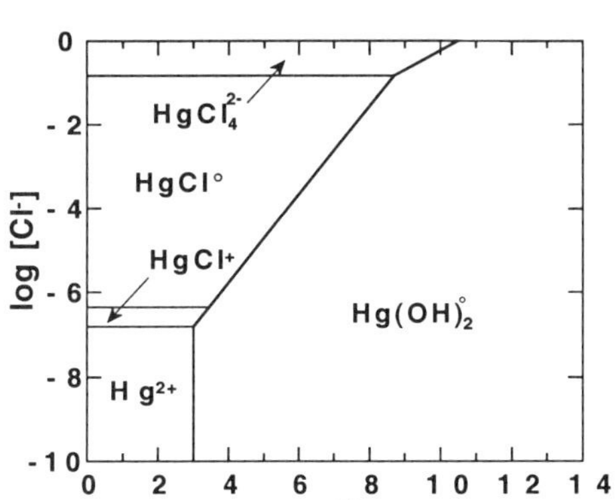

Figure 12.6. Fields of predominance of the chloride and hydroxyl complexes of mercury, calculated at 1 bar and 298 K from the stability constants given at the end of this chapter (see the book by Michard, 2001).

can be treated therapeutically through complexation by ligands such as 2,3-dimercapto-1-propanol. Among the numerous ligands used in chemistry, we will mention only the celebrated ethylene-diaminetetra-acetic acid (EDTA) whose stability constants are for instance about 10^{10} for Mg^{2+} and 10^{18} for Zn^{2+}. These differences are consistent with a general trend whereby the stability of complexes increases from alkaline to alkaline-earth and then to transition elements. (These constants in fact strongly varies with pH because of the four acid functions of EDTA, which are due to the CH_2COOH groups whose pk varies from 2.0 to 10.3.)

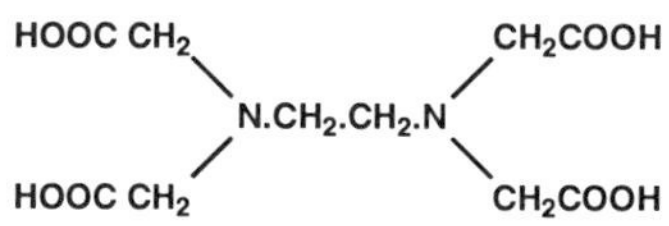

EDTA

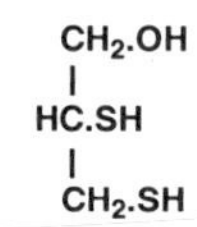

2,3 DIMERCAPTO-1-PROPANOL

12.3d SOLUBILITY PRODUCT

When forming complexes, the components associate but they remain dissolved in the solution. By precipitating in the form of a crystalline solid, like a salt, the components associate in a more intimate manner. Depending on many factors, this can happen at low as well as at high concentration of the species in solution. Denoting by μ^c the chemical potential of the solid precipitating from an electrolyte dissociating into ν_- anions and ν_+ cations, we have at equilibrium

$$\mu^c = \nu_+\mu_+ + \nu_-\mu_-. \tag{12.35}$$

Taking the pure crystal at the T and P of interest as a standard state, we write the equilibrium constant k_s

$$k_s = (m_+)^{\nu_+}(m_-)^{\nu_-}\gamma_\pm^{(\nu_+ + \nu_-)} = \exp(A^\circ/RT). \tag{12.36}$$

This constant k_s is called the *solubility product*. Of course, the presence of other ions in the solution must not be forgotten when calculating the ionic strength and the activity coefficients of (12.36) with Debye-Hückel, Pitzer or other models.

Like any phase equilibria determination, the determination of a solubility product in turns allow the activity coefficients or the standard affinity of precipitation to be determined. In this case one must make sure that the measured equilibrium product is the equilibrium value. Contrary to acid-base or complexation reactions, precipitation can be a very slow process because the creation and growth of a crystal nucleus involve surface energy (see 7.3c). In the same way as a liquid is supercooled when crystallization does not take place, a solution is *oversaturated* when the solubility product is exceeded. It is then metastable. (An aqueous solution can therefore be doubly metastable, being at the time supercooled and oversaturated.) The simplest way to characterize the *degree of oversaturation* is to take the ratio between the actual product of the activities of the relevant components and the solubility product of the precipitate.

The distinction between supercooling and oversaturation vanishes for ionic liquids where crystallization of a salt on the liquidus takes place through association of the ions of the solution. The relation (12.36) then becomes equivalent to the expression (11.72) of the Temkin model. In case of polymorphism, the reasons already advanced in 9.2c lead to the conclusion that solubility increases with the degree of metastability. This is why aragonite ($CaCO_3$), which is metastable at room T and P, is more soluble (by a factor of 2) than the stable polymorph calcite in sea water (which does not prevent many sea animals from synthesizing this aragonite to make their shells). The same effect is at work when cristallinity is imperfect, solubility being lower for a perfect crystal than for a poorly crystallized material.

Two different cases must be distinguished for species that have too low a concentration to precipitate on their own. In *coprecipitation*, the dilute species precipitate as a minor component of a solid solution. When writing the solubility product, one must take into account the activity-composition relationship and use is generally made of Henry's and Raoult's laws to calculate partition coefficients defined in the same way as in 10.5b. Another mechanism is *adsorption*, which is governed by surface forces, whereby the dilute species are deposited at the surface of the precipitate. This is specially common at the surface of ionic compounds, where the existence of non compensated electrical charges make the fixation of ionic species easy, and is a more efficient process when the surface area of the precipitate is large, i.e., when precipitation takes place as tiny crystals.

The kinetics of precipitation in a given solution strongly depends on which precipitate is considered so that the degree of oversaturation can greatly vary from one precipitate to another. An important solution is sea water. Wheras cations such as Fe^{2+}, Fe^{3+} and especially Mn^{2+} remain consistently oversaturated with respect to their salts, the solubility product of $CaCO_3$ is significantly exceeded only in shallow waters. Between the surface and a depth of 5000 m, the solubilities of calcite and aragonite double because the temperature decreases and the pressure and concentration of dissolved CO_2 increase. As sea water becomes undersaturated below 4500 m, the falling shells of planktonic animals then dissolve at greater depth. This is why limestone sedimentation takes place only on continental margins, whereas the eruptive rocks that have been formed along the mid-oceanic ridges are still laying on the deep oceanic floor.

12.4 REDOX REACTIONS

12.4a OXIDANTS AND REDUCTANTS

Still more than acid-base reactions, redox reactions constitute a very large theme that cannot be covered comprehensively in the rest of this chapter. Our goal will be modest. It will mainly be to establish connections between these reactions and thermodynamic potentials, to show how electrochemical measurements are used to determine Gibbs free energies of formation, and to state under which conditions of pH and electric potential a given species can predominate. We will turn to these points after a brief reminder of the main features of redox rections and a short description of electrochemical cells, a topic that has been deliberately left out in chapter 4.

As already mentioned, a redox reaction is due to the exchange of one or several electrons between an reducing agent, the *reductant*, which donates them, and a oxidizing agent, the *oxidant*, which accepts them. An element is thus oxidized if it donates electrons, and it is reduced if it gains electrons. In contrast to acid-base reactions which are always rapid, redox reactions can be very slow so that thermodynamic equilibrium is not necessarily attained. For instance, the reaction

$$C + O_2 \rightleftharpoons CO_2, \quad (12.37)$$

has a fast kinetic only at high temperature (otherwise both graphite and diamond could not be kept in air). In this case, the electrons donated by the oxidant are shared in the product (CO_2). As a result, the actual number of electrons exchanged between carbon and oxygen cannot be readily ascertained. In addition, the gaseous nature of the product and one reactant makes it difficult to let such reactions advance reversibly in electrochemical experiments. Noting that such equilibria are conveniently treated with the concepts presented in previous chapters, we will thus focus in this chapter on reactions that lend themselves more readily to electrochemical investigations. This will be mostly the case of reactions involving electrolytes and ionic substances.

The whole atmosphere is in metastable equilibrium because the reaction 5/2 O_2 + N_2 + $H_2O \rightleftharpoons 2\ NO_3^-$ + $2H^+$ does not proceed at 1 bar and 298 K in spite of its considerable affinity of 223 kJ (with respect to a nitrate solution of unit molality). If it were actually taking place, it should not only make all oxygen disappear but also oxidize an important fraction of nitrogen into nitrate. Likewise, the organic matter synthesized from water and carbon dioxide by the sophisticated photosynthetic machinery of vegetal cells could not build up if the kinetics of oxidation reactions were generally fast. It is therefore thanks to the slowness of such redox reactions that a biosphere can exist.

12.4b HALF REACTIONS

Consider the reaction

$$2\ HCl + 2\ Fe \rightleftharpoons 2\ FeCl_2 + H_2. \quad (12.38)$$

Although the exchange of electrons between oxidant and reductant is more easily recognized, it is useful to split this reaction into two *half reactions* to show more clearly the distinct role of the *conjugate* oxidants and reductants, viz.:

$$Fe \rightleftharpoons Fe^{2+} + 2\ e^-, \quad (12.39a)$$

$$2\ H^+ + 2\ e^- \rightleftharpoons H_2, \quad (12.39b)$$

where chlorine does not appear because it remains in the form of a Cl^- ion. In case the reaction would have a single product

$$2\ Na + Cl_2 \rightleftharpoons 2\ NaCl, \quad (12.40)$$

we would write in the same way

$$2\ Na \rightleftharpoons Na^+ + 2\ e^-, \quad (12.41a)$$

$$Cl_2 + 2\ e^- \rightleftharpoons 2\ Cl^-. \quad (12.41b)$$

Half reactions can provide an oversimplified picture of the reaction, however, because free electrons do not exist in aqueous solutions and because the actual redox mechanisms may be complex, especially when several electrons are exchanged.

Hence, half reactions are mainly a convenient way of describing the redox process. As such, they cannot be associated with an equilibrium constant.

In the same way as all acids and bases do not have the same facility to accept or donate protons, oxidants and reductants differ in their affinity for electrons. This is why they are also termed either strong or weak, and why a strong oxidant is conjugated to an inactive reductant, and vice versa. In the aforementioned examples, sodium is a strong reductant whose conjugate reductant Cl_2 is inactive, whereas ferrous iron (Fe^{2+}) is a weak oxidant. The capacity of an atom of to attract the electrons of another atom is characterized by its *electronegativity*. Although the electronegativity depends on the species in which the element is present, general scales have been established, for instance on the basis of systematics of bond energies.

In a somewhat arbitrary fashion, one then attributes the formal charge of the single or multiple bonding electronic doublets to the most electronegative atom. This charge is called the *oxidation number* because it increases in an oxidation and decreases in a reduction reaction. A given element can have different oxidation numbers in the same species, like carbon in hydrocarbons where the values -4 and 0 are attributed to C in CH_4 et $C{\equiv}C$ groups, respectively. The net electrical charge of a species is equal to the sum of the oxidation numbers of its various atoms. By convention, one distinguishes redox from acid-base and complexation reactions by the fact that only the former give rise to changes in oxidation numbers. Oxidation numbers also allow one to characterize simply *dismutation* reactions where a given element is at the same time oxidized and reduced. As an example, the oxidation number of oxygen changes from -1 to -2 and 0 when hydrogen peroxide yields water and oxygen

$$2\,H_2O_2 \rightleftharpoons H_2O + 1/2\,O_2. \qquad (12.42)$$

The terms oxidation numbers and valence are often used synonymously. Oxidation numbers have raised quite a few discussions. Their only unambiguous values are for pure elements, for which they are necessarily zero. By definition, this concept is approximate in other instances because it neglects the influence of chemical bonds on the external electronic orbitals of the element considered. When bonds are predominantly covalent, like in CO_2, it is a purely formal decision to attribute to carbon and oxygen the valences +4 and -2, respectively. In such cases, the electron density calculations made through quantum mechanical methods do show that the charge distributions are more regular than indicated by oxidation numbers. For species like NaCl and $FeCl_2$, a definite ionic character provides a good justification for giving the valences +1 and -1 to sodium and chlorine, respectively, and +2 to ferric iron. In any case, however, oxidation numbers must be taken only as empirical yardsticks, not as detailed pictures of the physical reality.

12.4c ELECTROCHEMICAL CELLS

When a redox reaction takes place spontaneously, the energy resulting from the transfer of electrons is usually dissipated as heat. It is the purpose of *electrochemical cells* to keep it as electrical energy by limiting the amount of heat produced by the displacement of electrons through the reaction medium (Fig. 12.7). This implies to separate materially the elements that donate and accept the electrons. In practice, this achieved

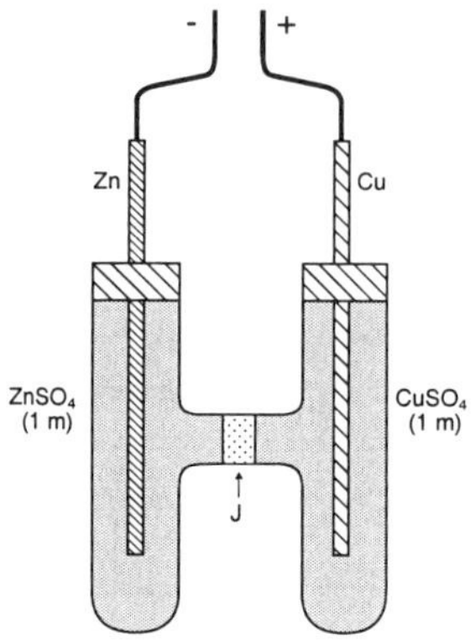

Figure 12.7. Schematic picture of a Daniell cell. In this case the junction J is made of a porous plug that closes the electrical circuit while limiting interdiffusion of the two solutions.

through the realization of two *half cells* in which the two half reactions proceed separately, but simultaneously. The electrons go from the oxidant to the reductant through two electrodes dipping each in a half cell, the oxidation occuring at the *anode*, and the reduction at the *cathode*. According to the nomenclature designed by Faraday (1791-1867), cations indeed move toward the cathode, anions move toward the anode and electrons from the anode to the cathode. To close the electrical circuit and allow electrical neutrality to be maintained in both half cells, an ionic diplacement takes place in the opposite direction through a *junction* which can be of many different kinds. Actually a cell can work in two different ways. In a *galvanic* cell, the redox reaction proceeds spontaneously and produces electrical energy. In an *electrolytic* cell, a current produced by some external source flows in the opposite direction and induces the reverse chemical process in the half cells.

Here we are not interested in cells as a means of getting electrical power, but mainly in their use for determining thermodynamic properties. The objective is then to measure the potential difference between the half cells under conditions that approach equilibrium as closely as possible. The current drawn must be negligible to avoid production of heat as a result of the internal resistance of the cell. In a reversible cell, any reaction can then advance forward or backward following a very small change in the direction of the current. By definition, the *electromotive force* (emf) of the cell is the potential difference between the half cells when no current is drawn through the electrical circuit. Experimentally, these conditions can be closely approached since measurements can be made with currents typically lower than 10^{-12} A.

As a simple example one often takes the Daniell cell. Its emf is that of the reaction

$$\mathrm{Zn} + \mathrm{Cu}^{2+} \rightarrow \mathrm{Zn}^{2+} + \mathrm{Cu}, \tag{12.43}$$

and its notation is

$$\mathrm{Zn(s)} \mid \mathrm{Zn}^{2+}(\mathrm{m}) \mid \mathrm{Cu}^{2+}(\mathrm{m}) \mid \mathrm{Cu(s)}.$$

In accordance with the convention for the initial and final states of a transformation, the notation of a cell indicates the reactants and the products on the left and right hand sides, respectively. For the same elements, the notation is then inverted depending on whether the cell is galvanic or electrolytic. One also specifies the nature of the phases present in the half cells (i.e., for the Daniell cell: s = solid, and m = aqueous solution of molality m) and insert vertical bars to indicate the limits between the phases.

From a practical standpoint, the most important limit is the junction which is a source of irreversibility that one tries to reduce. For the Daniell cell, one can for instance replace the porous plug by a *salt bridge* (a gel soaked with a concentrated salt solution) through which the Cu^{2+} or Zn^{2+} (along with SO_4^{2-}) ions will diffuse. A junction which has no emf is represented by a double bar. For the Daniell cell, we would thus write

$$\mathrm{Zn(s)} \mid \mathrm{Zn^{2+}(m)} \parallel \mathrm{Cu^{2+}(m)} \mid \mathrm{Cu(s)}.$$

Note, however, that a cell cannot be exactly reversible because the concentration gradient that necessarily exist in the neighborhood of the electrodes as a result of the electrochemical reaction is itself a source of irreversibility.

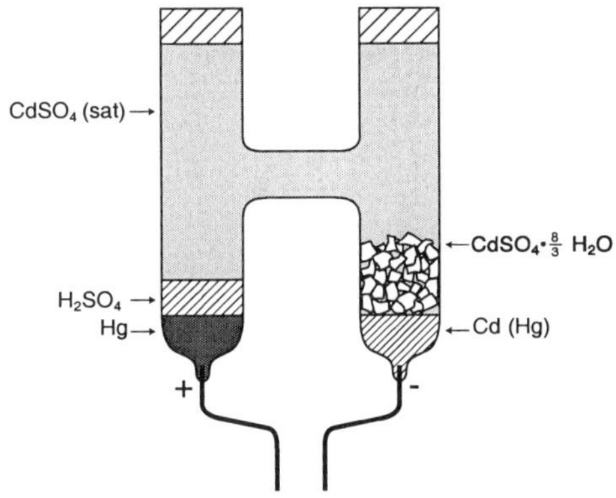

Figure 12.8. Scheme of a Weston cell, where a single electrolyte, the saturated cadmium sulfate solution, connects the two half cells. Redrawn from the book by Moore (1972).

An important example of a more complex reaction is provided by the Weston cell (Fig. 12.8), which has served for decades as a reference in thermodynamic measurements because of the remarkable constancy of its emf with time. Its reaction is

$$\mathrm{Cd(s)} + \mathrm{Hg_2SO_4\ (s)} + 8/3\ \mathrm{H_2O\ (l)} \rightleftharpoons \mathrm{CdSO_4} \cdot 8/3\ \mathrm{H_2O\ (s)} + 2\mathrm{Hg\ (l)}, \qquad (12.44)$$

where s and l designates solid and liquid phases, respectively. As the Weston cell is made up of a single electrolyte, it has the advantage of being deprived of a junction. Its notation is

$$\mathrm{Cd(Hg)} \mid \mathrm{CdSO_4} \cdot 8/3\ \mathrm{H_2O} \mid \mathrm{CdSO_4(sat)} \mid \mathrm{Hg_2SO_4} \mid \mathrm{Hg}$$

where Hg indicates an amalgam of Cd with Hg, and sat that the cadmium sulfate solution is saturated with respect to $Cd_2SO_4 \cdot$ 8/3 H_2O (and Hg_2SO_4 as well, whose solubility is very low). For various reasons it happens that the half cells of interest are chemically incompatible. One must then insert in between them a third electrolyte, which makes the design of the cell more difficult and, incidentally,its notation more complicated.

12.4d GIBBS FREE ENERGY OF REACTION AND NERNST'S EQUATION

We will now consider reversible conditions to relate the emf of a cell to the Gibbs free energy of the reaction ΔG. Let ν be the number of moles of electrons exchanged. When the reaction advances by $d\xi$, $\nu d\xi$ electrons are transferred and the variation of the Gibbs free energy is

$$dG = \Delta G d\xi. \tag{12.45}$$

At constant pressure and temperature, the electrical work done is equal to the product of the charge and the potential difference $\Delta V = E$ between them. If E is positive (galvanic cell), the electrical work must be counted as negative since if it done by the cell. If the work is infinitesimal, then the composition of the system does not change when the reaction advances by $d\xi$. Hence we have

$$dW = -\nu N_A e E d\xi = -\nu \mathcal{F} E d\xi, \tag{12.46}$$

where e is the charge of the electron, N_A the Avogadro number and $\mathcal{F} = N_A e$ represents one Faraday. From (2.27), the work is equal to the change in the Gibbs free energy of the system. We thus conclude from (12.45) and (12.46) that

$$\Delta G = -\nu \mathcal{F} E. \tag{12.47}$$

These reversible conditions do not correspond to the practical use of a cell to produce electrical work. As already noted, part of the chemical energy change is then transformed into heat because of the internal resistance of the cell. It remains, however, that the efficiency of the conversion of chemical energy into electrical work is much higher than if all the chemical energy change is first transformed into heat.

Let us then consider a general redox reaction of the form

$$\nu_1 \, \mathrm{M}_1 + \nu_2 \, \mathrm{M}_2 + \ldots \rightleftharpoons \nu_1' \, \mathrm{M}_1' + \nu_2' \, \mathrm{M}_2' + \ldots \tag{12.48}$$

whose Gibbs free energy is

$$\Delta G = \sum \nu_i \mu_i = \Delta G^\circ + RT \ln \prod \frac{a_i'^{\nu_i'}}{a_i^{\nu_i}}. \tag{12.49}$$

With (12.47) we can relate the emf of the cell to the activities of the species with the so-called *Nernst's equation*

$$E = -\frac{\Delta G^\circ}{\nu \mathcal{F}} - \frac{RT}{\nu \mathcal{F}} \ln \prod \frac{a_i'^{\nu_i'}}{a_i^{\nu_i}} = E^\circ + \frac{RT}{\nu \mathcal{F}} \ln \prod \frac{a_i^{\nu_i}}{a_i'^{\nu_i'}}, \tag{12.50}$$

where the activities are calculated with a model appropriate for the concentrations investigated (see 11.6). In passing, note that this equation may also be used for *concentration* cells, where both half cells are made up of the same solution of different concentration. When the activities of all products and reactants are unity, we obtain

$$E^\circ = -\Delta G^\circ / \nu \mathcal{F} = -\sum \nu_i \mu_i^\circ / \nu \mathcal{F}. \tag{12.51}$$

One calls E° the *standard* emf of the cell. It depends only on temperature.

Electrochemical measurements are especially useful because Gibbs free energy determinations made as a function of temperature also yield the entropy of the reaction with

$$\Delta S = -(\partial G/\partial T)_P = \nu\mathcal{F}(\partial E/\partial T)_P. \tag{12.52}$$

As $\Delta G \neq 0$, we remark that $\Delta H \neq T\Delta S$. The heat of the reversible reaction is by definition $Q = T\Delta S$. Hence it differs from ΔH, which must be calculated with

$$\Delta H = \Delta G + T\Delta S = -\nu\mathcal{F}E + \nu\mathcal{F}T(\partial E/\partial T)_P. \tag{12.53}$$

Although often difficult to make, electrochemical measurements are thus a very important source of thermochemical information.

12.4e RELATIVE ELECTRODE POTENTIALS

It is of course desirable to predict the emf of any redox reaction without having to measure it. By following a procedure that we have already used many times, one selects a half cell as a reference with which the emf of any other half cell will be compared. We will then have to subtract the *single electrode potentials* for the two relevant half reactions to obtain the emf of the reaction of interest. Since problem is similar to that discussed in 12.3b for defining single ion activities, it will not be a surprise that the hydrogen electrode has been selected as a reference. Its half reaction is the reduction of gaseous hydrogen under a 1 bar pressure (assumed unit fugacity) to protons in a solution of hydrogen acid in which the mean ionic activity $a_\pm$ is unity

$$H_2 \rightleftharpoons 2\,H^+ + 2\,e^-. \tag{12.54}$$

By definition the electrode potential E° of this reaction is set to 0. The sign of the relative electrode potential is the sign measured for the cell emf since the standard hydrogen electrode is always on the left and the particular electrode of interest on the right. From the extensive data tabulated in the literature, it is then possible to obtain directly the standard emf of any cell for which the electrode potentials E_1° and E_2° are known for both half cells, and thus the equilibrium constant of the redox reaction as a function of $E^\circ = E_1^\circ - E_2^\circ$ with

$$k = \prod \frac{a_i'^{\nu_i'}}{a_i^{\nu_i}} = \exp\left(\frac{\nu\mathcal{F}E^\circ}{RT}\right). \tag{12.55}$$

As examples we have listed in Table 12.3 the potentials for a few important reactions spanning the whole domain of observed values. The highest potentials correspond to the reduction of the most electronegative elements (the halogen elements), whereas the lowest are associated with the oxidation of the most electropositive elements (the alkaline elements).

The standard hydrogen electrode is made up of a platinum strip exposed to a current of hydrogen, which is partly immersed in an acid solution. The notation of the half cell is

Table 12.3. Electrode potentials (*Handbook of Chemistry and Physics*, 1998-99).

Redox reaction	$E°$ (V)
$XeF + e^- \rightleftharpoons Xe + F^-$	3.4
$F_2 + 2\,e^- \rightleftharpoons 2\,F^-$	2.866
$O_3 + 2\,H^+ + 2\,e^- \rightleftharpoons O_2 + H_2O$	2.076
$HClO + H^+ + 2\,e^- \rightleftharpoons Cl^- + H_2O$	1.482
$Cl_2\,(g) + 2\,e^- \rightleftharpoons 2\,Cl^-$	1.35827
$O_2 + 4\,H^+ + 4\,e^- \rightleftharpoons 2\,H_2O$	1.229
$Hg^{2+} + 2\,e^- \rightleftharpoons Hg$	0.851
$ClO^- + H_2O + 2\,e^- \rightleftharpoons Cl^- + 2\,OH^-$	0.81
$Ag^+ + e^- \rightleftharpoons Ag$	0.7996
$Hg_2^{2+} + 2\,e^- \rightleftharpoons 2\,Hg$	0.7973
$O_2 + 2\,H_2O + 4\,e^- \rightleftharpoons 4\,OH^-$	0.401
$Cu_2^{2+} + 2\,e^- \rightleftharpoons Cu$	0.3419
$Cu_2^{2+} + e^- \rightleftharpoons Cu^+$	0.153
$2\,H^+ + 2\,e^- \rightleftharpoons H_2$	0
$AgI + e^- \rightleftharpoons Ag + I^-$	0.15224
$Fe^{3+} + 3\,e^- \rightleftharpoons Fe$	-0.037
$Cd^{2+} + 2\,e^- \rightleftharpoons Cd$	-0.4030
$Cd^{2+} + 2\,e^- \rightleftharpoons Cd\,(Hg)$	-0.3521
$Fe^{2+} + 2\,e^- \rightleftharpoons Fe$	-0.447
$Zn_2^{2+} + 2\,e^- \rightleftharpoons Zn$	-0.7618
$2\,H_2O + 2\,e^- \rightleftharpoons H_2 + 2\,OH^-$	-0.8277
$Al^{3+} + 3\,e^- \rightleftharpoons Al$	-1.662
$H_2 + 2\,e^- \rightleftharpoons 2\,H^-$	-2.230
$Al(OH)_3 + 3\,e^- \rightleftharpoons Al + 3\,OH^-$	-2.310
$Mg^{2+} + 2\,e^- \rightleftharpoons Mg$	-2.372
$Na^+ + e^- \rightleftharpoons Na$	-2.71
$Li^+ + e^- \rightleftharpoons Li$	-3.0401

thus Pt | H_2 (1 bar) | H^+ ($a_\pm = 1$). As a matter of fact, the standard hydrogen electrode does not exist for the very reason that it relies on the ideal hypothetical standard solution of unit molality. The measurements are thus made at high dilution and extrapolated to this concentration as described in 11.2e for evaluating the properties of ions in their standard state. In practice, it is more convenient to use secondary reference electrodes which are of too many kinds to be reviewed in this section.

12.4f POTENTIAL-PH DIAGRAMS

The last point to be discussed is that redox potentials are not constant, but vary with the composition of the system through the changes in the activities of the ions that participate in half reactions. This effect has already been alluded to in the case of complexation reactions. The same effect is observed when a precipitate is formed, and we will investigate it here in the case of acid-base reactions. It will be at the basis of predominance diagrams, which we have already introduced in 12.3c. Here, we will indicate under which redox and pH conditions a given species can be the most abundant. These potential-pH representations are sometimes called Pourbaix diagrams, after the Belgian chemist who determined them in a systematic way in the 1950's. Because we have been mostly interested in aqueous solutions, water is the obvious starting point to introduce these diagrams.

Water is an oxidant whose conjugate reductant is hydrogen, and a reductant whose conjugate oxidant is oxygen. Since the two redox reactions of water involve H^+ and OH^- ions, they are necessarily coupled with the two acid-base reaction and we will determine in this section how they are mutually related. Let us first consider water as an oxidant. Under acidic conditions, the reaction of interest is

$$2\,H^+ + 2\,e^- \rightleftharpoons H_2, \tag{12.56}$$

and its redox potential is given by the expression

$$E = E^\circ_{H^+/H_2} + 0.03\log[a^2_{H^+}/a_{H_2}] = -0.06\,pH - 0.03\log a_{H_2}, \tag{12.57}$$

where the activity of water has been taken as unity and the factor 0.03 represents the term $RT/\nu\mathcal{F}$ at room temperature. Under basic conditions, we have

$$2\,H_2O + 2\,e^- \rightleftharpoons H_2 + 2\,OH^-, \tag{12.58}$$

with

$$\begin{aligned} E &= E^\circ_{H_2O/H_2} + 0.03\log[a^2_{H_2O}/a_{H_2}a^2_{OH^-}] \\ &= -0.828 - 0.03\log a_{H_2} + 0.06\,pOH. \end{aligned} \tag{12.59}$$

We also note that (12.7) states that the product $a_{H^+}a_{OH^-}$ is equal to k_w, the ion product of water, i.e., that $\log a_{H^+} + \log a_{OH^-} = 14$. From (12.57) and (12.59) we therefore conclude that $E^\circ_{H_2O/H_2} = E^\circ_{H^+/H_2} - 0.06\,pk_w$ (= -0.084 V at 298 K). For both reactions, the potential depends on the acidity of the solution through a -0.06 pH term. Water is thus most oxidant in most acidic solutions.

Let us then consider water as a reductant. Under acidic conditions we have

$$O_2 + 4\,H^+ + 4\,e^- \rightleftharpoons 2\,H_2O, \tag{12.60}$$

with

$$E = E^\circ_{O_2/H_2O} + 0.015\log[a_{O_2}\,a^4_{H^+}] = 1.229 + 0.015\log p_{O_2} - 0.06\,pH. \tag{12.61}$$

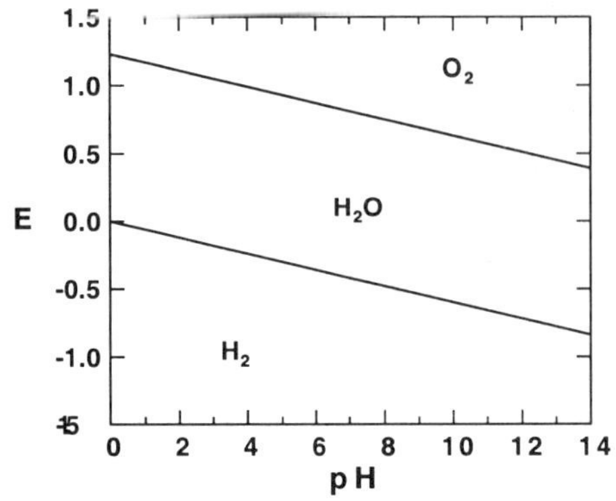

Figure 12.9. Potential-pH diagram of water.

Under neutral or basic conditions, we have

$$O_2 + 4\,H_2O + 4\,e^- \rightleftharpoons 4OH^-, \tag{12.62}$$

whose redox potential is

$$\begin{aligned} E &= E^\circ_{O_2/OH^-} + 0.015\log[a_{O_2}a^4_{H_2O}/a^4_{OH^-}] \\ &= 0.401 + 0.06\ pOH - 0.015\ \log p_{O_2}. \end{aligned} \tag{12.63}$$

and find that water is most reductant in basic solutions, when the pH is highest. These conclusions are conveniently summarized in the potential-pH diagram of Fig. 12.9 which shows the domains of predominance of a given species as a function of redox potential and pH. As already noted in 12.3c, the lines represent the conditions where the two most abundant species have the same concentration. Starting from low potentials, we first observe the domain where water is oxidized into hydrogen. At higher potentials when the pH is low, water then predominates. At the highest potentials, water is eventually reduced and liberates hydrogen.

The water diagram is the simplest one can plot. As a less obvious example, we show in Fig. 12.10 the potential-pH diagram of chlorine in aqueous solutions. This diagram has been calculated from the data of table 12.3 and the pk = 7.40 of hypochlorous acid (HClO/ClO$^-$ acid-base pair). Because the electric potential and pH depend on the concentration of dissolved species, we must also fix the total concentration of the element or species of interest. Arbitrarily, one often takes a molality of 10^{-2} mol/kg, which is the value used in Fig. 12.10. The very restricted range of predominance of molecular chlorine is apparent. In these diagrams, the vertical limits are determined by acid base reactions. By definition, the limit between two fields of predominance is given by the pk of the relevant acid-base pair. As to the horizontal or oblique limits, which have slopes generally ranging from 0 to -0.18, they are determined in the same way by the potentials of the successive redox reactions. Determining which are the predominant species at the ends of the diagram is thus immediate. With such starting points, as the reader may care to check, calculation of the limits within the diagram itself does not raise much difficulty.

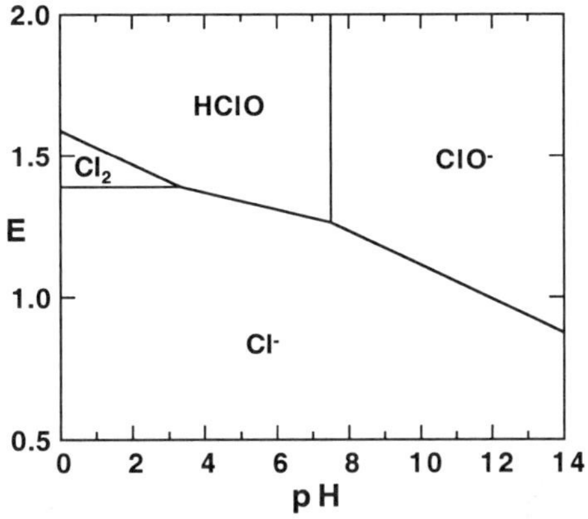

Figure 12.10. Potential-pH diagram of chlorine.

In the case of chlorine all species are present in the solution. For such homogeneous equilibria, it is not difficult to represent the conditions under which the two most abundant species have the same concentration. However, the presence of gaseous species in the potential-pH diagram of water shows that one must also deal with heterogeneous equilibria. Because one cannot equate concentrations in a liquid and gaseous phases, one arbitrarily fixes a pressure beyond which it is assumed that the gaseous species is predominant. One generally takes for this purpose the 1 bar standard pressure. The other case of heterogeneous equilibria to be mentioned is that of precipitation of one or several solid phases. The diagram is plotted in the same way as in the other cases once the total concentration of the element of interest has been fixed. The differences are that the solid phase will not precipitate if the selected concentration is too low, and that the boundary between the domains of predominance of two solid phases will then represent the only conditions where both solids can coexist.

12.4g ON THE MISUSE OF MATHEMATICS IN CHEMICAL THERMODYNAMICS

In conclusion to this discussion of chemical equilbria it might be useful to recall, on a different tone, that equations are a most precious tool, but that one should not overlook the physical reality they represent. This was aptly demonstrated in 1903 by a distinguished British physicist, Sir Arthur Schuster (1851-1934) in a fantasy (reprinted in the autobiography he published in 1932), whose gist will be summarized in the following paragraphs. As will be seen, a few redox and acid-base reactions allowed him to put forward, with a full mathematical rigor, a theory of matter that solved the fundamental issue of "the probable unity of chemical substances".

The starting point of Schuster's demonstration was the equality between hydrogen and chlorine atoms established from the classical reaction $H_2 + Cl_2 \rightleftharpoons 2\ HCl$, which can be rewritten as $HH + ClCl - 2\ HCl = 0$ to yield $(H - Cl)^2 = 0$, whence $H = Cl$. The next step is made with the other well-known redox reaction describing the production of hydrogen from sulfuric acid and zinc : $Zn + H_2SO_4 \rightleftharpoons H_2 + ZnSO_4$. An anologous mathematical transformation yields $(Zn - H_2) = SO_4\ (Zn - H_2)$, from which one finds $1 = SO_4$ or $S = 1/O_4$. One arrives in this way at the original concept of the inverse of a chemical atom, whose fertility is illustrated by the reaction $K + H_2O \rightleftharpoons HKO + H$. One directly obtains from it OH

(H - K) = H - K, and thus proves that O = 1/H. With a single assumption, "the very plausible one that chemical formulae are subject to the laws of algebra", Schuster thus demonstrated that H = Cl, O = 1/H = 1/Cl, and S = $1/O_4 = H_4 = Cl_4$.

One can in fact go one step further with the combustion reaction of carbon, C + $O_2 \rightleftharpoons CO_2$. By dividing the two sides of this equation by CO_2 Schuster obtained the more fundamental relationship $1/O_2 + 1/C = 1$. He could then state that "the absolute unit of chemical atomicity is the sum of the inverse carbon atom and the inverted double oxygen atom", and add that the equation teach us something more since "we see that the double oxygen atom and the carbon atom bear the same relation to each other as the conjugate foci in a spherical mirror, the principal focal length of which is unity". By avoiding in this short notice "anything beyond a strict statement of facts and indisputable arguments, Schuster concluded, I have assumed that a strictly scientific discussion will be more acceptable [...] than daring speculations which, more brilliant but less illuminating, remind us of the flickering efforts of a painting and overworked dynamo." Indeed, what could be added to such a neat demonstration?

Problems

1. Calculate the *p*H of a solution of HCl (molality 10^{-3}) and a weak base as a function of the *pk* and molality of the latter.

2. The *p*H of a NaClO solution of molality 0.01 is 9.7. What is the *pk* of the HClO/ClO^- pair?

3. In aqueous solutions, ammonium nitrite NH_4NO_2 has two acid-base functions (HNO_2/NO_2^-, $pk_1 = 3.4$) and (NH_4^+/NH_3, $pk_2 = 9.2$). Under which conditions is the *p*H independent of the molality *m* of NH_4NO_2? What is this *p*H at room *T* and *P*?

4. From the data given in 12.1c and 12.2c, calculate at 298 K the *p*H of a 10^{-3} molal solution of lysine, and then redo the calculation at 310 K after having determined the *pk*'s at this temperature from the enthalpies of the various acid-base reactions that are also reported.

5. One dissolves 10 g of NH_4Cl in 1 kg of water. What is the *p*H of the solution at room pressure and temperature? The *pk* of the NH_4^+/NH_3 pair is x9.24. A weak acid is then added to the solution. How does the *p*H of the solution vary as a function of the molality and *pk* of the acid?

6. Calculate the diagram of predominance of the chloride and hydroxyl complexes of mercury (Fig. 12.5). The stability constants are as follows:

   $Hg^{2+} + Cl^- \rightleftharpoons HgCl^+$ $\log k_1 = 6.74$
   $HgCl^+ + Cl^- \rightleftharpoons HgCl_2^\circ$ $\log k_2 = 6.40$
   $HgCl_2^\circ + Cl^- \rightleftharpoons HgCl_3^-$ $\log k_3 = 0.72$
   $HgCl_3^- + Cl^- \rightleftharpoons HgCl^+$ $\log k_4 = 1.10$
   $Hg^{2+} + OH^- \rightleftharpoons HgOH^+$ $\log k_1' = 10.41$
   $HgOH^+ + OH^- \rightleftharpoons Hg(OH)_2^\circ$ $\log k_2' = 11.44$

   From these values, justify that the complex $HgOH^+$ is never predominant.

7. Experimentally, one observes that the solubility *s* of silver chloride in a nitric acid solution is related to the ionic strength *I* of the solution by the relation $\log s = -4.895 + 0.512I1/2$. What are the solubility product of AgCl and its mean ionic activity coefficient?

8. The equilibrium SiO_2 (q) + 2 $H_2O \rightleftharpoons H_4SiO_4$ limits the solubility of quartz in water. The constant of this equilibrium is 10^{-4} at 298 K. For amorphous silica, which will be assumed to be similar to SiO_2 glass, the equilibrium constant of the analogous equilibrium with

H_4SiO_4 is 0.002. Calculate the Gibbs free energy difference between the two forms of SiO_2. In a more general way, does the existence of a configurational entropy favor the solubility of a phase?

9. Calculate the extreme values of the Gibbs free energies of reaction corresponding to the standard potentials listed in Table 12.3. Are these values specially high or low? Calculate from the data of this Table the Gibbs free energies and the equilibrium constants of the various redox reactions reported in section 12.4.

10. Write the two half reactions of the Weston cell and calculate its emf from the data of Table 12.3. As a function of temperature, the emf can actually be represented by the expression E (V) = 0.9454 +51.65 10^{-5} T -9.5 10^{-7} T^2. What are the entropy and enthalpy changes of the redox reaction (12.44)?

13
BASICS OF STATISTICAL MECHANICS

This chapter is an introduction to statistical mechanics. To get a glimpse at the microscopic meaning of temperature and pressure, kinetic models of classical mechanics will first serve to determine the equation of state and the velocity distribution of molecules of ideal monoatomic gases. We will then describe how the quantum state of a system is specified at a microscopic level and take as an example the case of particles in translation. The Maxwell-Boltzmann energy distribution will be derived and the basic importance of partition functions for calculations of thermodynamic functions stressed. We will finish with a closer examination of entropy, temperature and heat capacity before introducing the Fermi-Dirac and Bose-Einstein quantum statistics and discussing the properties of black-body radiation.

13.1 A PRIMER WITH IDEAL GASES

13.1a PREAMBLE

Why is the heat capacity of a crystal higher than that of the vapor phase of the same substance, and why does C_p vary from a substance to another? Classical thermodynamics cannot answer such questions because its power lies in the general applicability of its basic tenets, which implies considering matter as a black box. Different substances have different properties because interactions between molecules, ions or atoms vary from one system to another. But all such interactions can be described in terms of potentials between pairs, triplets, etc., of atoms so that the laws of thermodynamics eventually fall within the realm of mechanics. Application of mechanical laws to macroscopic systems is, however, impossible, even for the smallest amounts of matter that can be investigated experimentally (i.e., about 1 μm^3, or $\sim 10^{-9}$ g of matter), where atoms number more than 10^{12}. As will be described in the next chapter, the most powerful computing means makes detailed calculations possible, for extremely short time periods, for systems made up of only a few thousand atoms. There is thus no way to determine how the state of one mol of ideal gas may change 10^{32} times in one second or how, owing to short-lived fluctuations, the density and the entropy may vary at a small scale in a liquid.

Such limitations are not serious, for statistical methods make following any particle of a macroscopic system at any time unnecessary. The main simplification results from the fact that we do not need to know which particle is in which state, but just how many particles are in a given state. Thanks to the law of large numbers, the great number of particles considered gives rise to well-defined physical laws and implies that the properties of any macroscopic system are specified only by their most probable values, which do not vary with time. This is why, as expressed by the Duhem postulate, the state of a macroscopic system made up of billions of billions of particles can be specified by only two variables. Here, the important point will be to determine how energy is distributed between the particles of a system, and how average values may be calculated. Although energy is actually linked to electronic, translational, rotational or vibrational displacements, a simplifying feature is that, at this stage, we will not need to examine how the energy of molecules is distributed over these various *degrees of freedom*.

Statistical mechanics is itself a very broad field which cannot be summarized in a few chapters, even at an elementary level. We thus refer the reader to available textbooks for rigorous or comprehensive accounts. In these final chapters our aim will be to introduce new perspectives about some important topics that we have already discussed. Here attention will be focussed on the concepts of temperature and entropy, and on calculations of thermodynamic functions. Statistical mechanical models become increasingly complicated for complex systems. Hence we will be interested only in systems made up of *weakly interacting particles* because the total energy can then be obtained by direct summation of the energy of individual particles. Ideal gases represent the best known example, but we will see that crystals also belong to this category. To illustrate the goals and methods of statistical mechanics, which will be described in more detail in the following, we will begin with two classic models pertaining to ideal gases.

13.1b KINETIC MODEL OF AN IDEAL MONOATOMIC GAS

The birth of statistical mechanics can perhaps be associated with the kinetic model of gas conceived in 1738 by the Swiss mathematician Daniel Bernouilli (1700-1782), at a time where atoms were nothing but a bold assumption. In its modern form, this model applies to ideal monoatomic gases. Without any atomic interactions and at temperatures low enough that electronic energy can be neglected, the potential energy is zero and the internal energy reduces to the kinetic energy of translation of the atoms. Calculating the internal energy thus boils down to calculating the velocity distribution of the atoms. Before doing this in the next section, here we will relate the temperature and pressure of the gas to the velocities of the molecules.

Let N be the number of identical atoms of mass m in a cubic box of size a, and v_x, v_y and v_z be the three components of the velocity of an atom referred for convenience to Cartesian axes oriented parallel to the faces of the box (Fig. 13.1). The atoms move along straight lines until colliding elastically with either other atoms or with the walls of the box, exerting in this way a force equal to the sum of the individual momentum changes $\Delta(mv)$. The pressure P of the gas is the ratio between this force and the surface area of the walls. To calculate P, we thus have to integrate the changes in momentum over all atoms moving with different velocities in different directions.

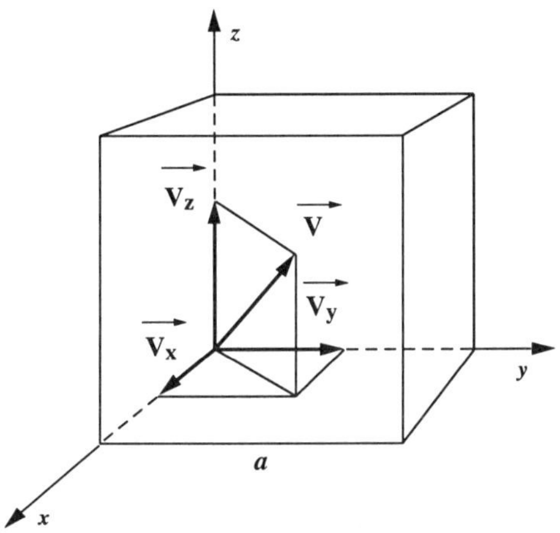

Figure 13.1. Components of the velocity of an atom.

Assuming the gas to be isotropic, we will denote by $n(v_x)$ the number of atoms *per unit volume* having a velocity between v_x and $v_x + dv_x$ along the x axis, irrespective of the velocities v_y and v_z along the two other axes; by $n(v_x, v_y)$ the number of atoms of velocities between v_x and $v_x + dv_x$ and v_y and $v_y + dv_y$, irrespective of v_z; and by $n(v_x, v_y, v_z)$ the number of atoms with velocities between v_x and $v_x + dv_x$, v_y and $v_y + dv_y$, and v_z and $v_z + dv_z$. For any gas, N is so large that n can be assumed to be a continuous function of the velocities, which does not fluctuate with time. The total number of atoms is then

$$
\begin{aligned}
N = \int_{-\infty}^{\infty} n(v_x)dv_x &= \int_{-\infty}^{\infty}\int_{-\infty}^{\infty} n(v_x, v_y)dv_x dv_y \\
&= \int_{-\infty}^{\infty}\int_{-\infty}^{\infty}\int_{-\infty}^{\infty} n(v_x, v_y, v_z)dv_x dv_y dv_z. \qquad (13.1)
\end{aligned}
$$

Consider the collisions on the wall of the box perpendicular to the x axis. An atom with a velocity v_x takes the time a/v_x to move along the x side of the box and it transfers a momentum $2mv_x$ to the wall when it rebounds on it. If all atoms were moving only along the x axis, the number of collisions of atoms with a velocity v_x would be $v_x n(v_x)/a$ per unit of time. But atoms move along all three directions of space so that we must consider that pressure (i.e, a force per unit of surface area) is exerted by all atoms with a velocity v_x that are enclosed within a volume element of length v_x and unit section. The relevant number of atoms is $[v_x n(v_x)/V]dv_x$, where V is the volume. The total pressure exerted on the wall is

$$P = \int_0^\infty \frac{2mv_x v_x n(v_x)}{V}\, dv_x = \frac{2m}{V}\int_0^\infty v_x^2 n(v_x)\, dv_x = \frac{m}{V}\int_{-\infty}^\infty v_x^2 n(v_x)\, dv_x. \qquad (13.2)$$

We will see in the next section how $n(v_x)$ can be calculated. For the time being, let us denote by $\overline{v_x^2}$ the mean of v_x^2 for the N atoms. Since $n(v_x) = n(-v_x)$, the integral (13.2) from 0 to ∞ is half that from $-\infty$ to ∞. We thus write for the pressure

$$P = \frac{m}{V}N\overline{v_x^2} = \frac{1}{3}\frac{m}{V}N\overline{v^2}. \qquad (13.3)$$

For reasons of isotropy, the mean of v_y^2 and v_z^2 along the other axes is equal to $\overline{v_x^2}$. Because $v^2 = v_x^2 + v_y^2 + v_z^2$, we have $\overline{v_x^2} = 1/3\ \overline{v^2}$ where $\overline{v^2}$ is called the *mean quadratic velocity*. The product PV is then

$$PV = \frac{2}{3}\left(\frac{1}{2}m\overline{v^2}\right)N = \frac{2}{3}U, \qquad (13.4)$$

and we have directly related it to the internal energy which is given by

$$U = \frac{1}{2}Nm\overline{v^2}. \qquad (13.5)$$

A comparison of these expressions with the ideal gas equation of state $PV = NkT$ finally gives

$$kT = \frac{1}{3}m\overline{v^2}, \qquad T = \frac{m\overline{v^2}}{3k}, \qquad (13.6)$$

and

$$U = \frac{3}{2}NkT. \qquad (13.7)$$

We have thus found that temperature is a measure of the mean quadratic velocity of the atoms of the gas. The other important conclusion is that the temperature and pressure of a gas have only a statistical significance. An isolated atom has no temperature and pressure, but just a given momentum and energy. Temperature and pressure can be defined only when the number of atoms is large enough that their values are time independent. For a given temperature, (13.6) states that the mean quadratic velocity is inversely proportional to the mass of the atoms. At room temperature, $(\overline{v^2})^{1/2}$ is for instance 1850 and 240 m/s for hydrogen and xenon atoms, respectively, which is of the same order of magnitude as the sound velocity in air. (Note that, being necessarily smaller than $(\overline{v^2})^{1/2}$, the mean velocities $< v >$ are slightly lower than these figures.) In passing, the importance of collisions as a way of ensuring an isotropic velocity distribution must be emphasized. If all atoms had initially the same speed along parallel directions, one would observe this perfectly "ordered" distribution to disappear and never to reappear spontaneously. Here, the irreversible nature of heat production would manifest itself by the well-defined isotropic velocity distribution that is eventually established. Finally, the ideal gas law is more closely approached by real gases at higher temperatures because kinetic energy progressively overwhelms the energy of molecular interactions, which depends only on interatomic distances.

At a microscopic level, the *mean free path* l between two collisions depends mainly on the molecular diameter d and the number of molecules per unit volume (c). For a pure, ideal gas, one shows that $l = 1/c\pi d^2$. At room pressure and temperature, l is of the order of 10^{-8} m for a molecule with a diameter of a few Å, which leaves only about 10^{-10} s between two collisions. For an ideal gas, viscosity (η) is entirely due to collisions between molecules. Hence it is from viscosity measurements that the mean free path is determined in low-density gases. For an ideal gas one finds that the viscosity does not depend on pressure and density, but only on temperature through the average velocity of the molecules and the law $\eta = m < v > /3\pi d^2$). Because the rate of collisions increases with temperature, the viscosity of gases actually increases, contrary to the viscosity of liquids which decreases with rising temperature.

13.1c VELOCITY DISTRIBUTION IN AN IDEAL GAS

In the previous section we did not try to determine the velocity distribution of the gas molecules. To solve this problem we will now reproduce part of the demonstration published by Maxwell in 1860 of the distribution law that bears his name. (Only the notation will be changed to make it consistent with that followed in this book.) Before doing so we will note that kinetic models of ideal polyatomic gases are more complex than described in the previous section because the energy of individual molecules is made up of rotational and vibrational contributions, in addition to the single translational energy that must be considered for monoatomic gases. As long as only translational energy is concerned, however, there is no difference with monoatomic gases so that Maxwell's model remains applicable.

Maxwell's goal was "to find the average number of particles whose velocities lie between given limits, after a great number of collisions among a great number of equal particles." His starting point was that, "If a great many equal spherical particles were in motion in a perfectly elastic vessel, collisions would take place among the particles, and their velocities would be altered at every collision; so that after a certain time the *vis viva* [the kinetic energy] will be divided among the particles according to some regular law, the average number of particles whose velocity lies between certain limits being ascertainable, though the velocity of each particle changes at every collision." To determine this law, we will now denote by $Nf(v_x)dv_x$ the number of particles for which v_x lies between v_x and $v_x + dv_x$. Similarly $Nf(v_y)dv_y$ and $Nf(v_z)dv_z$ will denote the number of particles for which v_y lies between v_y and $v_y + dv_y$, and v_z lies between v_z and $v_z + dv_z$, respectively, where f always stands for the same function. Let us then continue with Maxwell's words:

"Now the existence of the velocity v_x does not in any way affect that of the velocities v_y or v_z, since they are all at right angles to each other and independent, so that the number of particles whose velocity lies between v_x and $v_x + dv_x$, and also between

v_y and $v_y + dv_y$ and also between v_z and $v_z + dv_z$, is:

$$Nf(v_x)f(v_y)f(v_z)dv_x dv_y dv_z. \tag{13.8}$$

If we suppose the N particules to start from the origin at the same instant, then this will be the number in the element of volume $(dv_x dv_y dv_z)$ after unit of time, and the number referred to unit of volume will be

$$Nf(v_x)f(v_y)f(v_z). \tag{13.9}$$

But the directions of the coordinates are perfectly arbitrary, and therefore this number must depend on the distance $[r]$ from the origin alone, that is

$$Nf(v_x)f(v_y)f(v_z) = \phi(v_x^2 + v_y^2 + v_z^2). \tag{13.10}$$

Solving this functional equation, we find

$$f(v_x) = Ce^{Av_x{}^2}, \qquad \phi(r^2) = C^3 e^{Ar^2}. \tag{13.11}$$

If we make A positive, the number of particles will increase with the velocity, and we should find the whole number of particles infinite. We therefore make A negative and equal to $-1/\alpha^2$, so that the number between v_x and $v_x + dv_x$ is

$$NC\exp\left(\frac{-v_x^2}{\alpha^2}\right) dv_x. \tag{13.12}$$

Integrating from $v_x = -\infty$ to $v_x = +\infty$, we find the whole number of particles,

$$NC\sqrt{\pi}\alpha = N, \qquad C = \frac{1}{\alpha\sqrt{\pi}}, \tag{13.13}$$

$f(v_x)$ is therefore

$$\frac{1}{\alpha\sqrt{\pi}}\exp\left(\frac{-v_x^2}{\alpha^2}\right). \tag{13.14}$$

Whence we may draw the following conclusions:

1. The number of particles whose velocity, resolved in a certain direction, lies between v_x and $v_x + dv_x$ is

$$N\frac{1}{\alpha\sqrt{\pi}}\exp\left(\frac{-v_x^2}{\alpha^2}\right) dv_x \tag{13.15}$$

2. The number whose actual velocity lies between v and $v + dv$ is [see 13.3c]

$$N\frac{4}{\alpha^3\sqrt{\pi}}v^2\exp\left(\frac{-v^2}{\alpha^2}\right) dv. \tag{13.16}$$

3. To find the mean value of v, add the velocities of all particles together and divide by the number of particles; the result is

$$\text{mean velocity} = 2\alpha/\sqrt{\pi}. \tag{13.17}$$

4. To find the mean value of v^2, add all the values together and divide by N,

$$\text{mean value of } v^2 = 3/2\ \alpha^2. \tag{13.18}$$

This is greater than the square of the mean velocity, as it ought to be.

It appears from this proposition that the velocities are distributed among the particles according to the same law as the errors are distributed among the observations in the theory of the 'method of least squares'. The velocities range from 0 to ∞, but the number of those having great velocities is comparatively small."

This is how Maxwell concluded his discussion. His reasoning was in fact incomplete because the assumed isotropy of the gas does not necessarily imply the statistical independence of the variables along the different directions of space. Without trying to prove this point, we will just remark that Maxwell's demonstration illustrates well the results sought after in statistical mechanics in that it gives a distribution law from which average values of microscopic parameters can be calculated. This law is represented by the bell-shaped curves plotted in Fig. 13.2 for several temperatures. As expected from the increase of kinetic energy, the maximum of the velocity distribution shifts toward higher velocities when the temperature increases. The width of the curve also increases, in accordance with the fact that we have long stressed, namely, that entropy increases at higher temperatures correspond to widening of distributions.

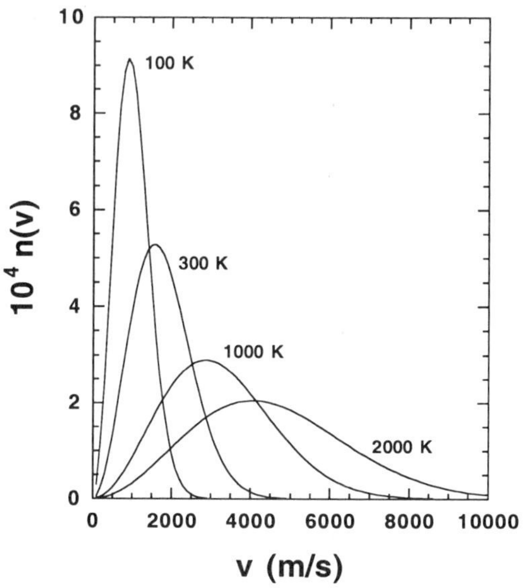

Figure 13.2. Maxwell distribution of velocities of hydrogen molecules (ideal gas) as calculated from (13.20) at the temperatures indicated. Note the shift of the maximum toward higher velocities (and energies) when T increases, which is accompanied by a strong widening of the distribution.

13.1d EQUIPARTITION OF ENERGY

Because the kinetic energy of a molecule is $E_c = 1/2\, mv^2$, the distribution law (13.16) states that the probability for a gas molecule to have a kinetic energy E_c is proportional to $\exp(-2E_c/\alpha^2 m)$. We will see in 13.4c that this exponential factor is not particular to kinetic energy and that it is actually $\exp(-E_c/kT)$. For the time being, we will thus assume that

$$\alpha = (2kT/m)^{1/2}. \tag{13.19}$$

This will allow us to rewrite Maxwell's distribution law as

$$n(v) = 4\pi N \left(\frac{m}{2\pi kT}\right)^{3/2} v^2 \exp\left(-\frac{mv^2}{2kT}\right), \tag{13.20}$$

which is the expression that has been used for the plots of Fig. 13.2.

We have seen in previous sections that it is useful to define several average velocities. From (13.20), the mean and mean quadratic velocities given by (13.17) and (13.18), respectively, are

$$< v >= \left(\frac{8kT}{\pi m}\right)^{1/2}, \qquad \overline{v^2} = \frac{3kT}{m}, \qquad v_{pr} = \left(\frac{2kT}{m}\right)^{1/2}. \tag{13.21}$$

To complement these relations we have added the most probable velocity v_{pr} as determined by the maximum of the distribution law, $[\partial n(v)/\partial v]_T = 0$. Consistent with (13.7), we derive the mean kinetic energy per molecule from the expression for the mean quadratic velocity $\overline{v^2}$

$$< E_c >= 3/2\, kT. \tag{13.22}$$

In the same way the most probable velocity corresponds to a kinetic energy kT. Because the gas is isotropic and at rest as a whole, the mean quadratic velocity is the same in the three dimensions of space for all molecules. We have thus

$$< E_{cx} >=< E_{cy} >=< E_{cz} >= 1/2\, kT. \tag{13.23}$$

This is the best example of *equipartition* of energy since each translational degree of freedom contributes $kT/2$ and $k/2$ to the internal energy and heat capacity of the system, respectively. In classical mechanics it has long been assumed that this was also holding true for rotational and vibrational energies. As will be shown in next chapter, this is incorrect for rotational degrees of freedom at low temperatures, and for vibrational degrees of freedom at any T. Equipartition of energy is thus always valid for translations only.

In classical mechanics one vibrational degree of freedom contributes kT to the internal energy (see 14.3c). In contrast to translational and rotational energies (see 14.1), the energy of an oscillator includes a potential energy term (which is proportional to the square of the displacement) along with a kinetic energy term (which is proportional to the square of the momentum). A correct statement of the principle of equipartition is therefore that a heat capacity $k/2$ is associated with each quadratic term of the energy.

13.2 STATE OF A MICROSCOPIC SYSTEM

13.2a PHASE SPACE, ERGODICITY

We must now define the state of a system at a microscopic scale where macroscopic characterization by only two variables obviously cannot work. We will do this here within the framework of classical mechanics before examining the quantum mechanical situation in the next section. Because energy is ultimately mechanical, in classical mechanics the obvious way to characterize each particle is by its position and momentum. The *microstate* of a system is then specified by the coordinates q_i and conjugate momenta $p_i = m_i v_i$ of all particles. For a system of N particles, it is represented at any moment by a point in a $2N$-dimensional space, the *phase space*. As a function of time, this point thus follows a path that could in principle be calculated from the laws of mechanics.

For a system of given energy $E \pm \Delta E$, the fundamental assumption made is that all states compatible with this energy have the same probability of occurrence. Over a long enough period of time the system will eventually pass through every state having the same energy and, on average, will spend the same time in each. This is the *ergodic* hypothesis. Because the accuracy with which the positions and momenta are known depends only on the limitations of the measuring instruments, physical laws are strictly deterministic. As illustrated by the two models presented at the beginning of this chapter for monoatomic gases, the probabilistic approach arises simply from the fact that problems are much too complex to be treated in an exact manner.

13.2b WAVE FUNCTION

In quantum mechanics, this conclusion no longer holds true because there is an intrinsic limit to the accuracy of measurements. As embodied by the uncertainty principle due to Heisenberg (1901-1976), any improvement in the knowledge of the position of a particle is counterbalanced by a greater uncertainty on the momentum

$$\delta q_i \, \delta p_i \geq h/2\pi, \tag{13.24}$$

where δ designates an uncertainty and h is a fundamental constant, the Planck constant. It is the smallness of h (= 6.63 10^{-34} J.s) that makes this principle relevant only at a microscopic scale.

Accordingly, the position of a particle is described as a probability of presence in a given portion of space which is mathematically represented by the *wave function* Ψ. Let x_i, y_i and z_i be the Cartesian coordinates of particle i of a system. The probability p_p of finding the particle between x_i, y_i and z_i and $x_i + dx_i$, $y_i + dy_i$ and $z_i + dz_i$ is then represented by the square of the amplitude of its wave function

$$p_p = |\Psi_i^2| \prod dx_i dy_i dz_i. \tag{13.25}$$

We will see in 13.2d that a state is associated with a volume element $(2\pi)^3/V$ in the wave vector space, and in 13.2c that the momentum is related to the wave vector K by $p = hK/2\pi$.

Hence we can also associate a state with a volume element h^3/V in a momentum space. At a given time, the microstate of a system is represented not by a point in the phase space, like in classical mechanics, but by a volume element h^3/V. Strictly speaking, quantization of energy (and momentum) prevents a continuous pathway from being followed as a function of time in the phase space. Any summation over accessible states of a system should thus be made in a stepwise manner. When dealing with the great many different degrees of freedom relevant to real systems, however, the spacing of energy levels is often small enough that summations over quantum states reduce to integrations along continuous "classical" pathways.

13.2c SCHRÖDINGER EQUATION

The notion of the wave equation originates in the wave-particle duality expounded by de Broglie (1892-1987) in 1923, according to which any particle of mass m is associated with a wave of wavelength

$$\lambda = \frac{h}{p} = \frac{h}{mv}. \tag{13.26}$$

Because the wave vector K is related to the wavelength by $K = 2\pi/\lambda$, it follows that

$$p = \frac{hK}{2\pi}. \tag{13.27}$$

If a moving particle has a wave-like character, then the waves associated with it must be in phase otherwise these would produce destructive, and not constructive interferences. As will be seen below, this condition of a standing wave in a finite volume imposes that the particle wavelength and energy can only take on discrete values. Denoting by E and V (not to be confused here with the volume) the energy and potential energy of the system, respectively, one determines the wave function and the associated energy levels by solving the equation derived by Schrödinger (1887-1961)

$$-\sum_i \frac{h^2}{8\pi^2 m}\left(\frac{\partial^2\Psi}{\partial x_i^2} + \frac{\partial^2\Psi}{\partial y_i^2} + \frac{\partial^2\Psi}{\partial z_i^2}\right) = (E - V)\Psi. \tag{13.28}$$

The wave functions and energies that satisfy (13.28) are the *eigenfunctions* and *eigenvalues* of the wave equation, which are associated with the *stationary states*. The function Ψ and its derivative must be finite, continuous and single-valued at any point of the system. For the wave function to be a probability density, it must also satisfy

$$\int_{-\infty}^{\infty} |\Psi|^2 \prod dx_i dy_i dz_i = 1, \tag{13.29}$$

for each particle i, which implies that Ψ tends rapidly enough to 0 when the coordinates x_i, y_i and z_i tend to infinity. This set of conditions actually implies that the solutions to (13.28) correspond to discrete values of the energy which are characterized by specific *quantum numbers*. The difference between two consecutive energy levels represents one *quantum* of energy, whose value varies enormously with the kind of energy considered. Typically one quantum is 10^{15} times higher for electronic than for translational degrees of freedom. As will be seen below, it frequently happens that different states have the same energy. If there are n such states, then the energy level is said to be n-fold *degenerate*.

The connection between matter waves and the Schrödinger equation can be cursorily described as follows. In one dimension, the propagation equation for a wave of velocity c is

$$\frac{\partial^2 \psi}{\partial x^2} = \frac{1}{c^2}\frac{\partial^2 \psi}{\partial t^2}, \tag{13.30}$$

where ψ is the displacement. To separate the space and time variables in this equation, let us write $\psi(x,t) = \Psi(x)\sin(2\pi\nu t)$ and replace ψ by this expression. We find

$$\frac{\partial^2 \Psi}{\partial x^2} + \frac{4\pi^2\nu^2}{c^2}\Psi = 0. \tag{13.31}$$

Before applying this equation to matter waves, we note that the energy E is the sum of potential (V) and kinetic ($p^2/2m$) energies. We thus have $p = [2m(E-V)]^{1/2}$ and $\nu = cp/h = h[2m(E-V)]^{1/2}$. Replacing ν by this relation in (13.31) we derive in this way the one-dimensional Schrödinger equation

$$\frac{\partial^2 \Psi}{\partial x^2} + \frac{8\pi^2 m}{h^2}(E-V)\Psi = 0. \tag{13.32}$$

13.2d WAVE FUNCTION CALCULATION: PARTICLE IN TRANSLATION

For illustration purposes, consider the case of a single particle in translation. Let it have a mass m and move along an axis x in a box of length a without being subjected to any other force than gravitation. Because the gravitational energy of the particle is negligible with respect to its kinetic energy, the Schrödinger equation (13.28) reduces to the single term

$$\frac{\partial^2 \Psi_x^{tr}}{\partial x^2} = -\frac{8\pi^2 m}{h^2}E_x^{tr}\Psi_x^{tr}. \tag{13.33}$$

The solutions to this well-known differential equation have the form $\Psi_{x0}\sin(bx+c)$. The boundary conditions impose that the probability of finding a particle outside the box is zero. For $x < 0$, we thus have $\Psi_x = 0$, which implies $c = 0$. Because Ψ_x^{tr} is also zero for $x > a$, the constant b must be an integral multiple of π/a

$$b = \frac{\pi}{h}(8mE_x^{tr})^{1/2} = \frac{n_x\pi}{a}. \tag{13.34}$$

It is in this way that the translation quantum number n_x is introduced. The wave function is then

$$\Psi_x^{tr} = \Psi_{xo}^{tr}\sin\left[\frac{\pi x}{h}(8mE_x^{tr})^{1/2}\right], \tag{13.35}$$

and the kinetic energy is

$$E_x^{tr} = n_x^2\frac{h^2}{8ma^2}. \tag{13.36}$$

Because n_x is an integer the energy of the particle is quantized. Owing to the smallness of h, quantization is imperceptible even in the case of very light atoms in translation. Like Ψ_x^{tr}, translational energy is a single valued function of n_x so that it is not degenerate.

13.2e FACTORIZATION OF THE WAVE FUNCTION

For a system made up of N particles the Schrödinger equation is a function of $3N$ variables. Even for a single atom with a moderate number of electrons, solving it is a formidable task which in general must be accomplished numerically. Hence the goals of statistical thermodynamics can be reached solely because it is possible to separate the wave function and energy of a microscopic system into mutually independent components. This is performed in two different ways. For a system made up of independent particles, the total wave function is the product of the wave functions of the individual particles. For a given particle, the factorization can in addition be performed for other mutually independent degrees of freedom. A wave function is then the product of the wave functions for the relevant degrees of freedom. As will be seen in 14.2, this factorization is especially useful for treating separately the electronic, translational, rotational and vibrational energies of a molecule. The Schrödinger equation is then also split into various differential equations of lower degree which can be solved, at least in a numerical manner.

As an example, we will again take translations by considering a displacement in any direction of space. Since any translation can be decomposed into independent translations along three directions, it is tempting to express the translation wave function as the product of three independent functions

$$\Psi^{tr}(x,y,z) = \Psi_x^{tr}\,\Psi_y^{tr}\,\Psi_z^{tr} \tag{13.37}$$

which would have all the same analytical form (13.35). By factorizing in this way Ψ in the Schrödinger equation (13.28), we actually find that the translational kinetic energy is the sum of the eigenvalues of the three wave functions

$$E^{tr} = E_x^{tr} + E_y^{tr} + E_z^{tr} = (n_x^2 + n_y^2 + n_z^2)\left(\frac{h^2}{8ma^2}\right). \tag{13.38}$$

Although translational energy is not degenerate for a given direction of space, the total energy is degenerate in a cubic box where the same energy can result from the many different ways of distributing $n = n_x + n_y + n_z$ translational quanta over the three directions of space. This kind of counting is of fundamental importance in statistical mechanics. We will thus calculate the number of ways in which N indistinguishable elements (particles or energy quanta) can be divided into groups of N_i, such that $\Sigma N_i = N$. The total number of permutations between the N elements is of course $N!$ but, in each group there are $N_i!$ indistinguishable arrangements. The number sought after is

$$\Omega = \frac{N!}{N_1!\,N_2!\,...N_i!} = \frac{N!}{\prod N_i!}. \tag{13.39}$$

This formula rerpesents the number of combinations as obtained in probability theory, and it is the result we found in 11.1a when calculating the configurational entropy of a solid solution. For a translator in a cubic box, it therefore indicates that energy is $n!/(3!)^3$ degenerate.

13.3 COUNTING OF STATES

On the basis of quantum states and energy quantization we are now poised to calculate averages from which we will derive statistical analogs of thermodynamic variables. For calculating such averages, however, we need to know three important things: (i) what are the accessible states for the relevant degrees of freedom; (ii) what is the energy of each state; (iii) what is the law that relates the probability of occurrence of a given state to its energy. It is the first two points that will be addressed in the following sections.

13.3a LOCALIZED AND NONLOCALIZED PARTICLES

First of all, from a practical standpoint an important distinction must be made between *localized* and *nonlocalized* particles. Like atoms in crystals, the N localized particles of a system of volume V are confined to a volume element of the order of V/N; their energy levels are widely spaced, and only those states with the lowest energy tend to be occupied. This makes it possible to count the quantum states one by one as given (along with their energy levels) by the solutions to the Schrödinger equation. Like gas molecules, nonlocalized particles occupy all the volume V available to the system; in view of weak interactions, their energy levels are closely spaced and highly degenerate, and the associated quantum states are densely populated. It is then more convenient to assume that they consitute a continuum, and we will evaluate the number and energy of the states from the "volume" and *density of states* of this continuum. In both cases, note that knowledge of the wave function itself is actually unnecessary. We just need to know all the relevant energy levels along with their degeneracy. Their discreteness often makes counting of the microstates easier in quantum than in classical mechanics.

13.3b DISTINGUISHABILITY, INDISTINGUISHABILITY

Another point is that quantum states are not counted in the same manner depending on whether or not the particles are distinguishable. We have already been faced with this question about mixing of ideal gases when presenting Gibbs' paradox in 8.2c. If we just know that the contents of two bottles are being mixed we cannot decide whether or not the terms $-R\Sigma x_i \ln x_i$ and $RT\Sigma x_i \ln x_i$ representing entropy creation must be included in the entropy and Gibbs free energy of the system, respectively. Mixing actually takes place only if two gases are different, i.e., if we have a means of distinguishing one kind of molecules from the other.

If several particles can share the same microstates this distinction has an important consequence for the counting of states for the system. For a system of N indistinguishable particles, $N!$ states are identical since permutation of the atoms leads to microstates that cannot be distinguished from each other. This is obviously the case for gases. The situation is however different in a crystal. Although atoms of the same element remain identical, their fixed position in a lattice allows them to be identified without any ambiguity. They are then localized particles, for which the number of states must be counted as if they were distinguishable.

13.3c DENSITY OF STATES

We already had a first glimpse at a density of states with the implicit counting made in the Maxwell model of classical mechanics. In the derivation of the probability (13.16) of a molecule having a velocity v, the reader will have noticed that Maxwell did not simply cube the probability (13.17) obtained for a velocity v_x, but also introduced in his expression a factor $4\pi v^2$. When passing from one to three dimensions, the number of states characterized by a velocity v must be counted along all directions of space. For a sphere of radius v, a given velocity change thus corresponds to a linear element dV_x in a single direction, but to a volume element $4\pi v^2 dv$ in the three directions of space. This is why a factor $4\pi v^2 dv$ must be included in the probability calculation when passing from one to three dimensions.

In quantum mechanics the Schrödinger equation determines the states associated with the various degrees of freedom. As seen in 13.2d, the finite volume that is available to nonlocalized particles implies that the translation wave function be zero at the boundaries of the system. For any standing wave, this periodic boundary condition imposes that the wavelength is an integer multiple of the length of the system. Along a single direction of space, a wave function Ψ of the form $\exp(iKx)$ is a standing wave if it satisfies the condition $\exp(iK0) = \exp(n\pi/a)$ where n is an integer and a the length of the box. It follows that the wave vector K can take only the discrete value $n\pi/a$. The same happens in three dimensions, where the components K_x, K_y and K_z of the wave vector K can take on solely the discrete values $n\pi/a$, $n\pi/b$ and $n\pi/c$, respectively, where a, b and c are the three dimensions of the system.

We can thus represent any state by the three integers n_x, n_y and n_z which characterize its wave number K. In the *wave vector space* , the various possible states define an array whose points are spaced by π/a, π/b or π/c depending on the direction of space considered. Because this spacing is very small in macroscopic systems, where a, b and c are large, we will not try to count the states one by one. We will assume instead that they are uniformly distributed in the wave vector space in which each of them is associated with a volume element $(2\pi)^3/abc = (2\pi)^3/V$. We will then count the number of states for which the components of the wave vector K range from K_x to $K_x + \delta K_x$, K_y to $K_y + \delta K_y$, and K_z to $K_z + \delta K_z$. This number is readily obtained from the *density of states* g, which is the reciprocal of the volume element $(2\pi)^3/V$:

$$g(K_x, K_y, K_z)\delta K_x \delta K_y \delta K_z = V/(2\pi)^3\ \delta K_x \delta K_y \delta K_z. \qquad (13.40)$$

To quote only applications discussed in this book, the parameters a, b and c can be the dimensions of a box where an atom is in translation, those of a cavity where photons are emitted, or the lattice parameters of crystals in which acoustic waves are propagating. In all these problems we have to determine how the number of states varies with the wave vector. Actually, we are not interested in the direction of the wave vector, but solely in its magnitude, so that we must introduce another density of states which indicates how many new states are encompassed when the wave vector changes from K to $K+\delta K$. The calculation is straightforward, but we must just be cautious to select a value of δK that is neither too large or too small to ensure that $g(K)$ is a smooth, continuous function of K. Since the volume difference between spheres with radii K and $K+\delta K$ is $4\pi K^2\delta K$, the variation of the density of states is

$$g(K)\delta K = \frac{4\pi K^2}{8\pi^3/V}\delta K = \frac{K^2 V}{2\pi^2}\delta K. \tag{13.41}$$

The next step is to transform the density in wave vector space into an energy density. For transforming the K^2 factor and the range δK, we will make use of the *dispersion relation*, that is, the relationship between the wave vector and energy. Again, the translator will be taken as an example to illustrate this transformation. From the relationship (13.27) between the wave vector and the momentum, the kinetic energy is

$$E = p^2/2m. \tag{13.42}$$

The expression (13.38) for kinetic energy then leads to

$$E = \frac{h^2K^2}{8\pi^2 m}. \tag{13.43}$$

To obtain the density of states we first differentiate this relation

$$dE = \frac{h^2KdK}{4\pi^2 m}. \tag{13.44}$$

With the expression of K drawn from (13.43)

$$K = \left(\frac{8\pi^2 mE}{h^2}\right)^{1/2}, \tag{13.45}$$

we finally obtain the energy density of the translator from (13.41)

$$g(E)\delta E = 2\pi V(2m/h^2)^{3/2}E^{1/2}\delta E. \tag{13.46}$$

Equations (13.41) and (13.46) carry the same physical information. This is why both functions share the same name (density of states) and symbol (g) although they have different dimensions. A density is a dimensionless variable. We see that this holds true when $g(K)$ and $g(E)$ are multiplied by their ranges ∂K and δE, respectively.

13.4 MAXWELL-BOLTZMANN STATISTICS

We now know how to count states for both localized and nonlocalized particles. The last step is thus to determine the law that relates the probability of occurrence of a given state to its energy. Because this law depends on the conditions considered, we

Table 13.1. Main features of a few ensembles

Ensemble	Microcanonical	Canonical	Grand canonical
Variables kept constant	N, V, E	N, V, T	μ, V, T
Characteristic function	Ω	Z	$\Theta(T, \mu, V)$
Probability of state p_i	$1/\Omega$	$e^{-\beta E_i}/Z$	$e^{-(\beta E_i - \mu n_i}/\Theta$
Statistical analog	$S = k \ln\Omega$	$F = -kT\ln Z$	$J = -kT\ln\Theta$

will deal with the distribution law relevant for most problems dealt with in this book, namely that of Maxwell-Boltzmann. The other *statistics* of Fermi-Dirac and Bose-Einstein will be introduced at the end of this chapter. Unfortunately, some problems may arise in the following sections because of the standard notation we follow. To limit the risk of confusion, we thus stress that the variable p will now designate a probability instead of a momentum.

13.4a ENSEMBLES

Let a system be made up of a large number N of weakly interacting particles. To determine its properties, we can simply make a very large number n of observations and calculate with $p_i = n_i/n$ the probability of finding the system in the state i. The average value of a given property Y then is $< Y >= \Sigma\ p_i Y_i$, where Y_i is the Y value for state i. This average must be calculated over all states that are actually accessible, which implies not to take into account the metastable states of the system. This is the procedure we will adopt in the next section for determining the statistical analog of entropy.

Instead of taking such *time* averages, one can alternatively follow an approach proposed by Gibbs whereby one imagines that the system of interest interacts with a great many identical systems with which it constitutes an *ensemble*. As will become clear below when determining the Maxwell-Boltzmann distribution law, this abstraction will allow us to derive averages over all possible microstates, as exhibited by a whole set of identical systems, without knowing in which microstate the investigated system actually is. For this purpose we will accept the postulate that ensemble and time averages are equivalent. As a matter of fact, this is another way of stating the ergodic hypothesis.

Depending on which conditions are imposed it is useful to distinguish several kinds of ensembles. For convenience, the most commonly used are summarized in Table 13.1. It the system is mechanically and thermally isolated, the number of particles N, the volume V and the internal energy U are constant and the ensemble is said to be *microcanonical*. We have seen in chapter 2 that choosing a given pair of independent variables for describing the state of a system leads to a specific thermodynamic potential. In the same way, each ensemble leads naturally to the introduction of specific functions and statistical analogs of thermodynamic properties. These are also included in Table 14.1. For instance, it should not be a surprise that entropy shows up for conditions of constant volume and energy, or the Helmholtz free energy for constant volume and temperature. For reasons of space and scope, we will not attempt a rigorous discussion of the analogies between macroscopic and statistical variables. Likewise, we will not discuss either the analogs derived from other ensembles as applications to real systems are not straightforward, although the *grand partition function* Θ derived from the grand canonical ensemble is important for treating open systems.

13.4b ANALOG OF ENTROPY

As made in chapter 2, we will begin with an isolated system and thus consider the constant V and U of the microcanonical ensemble. Since all microstates have the same energy, the ergodic postulate states that they have the same probability p_i of occurrence. If the number of accessible states of the system is Ω, then their probability is

$$p_i = 1/\Omega, \tag{13.47}$$

and the average of any property Y is

$$< Y >= 1/\Omega \sum Y_i. \tag{13.48}$$

That a constant value is observed when a macroscopic measurement is made thus indicates that an adequate sampling of the states is made during the timescale of the experiment.

We have already discussed extensively in 11.1 the connection between Ω and entropy. In view of (13.47), we will thus directly study the properties of the variable S_s defined by

$$S_s = -k \sum p_i \ln p_i, \tag{13.49}$$

which is necessarily positive because all p_i range from 0 to 1. As will be followed at the end of this chapter, the standard procedure is to show, with the help of Lagrange multipliers, that S_s is maximum at equilibrium. As it is simpler, we will adopt the treatment of Denbigh's book and assume that a spontaneous transformation takes place in the system and changes by dp_j and dp_k the probabilities of only two states j and k. Because $\Sigma p_i = 1$, we have $dP_j = -dP_k$ and the variation of S_s is

$$\begin{aligned} dS_s &= -k \sum (\ln p_i + 1)\, dp_i = -k[(\ln p_j + 1)dp_j + (\ln p_k + 1)dp_k] \\ &= -k(\ln p_j - \ln p_k)dp_j. \end{aligned} \tag{13.50}$$

We find that dS_s/dp_i is 0 if $p_j = p_k$. This is actually verified if the system is at equilibrium. Because this reasoning can be applied to a change affecting any pair of states, we conclude that equilibrium corresponds to an extremum of S_s, and that this extremum is a maximum because

$$\frac{d^2 S_s}{dp_j^2} = -k \left(\frac{1}{p_j} + \frac{1}{p_k} \right) \tag{13.51}$$

is negative when $p_j = p_k$. For an isolated system the variable S_s thus behaves as the entropy. From (13.47) and (13.49) we write

$$S_s = -k \sum 1/\Omega \ln(1/\Omega) = k \ln \Omega. \tag{13.52}$$

With k = Boltzmann constant, S_s is therefore the statistical analog of entropy and (13.52) states that any spontaneous transformation in an isolated system will cause an increase of the number of accessible states. Whereas the macroscopic equilibrium criterion of maximum entropy is of little practical use, we have already seen in chapter 11 how one uses (13.52) to calculate configurational and mixing entropies. From a theoretical standpoint, however, the microcanonical ensemble does not allow the temperature to be dealt with for the simple reason that temperature is not an explicit variable for systems of constant energy. This is why we must now turn to systems at constant T.

13.4c MAXWELL-BOLTZMANN DISTRIBUTION LAW

Let us consider a canonical ensemble and denote by e the ensemble, s the system of interest and r the rest of the ensemble. Since s and r are mutually independent, so are the probabilities $p_s(E_i)$ that s be in a state of energy E_i and $p_r(E_j)$ that r be in state of energy E_j. If we assume that the ergodic hypothesis is obeyed, then we have

$$p_e(E_i + E_j) = p_s(E_i) p_r(E_j), \tag{13.53}$$

regardless of the values i and j. Furthermore, let us suppose that the energy levels of both s and r are spaced closely enough that E_i and E_j can be taken as continuous functions, and the probabilities p as differentiable functions. Since E_i does not depend on E_j, we obtain through logarithmic differentiation of (13.53) with respect to E_j

$$\frac{\partial \ln p_e(E_i + E_j)}{\partial E_j} = \frac{\partial \ln p_r(E_j)}{\partial E_j}. \tag{13.54}$$

This equation can be rewritten as

$$\frac{\partial \ln p_e(E_i + E_j)}{\partial (E_i + E_j)} \frac{\partial (E_i + E_j)}{\partial E_j} = \frac{\partial \ln p_e(E_i + E_j)}{\partial (E_i + E_j)} = \frac{\partial \ln p_r(E_j)}{\partial E_j}. \tag{13.55}$$

Likewise, we find for the other partial derivative

$$\frac{\partial \ln p_e(E_i + E_j)}{\partial (E_i + E_j)} = \frac{\partial \ln p_s(E_i)}{\partial E_i}. \tag{13.56}$$

If follows that the two partial derivatives are equal

$$\frac{\partial \ln p_r(E_j)}{\partial E_j} = \frac{\partial \ln p_s(E_i)}{\partial E_i} = -\beta. \tag{13.57}$$

Because p_r and p_s are mutually independent, this equality can be satisfied for any values of E_i and E_j only if both partial deriviatives are equal to the same constant, which we have denoted by -β. Integrating this equation, we readily find

$$p_s(E_i) = 1/Z \exp(-E_i\beta) \tag{13.58}$$

where only the discrete values of energy are relevant and $1/Z$ stands for the integration constant.

As will be described below, Z is the *partition function* of the system. The choice of the system s within the ensemble has been arbitrary. The same result would therefore be obtained for any other chosen system, so that the parameter β is an intensive parameter characteristic of the ensemble. In fact β is necessarily positive otherwise the probability of having a state of infinite energy would not be zero. Without waiting to present in 13.4e a few arguments in favor of $\beta = 1/kT$, let us make use of this relation to rewrite (13.58) in the form

$$p_e(E_i) = 1/Z \exp(-E_i/kT). \tag{13.59}$$

We then obtain the Maxwell-Boltzmann distribution law in a more general way than derived in 13.1c for the ideal monoatomic gas.

13.4d ANALOG OF HELMHOLTZ FREE ENERGY

We do know that the equilibrium criterion for isothermal, isochoric transformations is the minimum of the Helmholtz free energy (see 2.3e). We will thus study the properties of the statistical analog of $F = U - TS$. Whereas the internal energy is given by $U = \sum p_i E_i$, a statistical analog of the entropy is (13.49) and, as suggested by the Maxwell-Boltzmann law (13.59), that of temperature is $\beta = 1/kT$. Hence the analog F_s of the Helmholtz-free energy should be written

$$F_s = U - S_s/k\beta = \sum p_i E_i + (\sum p_i \ln p_i)/\beta = 1/\beta \sum p_i \ (\beta E_i + \ln p_i). \tag{13.60}$$

We will repeat the reasoning made in 13.4b to examine how F_s varies when a transformation affecting only a pair of states j and k takes place at constant β. Differentiating F_s and applying the condition $dp_j + dp_k = 0$, we find

$$dF_s/dp_j = 1/\beta \ (\beta E_j + \ln p_j - \beta E_k - \ln p_k). \tag{13.61}$$

At equilibrium, the Maxwell-Boltzmann distribution law applies. Because both probabilities refer to the same partition function Z, we write

$$p_j = \exp(-\beta E_j)/Z, \qquad p_k = \exp(-\beta E_k)/Z. \tag{13.62}$$

With these expressions (13.61) reduce to the expected result $dF_s/dp_j = 0$. As already made in 13.4b, one finds that the second derivative of F_s

$$\frac{d^2F_s}{dp_j^2} = -\frac{1}{\beta}\left(\frac{1}{p_j} + \frac{1}{p_k}\right) \tag{13.63}$$

is negative. At equilibrium the extremum of F_s is therefore a minimum. Replacing the probabilities p_j and p_k by their expressions (13.62) in (13.60), we can finally simplify the expression of F_s

$$F_s = 1/\beta \sum[p_i(\beta E_i - \beta E_i - \ln Z)] = -\ln Z/\beta \sum p_i = -\ln Z/\beta. \tag{13.64}$$

13.4e TEMPERATURE AND ENTROPY

Before reviewing how statistical analogs serve to calculate the properties that have been extensively dealt with in the rest of this book, we must re-examine the meaning of temperature and entropy in the light of the preceding results. Of particular importance is the equality

$$\beta = 1/kT, \tag{13.65}$$

which is suggested by the analogy between microscopic and macroscopic variables, and especially by the fact that β is an intensive parameter that characterizes the thermal state of sytems belonging to a canonical ensemble. We can have a closer look at the connection between β and entropy by considering the statistical analog of the macroscopic relation $dS = dQ/T$ for a reversible transformation. We thus begin with the differential of the statistical internal energy

$$dU = \sum E_i\, dp_i + \sum p_i\, dE_i. \tag{13.66}$$

Since $dU = dQ + dW$, it is tempting to write

$$dQ = \sum E_i\, dp_i, \qquad dW = \sum p_i\, dE_i. \tag{13.67}$$

The first term represents variations of the occupancy of the energy levels, which do not change. According to the Born-Oppenheimer approximation (see Appendix E), it corresponds to conditions of constant volume under which no work can be done. The second term, work, therefore represents changes in the energy levels whose occupancy does not vary.

Let us then differentiate the entropy expression (13.49) to obtain

$$dS_s = -k \sum(\ln p_i + 1) dp_i. \tag{13.68}$$

If the system is in equilibrium with a thermostat, the probabilities p_i are related by (13.58) to the partition function. Noting that $\sum dp_i = 0$ for a system in equilibrium, we write

$$dS_s = -k \sum(-\beta E_i - \ln Z + 1) dp_i = k\beta \sum E_i dp_i. \tag{13.69}$$

Comparing this expression with (13.67) we thus find that the relation

$$dS_s = k\beta dQ \tag{13.70}$$

is consistent with both $dS = dQ/T$ and $\beta = 1/kT$ so that we will now denote by S (instead of S_s) the statistical entropy.

For the translational energy of a monoatomic gas, we saw in 13.1b and 13.1c that temperature and entropy are measures of the mean kinetic energy of the atoms and of the width of the velocity distribution, respectively. The Maxwell-Boltzman distribution law and the statistical expression of entropy extend these conclusions to other forms of energy. More precisely, temperature is a measure of the occupation of the quantum states of a system as a function of their energy. It is an intensive variable because it does not depend on the number of accessible states, but is determined only by their rates of occupancy. Entropy, in contrast, is an extensive variable because it depends on the logarithm of the number of accessible states, which increases linearly with the size of the system.

In the very first chapter we indicated that entropy could be split into thermal and configurational parts. Since we did not make any assumption here as to the nature of the quantum states of the system, the two entropy expressions $S = -k\sum p_i \ln p_i$ and $S = k \ln \Omega$ are equally valid and appliy to thermal as well as to configurational contributions. From a practical standpoint, however, counting thermal states is generally not feasible; as will be seen in 13.5b, one then evaluates thermal entropy with the former expression through calculation of the partition function. As shown in chapter 11, the latter expression can in contrast be very convenient for counting configurations and thus calculating configurational entropies. Besides, one can often assume that the thermal and configurational states are independent. The number of configurations is then

$$\Omega = \Omega^{conf} \Omega^{th}. \tag{13.71}$$

Entropy is created when heat is transferred and matter is mixed because the number of quantum states between which energy quanta and particles can be distributed increases considerably. This is another way of saying that energy and particles flow to maximize the number of configurations when heat and matter are exchanged. As already alluded to in 1.5d, this is the reason for the irreversibility of these processes. But the manner in which this widening takes place depend on the energy levels (and their degeneracy) of the system of interest. Knowing that two phases have the same temperature does not tell us anything about their respective energy levels. In contrast, knowing that many more energy levels are available for one phase, either because they are more numerous for a given energy or because they are more closely spaced, will allow us to predict that this phase has a higher entropy and heat capacity.

As a matter of fact, (13.71) is implicitly assumed to hold true whenever entropies of mixing are considered to be purely configurational. For example, this is the case for multisite solutions where thermal entropy depends little on the distribution of atoms on the crystallographic sites because, as a whole, vibrational frequencies are little affected by such substitutions.

13.4f NEGATIVE TEMPERATURES

The temperature of a system would be zero if all particles were in their fundamental energy levels, in which case the entropy would be also zero because there would remain just one accessible state. On the other hand, the entropy would reach its highest possible value and the temperature would be infinite if all energy levels were equally populated. This could happen only if the number of available quantum states were finite. It is impossible to set an upper limit to the number of states, as shown in 6.1b by the decomposition of molecular hydrogen into atomic hydrogen, and then into a plasma (before thermonuclear reactions are triggered at still higher temperatures). The states of lower energy therefore remain the most populated. The temperature thus remains finite and the entropy does not reach the maximum possible value corresponding to an even occupation of states.

However, certain degrees of freedom have a limited number of energy levels and are weakly coupled to the other degrees of freedom, which makes it possible to assume that they are largely independent. This is the case of the magnetic energy of atomic nuclei. A nucleus of spin I can have only $2I+1$ possible orientations in an external field, each corresponding to a different energy. For a unit spin, these are parallel, perpendicular and or antiparallel to the field (in increasing order of energy). Consider different chemical elements with different spins in a crystal. At equilibrium, the distributions of thermal energy within the lattice and of magnetic energy within the spin systems are both set by the Maxwell-Boltzmann law with a common temperature T. Now, the weak coupling between the lattice and the spin systems allows energy to be accumulated in the latter by appropriate radiofrequency pulses. Without significant spin-lattice relaxation, the spin temperature differs from the lattice temperature and can even vary from one spin system to another within the same crystal.

More interestingly, this weak coupling makes it possible to accumulate magnetic energy in a spin system until an even distribution of the spin orientation is achieved. For $I = 1$, one finds one third of the spins in each of the three possible orientations. This corresponds to an infinite temperature at which the entropy is maximum. The energy is not maximum, however, since only one third of the spins has the antiparallel orientation. If we keep delivering the right radiofrequency pulses to the spin system, then the magnetic energy increases further whereas the entropy decreases since the high-energy level become more populated than the low-energy levels. This situation corresponds to *negative* temperatures.

The striking fact is that T changes directly from $+\infty$ to $-\infty$ without going to 0 K. The temperature then rises and eventually tends to the zero temperature at which entropy would also be zero because only the highest energy state would be populated. Negative temperatures are thus possible when the number of states is finite, and they characterize real, albeit metastable, states of such systems. Another case of practical importance is that of laser cavities where an inversion of the population of electronic energy levels is produced, by optical pumping, in order that coherent radiation be emitted; contrary to those of spin systems, however, the lifetime of such population inversions is extremely short. Finally, the fact that the sign of $\beta = 1/kT$ changes without any discontinuity suggests that it is a more fundamental parameter than T, which diverges when the entropy begins to decrease. (From a theoretical standpoint one notices that temperature is always combined to k. It has therefore been proposed to use a scale where temperatures would have the dimension of energy, but such a change is of course not going to be accepted!)

13.4g ENTROPY AND INFORMATION

An interesting corrolary of the link between entropy and probability is the modern theory of information. The starting point was a speculation published in 1871 by Maxwell who imagined that a "vessel is divided into two portions, by a division in which there is a small hole, and that a being, who can see the individual molecules, opens and close this hole so as to allow only the swifter molecules to pass from A to B, and only the slower ones to pass from B to A. He will thus, without the expenditure of work raise the temperature of B and lower that of A, in contradiction to the second law of thermodynamics". This small being, whose "faculties are so sharpened that he can follow every molecule in its course", was shortly thereafter dubbed "demon" by Kelvin, and he has raised much debate and controversy ever since.

To explain why the demon cannot act as imagined by Maxwell, the physicist L. Szilard remarked in 1929 that, if one could know the state of all molecules of a gas, then one could build an engine which would process the molecules one by one to transform their kinetic energy into work. As the kinetic energy would then be replenished by impact of the "cold" molecules on the hot walls of the vessel, heat could be entirely transformed into work, which is of course impossible. Szilard thus concluded that the action of measuring the position of a molecule would raise the entropy by an amount greater than the decrease due to the conversion of heat into work. L. Brillouin and D. Gabor later emphasized that a molecule could be detected only if it scattered at least one photon. In order for it to be distinguishable, this photon had to have an energy higher than that of the photons of black-body radiation (see 13.6f), so that conversion of its energy into heat would result in a net increase of the entropy of the system.

Regardless of possible reversible ways to measure the position of molecules, the basic fact is that there exists a close connection between entropy and information. In the same manner as one can distribute n distinguishable particles between the two compartments of a vessel in 2^n ways, an n bit computer memory has 2^n different states and each bit of information must be physically associated with an entropy $k\ ln\Omega = k\ \ln 2$. When one bit of information is erased, the probability 1/2 that the memory was in a state 0 or 1 becomes a certainty that it is in a state 0. We conclude that the memory must dissipate some heat (which represents a negligible fraction of that it actually produces) to compensate the entropy decrease due to its erasing.

In a general way, one can describe any change in information by the *information entropy* defined in 1948 by C. Shannon as $S(Q/X) = -K\sum p_i \ln p_i$, where K is an arbitrary scale factor, Q a well-defined question (e.g., what is the state of the memory i?) and X its possible answers as characterized by a set of probabilities p_i (i.e., for the two states of the memory, p_i = 1/2 and 1/2 before erasing, and 1 and 0 after). The message has then an information content that can be characterized by the change in $S(Q/X)$ it induces (i.e., $K\ \ln 2$ for that that announces the erasing of the memory). But what about Maxwell's demon? The relevance of these ideas to his job is that he must forget the result of his previous observation to make a new one. By doing so, he creates the entropy required in order not to violate the second principle. In case a very large memory would store the results of his observations, its size would remain finite so that it would to have to be erased periodically anyway to let the demon continue indefinitely his observations.

13.5 APPLICATIONS OF THE MAXWELL-BOLTZMANN STATISTICS

13.5a PARTITION FUNCTION

To begin with, we must make use of the analogy between T and β as embodied by $T = 1/k\beta$ to evaluate the partition function Z introduced in 13.4c as a characteristic property of the system for conditions of *constant temperature and volume*. For this purpose, we first remark that

$$\sum_i p(E_i) = \sum_i 1/Z\ \exp(-E_i/kT) = 1. \tag{13.72}$$

We thus have

$$Z = \sum_i \exp(-E_i/kT), \tag{13.73}$$

where the summation is made over all *quantum states* accessible to the system. If the summation is performed over the energy levels, then the degeneracies g_i have to be taken into account with

$$Z = \sum_i g_i \exp(-E_i/kT). \tag{13.74}$$

Turning back to probabilities, we rewrite (13.58) as

$$p(E_i) = \frac{exp(-E_i/kT)}{\Sigma_i \exp(-E_i/kT)}. \tag{13.75}$$

For a system in equilibrium at temperature T, the probability of state i is proportional to the term in Z to which it corresponds. The particles of the system are therefore partitioned into their possible states in the same ratios as the Boltzmann factors in Z. This is the origin of the name of *partition function* given to Z, which might otherwise seem less evocative than the old name of *sum of states*.

From the translational energy levels derived in 13.2d, we will calculate in 14.2b that the partition function of an ideal monoatomic gas is

$$Z = V\left(\frac{2\pi mkT}{h^2}\right)^{3/2}. \tag{13.76}$$

The reader may care to obtain the same result from the density of states (13.46) by writing the partition function

$$Z = \int_0^\infty g(E)\exp(-E/kT)dE. \tag{13.77}$$

13.5b THERMODYNAMIC PROPERTIES

For systems of constant composition we will now derive from the results obtained in previous sections the statistical expressions of the main thermodynamic properties. Starting from the Helmholtz free energy expression (13.64), we rewrite it as

$$F = -\ln Z/\beta = -kT\ln Z. \tag{13.78}$$

For the internal energy, we write

$$U = \sum p(E_i)E_i = E_i \exp(-E_i/kT)/Z. \tag{13.79}$$

Because we have

$$\frac{dZ}{dT} = \sum \frac{E_i \exp(-E_i/kT)}{kT^2} = \frac{ZU}{kT^2}, \tag{13.80}$$

U is related to the partition function by

$$U = \frac{kT^2}{Z}\left(\frac{\partial Z}{\partial T}\right)_{V,n_i} = kT^2\left(\frac{\partial \ln Z}{\partial T}\right)_{V,n_i}. \tag{13.81}$$

From these results, we then derive the entropy

$$S = \frac{U-F}{T} = k\ln Z + kT\left(\frac{\partial \ln Z}{\partial T}\right)_{V,n_i}. \tag{13.82}$$

In the same way, we obtain

$$P = -\left(\frac{\partial F}{\partial V}\right)_T = kT\left(\frac{\partial \ln Z}{\partial V}\right)_T, \tag{13.83}$$

$$H = U + PV = kT^2\left[\frac{\partial \ln Z}{\partial T}\right]_{T,n_i} + kTV\left(\frac{\partial \ln Z}{\partial V}\right)_{T,n_i}, \tag{13.84}$$

and we may derive $G = H - TS$ by combining the preceding equations. Note finally that the calculated heat capacity is C_v

$$C_v = \left(\frac{\partial U}{\partial T}\right)_{V,n_i} = 2kT\left(\frac{\partial \ln Z}{\partial T}\right)_{V,n_i} + kT^2\left(\frac{\partial^2 \ln Z}{\partial T^2}\right)_{V,n_i}. \tag{13.85}$$

As will be seen in chapter 14, the standard practice is to calculate C_v by a direct differentiation of the expression obtained for U from a specific model, rather than evaluating the two differentials of Z that appear in (13.85). In any case, comparisons of experimental and theoretical values require either the subtraction of $TV\alpha^2/\beta_T$ from the measurements, or the addition of this term to the calculations. When one is interested in molar properties, it suffices to to consider molar energies and replace the Boltzmann constant k by the gas constant R in the above equations.

13.5c FACTORIZATION OF PARTITION FUNCTIONS

For particles having only translational energy, the summation in the partition function must be made over all different quantum states

$$Z = \sum_i \exp(-E_i/kT) = \sum_{j,k,l} \exp[-(E_{xj} + E_{yk} + E_{zl})/kT]. \tag{13.86}$$

Now the various quantum levels j, k and l of the translations along the x, y and z directions are occupied in a completely independent manner. This summation can therefore be split into three independent terms

$$Z = \sum_j \exp[-E_{xj}/kT] \sum_k \exp[-E_{yk}/kT] \sum_l \exp[-E_{zl}/kT] = Z_x Z_y Z_z. \tag{13.87}$$

In a more general way, each time a wave function can be factorized in several independent parts

$$\Psi = \prod \Psi_j, \tag{13.88}$$

energy is the sum of the eigenvalues of the various wave functions

$$E = \sum E_j, \tag{13.89}$$

and the partition function is the product of the associated partition functions

$$Z = \prod Z_j. \tag{13.90}$$

Of particular interest is this factorization for an assembly of N non interacting particles. Let us denote by z the partition function of a single particle and by Z that of the whole system. Two cases must then be identified depending on whether or not the particles are distinguishable. In the former (that of a crystal, for instance), we simply write

$$Z = z^N. \tag{13.91}$$

In the latter, such as in an ideal gas, the indistinguishability of particles has the consequence that there are $N!$ ways of achieving the same microstate. We thus have

$$Z = z^N / N! \tag{13.92}$$

Note that the presence of the constant factor $1/N!$ in the partition function for undistinguishable particles affects the entropy, but not the internal energy.

13.5d HEAT CAPACITY OF TWO-LEVEL SYSTEMS

These results allow us to have a fresh look at heat capacities and to examine for a more usual thermodynamic situation the consequences of a finite number of energy levels discussed in 13.4f for spin systems. We will consider the case of a two-level system, namely, that of a system with a single energy level above the ground state where the energy difference between them is ΔE. Let N_0 and N_1 be the fractions of the occupied groundstate and excited levels, respectively ($N_0 + N_1 = 1$). The Maxwell-Boltzmann distribution law states that

$$N_1 = \frac{\exp(-\Delta E/kT)}{Z} = \frac{\exp(-\Delta E/kT)}{1+\exp(-\Delta E/kT)} = \frac{1}{1+\exp(\Delta E/kT)}. \tag{13.93}$$

The internal energy is thus

$$U = N_1 \Delta E = \frac{\Delta E}{1+\exp(\Delta E/kT)}. \tag{13.94}$$

With regard to the heat capacity we find for a total of N particles

$$C_v = \left(\frac{\partial U}{\partial T}\right)_V = Nk\frac{(\Delta E/kT)^2 \exp(\Delta E/kT)}{[1+\exp(\Delta E/kT)]^2}. \tag{13.95}$$

The values of N_1 and of C_v are plotted in Fig. 13.3 against $kT/\Delta E$. From $T = 0$, we observe that both N_1 and C_v first increase very rapidly. When N_1 tends to level off, C_v then drops off and tends to 0. The high-temperature limit is not characterized by $N_1 = 1$, but by $N_1 = N_0$ otherwise the energy of the system would keep increasing whereas its entropy would decrease. Once more, we observe a competition between entropy and energy factors. The assymmetry between high and low temperatures, where $N_0 \sim 1$ and $N_1 \sim 0$, is due to the fact that ΔE becomes relatively smaller at higher T and thus negligible with respect to the term $-RT\Sigma N_i \ln N_i$. When the temperature increases, the fact that heat capacities seldom decrease, and never tend to 0, in contrast indicates that there always exist states of higher energy. As discussed in 13.4f, high temperatures thus drives all systems to states of higher energy and entropy.

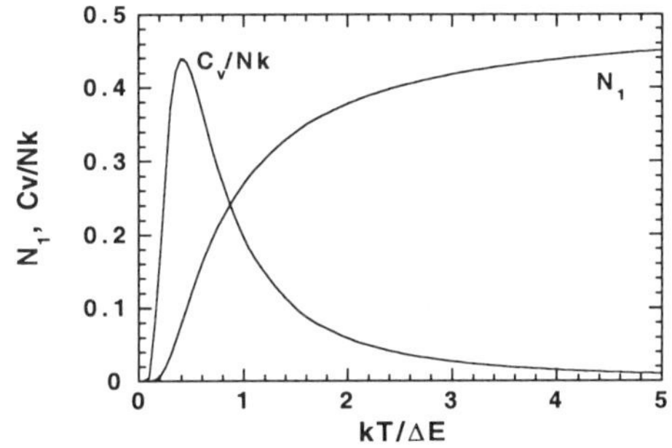

Figure 13.3. Heat capacity for a two-level system of energy and fraction N_1 of particles in the excited state against $kT/\Delta E$.

In various instances two-level systems nonetheless manifest themselves as *Schottky anomalies*, which are superimposed on normal C_p curves. For instance, one observes these anomalies when the groundstate of paramagnetic ions is split in a crystal field. It also happens that ΔE is so small that only the high-temperature tail of the anomaly can actually be observed.

13.5e EQUILIBRIUM BETWEEN IDEAL GASES

The calculation of equilibrium constants for reactions between ideal gases represents a direct application of the preceding results. Consider the general reaction

$$\sum \nu_i M_i = 0, \tag{13.96}$$

where ν_i is the stoichiometric coefficient of species M_i. Because the gases are ideal all molecules are independent in the volume V occupied by the mixture. The factorization of partition functions allows us to write for the system

$$Z = \frac{\prod q_i^{N_i}}{\prod N_i!}, \tag{13.97}$$

where N_i the number of molecules of species i. Let us keep the volume and temperature constant so that the equilibrium criterion is the minimum of the Helmholtz free energy. For the gas mixture, F is

$$F = -kT \ln Z = -kT \sum (N_i \ln q_i - \ln N_i!). \tag{13.98}$$

At equilibrium we have

$$\left(\frac{\partial F}{\partial N_i}\right)_{T,V} = 0. \tag{13.99}$$

Denoting by $d\xi = dN_i/\nu_i$ the advancement of the reaction as given by (7.2), we rewrite this equation with the help of Stirling's approximation as

$$\left(\frac{\partial F}{\partial \xi}\right)_{T,V} = kT \sum \nu_i (\ln q_i - \ln N_i) = 0. \tag{13.100}$$

We thus have

$$\prod N_i^{\nu_i} = \prod q_i^{\nu_i}. \tag{13.101}$$

Let us introduce the molarity $c_i = N_i/V$ to express the chemical composition of the gas mixture in terms of intensive variables. Instead of inserting a volume dependent term in the right-hand side of this expression it is simpler to replace q_i by a partition function per unit volume q_i^V. We then write the equilibrium constant as a function of such partition functions

$$k_c = \prod c_i^{\nu_i} = \prod (q_i^V)^{\nu_i}. \tag{13.102}$$

We emphasize that the partition functions in (13.102) refer to a common zero point of energy. The problem is that the fundamental level, which is selected as the zero point of energy for each molecule, varies from one species to another except for isotopic molecules (see chapter 15). For each species we must therefore take out of the right-hand side of this equation the difference E_{i0} between a common zero point of energy and the fundamental energy level of the species i. We thus write

$$\prod c_i^{\nu_i} = \prod (q_{i0}^V)^{\nu_i} \exp(-\nu_i E_{i0}/kT), \tag{13.103}$$

where q_{i0} is the partition function referenced to the fundamental energy level of species i. The term $\exp(-\sum \nu_i E_{i0}/kT)$ represents the energy change of the reaction at 0 K and can be obtained from the dissociation energies of the molecules (see Appendix E). We will see in chapter 14 how the partition functions can be calculated for diatomic and polyatomic molecules from spectroscopic information. By combining these two kinds of data, one can thus calculate equilibrium constants for reactions between ideal gases from purely microscopic observations.

13.6 QUANTUM STATISTICS

13.6a PROBABILITY LAWS

The Maxwell-distribution law is in fact valid only if the number of available states is much greater than the number of particles, in which case all significant terms of the partition function are associated with particles in different energy states. In practice, this generally corresponds to high-temperature conditions under which the Maxwell-Boltzmann statistics represent the limit of quantum statistics that will be determined in the following sections. Starting with the common postulate that states of equal energy have the same probability of occurence, we will examine how the different ways in which N identical particles distribute over known energy levels E_i, with known degeneracies g_i, give rise to different probability laws. Two cases will be envisaged depending on whether different particles can or cannot occupy the same quantum state.

The procedure will be to count the number N^s of different states and find how the probability of each state must be related to its energy in order that N^s be maximum for a constant number of particles and a constant energy. If n_i is the number of particles with energy E_i, these conditions are

$$N = \sum n_i, \qquad \sum dn_i = 0, \tag{13.104}$$

and

$$E = \sum n_i E_i, \qquad \sum E_i dn_i = 0. \tag{13.105}$$

As already mentioned in 8.6d, there is a standard method for finding the maximum of a function subjected to such conditions. It consists of differentiating the function $N_d = f(n_i)$ with respect to dn_i, and of incorporating the constraints (13.104)-(13.105) through the Lagrange multipliers α and β

$$f(n_i)dn_i - \alpha dn_i - \beta E_i dn_i = 0. \tag{13.106}$$

It is the relation drawn from this equation between n_i and the Lagrange multipliers that will yield the desired distribution law.

13.6b FERMI-DIRAC STATISTICS

We will begin with the case where particles obey the Pauli exclusion principle, according to which two different particles cannot be in the same quantum state. This is the case for electrons, protons and neutrons, and consequently of nuclei with an odd number of nucleons. Let n_i such particles distribute over a given energy level E_i having a degeneracy g_i. To calculate the number of ways the distribution can be made, we note that n_i must be smaller than g_i. If n_i states are occupied then $g_i - n_i$ states are unoccupied so that the counting is directly given by (13.39)

$$N_i^{FD} = \frac{g_i!}{n_i!(g_i - n_i)!}. \tag{13.107}$$

To derive the total number of states (N^{FD}), we consider all states by taking the product of N_i^{FD} over all energy levels

$$N^{FD} = \prod N_i^{FD} = \prod \frac{g_i!}{n_i!(g_i - n_i)!}. \tag{13.108}$$

With Stirling's approximation we have

$$\ln N^{FD} = \sum [(n_i - g_i)\ln(g_i - n_i) - n_i \ln n_i + g_i \ln g_i]. \tag{13.109}$$

By differentiating this expression, where the degeneracies g_i are constant, we obtain

$$d\ln N^{FD} = \sum [(\ln n_i - \ln(g_i - n_i)]. \tag{13.110}$$

Let us make use of (13.106) and Lagrange multipliers to introduce the constraints of total energy and constant number of particles on $d\ln N^{FD} = 0$. We then write

$$\sum [\ln \frac{n_i}{g_i - n_i} + \alpha + \beta E_i] dn_i = 0. \tag{13.111}$$

For this expression to be zero, all coefficients must vanish

$$\ln \frac{n_i}{g_i - n_i} + \alpha + \beta E_i = 0. \tag{13.112}$$

It follows that

$$n_i^{FD} = \frac{g_i}{\exp(\alpha + \beta E_i) + 1}. \tag{13.113}$$

This distribution law defines Fermi-Dirac statistics. Particles obeying it are called *fermions*.

In conductors on can consider that conducting electrons constitute a gas of free electrons within the crystal lattice. As described in solid state physics textbooks, an important application of Fermi-Dirac statistics is thus the calculation of the electronic contribution to the heat capacity of conductors. In (13.113), the parameter α (which is then noted μ and called the *Fermi level*), represents the chemical potential of electrons. Another classical application of this statistics concerns ^{3}He, not in its gaseous state where deviations from Maxwell-Boltzmann statistics are small even at the lowest temperatures, but in the liquid phase. Although atoms in a liquid are not weakly interacting particles, one can consider that in the case of ^{3}He they constitute fermions to which their mutual interactions give an effective mass from 3 to 6 times greater than their real mass. Another peculiarity of ^{3}He is that the melting curve has a negative slope. Contrary to that of ice, this is not due to a volume but to an entropy effect. Whereas spins are completely disordered in the solid above 2 mK, because of the Fermi statistics they are not in the liquid which is thus the phase with the lowest entropy.

13.6c BOSE-EINSTEIN STATISTICS

Contrary to fermions, *bosons* like photons and nuclei with an even number of nucleons can occupy all possible quantum states without any restriction. For a given energy level, the problem is to count the number of ways n_i indistinguishable elements can be distributed over g_i distinguishable boxes. We show below that this number is

$$N_i^{BE} = \frac{(n_i+g_i-1)!}{n_i!(g_i-1)!}. \tag{13.114}$$

For bosons, the number of quantum states is generally very great compared to the number of particles because either the degeneracy is very large or there are few particles to populate each energy level. Under these conditions, we can make the approximation $g_i - 1 \simeq g_i$, which allows us to write the total number of states as

$$N^{BE} = \prod N_i^{BE} = \prod \frac{(n_i+g_i)!}{n_i!g_i!}. \tag{13.115}$$

With Stirling's approximation, we obtain

$$\sum[\ln(n_i+g_i) - \ln n_i - \alpha - \beta E_i]dn_i = 0. \tag{13.116}$$

As made in the preceding section, we derive through differentiation of this equation and use of the Lagrange multipliers α and β another distribution law

$$n_i^{BE} = \frac{g_i}{\exp(\alpha+\beta E_i)-1}, \tag{13.117}$$

which defines the Bose-Einstein statistics.

Here we must count the number of ways of placing, without any restriction, N indistinguishable balls in G distinguishable boxes. Let us assume that the boxes are aligned and mutually separated by walls. Starting with a given arrangement of the balls in the boxes, we can obtain the other arrangements by displacing stepwise the walls to make the balls switch from one box to another one by one. Since we have $G-1$ walls, the total number of states is the number of arrangements of $N+G-1$ elements, which is $(N+G-1)!$. Because the balls are indistinguishable, this number must be divided by $N!$ In addition, the walls are also indistinguishable although the boxes are not. All states that differ only by permutations of walls are therefore the same, which is why we must also divide the total number of states by $(G-1)!$. In this way (13.114) is obtained.

13.6d COMPARISON WITH MAXWELL-BOLTZMANN STATISTICS

For comparison purposes, we will first redetermine the Maxwell-Boltzmann law in a similar, quantum manner. We now consider a set of N distinguishable particles and assume that energy is not degenerate. If the levels 1, 2, etc. were occupied by n_1, n_2, etc. particles, the number of states would be $N^{MB} = N!/\prod n_i!$. Owing to the existence of degeneracy, the actual number of states is actually g_i times higher for each level so that the total number of states is

$$N^{MB} = N! \prod \frac{g_i}{n_i!}. \tag{13.118}$$

Since the particles are indistinguishable, we have in fact

$$N^{MB} = \prod \frac{g_i}{n_i!}. \tag{13.119}$$

With Stirling's approximation, we find after differentiation and use of Lagrange multipliers

$$\ln \frac{n_i}{g_i} + \alpha + \beta E_i = 0, \tag{13.120}$$

from which we obtain the Maxwell-Boltzmann law

$$n_i^{MB} = \frac{g_i}{\exp(\alpha + \beta E_i)}. \tag{13.121}$$

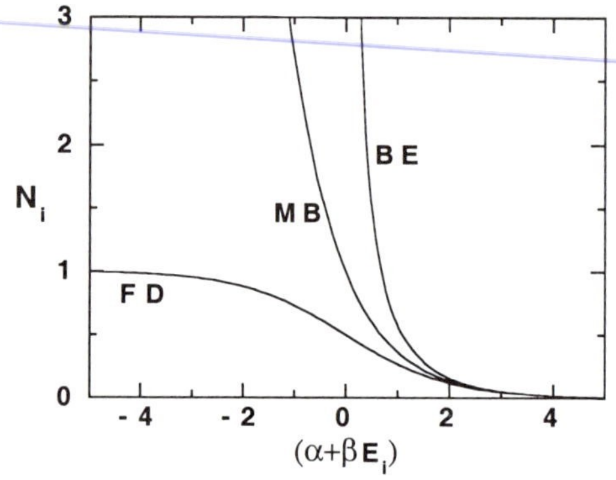

Figure 13.4. Comparison of the average numbers of particles per state as given by Fermi-Dirac (FD), Bose-Einstein (BE) and Maxwell-Boltzmann (MB) statistics. Note the very marked contrast between bosons and fermions.

The numbers of particles per state calculated with Maxwell-Boltzmann statistics are compared in Fig. 13.4 with those given by Fermi-Dirac and Bose-Einstein statistics. Consistent with the fact that quantum states of the same energy are equiprobable, the three statistics predict that the number of particles having a given energy E_i is proportional to the degeneracy g_i of this level. Besides, we observe that the laws (13.113) and (13.117) tend to the Maxwell-Boltzmann law (13.121) when $\exp(\alpha + \beta E_i)$ becomes much greater than unity (Fig. 13.4). This common limit implies that the constant β is the same in all three cases. Like for Maxwell-Boltzmann statistics, we thus have $\beta = 1/kT$. In contrast, the constant α is not the same. As made in 13.5a for Maxwell-Boltzmann statistics, one can determine α from the total number of particles of the system, but we will not perform the calculation here because it is more involved for both Fermi-Dirac and Bose-Einstein statistics. Denoting e^{α} by B, we can nonetheless write the three laws in the same form

$$f_i = \frac{1}{B\exp(E_i/kT) + I}, \tag{13.122}$$

where $I = 1$ (Fermi-Dirac), 0 (Maxwell-Boltzmann) and -1 (Bose-Einstein).

The Maxwell-Boltzmann statistics correspond to the "classical" case where $n_i << g_i$. Under these conditions, the Pauli exclusion principle is no longer relevant and the differences between fermions and bosons vanish since both n_i^{FD} and n_i^{BE} tend to $g_i!/n_i!$, the value of the Maxwell-Boltzmann distribution. For gases, this happens at rather low temperatures at low densities. This is why Maxwell-Boltzmann statistics would break down well below the boiling point and why the ideal gas law, which is valid solely within the framework of this statistics, remain applicable at low densities. It is only for ^{3}He and ^{4}He, which have very low boiling points of about 4 K, that quantum effects become observable although they remain minor.

13.6e BLACK-BODY RADIATION

The important problem of black-body radiation will allow us to make good use of several of the results obtained in previous sections. Before doing so, it is appropriate to recall that quantum mechanics owes a great deal to thermodynamics because it is while trying to solve the problem of black-body radiation that Planck (1858-1947) first assumed in 1900 that energy was quantized. Soon after, Einstein (1879-1955) had also to assume that energy was quantized to account for both the photoelectric effect and the heat capacity of solids. Then, it became apparent that energy quantization was not just a mathematical trick but a feature of profound physical significance.

In daily life black-body radiation manifests itself as a glow observed when a coal is heated at about 1000 K. On further heating the light emitted turns red, yellow and then white at around 2000 K. At the same time the light intensity grows so strongly that the light becomes blinding. For a given wavelength, both the intensity and the relative intensity of the emitted light thus changes with temperature. Before examining this point in more detail we will examine a few other important features for the case of an isothermal radiating cavity. A perfect *black body* absorbs everything and reflects nothing. The radiation in it must be isotropic and unpolarized and it must have the same intensity for a given frequency everywhere otherwise one could exploit a heterogeneity to produce work from such a heat reservoir whose temperature is constant. For the same reason, the radiation will not be affected by the shape of the cavity or by the insertion of another body at the same temperature. If follows that there must exist a universal function of the frequency ν and the temperature T, which gives the energy density of heat radiation per unit frequency range at a given temperature.

We then note that the energy of the radiation varies with the volume V of the cavity. As energy is an extensive property, that of the radiation is proportional to V at constant T. If the cavity has energy it follows that it has also entropy and, like for any thermodynamic system, the entropy of the cavity varies with V

$$\left(\frac{\partial S}{\partial V}\right)_T = \left(\frac{\partial P}{\partial T}\right)_V. \tag{13.123}$$

Without knowing anything about Maxwell's theory of electromagnetism, we can conclude from this relation that a pressure P is exerted by the radiation on the walls of the cavity, and that it increases with T. As a matter of fact, electromagnetic theory shows that

$$P = U/3V, \tag{13.124}$$

where U is the energy of the radiation. From (13.124) and the thermodynamic equation of state (3.26)

$$T\left(\frac{\partial P}{\partial T}\right)_V = P + \left(\frac{\partial U}{\partial V}\right)_T, \tag{13.125}$$

we obtain

$$T\left(\frac{\partial P}{\partial T}\right)_V = 4P. \tag{13.126}$$

Denoting by A a constant, we find by integration

$$P = AT^4, \tag{13.127}$$

which shows that the pressure of radiation P is proportional to T^4, but that is independent of volume. This is the pressure that would prevail in a cavity in which a very high vacuum would be made. In other words, radiation in thermal equilibrium can be viewed as a gas of photons, for which we can also calculate the variation of energy with temperature. For this purpose, let $u(T)$ denote the energy density, which is the proportionality constant between U and V :

$$U(T,V) = u(T)V. \tag{13.128}$$

Making use of (13.124) to eliminate P, this equation allows us to write the thermodynamic equation of state in a form that can be directly integrated

$$\left(\frac{\partial u}{\partial T}\right)_V = \frac{4u(T)}{T}, \qquad u(T) = u_0 T^4. \tag{13.129}$$

This one form of the Stefan-Boltzmann law, where u_0 is a constant.

13.6f PLANCK'S DISTRIBUTION LAW

We now want to determine how the energy of a black body varies with frequency. For this purpose, we need to know only three things: how the energy of a photon depends on frequency; how many states are available in a given frequency range $\delta\nu$; and what is the probability of occupation of these states as a function of energy. The energy distribution will then be the product of these three terms.

According to the assumption made by Planck that energy is quantized, the first term is simply given by

$$E = nh\nu, \tag{13.130}$$

where n is an integer and the Planck constant h is the proportionality factor between the energy of a photon and its frequency. The second term is the density of states.

We almost already know it from the calculation made in 13.3c. The only change to make to (13.41) originates in the fact that electromagnetic waves have two (transverse) polarizations. The number of states is thus twice that we derived, namely

$$g(K)\delta K = 2\frac{K^2 V}{2\pi^2}\delta K = \frac{K^2 V}{\pi^2}\delta K. \qquad (13.131)$$

The transformation to the energy density is readily made with the relationship $\nu = cK/2\pi$, whence

$$g(\nu)\delta\nu = 8\pi^2 V\delta\nu/c^3. \qquad (13.132)$$

Note finally that photons are peculiar bosons because they are massless. Since they can be created and annihilated, their number is not fixed and the conservation equation (13.104) does not apply to them. Without this constraint, the term α disappears from the Bose-Einstein distribution law. We thus have

$$n_i = \frac{g_i}{\exp(E_i/kT) - 1}. \qquad (13.133)$$

By combining (13.130), (13.131) and (13.133), we obtain directly the celebrated Planck distribution law

$$\frac{E(\nu)\delta\nu}{V} = \frac{8\pi h}{c^3}\frac{\nu^3}{\exp(h\nu/kT) - 1}\delta\nu, \qquad (13.134)$$

which we have expressed for a unit volume. It happens that this spectral distribution of energy does not lend itself to an experimental determination. With a spectrometer, it is in contrast easy to collect the light emitted by the cavity through a hole (which is small enough that the thermal regime of the cavity is not affected), and to measure the energy flux E_T as a function of the frequency. By integrating the energy radiated for an element of black surface of unit area in half the directions of space during a time dt, one finds that E_T is related to the spectral distribution by

$$E_T = \frac{c}{4}\frac{E(\nu)}{V} = \frac{2\pi h}{c^2}\frac{\nu^3}{\exp(h\nu/kT) - 1}. \qquad (13.135)$$

This distribution is represented in Fig. 13.5 at several temperatures. The marked shift of the spectrum toward high frequency and the very strong increase of the radiated flux are obvious. Once more we observe that an entropy increase translates into the widening of a distribution. The maximum is in fact proportional to T^3 whereas one finds the same curve when doubling the temperature by doubling the abscissa scale and multiplying by 2^5=32 the ordinate scale.

By integrating (13.134) from 0 to infinity, let us finally calculate the total energy radiated per unit volume

$$\frac{U}{V} = \frac{8\pi h}{c^3}\int_0^\infty \frac{\nu^3}{\exp(h\nu/kT) - 1}d\nu = \frac{8\pi h}{c^3}\left(\frac{kT}{h}\right)^4 \int_0^\infty \frac{x^3}{\exp(x) - 1}dx. \qquad (13.136)$$

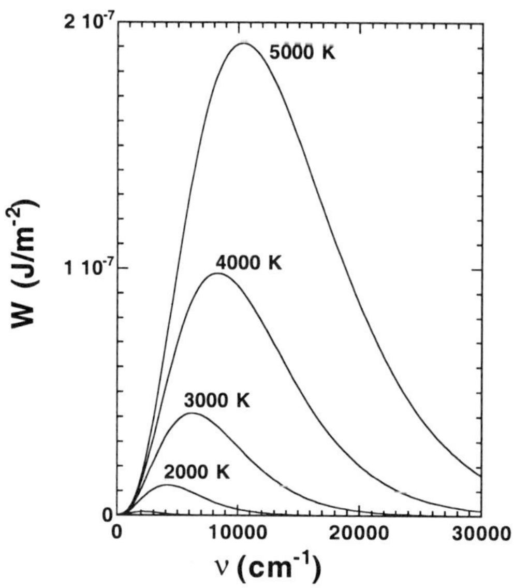

Figure 13.5. Black-body radiation at the temperatures indicated. The very small bump at the bottom of the figure is the calculation for T = 1000 K. The visible part of spectrum is between 12,500 and 25,000 cm^{-1}.

Because the definite integral of this equation is equal to $\pi^4/15$, the energy density is

$$u(T) = \frac{U}{V} = \frac{8\pi^5}{15}\frac{(kT)^4}{(hc)^3}. \tag{13.137}$$

This result allows us to find the emitting power W of the surface of the cavity

$$W = \frac{c}{4}u(T) = \frac{2\pi^5 k^4}{15h^3c^2}T^4 = \sigma T^4. \tag{13.138}$$

With (13.129), this expression constitute the Stefan-Boltzmann law and it allows us to express the Stefan-Boltzmann constant σ (=$u_0c/4$) in terms of h, c and k.

Because black-body radiation depends only on temperature, it represents a primary temperature standard (above 1235 K in the 1990 international practical temperature scale) under carefully controlled conditions. Conversely, it can serve for temperature measurements when use of thermocouples becomes less easy above 1800 K. By *optical pyrometry*, one measures the ratios between the intensities at different temperatures for the same wavelength. A real body is not black, however, but *grey*, and it radiates less light than predicted by (13.135). One thus introduces in this equation a parameter specific to the body, the *emissivity* $\varepsilon(T,\nu)$, whose value ranges from 0 to 1 and which can depend on both temperature and frequency. In case the emissivity would depend only on temperature, we would have

$$W = \varepsilon(T)\,\frac{2\pi h}{c^2}\frac{\nu^3}{\exp(h\nu/kT)-1}. \tag{13.139}$$

The "color" temperature obtained from the intensity at a given frequency differs the most from the actual thermodynamic temperature when the emissivity is smallest. Hence measurement are often made at two different wavelengths to reduce the errors.

Problems

1. Calculate the de Broglie wavelengths of a riffle bullet of 10 g moving at 1000 m/s, and of a neutron of energy 0.04 Mev ("thermal" neutron at 300 K), and state whether quantum effects could be observed for either particle.

2. When the sunlight is focussed with a magnifier one observes that the temperature at the focus point increases when the diameter of this point decreases. Is there a limit to the temperature that can be reached when the diameter decreases? If so, what is this limit?

3. A vacuum acquires a mass when it gets hot and full of radiating energy. By combining Einstein's relation $E = mc^2$, where c is the velocity of light and the Stefan-Boltzmann law (13.129), calculate the density (mass of 1-cm^3) of a cavity as a function of temperature from 10^5 to 10^{11} K.

4. A particle can occupy only two energy levels of which one is doubly degenerate. Calculate its heat capacity as a function of temperature and compare the result with that obtained in 13.5d for the nondegenerate two-level case.

5. The partition function of one mole of a particle is represented by $Z = A + B/RT$. Calculate the molar internal energy, entropy and heat capacity of the particle as a function of temperature.

6. The sun may be considered as circular black body with a surface temperature of 5780 K. Its radius is 7 10^8m and average distance from the earth is 1.5 10^{11} m. What is the power received per m^2 at the earth's equator? By assuming the the earth is also a black body, which reemits all the energy it absorbs, calculate what would be its surface temperature.

7. At what temperature will the pressure of black body radiation be 1 bar?

8. Among the equations derived in (13.5e), which is that that represents the equation of state of a gas of photons? What is its particularity ? Calculate the isochoric and isobaric heat capacities, and the enthalpy, entropy and Gibbs free energy of a gas of photons. Why are C_p and G zero?

9. From the Planck distribution law for black-body radiation, derive the limiting laws of Wien, $E(\nu)\delta\nu/V = [8\pi h\nu^3/c^3]exp(-h\nu/kT)$, and of Rayleigh-Jeans, $E(\nu)\delta\nu/V = (8\pi\nu^2/c^3)kT$, which had been established earlier for high and low frequencies, respectively. Explain why extrapolation of the latter to high frequencies led to the *ultraviolet catastrophe*. Show that the Rayleigh-Jeans law can be derived from the density of states (13.131) and the assumption of equipartition of energy (i.e., black body radiation is made up of oscillators whose energy is equal to kT). Explain why this is incorrect.

10. When a gas is adsorbed on a surface, the fraction f of the surface occupied by the gas molecules increases with the gas pressure P and decreases when the temperature T increases. Experimentally, one often finds that $P = Cf/(1-f)$, a result known as Langmuir adsorption isotherm, where C is a temperature dependent parameter.
1. One introduces N molecules of an ideal gas in a box of volume V in which a surface offers B equivalent sites for the adsorption of M gas molecules. Show that the partition function of the box-surface system is $Z = [B!z_a^M/M!(B-M)!]\ [z_g^{N-M}/(N-M)!]$, where z_a and z_g are the molecular partition functions of the adsorbed and free gas molecules, respectively. (Recall that the only latter are distinguishable.)
2. The equilibrium value of M is that which maximizes Z. Show that it is given by $(B-M)z_a/M = z_g/(N-M)$.
3. Writing $z_g = z_g'V$, where z_g' depends on T but not on V, demonstrate from the ideal gas

equation of state the equilibrium expression $P = (kTz'_g/z_a)f/(1-f)$ of the Langmuir adsorption isotherm. What information on the adsorption mechanisms can be drawn from this equation?

4. Assume now that the adsorbed molecules are no longer fixed on the surface, but that they can move as a bidimensional gas. Show that the new partition function of the system is $Z = z_a^M z_g^{N-M}/M!(N-M)!$, and that the equilibrium pressure is given by $P = kTMz'_g/z'_aS$ where S is the available surface and z'_a a factor of z_a that depends only on T.

5. To derive an expression analogous to the Langmuir isotherm let us arbitrarily set $f = aM/S$, with a = proportionality constant. Show that f is related to the pressure by $f = aP^n z'_a/kTz'_g$. Experimentally, this equation is often followed when the original Langmuir isotherm is not obeyed. Why?

14

THEORETICAL CALCULATIONS OF THERMODYNAMIC PROPERTIES

In this chapter calculations of thermodynamic functions will illustrate the practical usefulness of statistical mechanics. The partition and thermodynamic functions of translators, rotators and harmonic oscillators will be first evaluated. Assuming gaseous molecules to be rigid rotators and harmonic oscillators, we will apply these results to diatomic and polyatomic ideal gases for which the relevant molecular constants are obtained from spectroscopic data. Anharmonicity, due to the asymmetry of interatomic potentials, will be the main correction to the models that will be presented. The case of crystals will then be discussed with Einstein and Debye models and with lattice dynamics. We will also describe how the static and dynamic properties of systems made up of a few thousand atoms are simulated with molecular dynamics and Monte-Carlo methods. We will finally mention how, through resolution of the Schrödinger equation, the properties of simple systems can be evaluated in a purely theoretical manner.

14.1 ENERGIES AND PARTITION FUNCTIONS

14.1a PREAMBLE

Calculation of thermodynamic functions of ideal gases has long been one of the most important practical applications of statistical mechanics. For not too complicated molecules, this is the procedure used for determining these functions in the standard state. For instance, the heat capacities plotted in Fig. 4.3 have been obtained in this way. In real gases molecular interactions are not so easily accounted for. This is not a practical problem because knowledge of the equation of state allows the heat capacity and entropy of a real gas to be derived from those of the ideal gas with the aid of relations such as (5.31). Of particular importance in this semi-empirical procedure is that the statistical mechanical calculations yield the absolute entropy of gases. This is especially valuable since an experimental determination of this entropy could otherwise be problematic, as described in 9.3f for steam from a long entropy cycle that begins at 0 K with a crystal, ice, which does not follow Nernst's law.

We will thus focus in this chapter on the calculation of partition and thermodynamic functions, denoting by z the partition function of a single particle, and by Z that of a system of N particles. Along with gases, we will mainly deal with crystals and leave liquids out as there is currently no means for performing such calculations for the latter phases with an accuracy comparable to that of experimental determinations. The difficulty is that even for water, by far the most extensively studied substance, modeling the effects of configurational changes in the liquid state is still a formidable challenge. It is in fact through atomistic simulations, which will be described at the end of this chapter, that this challenge is being met.

For ideal gases, the simplest approximation considers the molecules as rigid rotators and harmonic oscillators in free translation. Although the two assumptions are of course inconsistent, for a vibrating molecule cannot be rigid, they work fairly well so that we will begin with the statistical mechanics of single translators, rotators and harmonic oscillators. The corrections to be made to these models when dealing with real molecules will be described before treating the case of crystalline insulators for which only vibrational energy has to be considered. As calculations of wave functions and eigenvalues are beyond the scope of this book, we refer the reader interested in these topics to standard textbooks of quantum mechanics.

For rotations and vibrations, frequencies (ν) are the relevant input information. They are usually expressed in the form of wavenumbers, $\omega = \nu/c$ (c = speed of light) in units of cm^{-1}, although wavenumbers and frequencies are often used synonymously. More important, frequencies are directly related to internuclear forces and distances. In passing, we will thus emphasize that the geometry and internuclear forces of gaseous molecules are known thanks the energy changes that manifest themselves in rotation and vibration spectra. We finally note that 3 coordinates are used to describe the translations of a molecule as a whole, in the three directions of space, with respect to the center of mass. For a nonlinear molecule three others characterize the rotations, whereas this number reduces to only two for a linear molecule because rotations around the molecule axis do not give rise to distinguishable positions in space. Hence there remain $3N-5$ and $3N-6$ vibrational degrees of freedom for linear and nonlinear molecules, respectively, for a total number of N atoms. A macroscopic crystal has also 3 translational and rotational degrees of freedom and $3N-6$ vibrational of freedom, but the number N of atoms is then so large as to make translations and rotations irrelevant.

Rotation frequencies are determined from spectroscopic measurements made at far infrared or microwave wavelengths. With fundamental frequencies ranging from a few ten to a few thousand cm^{-1}, vibrational transitions are observed in the infrared mostly by infrared and Raman spectroscopies (Fig. 14.1). With the former, the investigated substance is

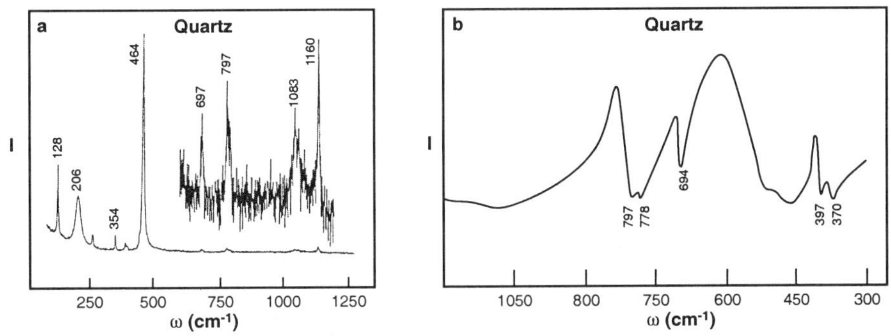

Figure 14.1. Raman (a) and infrared (b) spectra of quartz, indicating the wavenumbers of the observed modes (Sato and McMillan, 1987). Remark the very low intensity of the high-frequency Raman bands above the background.

subjected to a polychromatic infrared light source. The vibrational frequencies correspond to maxima in the absorption spectrum; the observable (*active*) vibrational modes are those which induce changes in the dipolar moment. In Raman spectroscopy, the active modes are those modes which induce changes in polarizability. A monochromatic visible (laser) light source is used and one records the frequency changes of this radiation caused by the absorption or the emission of photons by the oscillators of the substance. With respect to the frequency of the exciting radiation, a Raman spectrum thus consists of two symmetric series of bands called Stokes and anti-Stokes on the low and high-frequency sides, respectively. Because of the Boltzmann factor, however, more photons are absorbed than emitted so that the Stokes bands are more intense than their anti-Stockes counterparts. This is why one generally records only the former. Note that a variety of modes induce no change in either polarisability or dipolar moment: they are Raman and infrared *inactive*.

14.1b TRANSLATIONS

We have found in 13.2d that the kinetic energy of a translator moving along the x direction in a cubic box of size a is

$$E^{tr} = n_x^2 \frac{h^2}{8ma^2}, \tag{14.1}$$

where n_x is a quantum translation number. The energy is quantized but, owing to the smallness of $h = 6.626\ 10^{-34}$ J s, quantization is in fact imperceptible. The translation quantum would be lower than 10^{-25} J for a hydrogen atom moving in a 1Å-box and would become significant only for values of a of the order of 10^{-25} m, which are so small with respect to the size of atoms that they are irrelevant.

For a single degree of freedom, the partition function of a translator is

$$z^{tr} = \sum \exp(-E_i^{tr}/kT). \tag{14.2}$$

For any practical purpose, the translational energy levels cannot be distinguished from the continuous values of classical mechanics. Replacing the summation in (14.2) by an integral we write

$$z^{tr} = \int_0^\infty \exp\left[-n_x^2 \frac{h^2}{8ma^2}\right] dn_x. \tag{14.3}$$

Between 0 and ∞ the integral of $\exp(-Ax^2)$ is $1/2\ (\pi/A)^{1/2}$. Hence the translation partition function is

$$z^{tr} = \frac{a}{h}(2\pi mkT)^{1/2}. \tag{14.4}$$

From (13.81) and this expression and the internal energy and isochoric heat capacity of a translator are readily obtained as

$$U^{tr} = kT^2(d\ln z^{tr}/dT) = kT/2, \tag{14.5}$$

$$C_v^{tr} = (\partial U^{tr}/\partial T)_V = k/2. \tag{14.6}$$

The heat capacity is thus independent of temperature. For one mole of translators we find $C_v^{tr} = R/2$.

14.1c RIGID ROTATOR

The simplest case to be dealt with here is that of linear molecules with a moment of inertia I_e. Rotations take place around only two directions of space orthogonal to the molecule axis at the center of mass of the molecule. The problem reduces to that of a point mass m at a distance r_e from the center of rotation having the same moment of inertia I_e as the molecule. For the wavefunction to be finite, continuous and single-valued, one finds that the energy must be quantized

$$E^{rot} = J(J+1)\frac{h^2}{8\pi^2 I_e} = J(J+1)hcB_e, \tag{14.7}$$

where $B_e = h/8\pi^2 cI$ is the *rotational constant* and J, the *rotational quantum number*, takes integer values. As noted below $2J+1$ different wave functions have the same energy. The energy is therefore $2J+1$ degenerate.

Taking into account the degeneracy, we write for the partition function of a rotator

$$z^{rot} = \sum_J (2J+1)\exp\left[-\frac{J(J+1)B_e hc}{kT}\right]. \tag{14.8}$$

For low temperatures and large values of B_e (i.e., small mass and short rotator length), the rapid convergence of z^{rot} allows it to be calculated by direct summation. At 100 K, for instance, (14.8) is expanded as $1+0.8+9\ 10^{-2}+2\ 10^{-3}+10^{-5}$ for $B_e = 45.66$ cm^{-1} like for the HD molecule (D = deuterium). For high temperatures and low values of B_e, convergence is in contrast very slow. For $B_e = 0.3\ \text{cm}^{-1}$ like for Cl_2, the 180th term of the summation at 1000 K is still equal to $3\ 10^{-4}$. In this case the sum in (14.8) can be replaced by an integral. Since $2J+1 = d[J(J+1)]/dJ$, we find immediately

$$z^{rot} = kT/hcB_e. \tag{14.9}$$

This is the classical value of the rotational partition function. In between these two cases, one obtains a series expansion

$$z^{rot} = 1/\sigma(1+\sigma/3+\sigma^2/15+4\sigma^3/315+\ldots), \tag{14.10}$$

where $\sigma = hcB_e/kT$.

The partition functions (14.9) and (14.10) accounts for two rotational degrees of freedom. For only one degree, the high-temperature partition function would therefore be

$$z^{rot} = (kT/hcB_e)^{1/2}. \tag{14.11}$$

The internal energy is

$$U^{rot} = kT^2(d\ln z^{rot}/dT) = kT/2. \tag{14.12}$$

It follows that the heat capacity is constant

$$C_v^{rot} = (\partial U^{rot}/\partial T)_V = k/2, \tag{14.13}$$

and equal to that given by (14.6) for one translational degree of freedom.

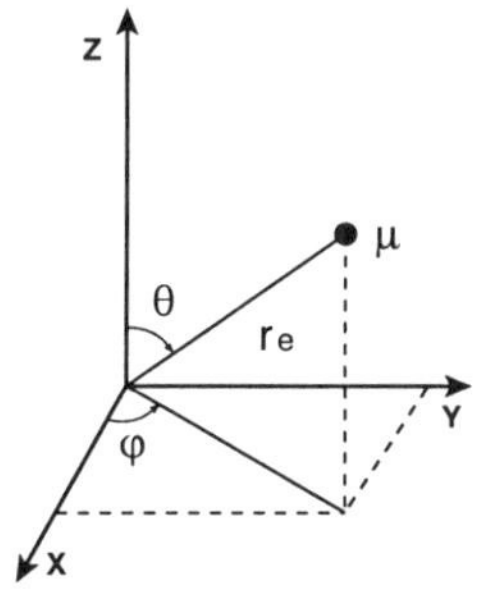

Figure 14.2. Spherical coordinates for the rigid rotator.

The rotational frequency $\nu = cB_e$ is a purely quantum mechanical feature. In classical mechanics the energy of a rotator varies continuously with the angular velocity ω

$$E^{rot} = 1/2\, I_e\omega^2 = P^2/2I_e, \tag{14.14}$$

so that the angular momentum P is $h/2\pi\, [J(J+1)]^{1/2} \sim h/2\pi\, J$. The integer J is thus an approximate measure of the angular momentum expressed in units $h/2\pi$. In quantum mechanics it is with the set of spherical coordinates ϕ, θ and r_e (Fig. 14.2) that the Schrödinger equation is most conveniently solved. Since r_e is constant, the wave function is the product of two terms depending on the angles ϕ and θ respectively

$$\Phi^{rot} = \Phi(\phi)\Theta(\theta) = A\, e^{im\phi}\, \Theta(J,m) \tag{14.15}$$

where $\Theta(J,m)$ is a Legendre polynomial and m is another integer. Although Φ^{rot} has actually a very complicated expression, the important point is that we do not need to know it. If suffices to know that Φ^{rot} is defined only if $8\pi^2 I_e/h^2 = J(J+1)$ and that m takes on values from J to $-J$ and represents, in units $h/2\pi$, the component of J along the axis of the rotator. This is why $2J+1$ rotational states have the same energy.

14.1d HARMONIC OSCILLATOR

Let r_e be the equilibrium length of an oscillator and $x = r - r_e$ its length change while vibrating. If the force exerted is proportional to r, then the potential V is said to be *harmonic* and can be written

$$V = 1/2\ k_f(r - r_e)^2 = 1/2\ k_f x^2, \tag{14.16}$$

where k_f is the force constant (Fig. 14.3). Although interatomic potentials are actually not harmonic for large vibrational amplitudes, they can be generally considered so in the vicinity of their minimum. In classical mechanics the problem is readily solved and one finds that the length of the oscillator varies as a function of time t as

$$x = x_0 \cos(2\pi\nu t + \varepsilon), \tag{14.17}$$

where the vibrational frequency is related to the force constant and mass m of the oscillator by

$$\nu = \frac{1}{2\pi}\left(\frac{k_f}{m}\right)^{1/2}. \tag{14.18}$$

Like the amplitude of vibration x_m, the vibrational energy $E^{vib} = 1/2\ k_f x_m^2$ thus varies continuously. As sketched in 14.3c, evaluation of the classical partition function yields a constant value k for the isochoric heat capacity of one oscillator, and thus R for one mole of oscillators, regardless of the value of the force constant.

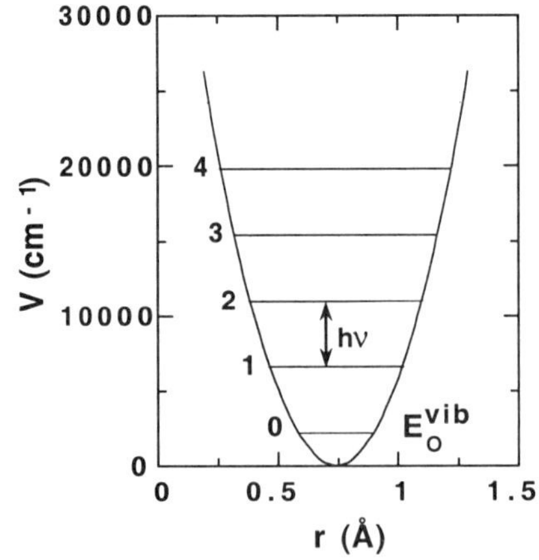

Figure 14.3. Harmonic potential and even spacing of vibrational energy levels for the ground state of H_2 (r_e = 0.7416 Å).

In quantum mechanics, solving the Schrödinger equation for an harmonic oscillator is much more tedious. For our purpose the essential feature is that the vibrational energy has the simple expression

$$E^{vib} = (n+1/2)h\nu = (n+1/2)hc\omega. \qquad (14.19)$$

The vibrational energy levels are thus equally spaced by an amount $h\nu$ (Fig. 14.3). Of special interest is the fact that the vibrational energy is not zero when $n = 0$. The existence of this *zero-point energy* $E_0^{vib} = h\nu/2$ indicates that a quantum oscillator is never at rest otherwise Heisenberg's principle (13.24) would be violated in that one would know without any uncertainty both the position (as determined by the bottom of the potential energy well) and the (then zero) momentum of the oscillator.

In quantum mechanics the classical formula (14.18) still relates the harmonic vibrational frequency ν to the force constant k_f, but the wave function is given by the complicated expression

$$\Psi^{vib}(x) = \left[\left(\frac{\alpha}{\pi}\right)^{1/2}\frac{1}{2^n n!}\right]^{1/2}\exp(-\frac{\alpha x^2}{2})H_n(\alpha^{1/2}x), \qquad (14.20)$$

where $\alpha = 4\pi^2 m\nu/h$ and H_n is a Hermite polynomial of degree n which, for the variable x, satisfies the recurrence relation

$$H_{n+1}(x) - 2xH_n(x) + 2nH_{n-1}(x) = 0, \qquad (14.21)$$

with $H_0 = 1$ and $H_1 = 2\alpha^{1/2}x$. We again find a marked contrast between the extremely simple expression of the energy, in which we are interested, and the very complicated form of the wave function. Since we are not concerned by the variations of Ψ^{vib} with the quantum vibrational number n, we will just note that the probability for the oscillator to have a certain length decreases when the vibration amplitude increases, like in classical mechanics, but that it is zero only for an infinite length.

The other important feature is that the quantization of vibrational energy levels can never be neglected. In the vibration partition function

$$z^{vib} = \sum \exp[-(n+1/2)h\nu/kT], \qquad (14.22)$$

the summation has thus to be performed over the discrete energy levels. The calculation is in fact easy because z^{vib} is a geometrical series with a ratio $\exp(-u)$, where u stands for $h\nu/kT$, whose sum is

$$z^{vib} = \frac{e^{-u/2}}{1-e^{-u}}. \qquad (14.23)$$

In this expression the numerator $e^{-u/2}$ represent the contribution of zero-point energy whereas the denominator $(1-e^{-u})$ accounts for the excited vibrational energy levels. Only the latter contributes to the heat capacity. For a single oscillator, the internal energy is

$$U^{vib} = kT^2\frac{d\ln z^{vib}}{dT} = \left[\frac{h\nu}{2} + \frac{h\nu e^{-u}}{1-e^{-u}}\right] = kT\left[\frac{u}{2} + \frac{u}{e^u - 1}\right]. \qquad (14.24)$$

Differentiation of this expression yields the heat capacity

$$C_v^{vib} = \left(\frac{\partial U^{vib}}{\partial T}\right)_V = \frac{h\nu(h\nu/kT^2)e^{-u}}{(1-e^{-u})^2} = \frac{ku^2e^u}{(e^u-1)^2}. \tag{14.25}$$

In contrast to the translational and rotational heat capacities, the vibrational C_v is not constant. Tending to 0 toward 0 K, it follows a familiar s-shaped curve before reaching the high-temperature limit of k for one oscillator. Calculation of the entropy and Helmholtz free energy is also straightforward

$$S^{vib} = k\ln z^{vib} + kT\frac{d\ln z^{vib}}{dT} = k\left[\frac{u}{e^u-1} - \ln(1-e^{-u})\right], \tag{14.26}$$

$$F^{vib} = -kT\ln z^{vib} = kT\left[\frac{u}{2} + \ln(1-e^{-u})\right]. \tag{14.27}$$

14.2 IDEAL GASES

14.2a INTERNAL AND ELECTRONIC DEGREES OF FREEDOM

For calculating thermodynamic functions of ideal gases the starting point is the split of molecular energy into translational and internal energies, $E = E^{tr} + E^{int}$, which is associated with the factorization of the corresponding wave functions and partition functions. Thanks to the Born-Oppenheimer approximation (Appendix E), the electronic and rotation-vibration energies are separated further. To each electronic state is associated a vibrational potential and specific equilibrium interatomic distances, and thus a rotation-vibration partition function z^{rv}. For many molecules it happens that the first excited electronic state is at least 10 000 cm^{-1} above the ground state. This energy is so high that the Boltzmann factor $\exp(-hcE^{el}/kT)$ is negligibly small at all but very high temperatures, where the molecule would no longer be stable anyway. Practically, the actual sum of states over the various electronic energy levels $E_0^{el}(=0), E_1^{el}$, etc.

$$z = \sum\exp(-E_{i0}^{rv}/kT) + \exp(-E_1^{el}/kT)\sum\exp(-E_{i1}^{rv}/kT) + ... \tag{14.28}$$

then reduces to the first term. If not, one has just to calculate the partition function for all electronic states whose contribution to the internal partition function is not negligible. A well-known example is O_2 whose first two electronic states have the rather low energies $E^{el} = 7918$ and 13 195 cm^{-1}. At 1600 K, $\exp(-E^{el}hc/kT) =$ 8 10^{-4} and 5 10^{-6}, respectively. Even in this case only the small contribution of the first excited state thus needs to be taken into account at this high temperature.

14.2b MONOATOMIC GASES

In addition to its electronic energy, a monoatomic gas has only translational energy. Because translations along the three directions of space x, y and z are mutually independent, so are the three components of the translational kinetic energy (see 13.2)

$$E^{tr} = E_x^{tr} + E_y^{tr} + E_z^{tr}. \tag{14.29}$$

For one degree of freedom, the translational partition function is given by (14.4). For a single gas molecule in a cubic box of volume $V = a^3$ the total translation partition function is thus

$$z^{tr} = z_x^{tr} z_y^{tr} z_z^{tr} = V \left(\frac{2\pi mkT}{h^2} \right)^{3/2}. \tag{14.30}$$

Because the atoms of the gas are indistinguishable, from (13.92) the partition function Z of N atoms thus includes a factor $1/N!$

$$Z = \frac{z^N}{N!} = \frac{1}{N!} V^N \left(\frac{2\pi mkT}{h^2} \right)^{3N/2}. \tag{14.31}$$

For one mole of ideal gas of molecular weight $M = mN_A$, where N_A is the Avogadro number, we find through standard calculation and use of the Stirling approximation that

$$U = kT^2 (\partial T^{3N/2} / \partial T)_{V,N} = 3/2\, RT. \tag{14.32}$$

The enthalpy and heat capacities of an ideal monoatomic gas follow immediately

$$H = U + PV = 5/2\, RT, \tag{14.33}$$

$$C_v = 3/2\, R, \tag{14.34}$$

$$C_p = 5/2\, R. \tag{14.35}$$

As shown in Fig. 4.3 for the isobaric heat capacity, noble gases are of course the most important substances to which these equations apply.

We finally derive the Helmholtz and Gibbs free energies along with the entropy

$$F = -kT \ln Z = -nRT \left[\ln \left(\frac{V M^{3/2} T^{3/2}}{n} \right) + 1 + \ln \left(\frac{(2\pi k)^{3/2}}{h^3 N_A^{5/2}} \right) \right], \tag{14.36}$$

$$G = F + PV = -nRT \left[\ln \left(\frac{V M^{3/2} T^{5/2}}{n} \right) + \ln \left(\frac{(2\pi k)^{3/2}}{h^3 N_A^{5/2}} \right) \right], \tag{14.37}$$

$$S = \frac{H - G}{T} = nR \left[\ln \left(\frac{M^{3/2} T^{5/2}}{P} \right) + \frac{5}{2} + \ln \left(\frac{2\pi}{N_A} \right)^{3/2} \frac{k^{5/2}}{h^3} \right]. \tag{14.38}$$

Only the entropy equation warrants a specific comment because it diverges to $-\infty$ instead of tending to 0 when T tends to 0. The reason is that Boltzmann statistics breaks down at very low temperatures, see 13.6d. Since gases are not stable phases in the vicinity of 0 K, however, this divergence does not prevent their absolute entropy from being calculated in the temperature intervals of practical interest.

14.2c DIATOMIC GASES

The vibrational and rotational energies must now be considered in addition to translational and electronic energies. In a diatomic molecule, the chemical bond acts as a spring exerting a force whose direction is opposite to that of the atomic displacements from the equilibrium position r_e. In the harmonic approximation this force $F = -k_f(r - r_e)$ is thus a direct measure of the bond strength. For both vibrational and rotational degrees of freedom, a molecule made up of two atoms of mass m_1 and m_2 is equivalent to an oscillator and a rotator having a *reduced mass*

$$\mu = \frac{m_1 m_2}{m_1 + m_2}, \tag{14.39}$$

a vibrational frequency

$$\nu = \frac{1}{2\pi}\left(\frac{k_f}{\mu}\right)^{1/2}, \tag{14.40}$$

and a moment of inertia

$$I = \frac{m_1 m_2}{m_1 + m_2} r_e^2 = \mu r_e^2. \tag{14.41}$$

Assuming as a further approximation that the vibrations and rotations are mutually independent, we obtain directly the internal partition function of one molecule from (14.9) and (14.23). Before doing so, however, we need to take into account the fact that a π rotation of a linear, symmetric molecule around its center of mass yields a quantum state that cannot be distinguished from the starting one. In this case the rotational partition function must be divided by 2 so that another parameter, the *symmetry number s* is introduced in the expression of the rotation partition function. We thus write for the total partition function

$$Z = \frac{1}{N!} z^{tr} z^{rot} z^{vib} = \frac{1}{N!} V \left(\frac{2\pi mkT}{h^2}\right)^{3/2} \frac{1}{s} \frac{kT}{hcB_e} \frac{e^{-u/2}}{1 - e^{-u}}. \tag{14.42}$$

For *homonuclear* molecules like H_2 ou O_2, $s = 2$; for *heteronuclear* molecules like HD or CO it is unity.

Of course (14.9) is valid only when rotations conform to the high-temperature limit, i.e., when the spacing of rotational energy levels is negligible. Except for light H-bearing molecules, this is always the case above room temperature. The reader may care to repeat the calculations made in the previous section for a monoatomic gas and derive analytically the thermodynamic functions of a diatomic gas. We will just give the isochoric heat capacity as determined by translations, rotations and harmonic vibration

$$C_v = \frac{5}{2} R + \frac{ku^2 e^u}{(e^u - 1)^2}. \tag{14.43}$$

The most important correction to be made to the rigid rotator-harmonic oscillator model concerns anharmonicity. Instead of being constant, the spectroscopically observed spacing of vibrational energy decreases when the vibrational quantum number increases. This reveals that the interatomic potential has an asymmetric shape. That atoms are bonded is of course direct evidence for attractive forces between them, but the equilibrium bond length is determined by a balance with the mutual repulsion exerted by nonbinding inner electrons. On one side, the repulsion becomes much stronger than indicated by an harmonic potential when the atoms are getting closer. On the other side, instead of increasing continuously the potential tends to the finite value D_0 of the dissociation energy when the atoms are getting farther apart (Fig. 14.4). Although even better agreement with spectroscopic observations is achieved when the potential energy is expanded in powers of $(r - r_e)$, a simple but realistic model is the Morse potential (Fig. 14.4)

$$V = D\{1 - \exp[-a(r - r_e)]\}. \tag{14.44}$$

which has only three adjustable parameters, D, a et r_e.

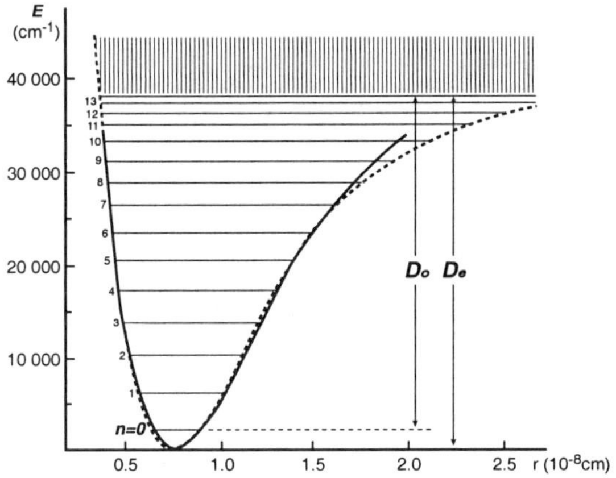

Figure 14.4. Comparison between harmonic and Morse potentials for the electronic groundstate of H_2. The energies of dissociation from the bottom of the potential energy well and from the zero-point energy are also indicated. Redrawn from the book by Herzberg (1950).

Through resolution of the Schrödinger equation, one finds that the energy levels of a rotating Morse oscillator are given by

$$\begin{aligned} E(J,n)/hc = \omega_e(n+1/2) &- \omega_e x_e(n+1/2)^2 + B_e J(J+1) \\ &- \alpha_e(n+1/2)J(J+1) - D_e J^2(J+1)^2, \end{aligned} \tag{14.45}$$

where the variations of the moment of inertia of the molecule due to vibrations and the centrifugal distortion due to rotation are dealt with by the terms involving the molecular constants α_e and D_e, respectively. These constants are related to the parameters of (14.44) by $\omega_e = a/2\pi\ (2D/\mu)^{1/2}$, $\omega_e x_e = hc\omega_e/4D$, $\alpha_e = 6(\omega_e x_e B_e^3)^{1/2}/\omega_e - 6B_e^2/\omega_e$ and $D_e = 4B_e^3/\omega_e^2$. To compare their importance, we list them in Table 14.1 for the electronic ground states of H_2, O_2 and Cl_2.

Appropriate summations or integrations of the extra terms of (14.45), after factorization of $z^{vib}$and $z^{rot}$for the rigid rotator-harmonic oscillator, yield other partition functions accounting for anharmonicity of the vibrations (z^{anh}), rotation-vibration interactions (z^{rvi}) and centrifugal distortion (z^{cd}), viz.

$$z^{anh} = \exp\left(\frac{-hc\omega_e x_e}{4kT}\right)\frac{1 - 2x_e u e^{-u}}{(1 - e^{-u})^2}, \tag{14.46}$$

$$z^{rvi} = 1 + \frac{\alpha_e/B_e}{e^u - 1}, \tag{14.47}$$

$$z^{cd} = 1 + \frac{hcD}{kT\sigma^2} + 12\left(\frac{hcD}{kT\sigma^2}\right)^2 + \ldots \tag{14.48}$$

Table 14.1. Molecular constants of H_2, O_2 and Cl_2 (cm^{-1})[a]

	ω_e	$\omega_e x_e$	B_e	α_e	D_e
H_2	4401.12	121.28	60.8380	3.0258	0.0463
O_2	1580.36	12.07	1.4456	0.1593	$5.\ 10^{-6}$
Cl_2	559.72	2.67	0.2439	0.0015	

[a] Data of Buijs and Gush (1971), Babcock and Herzberg (1948) and Clyne and Coxon (1970). The third vibrational term (yielded by a series expansion of the potential) is $\omega_e y_e (n + 1/2)^3$, with $\omega_e y_e = 0.804$ and $0.047\ cm^{-1}$ for H_2 and O_2, respectively.

The reader may care to compare their importance for the molecules of Table 14.1 and check that the effects of rotation-vibration and centrifugal distortion are no longer negligible at high temperatures. They remain small compared to those of anharmonicity, however, which represent themselves minor corrections to the partition function of the rigid rotator-harmonic oscillator. Note finally that the contribution of nonzero nuclear spins to partition functions should in principle be taken into account. In practice it can be neglected because the energy differences between the various orientation states of nuclear spins are extremely small. In addition, chemical reaction leaves nuclear spins unchanged so that the entropy that results from the degeneracy of the nuclear spin wave function exactly cancels out whenever differences in thermodynamic quantities are calculated.

14.2d POLYATOMIC GASES

Polyatomic molecules will be treated in less detail. Although they show qualitatively the same features as diatomic molecules, the vibrational modes and the expressions of the internal energy and partition functions become increasingly complex as the number of atoms increases.

For rotations, the simplest case is that of the two degrees of freedom of linear molecules, which are treated in the same way as for diatomic molecules. The symmetry number is 1 for nonsymmetric molecules like HCN and N_2O (N-N-O), and 2 for symmetric ones like CO_2 (O-C-O). For nonlinear molecules, any rotation can be expressed as the combination of rotations around the three principal axes of inertia which are defined as follows (Fig. 14.5). For a molecule fixed in a given orientation, the moment of inertia varies with direction and can be represented by an ellipsoid cen-

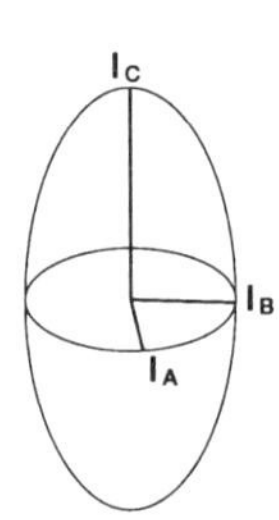

Figure 14.5. Ellipsoid of inertia moment for an asymmetric top ($I_A \neq I_B \neq I_C$). For a symmetric top, $I_B = I_C$, whereas $I_A = I_B = I_C$ for a spherical top.

tered on the center of mass. The minimum and maximum values define two principal axes and moments of inertia (I_A and I_C) whereas the third moment (I_B) is that on the axis of the ellipsoid orthogonal to the first two. With the notation $\sigma_i = h^2/8\pi^2 kTI_i$, one demonstrates that the rotation partition function is

$$z^{rot} = \frac{1}{s}\left(\frac{\pi}{\sigma_A\sigma_B\sigma_C}\right)^{1/2} \left[1 + \frac{1}{12}\left(2\sigma_A + 2\sigma_B + 2\sigma_C - \frac{\sigma_A\sigma_B}{\sigma_C} - \frac{\sigma_A\sigma_C}{\sigma_B} - \frac{\sigma_B\sigma_C}{\sigma_A}\right) + \ldots\right], \quad (14.49)$$

where the symmetry number s is again introduced. For symmetric bended molecules like H_2O and SO_2, s = 2, whereas $s = 3$ for pyramidal NH_3, 12 for tetrahedral CH_4, etc. Except for light species, the term in brackets reduces to 1 above 300 K.

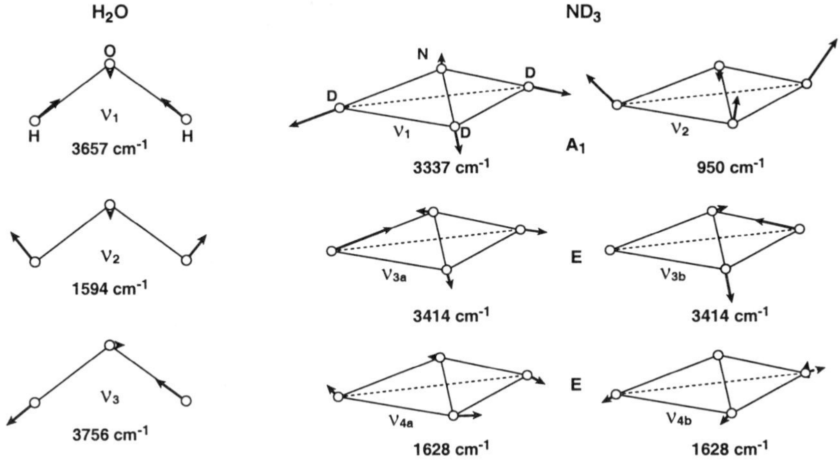

Figure 14.6. Normal modes of vibration of water and ammonia. Note the two doubly degenerate vibrations of ammonia. The arrows represent the maximum atomic displacements for the various modes. To make them more visible for nitrogen in ammonia, we have in fact shown the normal modes of the "heavy" species ND_3 (D = deuterium). From the book by Herzberg, 1945.

For polyatomic molecules the actual vibrational motions may be extremely complex. Fortunately, in the harmonic approximation it is possible to express any vibration as the sum of $3N - 5$ (for linear molecules) or $3N - 6$ (for nonlinear molecules) independent vibrational modes (see Appendix E). In each mode all atoms vibrate in phase around their equilibrium position with the same frequency but with an amplitude that decreases with the mass of the nucleus. The atomic displacements involved in the *normal modes* of vibration of water and ammonia are shown in Fig. 14.6. Vibrational frequencies may be calculated once a force field has been set up. This can be made in different ways.

The simplest interatomic potentials rely on the assumption of *central* forces, namely, of pairwise interactions between the atoms of the molecule. In the harmonic approximation, we thus write

$$V = 1/2 \sum_{i,j} k_{ij} r_{ij}^2 \tag{14.50}$$

where k_{ij} and r_{ij} are the force constant and distance between the atoms i and j. Such a potential is approximate because it does not take into account the actual bonds existing within the molecule. A better potential is thus provided by *valence* force fields, where one considers variations of potential energy arising from changes in bond lengths (d_i) and angles (a_j)

$$V = \sum k_i d_i^2 + \sum k_j a_j^2. \tag{14.51}$$

With a given potential and set of coordinates q_i one looks for solutions of the form $q_i = A_i \cos(2\pi\nu_i t + \varepsilon)$ when solving the equations of motion. For the vibrating atoms the potential and kinetic energies are written in the generalized Lagrangian form (Appendix E). These equations are considerably simpler when the coordinates used take into account the symmetry of the molecule. The adequate set of *symmetry coordinates* is thus determined from a group theory analysis of the symmetry of the molecule, which also yields the kind of atomic displacement involved in the various normal modes of vibration (Fig. 14.6). By solving the equation of motion, one then finds the frequencies of the normal modes of vibration and the actual vibrational amplitudes. The frequencies of the $3N-5$ or $3N-6$ normal modes of vibration are thus calculated within the framework of classical mechanics.

In quantum mechanics, the mutual independence of the normal modes implies that the wave equations and quantized energies of the modes are also independent so that the partition function (14.23) remains valid. That the Schrödinger equation of a polyatomic molecule is generally too complex to be solved therefore does not prevent the vibrational partition function from being evaluated. In the harmonic approximation, it suffices to write the product, over the $3N-5$ or $3N-6$ modes, of the partition functions of all modes

$$z^{vib} = \prod \frac{e^{-u_i/2}}{1-e^{-u_i}}. \tag{14.52}$$

For an asymmetric molecule, the total partition function is obtained from (14.4), (14.49) and (14.52). If only the first term of (14.49) is significant, we have

$$z = z^{tr} z^{rot} z^{vib} = V\left(\frac{2\pi mkT}{h^2}\right)^{3/2} \frac{1}{s}\left(\frac{\pi}{\sigma_A \sigma_B \sigma_C}\right)^{1/2} \prod \frac{e^{-u_i/2}}{1-e^{-u_i}}. \tag{14.53}$$

Several corrections must be made to reproduce the fine features of rotation-vibration spectra. In calculations of thermodynamic functions the smallness of their effects will be illustrated by steam. As mentioned in 9.3f about the residual entropy of ice, the calculated entropy of water vapor at 298 K used in 1936 was 188.66 J/mol K, which is in excellent agreement with the modern value of 188.84 J/mol K derived from a considerably more comprehensive analysis. These calculations are thus especially useful as they are not too sensitive to the actual energy levels of a molecule.

As for diatomic molecules, the corrections to be made to the harmonic oscillator-rigid rotator model concern anharmonicity, rotation-vibration interactions and centrifugal distortion. They are more complex because these features induce coupling between the vibrational modes, which are no longer mutually independent. As they do not play a major role, we will briefly introduce here only the most significant one, which deals with anharmonicity. To account for the coupling between the vibrational modes caused by anharmonicity, one expands the internuclear potential in a Taylor series of the relevant symmetry coordinates s_i, viz.

$$\begin{aligned} V &= V_0 + \sum_{i=1}^{3N-6} \left(\frac{\partial V}{\partial s_i}\right)_0 s_i + \frac{1}{2!} \sum_{i,j=1}^{3N-6} \left(\frac{\partial^2 V}{\partial s_i \partial s_j}\right)_0 s_i s_j \\ &+ \frac{1}{3!} \sum_{i,j,k=1}^{3N-6} \left(\frac{\partial^3 V}{\partial s_i \partial s_j \partial s_k}\right)_0 s_i s_j s_k + \ldots = V_0 + k_{ij} s_i s_j + k_{ijk} s_i s_j s_k + \ldots \end{aligned} \tag{14.54}$$

where k_{ij} and k_{ijk} are the harmonic and anharmonic force constants, respectively. The energy levels are found to be

$$\begin{aligned} E^{vib}(n_1, n_2 \ldots)/hc &= \sum_i \omega_i (n_i + 1/2) + \sum_{i \leq j} x_{ij} (n_i + 1/2)(n_j + 1/2) \\ &+ \sum_{i \leq j \leq k} x_{ijk} (n_i + 1/2)(n_j + 1/2)(n_k + 1/2) + \ldots \end{aligned} \tag{14.55}$$

From this expression one calculates an anharmonic partition function which involves the anharmonic constants x_{ij} and x_{ijk} and resembles the expression (14.46) derived for diatomic molecules.

14.3 CRYSTALS

14.3a VIBRATIONAL DENSITY OF STATES

The energy of a crystal is the sum of its energy at 0 K, the *lattice energy*, and of its thermal energy. In conductors, such as metals, electrons appreciably contribute to thermodynamic properties but the calculation of these effects within the framework of Fermi-Dirac statistics is beyond the scope of this book. We will thus restrict this discussion to insulators for which only internal vibrational energy must be taken into account. For the heat capacity, we write

$$C_v = (\partial U/\partial T)_V = (\partial U^{vib}/\partial T)_V. \tag{14.56}$$

In a crystal, atoms do not vibrate individually, but collectively through particle-like excitations called *phonons*, which have a quantized energy. A crystal with N atoms in its unit cell has $3N$ vibrational modes of which 3 are acoustic and $3N-3$ optic (Fig. 14.7). In acoustic modes (see Appendix C), the atoms within a single unit cell are displaced in the same direction, which is that of the wave vector in the compressional (or longitudinal) mode, or is perpendicular to that of the wave vector in the two shear (or transverse) modes. In the optic modes, the atoms of a given unit cell vibrate along different directions. For any mode, the periodicity of the crystal lattice implies that the wave vector varies within definite bounds. When the wave vector is zero, all unit cells of the crystal vibrate in phase. When the wave vector is maximum, analogous atoms in contiguous unit cells vibrate out of phase. The variation of the frequency with the wave vector constitutes the dispersion relation of a vibrational mode.

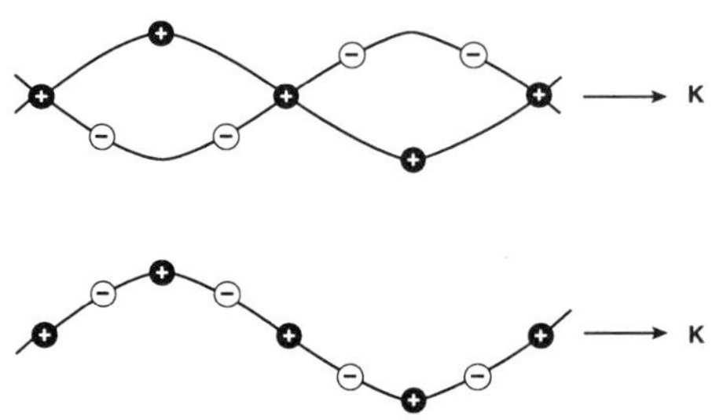

Figure 14.7. Acoustic (bottom) and optic (top) modes for a linear chain of atoms (from the book by Kittel, 1996).

In the harmonic approximation, the lattice energy of the crystal, U_0, is constant at constant volume. Regardless of its acoustic or optic nature, each mode contributes to thermodynamic functions as described in 14.1d. For the ith mode, the heat capacity thus is

$$C_{vi} = 3N\frac{ku^2e^u}{(e^u-1)^2}. \tag{14.57}$$

As already seen in 13.3c we need to know a density of states to calculate thermodynamic functions. The vibrational density of states, $g(\omega)$, which is relevant here, gives the number of modes with a wave number lying between ω and $\omega + d\omega$, from $\omega = 0$ to the highest wave number ω_m. It would be readily calculated if dispersion relations were known for all modes as one would obtain C_v through:

$$C_v = N\int_0^{\omega_m}\frac{ku^2e^u}{(e^u-1)^2}\,g(\omega)\,d\omega, \tag{14.58}$$

where $u = hc\omega/kT$.

Experimentally, one can determine $g(\omega)$ from measurements of the inelastic scattering of neutrons by phonons. These experiments are made on rather large amounts of material ($\sim$ 20 g) and require in addition complex lattice dynamical calculations for determining the complete density of states so that few crystals have been investigated yet. The good results obtained from simpler modeling suggests, however, that these detailed calculations of $g(\omega)$ are not necessarily needed for calculations of thermodynamic functions. In the following sections we will thus restrict ourselves to a few models for the density of states. Before doing so, we note that, strictly speaking, defects and impurities contribute to the heat capacity and other thermodynamic properties of crystals. Fortunately, their contribution is generally small enough to be safely neglected. Important crystal-field and magnetic contributions can exist, especially for compounds containing transition elements (see 6.1b), but it is beyond our scope to discuss these effects. In the following sections we will be primarily interested in low-temperature conditions where the differences between C_p and C_v can generally be neglected. Although the models that will be presented actually deal with C_v, they will be compared with experimentally observable C_p.

14.3b EINSTEIN'S MODEL

This model proposed for crystals by Einstein in 1907 is the simplest that can be thought of. Neglecting any dispersion, it assumes that all atoms vibrate independently, harmonically and isotropically with the same frequency ν around their equilibrium positions. This is a very rough approximation, even for monoatomic crystals, since one considers atoms as weakly interacting localized particules to calculate the energy of the crystal as a whole as the sum of the energies of the individual atoms. The Einstein model is of great historical interest, however, because it introduced the quantization of vibrational energy levels. In addition, it maintains a fundamental significance because, apart from zero-point energy (which was later discovered through resolution of the Schrödinger equation), it gives exact thermodynamic properties for an harmonic oscillator of given frequency. The quantization of vibrational energy assumed by Einstein,

$$E = n\,h\nu \tag{14.59}$$

indeed yields the correct expression (14.25) for C_v (Fig. 14.8). For N atoms, there are $3N$ vibrational modes so that the heat capacity is

$$C_v = 3N\frac{ku^2e^u}{(e^u-1)^2} \tag{14.60}$$

where $u = h\nu/kT$. It is shown in Fig. 10.8 as originally calculated by Einstein for diamond and compared to the measurements then available (in one of the two figures he ever published).

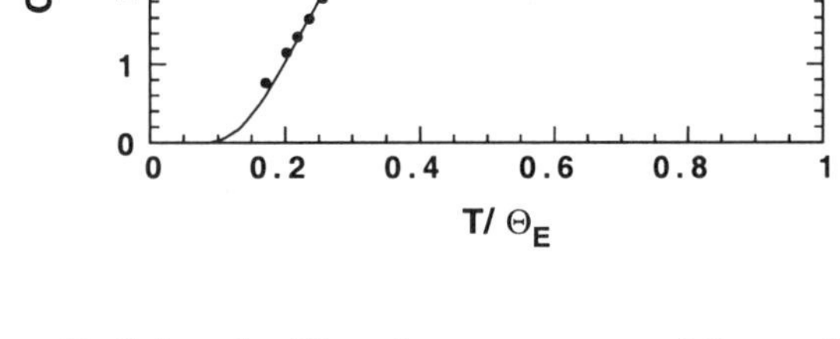

Figure 14.8. Heat capacity of diamond. The values calculated with (14.60) and $\theta_E = 1300$ K are plotted against θ_E and compared to the measurements made by Weber (1875) up to 1259 K. As the measurements begin only at 220 K, the poor results yielded by the model at very low temperatures are not apparent.

Defining the Einstein temperature of the oscillator as $\theta_E = h\nu/k$, we can rewrite this equation in a reduced form as

$$C_v = N\frac{k(\theta_E/T)^2\,exp(\theta_E/T)}{(\exp(\theta_E/T)-1)^2}. \tag{14.61}$$

As already stressed, the high-temperature limit of (14.60) is k for one oscillator. For a crystal made up of N_A atoms (N_A = Avogadro number), the high-temperature limit of C_v is thus $3N_Ak = 3NR$, which is the so-called Dulong and Petit limit (see 14.3c). In agreement with this law, Einstein's model states that C_v tends to $3R$/g atom K but it also accounts for the deviations from this law by predicting that C_v tends to 0 at 0 K (Fig. 14.8). It does so too rapidly, however, because the assumption of a single vibrational frequency implies that all oscillators are frozen out when the first energy level $h\nu$ becomes too great with respect to kT. Although Einstein's model gives the correct expression for the heat capacity of a quantum oscillator, it is inappropriate because it assumes that the density of states is zero except for a single frequency ν. It thus denies the existence of low-frequency oscillators that can be excited for low values of kT.

A noteworthy feature of the harmonic oscillator is that its average length remains constant, regardless of the vibrational amplitude, so that thermal expansion is zero. Contrary to a commonly made assumption, however, this does not necessarily hold true for a three-dimensional set of oscillators. When a crystal is made up of stiff polyhedral units, sharing atoms at their corners, vibrational modes representing rotational oscillations of the polyhedra can induce contraction of the crystal lattice, and thus a negative thermal expansion coefficient. For instance, this is observed for β-quartz and β-cristobalite, the two high-temperature polymorphs of silica which are made up of a three-dimensional framework of SiO_4 tetrahedra. Conversely, that thermal expansion coefficients are not negative, but generally positive, is thus compelling evidence for the anharmonic nature of interatomic potentials.

In passing an effect of anharmonicity must be noted. A vibrational band observed by Raman or infrared spectroscopy, for instance, is actually the result of all the transitions that take place between different energy levels. Since anharmonicity induces a closer spacing of these levels when the quantum number increases (Fig. 14.4), the mean vibrational frequency, which is actually observed, thus decreases when the mean vibrational number increases, i.e., when the temperature increases. The mean quantum number $<n>$ is thus another important concept because it gives the average energy of a set of oscillators at a given temperature. Given that $\sum ne^{-nu} = -\sum d(e^{-nu})/du$, we readily calculate that

$$<n>=\sum \frac{ne^{-nu}}{1/(1-e^{-u})} = \frac{h\nu/e^u - 1}{h\nu} = \frac{1}{e^u - 1}. \qquad (14.62)$$

This number increases with temperature and markedly depends on vibrational frequency. For ν =100 cm^{-1}, $<n>$ increases from 0.3 at 100 K to 10 at 1500 K. For ν = 1500 cm^{-1}, it changes from 0 to 0.3 between these temperatures. Vibrational amplitudes remain small in most cases, which is the practical justification for the harmonic approximation. In this respect, note that the first microscopic criterion for melting, proposed in 1910 by Lindemann, was that a crystal becomes unstable when the average vibrational amplitude reaches a critical fraction of bond lengths, which is actually about 0.1-0.2. Although the Lindemann criterion is of course a rough approximation, it remains of qualitative interest.

14.3c DULONG AND PETIT'S LAW

In 1819 Petit (1791-1820) and Dulong (1785-1838) reported their soon to be famous law stating that "the atoms of all simple bodies [elements] have exactly the same capacity for heat", with a value of $3R$ = 6 cal/mol K. Departures from this law were recognized, especially for diamond, well before low-temperature measurements could be made at the beginning of the

20th century. This did not prevent the law to be considered firmly established throughout the 19th century. For instance, it was used to check or even determine atomic weights which were often known apart from an integer factor.

Additional support to this law was provided by the fact that it could be demonstrated theoretically by the newly born statistical mechanics. In classical mechanics, where the state of a system is specified by the coordinates q_i and conjugate momenta p_i of all particles (see 13.2a), the partition function is

$$Z = 1/h^f \int \dots \int \dots \exp[-H(q_i, p_i)/kT] dq_i \dots dp_i \dots \tag{14.63}$$

where $H(q_i, p_i)$ is the sum of the potential and kinetic energies of particle i and the factor $1/h^f$ is introduced as the volume unit of a phase space of dimension f (see 13.2b). For a single harmonic oscillator, $H(p,q) = p^2/2m + k_f r^2/2$. Integration of (14.63) is straightforward and yields $(2m\pi kT)^{1/2}$ and $(2\pi kT/k_f)^{1/2}$ for the kinetic and potential energy parts, respectively, so that we have

$$z^{vib} = kT/h\nu, \tag{14.64}$$

and

$$C_v^{vib} = k \tag{14.65}$$

for one oscillator, or $3R$ for a set of $3N_A$ oscillators.

Now, we note that the difference between C_p and C_v is $TV\alpha^2/\beta_T$. Because this difference generally represents at most a few % of C_v for solids near room temperature, we predict that C_p is close to the $3R$ value of Dulong and Petit's law. As indicated by the expression (14.18) of the vibrational frequency for a diatomic molecule, this law is in fact the most closely followed when bonding is weak and atomic masses are high. In such cases, the vibrational frequencies are low and heat capacities rapidly reach their high-temperature limit. Diamond thus shows the strongest deviations from Dulong and Petit's law because of the exceptional strength of the C-C covalent bond.

14.3d DEBYE'S MODEL

That the vibrational density of states extends to very low-frequencies is indicated by the rather slow decrease of C_v near 0 K (Fig. 14.8). The model proposed by Debye in 1912 accounts for the existence of such a frequency distribution by considering that these vibrations are acoustic modes whose wavelengths (λ) are very long compared to interatomic distances. There is then no need to examine the details of coupling between atomic motions. The discontinous nature of matter appears simply through the constraint that, for N atoms, the total number of degrees of freedom is $3N$. In view of the very large magnitude of N, we may assume a continuous distribution of modes from a zero up to a maximum *cutoff* frequency ν_D such that

$$\int_0^{\nu_D} g(\nu) d\nu = 3N. \tag{14.66}$$

For determining the density of states $g(\nu)$, we make the further assumption that a crystal is an isotropic continuum in which acoustic waves of different frequencies propagate, at the same velocity c_a, with a linear relationship between the frequency and wave vector $K(-2\pi/\lambda)$

$$\nu = c_a K/2\pi = c/\lambda. \tag{14.67}$$

It is these dispersion relations (Fig. 14.9a), which are lacking in Einstein's model, that allow for modes of very low frequencies (or long wavelengths).

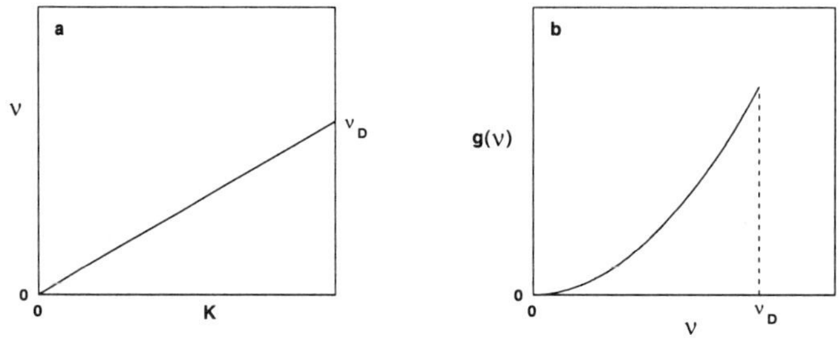

Figure 14.9. Dispersion relation (a) and vibrational density of states (b) for Debye's model.

As already indicated in 13.3c in a general way, we then note that the acoustic waves are subjected to periodic boundary conditions. For a standing wave within a cubic solid of length L, a solution to the wave equation of the form $\exp[i(Kr - \omega t)]$ imposes that $\exp[\mathrm{i}(K_x x + K_y y + K_z z) = \exp[\mathrm{i}(K_x(x+L) + K_y(y+L) + K_z(z+L)]$. Hence the components of the wave vector can only take on the values

$$K_x = K_y = K_z = \pm 2\pi/L, \pm 4\pi/L... \tag{14.68}$$

so that there can only be a single wave vector in a unit volume $(2\pi/L)^3 = 8\pi^3/V$. In a sphere of radius K the number of modes will be

$$N_\nu = \frac{4/3\ \pi K^3}{(8\pi^3/V)} = \frac{K^3 V}{6\pi^2}. \tag{14.69}$$

From (14.66) and (14.69), we write the density of states as

$$g(\nu) = \frac{dN_\nu}{d\nu} = \frac{4\pi V}{c_a^3}\nu^2, \tag{14.70}$$

which thus appears as a parabolic function of the frequency.

In a real crystal acoustic waves represent N longitudinal (compression) and $2N$ transversal (shear) modes with velocities c_l and c_t, respectively (see Appendix B). The total number of modes must thus be summed for the three polarizations

$$g(\nu) = \frac{12\pi V}{c_m^3}\nu^2, \tag{14.71}$$

where c_m is a mean sound velocity, which may be calculated from the rule

$$\frac{3}{c_m^3} = \left(\frac{1}{c_l^3} + \frac{2}{c_t^3}\right). \tag{14.72}$$

Finally, we calculate the cutoff frequency ν_D from the normalization equation (14.66) as

$$\nu_D = (3N/4\pi V)^{1/3} c_m. \tag{14.73}$$

In Fig. 14.9b we show a density of states calculated from these equations. Of particular interest is the fact that the factor $(3N/4\pi V)^{1/3}$ in (14.73) has the dimension of a length. As its value of a few Å is comparable to interatomic distances, it is a measure of the minimum wavelength of acoustic waves.

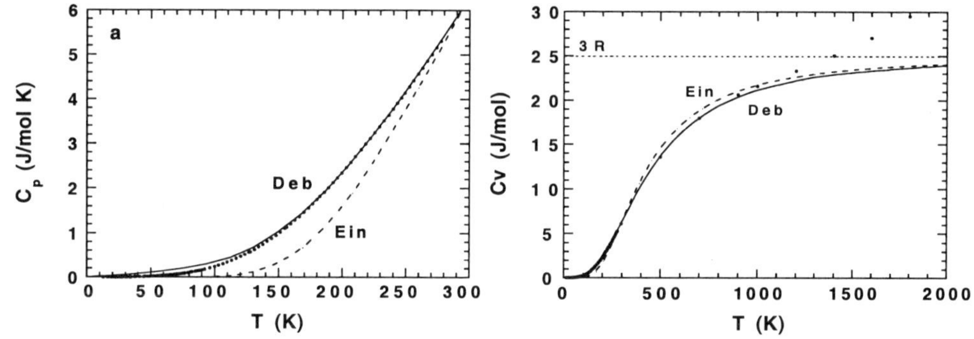

Figure 14.10. Heat capacity of diamond. At low temperature (a), the values calculated from Debye's model and θ_D= 2200 K (solid curve) are compared to the measurements of Desnoyers and Morrison (1958, circles) and to the values from an Einstein model with θ_E= 1300 K (dashed curve). At high temperatures (b), the apparent discrepancy above 900 K between the observations (Hultgren et al., 1979, squares) and the values calculated with Debye's model is due to the fact that the difference between C_p and C_v is no longer negligible. The 3R Dulong-and-Petit limit is included in (b) as a dotted line.

Knowing the density of states, we may calculate the internal energy and heat capacity as a sum over the Einstein functions (14.24) and (14.25). Defining the Debye temperature θ_D as

$$\theta_D = hc\nu_d/k \tag{14.74}$$

and denoting $hc\nu_D/kT$ by u_D we obtain

$$U = 3NkT\left(\frac{T}{\theta_D}\right)^3 \int_0^{u_D} \frac{u^3}{e^u - 1} du, \tag{14.75}$$

$$C_v = 3Nk\left(\frac{T}{\theta_D}\right)^3 \int_0^{u_D} \frac{u^4 e^u}{(e^u - 1)^2} du. \tag{14.76}$$

At high temperature, the limit of C_v is necessarily the same as in Einstein's model. At low temperatures, in contrast, the original feature of Debye's model is the prediction that C_v becomes proportional to T^3 when T tends to 0 with the limiting law $C_v = 12/5\ \pi^4 R(T/\theta_D)^3$. One finds this result because the integral present in (14.76) tends to $12\pi^4/5$ when expanding $(1\text{-}e^{-x})^{-2}$ in powers of e^{-x}. This low-temperature law generally works well for $T < \theta_D/15$. It predicts a decrease of C_v that is much slower than in Einstein's model because the existence of a frequency distribution ensures that, although fewer and fewer oscillators contribute to C_v, there always remain modes such that $h\nu$ is similar to kT. For diamond, the improvement with respect to Einstein's model is obvious, as shown in Fig. 14.10.

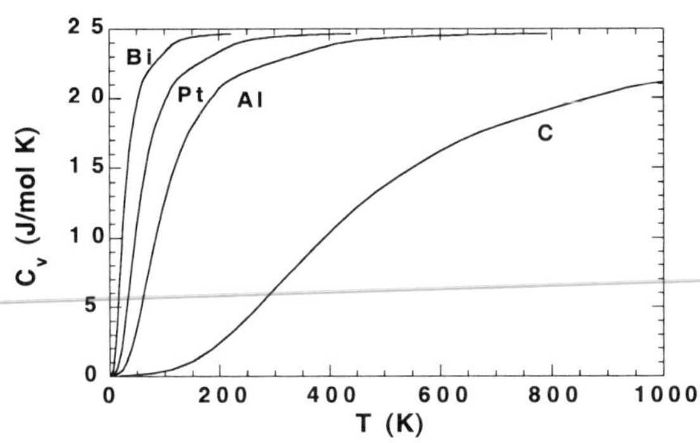

Figure 14.11. Heat capacities of several metals calculated with (14.76) and the thermal Debye temperatures of Fig. 14.12.

When expressed as a function of θ_D, the equations for the internal energy and heat capacity are universal. Plotting C_v for different materials against $\ln T$ (Fig. 14.11), we thus find that the curves have the same shape but are horizontally shifted by their difference in $\ln\theta_D$. These characteristic temperatures are proportional to the cutoff frequencies. Qualitatively they vary like Einstein temperatures since they increase with bond strength and decrease with increasing atomic masses.

However, a single Debye temperature does not allow all the experimental data to be reproduced to within their error margins. If one calculates the Debye temperature by fitting (14.76) to the individual experimental data, one observes that it is not constant, but that it varies with temperature. With a single fitting parameter θ_D, it will not be a surprise that Debye's model works best for monoatomic crystals. For more complex solids, better agreement may be found by expressing the heat capacity as a sum of three different Debye functions. But the heat capacities of crystals are seldom reproduced in this way to within their error margins over the whole interval of the measurements. The reason is that Debye's model has been derived for acoustic modes. As such it does not work well for modes with very low dispersion, which are especially frequent at high frequencies and are better represented by Einstein's model. Empirically, this is why a combination of Einstein and Debye functions can provide good fits to the low-temperature C_p measurements. Note finally that the equations (14.71) and (14.75) provide a direct link between elastic and thermal properties. If Debye's model were exact, sound velocities could be determined from calorimetric measurements and, conversely, heat capacities could be predicted from acoustic mea-

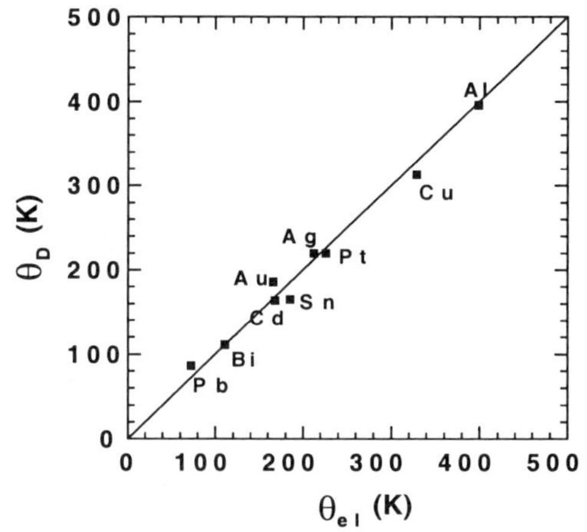

Figure 14.12. Comparisons between elastic and thermal Debye temperatures for some metals (Blackman, 1955).

surements. In practice, the agreement between the two kinds of measurements is good only for monoatomic crystals as a result of the limitations of the model (Fig. 14.12).

As long as calorimetric measurements could not be made routinely below about 50 K, the limiting Debye T^3 law was extensively used to extrapolate C_p mesurements down to 0 K. In determinations of the standard entropy $S_{298} - S_0$, this allowed the relative entropy $S_T - S_0$ to be calculated between the lowest temperature investigated T and 0 K. This law is not always followed, however. The heat capacities of amorphous substances are cases in point, usually showing large positive deviations from it (Fig. 14.13). This points to an excess of low-frequency excitations with respect to Debye's dispersion relations. In spectroscopic studies, this anomaly manifests itself as a low-frequency peak called the *boson* peak. Although its origin is not clearly understood, it is generally assigned to the existence of structural disorder in the substance. But it should be kept in mind that the C_v's of crystalline phases like quartz and especially cristobalite display similar deviations from Debye's limiting law (Fig. 14.13). Hence, structural disorder is not a prerequisite for these anomalies as the three-dimensional open structure of these crystals can also give rise to a number of vibrational modes of very low-frequencies.

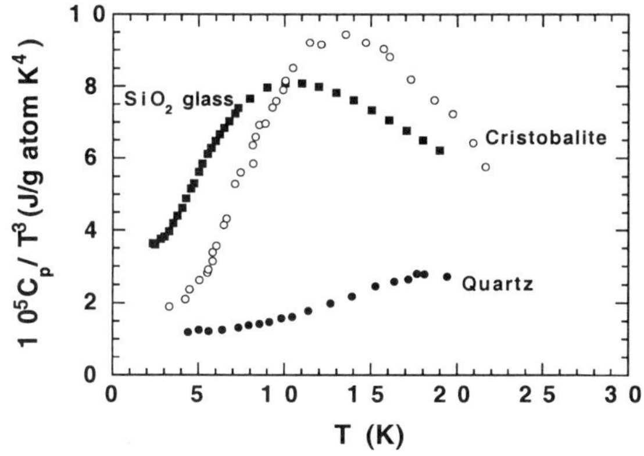

Figure 14.13. Excess heat capacities of cristobalite, quartz and SiO_2 glass with respect to horizontal lines that would indicate proportionality of C_v to T^3 at very low temperatures. Measurements by Bilir and Phillips (1975) and Flubacher et al. (1959).

14.3e DEBYE-EINSTEIN MODELS

We will sketch below how the vibrations of a crystal lattice are treated rigorously. When only heat capacities and entropies are to be calculated, however, simplified density of states can be appropriate. These have the advantage of being determined solely from spectroscopic observations, without the need to calculate the vibrational frequencies of all modes. A first simplification results from a distinction that can often be made between *lattice* and *internal* modes. For instance, the ionic compound $CaCO_3$ can be considered as made up of two ions, Ca^{2+} and CO_3^{2-}. The 9 internal modes are those of an isolated CO_3^{2-} ion. At lower frequencies, the 21 lattice modes (which include the three acoustic modes) involve Ca^{2+} and the carbonate ion CO_3^{2-} that can be taken as a rigid entity.

As already noted, an Einstein model is generally appropriate for optic modes whose dispersion is generally slight whereas a Debye model (or a variation on this model) is suitable for the acoustic modes. A density of states can then be set with the assumption that the $3N$ - 3 optical modes represent either discrete modes or continua, delimited by low- and high-frequency cutoffs, over which a uniform distribution of modes is assumed. With proper selection of the number of modes and boundaries of the optic continua, one usually reproduces the low-temperature calorimetric measurements when the relevant acoustic and spectroscopic data are known. Below 50 K, only the acoustic modes and the optic modes with wavenumbers lower than about 100 cm^{-1} contribute significantly to the heat capacity. At these temperatures, C_v is thus very sensitive to the lower end of the density of states, i.e., to the assumed low-frequency cutoff. This can result in errors of 10-20 % for C_v between 20 and 50 K, but these errors do not have too severe consequences because heat capacities in this interval have a relatively minor influence on standard entropies.

14.3f LATTICE DYNAMICS

The rigorous treatment of vibrations in crystals dates back to 1912 when Born (1882-1970) and von Karman (1881-1963) laid the foundations of *lattice dynamics*. The calculations were much too complex to be performed at that time, however, and they have become feasible only with modern computing means. The method aims at describing how atomic displacements propagate through a three-dimensional crystalline lattice in the form of phonons. If the force constants between the atoms are known, then one solves the equations of motion to determine the energy of the phonons as a function of the wave vector and the symmetry of the involved atomic displacements.

The calculations are generally made in the *quasi-harmonic* approximation. With it, one indirectly accounts for anharmonicity, which causes thermal expansion, by considering harmonic potentials whose force constants depend on the unit cell volume, and thus on temperature. As long as the temperature is not too high (for $T < \theta_D$), this approximation allows the variations with temperature of the mean vibrational frequencies (see 14.3b) to be accounted for simply. As already described in the case of polyatomic molecules, this is a semi-classical approach: the vibrational modes have a quantized energy, but the vibrational density of states is calculated in a classical manner once the empirical force constants of the interatomic potentials have been fitted to the experimentally determined vibrational frequencies. As an example the vibrational density of states of graphite is plotted in Fig. 14.14. Even for a monoatomic crystal, the actual density of states is much more complicated than suggested by a Debye model. In turn, the fact that the heat capacity of graphite is correctly modeled by a

Debye model indicates that a detailed knowledge of the vibrational density of states is not necessarily required for calculations of thermodynamic functions.

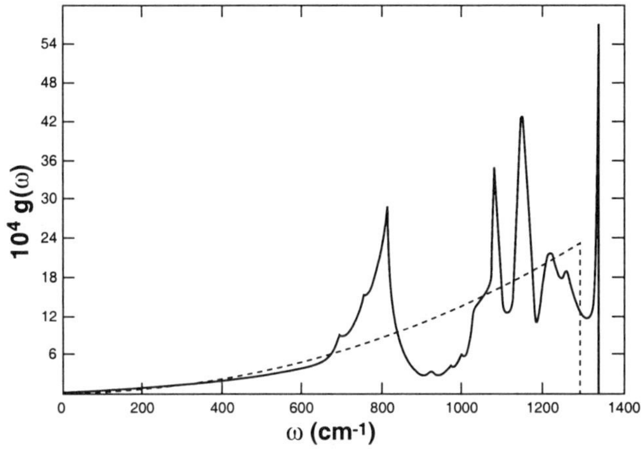

Figure 14.14. Vibrational density of states of diamond. Lattice dynamics calculation (solid curve) and Debye approximation (dashed curve). Calculations by Bottinga (1968).

As discussed in 14.4c, interatomic potentials deal with atomic interactions in terms of variations of bond angles and distances. With *rigid ion* potentials, one assumes that atoms are rigid spheres bearing a given electrical charge (which can be fractional). With *shell* potentials, one introduces a greater flexibility by considering the forces that are exerted separately on atomic nuclei and on massless electronic "shells". In both cases, Newton's equations of motion are expressed in terms of the atomic displacements $e_x(K), e_y(K), e_z(K)$ relevant for each normal mode of vibration. These displacements are represented by plane waves of wave vectors K :

$$m\omega^2(K)e_i(K) = D(K)e_j(K) \tag{14.77}$$

where m is the mass of a given atom and $D(K)$, the *dynamical matrix*, relates the variations of the potential with equilibrium distance (u_i, for atom i) to the interatomic distances (r_{ij}, for the pair i, j) through:

$$D(K) = \sum \left(\frac{\partial^2 U}{\partial u_i \partial u_j} \right) \exp(\mathrm{i}Kr_{ij}). \tag{14.78}$$

For a unit cell of N atoms, the frequencies of the $3N$ vibrational modes are the eigenvalues of the dynamical matrix calculated for a given value of K. The boundary conditions impose that analogous atoms on the opposite faces of a crystal vibrate in phase. Like for Debye's model, this implies that the wave vectors range from 0 to maximum values that depend on the crystallographic direction (the parameters of the *reciprocal lattice*, which define the first *Brillouin zone*). Because one cannot solve Newton's equations of motion as a function of K analytically, one repeats the calculation for discrete K values that adequately sample the first Brillouin zone. The vibrational density of states is then obtained as a histogram giving the number of modes lying in a given frequency interval (Fig. 14.14). Calculation of thermodynamic functions is then straightforward.

14.4 NUMERICAL SIMULATIONS

14.4a POTENTIALS, STRUCTURE AND PROPERTIES

The ideal gas or Debye's model of crystals are rare instances of problems that lend themselves to exact solutions in statistical mechanics thanks to the lack of interactions assumed between the particles of a system. In other instances like real gases under moderate pressures, analytical solutions of practical interest can be derived from adequate approximations of intermolecular potentials. For example, one can determine equations of state from the *virial* theorem which relates the kinetic and potential energies of the particles. As a demonstration, let F_x be the x component of the force exerted on particle of mass m. By multiplying by x both sides of Newton's equation of motion we obtain

$$mx\frac{d^2x}{dt^2} = \frac{m}{2}\frac{d^2(x^2)}{dt^2} - m\left(\frac{dx}{dt}\right)^2 = xF_x. \tag{14.79}$$

Doing the same thing for the y and z components, and summing up the three equations for one particle we find that

$$\frac{1}{2}mv^2 = \frac{m}{4}\frac{d^2(r^2)}{dt^2} - \frac{1}{2}(xF_x + yF_y + zF_z), \tag{14.80}$$

where v is the velocity of the particle $[v^2 = (dx/dt)^2 + (dy/dt)^2 + (dz/dt)^2]$ and r its distance from an arbitrary fixed point $(r^2 = x^2 + y^2 + z^2)$. If the center of mass of the system does not change we have $\sum d^2(r^2)/dt^2 = 0$. By summing (14.80) over all the particles of the system, we then derive

$$\sum 1/2\; mv^2 = -1/2 \sum(xF_x + yF_y + zF_z) = -V_r \tag{14.81}$$

where V_r is the *virial*. Finally, it is useful to split the force F into internal (intermolecular) and external (pressure) forces. Assuming that the former are central and represented by a function $\phi_i(r)$, where r is the distance to the particle i, one shows that (14.81) can be expressed as a function of the pressure P and volume V of the system by

$$\sum 1/2\; mv^2 = -1/2 \sum\; r\phi_i(r) + 3/2\; PV. \tag{14.82}$$

This is the equation mentioned in 5.4b which remains of fundamental importance for the study of not too dense fluids.

Unfortunately, no satisfactory approximation can be made about intermolecular forces so that (14.82) remains analytically intractable for liquids and dense fluids. It is to overcome such difficulties that numerical simulations have been attempted. Whereas the immense number of atoms in any macroscopic system only allows average properties to be determined, modern computers make it direct investigations possible at a microscopic level. This is achieved through what is called numerical experiments, to emphasize that these calculations are subjected to both systematic and statistical errors. The number of atoms considered must be high enough for time-independent results to be obtained, but not too high to maintain acceptable computing times. Regardless of the method used, the potentials of mutual interaction between atoms or ions are an essential feature that determine the structure and properties of the system. Initially made for liquids and dense fluids, these calculations have proven useful for calculating equations of state of gases, under conditions that are out of reach of experiments, or studying the dynamics of crystals and determining their structure. Even with the most powerful computers, however, one can deal with systems made up of a few thousand particles at most. These methods are thus well suited for crystals having a small unit cell, but they are less so for macromolecules like proteins, and they cannot account for the very large correlation lengths that are observed in the neighborhood of the critical point of fluids.

Three methods will be briefly reviewed in the rest of this chapter. With Monte Carlo simulations, one tries to determine the properties of a system from a random sequence of states generated along with a weighting scheme which rejects states whose energy is too high. With molecular dynamics, one calculates in a step-wise fashion the changes in the state of a system by solving the equations of motion for a set of particles. We will conclude with *ab initio* methods, where the only principles involved are those of quantum mechanics. Each of these approaches has its advantages and limitations. Before discussing them it is appropriate to describe interatomic potentials in more detail. An example will first illustrate how they are related to a macroscopic property. We will then present some of the most commonly used potentials.

14.4b COMPRESSIBILITY OF CRYSTALS

As an illustration of the direct link between interatomic potentials and compressibility we will follow Poirier (1999) to derive the Vinet equation of state (5.43) from an empirical expression for the binding energy of metals. To express the change in the volume V of the unit cell in terms of a single parameter, we define a radius r such as $V = 4/3\ \pi r^3$ and assume that it varies with pressure as $a = (r - r_0)/l$, where r_0 is the equilibrium radius at $P = 0$ and l is a scale length. We further assume that, at constant temperature, the changes in internal energy are represented by the following empirical expression

$$U = -U_0(1+a)e^{-a}. \tag{14.83}$$

Noting that $da/dV = 1/4\pi r^2$ and neglecting the differences between adiabatic and isothermal conditions, we readily derive the pressure with

$$P = -\frac{dU}{dV} = -\frac{dU}{da}\frac{da}{dV} = -\frac{U_0}{4\pi l r^2} a e^{-a}. \tag{14.84}$$

The bulk modulus follows

$$K = -V\frac{dP}{dV} = \frac{U_0}{12\pi l^2 r}(1-a)e^{-a}. \tag{14.85}$$

For $P = 0$, we have $a = 0$, $K = K_0$ and $r = r_0$, whence $K_0 = U_0/12\pi l^2 r_0$. Making use of this expression, one obtains the Vinet equation (5.43a)

$$P = \frac{3K_0[1-(V/V_0)^{1/3}]\ \exp\{\eta[1-(V/V_0)^{1/3}]\}}{(V/V_0)^{2/3}}, \tag{14.86}$$

where η stands for $3/2\ (K_0' - 1)$. Differentiating P with respect to V, we obtain the expression (5.43b) for the bulk modulus. Because crystals are of course the condensed phases with the best-known structures, this elementary calculation illustrates why their properties are particularly useful to adjust the parameters of the analytical expressions used for representing interatomic potentials.

14.4c INTERATOMIC POTENTIALS

For obvious ease of calculation, the first simulations have been made with the *hard sphere* potential

$$\phi_{ij}(r_{ij}) = 0 \text{ (if } r_{ij} < \sigma), \qquad \phi_{ij}(r_{ij}) = \infty \text{ (if } r_{ij} \geq \sigma), \tag{14.87}$$

where r_{ij} is the distance between atoms i and j and σ is the radius of the spheres. Despite its simplistic nature, this potential has allowed much fundamental insight to be gained about the liquid state (regarding in particular equations of state). For more realistic calculations the potential must describe interaction energies as a function of the distances and geometry of a given collection of atoms. In a general way, potentials may be expressed as a sum of several terms

$$\phi = \sum_{i=1}^{N} \phi_1(x_i) + \sum_{i<j} \phi_2(r_{ij}) + \sum_{i<j<k} \phi_3(r_{ij}.r_{jk}) + \ldots \tag{14.88}$$

where x_i designates the position of atom i. The first term, ϕ_1, represents external or density dependent forces, which generally need not be taken into account. The third term, ϕ_3, represents three-body interactions. It is usually neglected although it can be significant, especially when strong covalent bonding exits. It is thus the second term, ϕ_2, which represents pairwise interactions, that always predominates.

As seen in 14.1d for ideal diatomic gases, bonds are harder to compress than to extend. The resulting marked asymmetry of the interatomic potential is nevertheless well represented by simple expressions like the Morse potential (14.44). Regardless of the forces involved, these features are observed in more complex systems for which it is often useful to make a distinction between intra- and intermolecular forces. For a real gas this distinction is clear-cut at low pressure. Intramolecular bonds are generally so strong that they cannot be modeled classically, but this very strength results in low vibrational amplitudes which allow molecules to be considered as rigid entities with fixed bond lengths if not with constant bond and torsion angles. The distinction between intra- and intermolecular forces becomes less sharp when the pressure increases, especially with the onset of dimerization or other reactions, whereas it gets blurred in liquids and tends to vanish in crystals at still higher pressures. Depending on the existence or lack of such contrasts, systems are thus termed "molecular" or "atomic". In any case, the fundamental point remains that potentials are made up of attractive and repulsive parts which, to a first approximation, can be considered as pairwise. This is what makes it possible to determine the shape and the parameters of the potential energy "surface" of a complex system.

Once a suitable analytical form has been selected, empirical parameters are generally fitted to experimental structural and elastic data. As shown in previous section, these are direct, sensitive probes of potentials. Thermal data can also be used but calculations are much more tedious since, for instance, fitting a potential to heat capacity data first involves a calculation of the vibrational density of states. In addition, the information conveyed by a single scalar, C_p, is greatly reduced compared to that obtained from the numerous structural parameters and elastic constants of the same substance. Unfortunately, problems associated with extrapolation commonly remain because the experimental data used for fitting potentials are often unavailable for the P, T conditions under which simulations would be the most useful. Alternatively, it is possible to determine potentials from solutions to the Schrödinger equation for small clusters of atoms, see below. But these first-principles approaches do not remedy all the shortcomings of the potentials, especially since they are generally made for a temperature of 0 K. In any case, the approximate knowledge of the potentials is indicated by the fact that elastic and thermal properties of a phase are seldom predicted accurately with the same potential. Common reasons for these failures are the lack of directionality of potentials and frequent neglect of three-body interactions which would lead to considerable increases in computing time.

As examples, we will present a few well-known interatomic potentials. We will avoid molecular systems so as not to delve into the complications that result from the nonspherical shape of molecules. In any case, the goal is to find expressions as simple as possible to deal with complex forces. These originate in the diffuse nature of electron clouds surrounding the nuclei and represent short-range interactions associated with Pauli repulsion, between neighbouring charge clouds, and also short- and long-range components of van der Waals attraction. For ionic substances the strong, long-ranged Coulombic forces must be taken into account. Considering only atomic systems, we will begin with the celebrated "12-6" Lennard-Jones potential introduced in 1924 for noble gases

$$\phi_{ij}(r_{ij}) = 4\varepsilon\left[\left(\frac{\sigma_{ij}}{r_{ij}}\right)^{12} - \left(\frac{\sigma_{ij}}{r_{ij}}\right)^{6}\right], \tag{14.89}$$

where ε represents the depth of the potential energy well and σ_{ij} the position of the minimum. Illustrating nicely the balance between attractive (r^6) and repulsive (r^{12}) forces, this potential is still extensively used for fluids although it becomes too stiff at small distances.

For ionic substances, a good starting point is the Born-Mayer potential

$$\phi_{ij}(r_{ij}) = \frac{z_i z_j e^2}{r_{ij}} + A_{ij}\exp\left(-\frac{r_{ij}}{\rho_{ij}}\right). \tag{14.90}$$

The first term is the Coulombic attraction, $z_i e$ being the formal charge of ion i. The parameter A_{ij} characterizes repulsive forces and ρ_{ij} represents the "softness" of the ion pair i, j. More precise descriptions require more complicated expressions. With an additional term representing van der Waals attraction forces we obtain the Buckingham potential

$$\phi_{ij} = \frac{z_i z_j e^2}{r_{ij}} + A_{ij}\exp\left(\frac{-r_{ij}}{\rho_{ij}}\right) - \frac{C_{ij}}{r_{ij}^6}. \tag{14.91}$$

To get a glimpse at the many different forms given to potentials, we will finally mention the Born-Mayer-Huggins potential

$$\phi_{ij}(r_{ij}) = \frac{z_i z_j e^2}{r_{ij}} + A_{ij} \exp\left(\frac{\sigma_i + \sigma_j - r_{ij}}{\rho_{ij}}\right) - \frac{C_{ij}}{r_{ij}^6} - \frac{D_{ij}}{r_{ij}^8}. \tag{14.92}$$

More flexibility is introduced in the repulsive interaction, through the ion size parameter σ_i, and in the van der Waals dispersive energies which are split into dipole-dipole (C_{ij}) and dipole-quadrupole (D_{ij}) interactions.

At this point we must mention a practical difficulty that stems from the small number of particles dealt with in simulations. A few thousand atoms corresponds to cubic lattices with rows of only 10-20 atoms. As more than one third of the atoms would be at the surface for $N = 15^3 = 3375$ atoms, for instance, one often uses periodic boundary conditions for obviating surface effects. In all three directions of space, the cell actually investigated is then surrounded by its exact replicas. (In addition, this condition ensures constancy of the size and energy of the system since an atom moving outside the cell automatically re-enters it on the opposite side from one of the replicas.) For a pairwise potential ϕ_{ij}, the potential energy of the system is calculated with

$$E_p = \sum_{i<j} \phi_{ij}(r_{ij}), \tag{14.93}$$

where the summation involves the distances r_{ij} between atoms i and j within the cell. To limit the computing time one actually sets a cutoff distance, which is typically half the cell length, beyond which one generally assumes that the contribution to E_p may be replaced by a small, constant corrective term.

14.4d MONTE CARLO SIMULATIONS

The Monte Carlo method owes its name to the fundamental role played in it by random numbers. It was first applied in 1953 to atomistic simulations by Metropolis *et al.* for determining the properties of a system from a random sampling of its configurations. A perfectly random procedure would in fact be inadequate because it would sample too many configurations that are practically irrelevant because they have too high an energy. Rather than weighting a great many randomly chosen configurations by the Boltzmann factor $\exp(-E/kT)$, it is much faster to weight evenly a smaller number of configurations whose probability $\exp(-E/kT)$ is high. Hence, the trick consists in concentrating the sampling in regions of phase space that contribute the most to the partition function. Starting from a given configuration, one picks up a single atom at random and displaces it randomly along all coordinate directions (i.e., from x to $x + \alpha r_x$, where α is the maximum displacement allowed and r_x a random number ranging between -1 and 1, and so on for y and z). Assuming periodic boundary conditions, one calculates the energy change due to the move with the appropriate potential expression. If the energy decreases, the move is accepted because the new state is more probable than the former. If the potential energy increases, then the move is accepted if $\exp(-\Delta E/kT)$ is greater than a real number drawn randomly between 0 and 1. If it is smaller than this number, the move is rejected and the atom brought back to its former position.

Through the Boltzmann factor $\exp(-E/kT)$ the temperature is implicitly set to a constant value. If the volume and number of atoms are also kept constant, the calculations are made in the canonical ensemble. At each step the pressure can be obtained from the virial equation (14.82). The kinetic energy is constant, since the temperature is fixed, which implies that the total energy fluctuates as the potential energy. After m moves, the average value of any property Y is calculated as $\sum Y_m/m$ where all moves are counted, whether successful or not. The simulation is terminated when the variables of interest are calculated with the desired statistics. Without having actually to calculate the partition function it is thus possible to calculate not only thermodynamic, but also structural properties such as bond angles and distances or the number of nearest neighbors around a given atom.

The main advantages of Monte Carlo simulations are its simple computation scheme and relatively modest computation times. Although a great many steps may be required to calculate equilibrium properties, the discontinuous nature of the moves makes it unlikely that the system gets trapped in a metastable equilibrium state. Monte Carlo simulations can also be adapted for calculations of averages in any ensembles. Instead of the canonical ensemble, which was originally considered, one can impose isothermal-isobaric conditions and can even make simulations in the grand canonical ensemble by keeping chemical potentials constant while allowing the number of atoms to fluctuate. As to the main limitations of Monte Carlo simulations, they stem from the fact that the method yields only a series of snapshots. As no information on the pathways between two states is available, it is impossible to obtain time-dependent or transport properties like diffusion coefficients, thermal conductivity or viscosity.

In Monte Carlo simulations the random sampling is designed to ensure ergodicity. Thus the system moves from a given state to any other neighboring state with equal probability. This sampling rests on a *Markov chain* of states, whereby the outcome of the nth draw depends only on that of the $n-1^{\text{th}}$ and a finite set called the *state space* is defined by all outcomes. Practically, an important parameter is the maximum displacement allowed which determines the convergence of the Markov chain and must thus be adequately selected. If the maximum displacement is too small, the phase space is explored too slowly whereas too many states are rejected if it is too great. Note also that in some simulations all particles are moved simultaneously instead of a single one.

14.4e MOLECULAR DYNAMICS SIMULATIONS

Molecular dynamics are a more powerful tool because they yield not only static configurations and properties, but also their variations with time. The calculations involved aim at solving Newton's equations for motion for the N constituting particles of the system. Starting from a given configuration and a set of initial velocities, which fixes the sum of the potential and kinetic energies, one determines in a step-wise fashion the state of the system. The calculation is terminated when time-independent properties are obtained, which are assumed to characterize internal equilibrium. The conditions of constant volume, energy and number of atoms often imposed are those of the microcanonical ensemble, although other conditions can be selected. By equilibrating the system with a thermostat or a barostat, one can for instance perform the calculations at constant temperature or pressure.

Let us denote by f_i the resultant of the forces exerted on particle i

$$f_i = -\sum_{j \neq i}^{N} \frac{d\phi_{ij}}{dr_{ij}}. \tag{14.94}$$

Newton's equation of motion is then written as

$$md^2r_i/dt^2 = f_i(r_1, r_2, ..., r_n) \tag{14.95}$$

where r_i is the position of atom i at time t. For numerical calculations, this expression is frequently rewritten as a finite difference equation

$$r_i(t + \Delta t) = 2r_i(t) - r_i(t - \Delta t) + \Delta t^2 f_i(t)/m_i, \tag{14.96}$$

where m_i is the mass of atom i and Δt is a time increment of the order of 1 femtosecond (10^{-15} s). We thus stress that a rather extensive 10^5-step calculation will represent the evolution of the system over only 10^{-10} seconds. The temperature is given by

$$T = \frac{1}{3Nk} \sum m v_i^2, \tag{14.97}$$

where v_i is the velocity of particle i, whereas the pressure is obtained from the virial equation (14.82), which is rewritten as

$$P = \frac{NkT}{V} - \frac{1}{3V} \sum_{i,j} \frac{\partial \phi_{ij}}{(\partial r_{ij})} r_{ij} = \frac{NkT}{V} + \frac{1}{3V} \sum_{i,j} f_{ij} r_{ij}. \tag{14.98}$$

A great advantage of molecular dynamics is the detailed description of the system as obtained from the complete record of the positions and energies of particles as a function of time. By tracking the positions of the atoms of a given element, one can have a glimpse at the microscopic mechanisms causing configurational changes in liquids. One can also calculate transport properties such as self diffusion coefficients. That of element i is obtained from the mean of the velocities calculated with

$$D_i = \frac{1}{3} \int_0^{\infty} < \vec{v}_i(t) \cdot \vec{v}_i(0) > dt, \tag{14.99}$$

where $\vec{v}_i(t)$ is the velocity vector of i at time t.

As for Monte Carlo simulations, the calculations are only as accurate as the interatomic potentials used. Quantitatively, good results have been obtained for the equation of state of simple fluids or the melting curves of simple substances like rare gases for very high pressures and temperatures at which experiments are not yet possible yet. In view of their purely numerical nature, these calculations cannot yield functional relationships such as equations of state, which are not specified by thermodynamics. Like measurements, they just give numerical values of thermodynamic variables to which usual appropriate expressions are fitted. We will also note that simulations may have limited practical use in phase equilibria calculations, where the small differences between large enthalpy or entropy values that are involved are not necessarily calculated with the necessary accuracy.

The liquid-vapor coexistence curve of NaCl plotted in Fig. 14.15 nonetheless illustrates that good potentials yield results consistent with the experimental data and covering much wider P, T domains than the measurements. In this case, slight modifications of the potential parameters would allow the measurements to be very well reproduced. Because potentials lend themselves to principles of corresponding states, it suffices to note that rescaling through change of the critical parameters leads to the same result. More serious is probably that the curvature of the calculated coexistence curve is too great in the vicinity of the critical point. This is a direct consequence of the fact that the simulation is made with too few atoms to account for the very long range interactions involved in critical fluctuations.

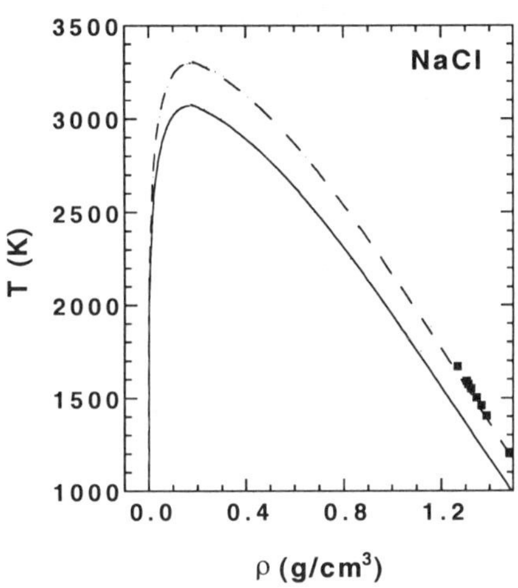

Figure 14.15. Liquid-vapor coexistence curve of NaCl. Measurements (squares) of Kirschenbaum et al. (1962) and molecular dynamics calculation (solid curve) by Guissani and Guillot (1994) for a set of 512 atoms with an empirical Born-Huggins-Mayer potential. The dashed curve is obtained from a corresponding states principle, by adjustment of the calculated critical temperature and density to 3300 K and 1.73 cm^3/mol.

Calculations are of course more difficult for more complex substances. Regardless of the fact that interatomic potentials tend to be less well known, a difficulty is that the number of particles cannot be increased at will because the computation time grows as the square of N. We must also stress that calculations can be made only if equilibrium is reached quickly. When configurational changes take place in the system (see 6.3b), relaxation times must thus be short for simulations to be feasible. Since each step represents a time interval of about 10^{-15} s, 6 10^{16} iterations would be necessary to reach equilibrium for a system in which the relaxation time is one minute! Because relaxation times decrease tremendously when T increases, simulations are thus better suited for high than for low temperatures. Likewise, kinetic transitions such as the glass transition (see 6.4a) are modeled at cooling rates that are orders of magnitude faster than experimentally attainable. (Although comparisons with measurements are not straightforward, these calculations can be used to study vitrification of substances, like argon, that do not form glass under laboratory cooling rates.) For crystals at high temperatures, calculations are especially useful for following phase transitions and determining the structure of new phases when the unit cell is unambiguously determined. When the quasi-harmonic approximation breaks down, molecular dynamics thus supersedes nicely lattice dynamics as anharmonicity is implicitly taken into account through the asymmetry of interatomic potentials. We will finally note that the presence of hydrogen atoms can raise special problems at lower temperatures where the breakdown of the Born-Oppenheimer approximation (Appendix E) implies that the potential energy of the system no longer depends only on the coordinates of the atomic nuclei.

14.4f AB INITIO CALCULATIONS

Electrons are indirectly dealt with in the aforementioned methods. With the Born-Oppenheimer approximation, their rapid motion is averaged out so that nuclei are considered to move within the fixed potential energy field set by a given electronic state. Since electronic energy is orders of magnitude higher than vibrational energy, the electronic groundstate alone is relevant for temperatures in which we are interested. But no attempt is made at solving the Schrödinger equation for the electrons and calculating the electronic wavefunctions and associated internuclear potential. One just assumes that this potential is well defined and tries to determine it by fitting adequate analytical expressions to a variety of experimental data.

In *ab initio* methods, the emphasis is in contrast put on electrons in order to determine directly the interatomic potential through resolution of the Schrödinger equation for a set of atoms representative of the system studied. As it is determined only by the kinetic energy of the nuclei, temperature is left out in these calculations of the groundstate electronic energy which thus refer to 0 K. Exact solutions to the Schrödinger equation cannot be found, however, because electrons interact in a complex manner with each other and with the ion cores of the nuclei. Fortunately, appropriate approximations and simplifications allow numerical solutions to be found in reasonable computing times. Once the potential energy is known, it is possible to calculate not only structural parameters (bond angles and distances) as determined from the energy minimum of the system, but also microscopic (charge density, etc.) and macroscopic (equations of state, etc.) properties. In practice, this can be made for about 100 atoms (or 1000 electrons) for crystals with a high symmetry, but only for a score of atoms for low-symmetry crystals and atomic clusters.

Since the method is parameter free, it has the great advantage of yielding results that should be as accurate at low as at very high pressures where measurements are scarce or even lacking. It could be less so for lower symmetry or bigger unit cells. As an illustration of the method we plot in Fig. 14.16 the volumes calculated up to 4 Mbar for the low- and high-pressure phases of iron. Except at low pressure for the ε phase, the agreement with experimental data is striking. As monoatomic substances, iron polymorphs have simple crystal structures but the existence of a partially filled 3d electronic band in transition elements raises particular difficulties, through strong n-body interactions, which thus appear to be overcome. On the other hand, the fact that calculations are performed at 0 K has the annoying consequence that finding the equilibrium configuration may be a kind of educated guess work, especially when one tries to calculate new structures not known from experiments. The purpose of *ab initio molecular dynamics* is thus to combine the advantages of both methods. In brief, interatomic potentials determined by *ab initio* calculations are used in molecular dynamics to optimize the structural parameters and calculate the forces resulting from the displacements of nuclei. In turn, these results obtained at temperatures different from 0 K serve as a starting point for a new round of *ab initio* calculations, and so on. In an iterative way, structural parameters are then optimized more efficiently and with less constraints than with *ab initio* calculations alone, whereas the effects of temperature are dealt with through the molecular dynamics simulations.

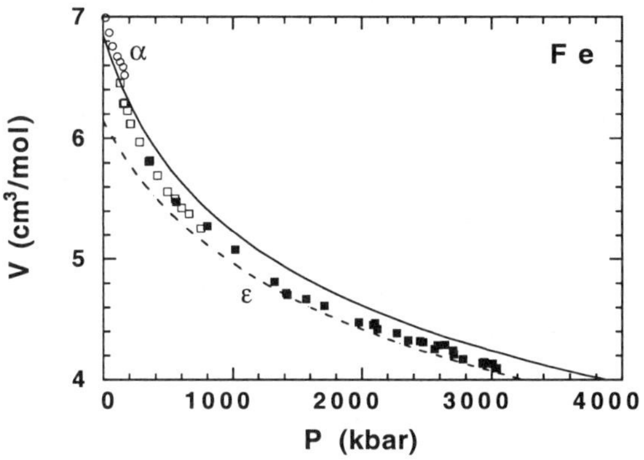

Figure 14.16. Compression of the α (body cubic centered) and ε (hexagonal close compact) phases of iron. Experimental data of Jephcoat et al. (1986, circles) and Mao et al. (1990, squares) and ab initio calculations of Stixrude et al. (1994).

In *ab initio* calculations one assumes that each electron is subjected to an effective potential representing the various interactions within the phase. The Schrödinger equation can then be solved self-consistently for a single electron, and the structural and electronic properties obtained by summation of the individual electronic contributions. But corrections must be made to this simple scheme because electrons are not actually subjected to average fields and also because, according to Pauli's exclusion principle, two electrons cannot have the same quantum state. These corrections represent *correlation* and *exchange* contributions to the potential, which are usually handled with the so-called density functional theory. This theory rests on the theorem that different groundstate wavefunctions result in different electron densities within the phase. One can thus let the electron density vary until the configuration with the lowest energy is found, which corresponds to the groundstate. As the actual form of the correlation-exchange functional is not known, one assumes in the *local density* approximation that the correlation-exchange potential is a function of electron density at a given coordinate position. Better results are in fact achieved in the *generalized gradient* approximation where the potential depends not only on electron density but also on its gradient. It is with this method that the results plotted in Fig. 14.16 have been obtained.

Problems

1. Why is the thermal expansion of a crystal tending to 0 at 0 K, whereas its compressibility does not vanish?

2. Let m_1 and m_2 the mass of the atoms and r_e the equilibrium distance in a diatomic molecule. Show that the molecule has the same moment of inertia as a mass $\mu = m_1 m_2/(m_1 + m_2)$ rotating at a distance r_e from the center of rotation, and that the molecule vibrates in the same way as the same mass μ suspended on a spring having the same force constant k as the molecule.

3. What are the high-temperature limits of C_v for the following gases: CO, CO_2 (linear molecule), H_2O, NH_3, CH_4 and C_3H_8?
4. Why is the heat capacity of a vapor systematically smaller than those of the solid and liquid phases of the same substance at the same temperature (provided that the temperature is not too low)?
5. Find the equation of state $PV = RT$ from the partition function of the ideal gas.
6. The partition function of a translator is calculated in two different ways in 13.5a and in 14.1d. Why is the same result found?
7. Taking the oxygen molecule as an example (Fig. E.1), calculate the work required to act against gravitational forces in order to bring the two atoms from their equilibrium internuclear distance to infinity and show that it contributes negligibly to the dissociation energy of the molecule.
8. One finds that C_v = 3/2 R for one mole of hydrogen just above the boiling point of 20 K. What are the degrees of freedom that do not contribute to C_v, and why are they not getting excited through molecular collisions? From the data of Tableau 14.1, calculate the temperatures at which $1/2\,kT$ (or kT) become higher than the spacings of the energy levels of these degrees of freedom.
9. Calculate C_v and C_p for hydrogen and oxygen (ideal gases) at 50, 200, 1000 and 1800 K from the data of Table 14.1 in the harmonic oscillator, rigid rotator approximation. Comment on the low-temperature results and compare the high-temperature values with the "exact" data plotted in Fig. 4.3.
10. With (14.5) and (14.6) and the rotational contants of Table 14.1, calculate the rotation partition functions of H_2 and Cl_2 as a function of temperature below 300 K. Why is (14.5) becoming valid at higher temperatures for H_2 than for Cl_2?
11. Find the development (14.6) of the rotation partition function of a diatomic molecule by using the Euler-MacLaurin formula

$$\Sigma_{n=a}^{\infty} f(n) = \int_a^{\infty} f(x)dx + 1/2f(a) - 1/12\ (df/dx)_{x=a} + 1/720\ (d^3f/dx^3)_{x=a} \quad (14.100)$$
$$-1/30240\ (d^5f/dx^5)_{x=a} + \ldots$$

Hint: write $f(x) = (2x+1)\exp[-\sigma x(x+1)] = gh$ and $f(n) = \Sigma_{p=0}^{n} C_n^p h^{n-p} g^p$.

12. For a Mie potential, $\phi(r) = -a/r^m + b/r^n$, show that the elastic Helmholtz free energy is

$$F(V) = \frac{9K_0V_0}{n-m}\left[-\frac{1}{m}\left(\frac{V}{V_0}\right)^{-m/3} + \frac{1}{n}\left(\frac{V}{V_0}\right)^{-n/3}\right],$$

and that the isothermal equation of state is

$$P = \frac{3K_0}{n-m}\left[-\left(\frac{V}{V_0}\right)^{-1+m/3} + \left(\frac{V}{V_0}\right)^{-1+n/3}\right].$$

15
ISOTOPIC EQUILIBRIA

For light elements like hydrogen or oxygen, species in chemical equilibrium generally have different isotopic compositions. In the last chapter, calculation of stable isotope fractionation will provide another illustration of the practical use of statistical mechanics. These calculations are made from basic spectroscopic data with the partition functions derived in the previous chapters. The quantum mechanical origin of isotope fractionation will be demonstrated and the importance of zero point energy emphasized. In conclusion to this book, the relevance of isotope fractionation theory to important environmental and cosmological problems will be pointed out.

15.1 ISOTOPIC REACTIONS

15.1a PREAMBLE

Soon after different isotopes of light elements were discovered in the 1930's, it became apparent that their relative abundances were not constant in phases at equilibrium, but that they varied with temperature and chemical composition. These variations are caused by isotope exchange reactions and they are strongest when the relative mass difference between isotopes is the highest. At room temperature, for example, differences of isotope ratios between coexisting species are less than 10 % for $^{18}O/^{16}O$ but can reach a factor of 4 for the ratio of deuterium (D) and hydrogen (H). Industrially such fractionation allows isotopes like D to be purified from starting materials having natural relative abundances as low as 10^{-4}. Furthermore, stable isotopes of light elements are routinely used as tracers to determine the mechanisms of chemical and

Table 15.1. Physical properties of isotopic waters[a]

	$H_2^{16}O$[a]	$D_2^{16}O$	$H_2^{18}O$
Volume at 298 K (cm^3/mol)	18.0686	18.1331	18.0428
Maximum density (kg/m^3)	999.972	1106.00	1112.49
Temperature of maximum density (K)	277.13	284.34	277.36
Melting temperature (K)	273.15	276.97	273.45
Enthalpy of melting (kJ/mol)	6.007 (4)	6.315 (6)	6.029 (4)
Boiling temperature (K)	373.15	374.57	373.30
Enthalpy of vaporization (kJ/mol)	40.657	42.339	40.734
Triple point temperature (K)	273.16	276.97	273.46
Critical temperature (K)	647.10	643.85	
Critical pressure (bar)	220.64	216.71	

[a] Data from Kell (1977), Jakli and Staschewski (1977), Kestin et al. (1984) and Nagano et al. (1993)

biochemical reactions and the variations of their relative abundances in natural fluids and minerals, represent tracers and thermometers of past or current geochemical processes. As such, they represent a most useful tool in environmental studies.

In view of their differing number of neutrons, the isotopes of a given element should be considered as different elements happening to have very similar physical and chemical properties. This is shown in Table 15.1 for the most important water *isotopic molecules* where variations in properties are clearly a function of molecule mass. Consistent with the nonidentical properties of isotopic molecules, one observes that the enthalpy of mixing is not zero, but a few tens of Joule (Fig. 15.1), when water (H_2O) and heavy water (D_2O) are mixed. This enthalpy is in fact associated with the formation of HDO through the reaction

$$H_2O + D_2O \rightleftharpoons 2\ HDO, \tag{15.1}$$

which illustrates the fact that isotopes of the same element participate in real chemical reactions.

From a theoretical standpoint, isotope mixtures are of interest because they approximate more closely ideal solutions than any other chemical systems. However, this holds true only at high enough temperatures. In contrast, at low temperatures isotopic mixtures have the peculiarity of being nonideal solutions in the absence of any molecular interactions! On these accounts, to which is associated the name of the chemist H. Urey (1893-1981), it is useful to know the rules that govern isotope partitioning under equilibrium conditions. These principles are those of statistical mechanics. The fact that the chemical properties of the various isotopes of the same element are almost identical has the fortunate consequence that, for simple substances, isotopic equilibrium constants can be calculated accurately. Along with those of thermodynamic func-

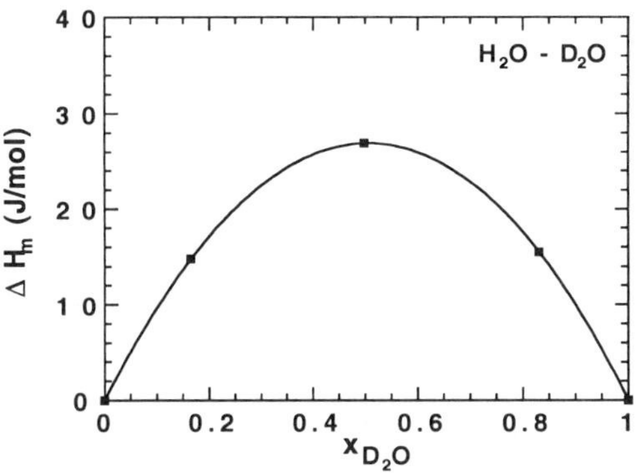

Figure 15.1. Enthalpy of mixing of liquid H_2O and D_2O at 523 K and 70 bar (Simonson, 1990).

tions, these calculations probably represent the most direct applications of the methods presented in the preceding chapters.

15.1b INTERNAL ISOTOPIC EQUILIBRIUM

The simplest case of isotopic reaction is that of an *internal isotopic equilibrium*. That of water, (15.1), has already been mentioned. For hydrogen, we have

$$H_2 + D_2 \rightleftharpoons 2\,HD. \tag{15.2}$$

As a starting point for studying this first kind of isotopic equilibrium, we will consider the situation where the constants of (15.1) and (15.2) correspond to random formation of H_2O, D_2O and HDO molecules for water, and H_2, D_2 and HD molecules for hydrogen. In the latter case, let a system be made up of N_D atoms of D and N_H atoms of H and denote by N_T the sum $N_D + N_H$. For the high numbers of atoms to be considered in a real system, the probability of forming a molecule D_2 is

$$\frac{N_D(N_D-1)}{N_T(N_T-1)} \sim \frac{N_D^2}{N_T^2}. \tag{15.3}$$

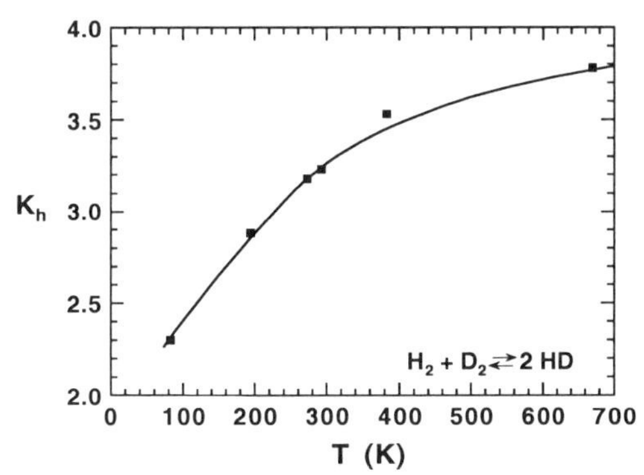

Figure 15.2. Internal isotopic equilibrium of hydrogen against temperature. Experimental points from the summary made by Pyper and Souers (1976) and calculated curve (Richet et al., 1977).

Likewise, that of forming a H_2 molecule is N_H^2/N_T^2 whereas that of forming a HD molecule is $2N_DN_H/N_T^2$ since we have two ways of making the latter, by taking first either a H atom or a D atom. Because the mole fractions are proportional to these probabilities, we conclude that a random distribution of isotopes would give for the equilibrium constant of (15.2)

$$K_h = \frac{(2N_DN_H/N_T^2)^2}{(N_H^2/N_T^2)(N_D^2/N_T^2)} = 4. \tag{15.4}$$

Now, experience shows that this value represents only the high-temperature limit of K_h. At 200 and 540 K, for instance (Fig. 15.2), the values of K_h are 2.8 and 3.6, respectively. Actually, there are more D_2 molecules and less HD molecules than if H and D were forming randomly the three isotopic molecules. Also note that the equilibrium constant of (15.1) is 3.8 at 75 K, indicating that HDO molecules would be similarly less abundant in steam than predicted from a random isotope distribution.

15.1c EXTERNAL ISOTOPIC EQUILIBRIUM

Let us now turn to an isotope exchange between two different chemical species and consider D/H exchange between hydrogen and water

$$HD + H_2O \rightleftharpoons H_2 + HDO. \tag{15.5}$$

Since hydrogen isotopes are not distributed at random within a given chemical species, there are even less reasons why they would distribute randomly between hydrogen and water. Indeed, the equilibrium constant K_{wh} of this reaction tends to unity, corresponding to a random distribution only in the high-temperature limit (Fig. 15. 3). It increases considerably at lower temperatures where it reaches for instance a high value of 3.4 at 300 K. Water is thus generally enriched in D with respect to hydrogen and, conversely, hydrogen is enriched in H with respect to water. The isotopes H and D thus fractionate, and the magnitude of fractionation increases when the temperature decreases.

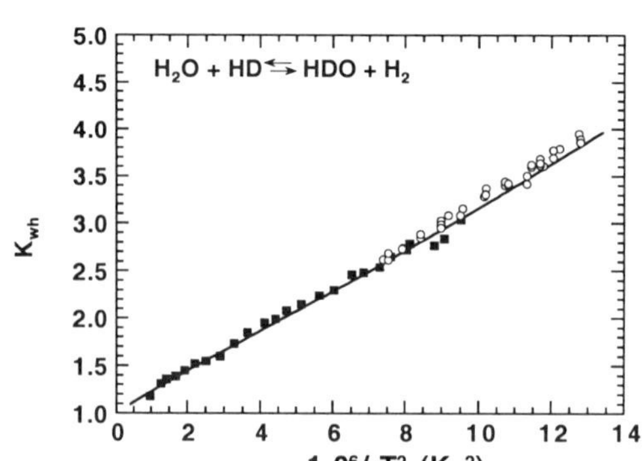

Figure 15.3. H-D isotope fractionation between hydrogen and water against $1/T^2$. Experimental points (Cerrai et al., 1954: squares; Rolston et al., 1976: circles) and calculated curve (Richet et al., 1977).

15.1d PARTITIONS FUNCTIONS AND EQUILIBRIUM CONSTANTS

The basic features of isotopic reactions are readily accounted for by statistical mechanics. These calculations rely on the fact that electronic wave functions (and orbitals) are essentially determined by the number of electrons and protons (see Appendix E). The internuclear potentials and molecular geometries are thus the same for the various isotopic molecules of a given species. Although they derive from the same internuclear distances and potentials, the vibration and rotation frequencies of the isotopic molecules differ because of mass differences between the molecules. With the exception of some H-D exchange reactions, the volumes of isotopic molecules are so similar that we can also consider volumes of reactions to be zero. As isotopic equilibrium constants do not depend appreciably on pressure, they are related to differences in Helmholtz free energies

$$K = \exp(-\Delta F/RT). \tag{15.6}$$

A further simplification originates from the fact that the total number of moles of a species does not change in isotopic reactions. Equilibrium constants can thus be expressed indifferently in terms of partial pressures or mole fractions. Experimentally, they can be determined from very accurate mass spectrometry measurements. By calorimetric methods, however, it is generally impossible to measure their very small ΔF which are typically of the order of a few tens of Joule at most (in the case of H-D exchange, see Fig. 15.1). To complement their experimental determination, equilibrium constants can alternatively be calculated from the statistical expression (13.78) of the Helmholtz free energies of the isotopic molecules

$$F = -k \log Z, \tag{15.7}$$

where k is the Boltzmann constant and Z the partition function of the investigated set of N molecules. Since we will be interested only in ratios between the partition functions of isotopic molecules, the factor $N!$ introduced in 13.3b to account for the indistinguishability of particles could be omitted. Hence we will directly write

$$Z = \sum \exp(-E_j/kT), \tag{15.8}$$

where, as defined in 13.5a, the summation is made over all accessible quantum states j. If we denote by Z^*/Z the ratio between the partition functions of two isotopic molecules, for example HD and H_2, or HD and D_2, it follows from (15.5)-(15.6) that K is expressed as the product of partition function ratios

$$K = \prod (Z^*/Z)^{\nu_i}, \tag{15.9}$$

where ν_i is the stoichiometric coefficient of species i in the isotopic reaction.

As in the preceding chapter, the energy will be split into independent translational, rotational and vibrational parts. As long as the temperature is lower than 1500 or 2000 K, the contribution of excited electronic energy levels to the partition function can be neglected (see Appendix E). Partition functions will thus be written

$$Z = Z^{trans} Z^{rot} Z^{vib}. \tag{15.10}$$

15.2 REDUCED PARTITION FUNCTION RATIOS

15.2a ELEMENT-SPECIES EQUILIBRIA

Before proceeding to evaluate the partition functions, let us establish relative scales of isotope partitioning by taking once more the elements as reference states. We thus consider reactions between isotopically pure species and elements, viz.

$$\mathrm{AX_n} + n\,\mathrm{X}^* \rightleftharpoons \mathrm{AX_n^*} + n\,\mathrm{X}, \tag{15.11}$$

where X and X* are the isotopes exchanged and A represents the rest of the chemical formula of the species. The equilibrium constants of these reactions will be denoted as $f_{\mathrm{AX_n}}$. For reasons that will become clear below, f (not to be confused with a fugacity) is called a *reduced partition function ratio*. Regarding the reaction

$$\mathrm{H_2O} + \mathrm{D_2} \rightleftharpoons \mathrm{D_2O} + \mathrm{H_2}, \tag{15.12}$$

we can thus express its equilibrium constant as

$$k = f_{\mathrm{H_2O}}/f_{\mathrm{H_2}}, \tag{15.13}$$

which is obviously more convenient than calculating anew equilibrium constants for such reactions between isotopically pure molecules.

From (15.9) $f_{\mathrm{AX_n}}$ is given by

$$f_{\mathrm{AX_n}} = \frac{Z_{\mathrm{AX_n^*}}/Z_{\mathrm{AX_n}}}{(Z_{\mathrm{X}^*}/Z_{\mathrm{X}})^n}, \tag{15.14}$$

and we note at once that, in the denominator, the partition functions refer to isolated atoms. Only the translational energy has thus to be taken into account for elements, so that, from (14.4), the ratio $Z_{\mathrm{X}^*}/Z_{\mathrm{X}}$ reduces to $(m_X^*/m_X)^{3/2}$ where m_X and m_X^* are the isotope masses. We will now evaluate the numerator of (15.14) for three different kinds of substances.

15.2b DIATOMIC GASEOUS MOLECULES

For reasons of simplicity, we first deal with diatomic molecules. In the harmonic approximation, the translation, rotation and vibrational partition functions are given by (14.4), (14.9) and (14.23) for a molecule with a mass $M = m + m'$, a moment of inertia I, a symmetry number σ, and a vibrational frequency ν. We thus write

$$f = \left(\frac{m}{m^*}\right)^{3n/2} \frac{\sigma}{\sigma^*} \left(\frac{M^*}{M}\right)^{3/2} \frac{I^*}{I} \frac{\mathrm{e}^{-u^*/2}(1-\mathrm{e}^{-u})}{\mathrm{e}^{-u/2}(1-\mathrm{e}^{-u^*})}, \tag{15.15}$$

where $u = h\nu/kT$, and the asterisks indicate the properties of an isotopically substituted molecule. This expression depends only on masses, vibrational frequencies and moments of inertia. It can be much simplified because the internuclear distance and vibrational force constant k_f are the same for isotopic molecules of the same species. Hence, I and ν are readily calculated for any isotopic molecule if both parameters are already known for a given one. We have indeed

$$I = \mu r^2, \qquad \nu = \frac{1}{2\pi}\left(\frac{k_f}{\mu}\right)^{1/2}, \qquad \mu = \frac{mm'}{m+m'}, \tag{15.16}$$

which gives

$$I^* = I\mu^*/\mu, \qquad \nu^* = \nu(\mu/\mu^*)^{1/2}, \tag{15.17}$$

and yields the relation

$$\frac{I^*}{I}\left(\frac{M^*}{M}\right)^{3/2}\left(\frac{m}{m^*}\right)^{3n/2}\frac{u}{u^*} = 1. \tag{15.18}$$

Making use of this expression in (15.15) we find

$$f = \frac{\sigma}{\sigma^*}\frac{u^*\,\mathrm{e}^{-u^*/2}}{u\,\mathrm{e}^{-u/2}}\frac{(1-\mathrm{e}^{-u})}{(1-\mathrm{e}^{-u^*})}, \tag{15.19}$$

where, from (15.17), $u^* = u(\mu^*/\mu)^{1/2}$. This result is remarkable in that it suffices to know only the harmonic vibrational frequency of a single isotopic molecule to calculate f.

15.2c POLYATOMIC GASEOUS MOLECULES

A polyatomic molecule is characterized by three principal moments of inertia I_A, I_B and I_C, and $3p-6$ (or $3p-5$ for a linear molecule) vibrational frequencies. Calculated in a similar fashion as for a diatomic molecule, the harmonic reduced partition function ratio is

$$f = \left(\frac{m}{m^*}\right)^{3n/2}\frac{\sigma}{\sigma^*}\left(\frac{M^*}{M}\right)^{3/2}\left(\frac{I_A^* I_B^* I_C^*}{I_A I_B I_C}\right)^{1/2}\prod\frac{\mathrm{e}^{-u_i^*/2}(1-\mathrm{e}^{-u_i})}{\mathrm{e}^{-u_i/2}(1-\mathrm{e}^{-u_i^*})} \tag{15.20}$$

where $u_i = h\nu_i/kT$. For polyatomic molecules, there exists a relation analogous to (15.18) between vibrational frequencies and moments of inertia, namely,

$$\left(\frac{I_A^* I_B^* I_C^*}{I_A I_B I_C}\right)^{1/2}\left(\frac{M^*}{M}\right)^{3/2}\left(\frac{m}{m^*}\right)^{3n/2}\prod\frac{u_i}{u_i^*} = 1, \tag{15.21}$$

which is known as the Teller-Redlich product rule. Dividing (15.20) by (15.21), we again derive an expression for f that depends only on the frequencies of the normal modes of vibration

$$f = \frac{\sigma}{\sigma^*}\prod\frac{u_i^*\mathrm{e}^{-u_i^*/2}(1-\mathrm{e}^{-u_i})}{u_i\mathrm{e}^{-u_i/2}(1-\mathrm{e}^{-u_i^*})}. \tag{15.22}$$

Here, however, the situation differs from that of diatomic molecules. Although the vibrational frequencies (and the geometry) of the various isotopic molecules are still determined by the same force field, their accurate calculation is not direct and requires instead a detailed analysis whose complexity increases rapidly with the number of atoms in the molecule (see 14.2d). Although vibrational frequencies must decrease when isotope masses increase, the isotope shift markedly depends on the nature of the normal mode. It is of course zero for modes in which the element considered does not participate. The variation of f with temperature is thus the result of a great many factors so that the ratio of f between two different species does not necessarily vary smoothly. (An example of this will be given in Fig. 15.6.)

15.2d CONDENSED PHASES

For a crystal only vibrational degrees of freedom have to be taken into account (see 14.3). Integrated between 0 and the upper limit ω_m of the vibrational density of states $g(\omega)$, the reduced partition function is

$$f = \left(\int_0^{\omega_m^*} g(\omega^*) \frac{e^{-u^*/2}}{(1-e^{-u^*})} d\omega^* \right) \Big/ \left(\int_0^{\omega_m} g(\omega) \frac{e^{-u/2}}{(1-e^{-u})} d\omega \right) \tag{15.23}$$

where $u = hc\omega/kT$. Of course, the difficulty lies in calculating $g(\omega)$ with the required accuracy. Except for the simplest substances, this cannot be made with lattice dynamics. The precision is still lower when one can estimate only simplified densities of states, for instance by neglecting the dispersion of optic modes. Calculations are thus considerably less accurate for solids than for gases.

The situation is still more difficult for liquids because one does not know how to calculate the effects of isotope substitutions on configurational degrees of freedom. Even with molecular dynamics or other simulation methods quantitative applications are still impossible for phases of practical interest like water or aqueous solutions. We can note, however, that the atomic degrees of freedom involved are qualitatively the same as for gases and crystals (translation, rotation, vibration) so that the fundamental laws determining isotope partitioning are also the same. It follows that the temperature dependence of equilibrium constants involving condensed phases follows the same laws as for gases. As indicated by (15.22)-(15.23), especially important in this respect is the fact that in all cases f tends to 1 at high temperatures, when all u's tend to 0.

15.3 ISOTOPIC EQUILIBRIA

15.3a ISOTOPE FRACTIONATION AND ZERO POINT ENERGY

Why, according to (15.22), translations and rotations do not contribute to isotopic equilibrium constants between gaseous species? For both degrees of freedom, we have derived the partition function in 14.1 by noting that the energy levels are closely enough spaced with respect to kT that it is possible to replace their summation by an integral. Implicitly, we have thus neglected the quantization of the translational and rotational energies and derived the expressions of classical mechanics. In contrast, quantization has to be taken into account in the vibrational partition function. Neglecting it, the partition function would be that of an harmonic oscillator in classical mechanics, which was shown in 14.3c to be

$$Z_c^{vib} = kT/h\nu. \tag{15.24}$$

Combining this expression with those previously derived for Z^{trans} and Z^{rot}, we find that the reduced partition function ratio in classical mechanics is such that

$$(\sigma^*/\sigma)f = 1. \tag{15.25}$$

Isotope fractionation is therefore purely quantum mechanical in origin. As a matter of fact, f can be defined as the ratio between the quantum mechanical and classical partition functions.

From inspection of (15.22) the low- and high temperature limits are readily identified. At low temperature u_i is large, $1 - e^{-u_i}$ differs little from 1 and f is essentially determined by the $e^{-u_i/2}$ terms, which represent the contribution of zero point energy to the partition function. The difference in zero-point energy is thus the main factor causing isotope fractionation. This is why even simple harmonic calculations can yield surprisingly accurate results. Since a zero-point energy represents the difference between the first energy level and the bottom of a potential energy well, it cannot be measured macroscopically and has to be determined from a complete calculation of vibrational frequencies. Its importance decreases when the temperature increases. At high enough T, the u_i have decreased to the point that f tends to the unity value of classical mechanics. The $h\nu_i$ are then small enough compared to kT that the quantization of vibrational energy levels is no longer an essential feature. Through the $1 - e^{-ui}$ terms the contribution of excited vibrational energy levels becomes significant before this high-temperature limit is reached, but its importance remains one order of magnitude smaller than that of the zero point energy.

An important point is that analyses made with mass spectrometry yield isotope ratios accurate to within 0.1 - 1‰ depending on the element analyzed. For calculating with the same accuracy isotopic equilibrium constants, one should know the zero point energy difference to better than about 1 cm^{-1}. This is possible only for simple gaseous substances through detailed vibrational analyses. For crystals, and especially liquids, calculations made with (15.23) are thus much less accurate than measured equilibrium constants. In these cases (15.23) is not useless, however. As described below, it allows better interpolations or extrapolations of experimental data to be made thanks to the analytical expression it gives for the temperature dependence of equilibrium constants.

15.3b ISOTOPIC EQUILIBRIUM CONSTANTS

The vibrational frequencies of a few gaseous molecules are listed in Table 15.2. For diatomic molecules, the $\omega_e x_e$ parameters are included with which anharmonic corrections can be made as described in 14.2c. The calculated constants plotted in Figs 15.1 and 15.2 for hydrogen internal equilibrium and water-hydrogen equilibrium illustrate the accuracy of the method, which is especially striking as there are no free parameters and the zero-point energy is not directly measurable. The reader may care to check that even simple harmonic calculation already give fairly good results.

Table 15.2. Vibrational frequencies of some gaseous molecules (cm^{-1})

	ω_e	$\omega_e x_e$ [a]		ω_1	ω_2	ω_3
H_2	4401.12	121.28	$H_2^{16}O$	3835.37	1647.59	3938.74
HF	4138.12	89.88	$HD^{16}O$	2823.19	1444.53	3888.63
$^{14}N_2$	2358.03	14.13	$D_2^{16}O$	2762.84	1206.72	2885.99
$^{16}O_2$	1580.36	12.07	$H_2^{18}O$	3827.59	1640.62	3922.69
$^{12}C^{16}O$	2169.82	13.29	$^{12}C^{16}O_2$	1353.64	672.62 [b]	2396.27
$^{14}N^{16}O$	1904.09	14.04	$^{13}C^{16}O_2$	1353.68	653.77 [b]	2328.04
$^{32}S_2$	725.68	2.85	$^{12}C^{18}O_2$	1275.98	662.32 [b]	2359.59

[a] Anharmonically, the vibrational energy of a diatomic molecule is $E = hc[\omega_e(n+1/2) - \omega_e x_e(n+1/2)^2]$, see 14.2c; the zero-point energy then is $E_0 = hc(\omega_e/2 - \omega_e x_e/4)$.
[b] Doubly degenerate vibrational mode

As another illustration of the general trends of isotope fractionation, the reduced partition functions for various exchanges are plotted in Fig. 15.4. All other things being equal, the decreasing magnitude of f with the relative mass difference of the isotopes is obvious. Because the mass difference between stable isotopes is generally 1, 2 or 3 units of atomic mass, regardless of the mass of the element, fractionation is hardly detectable for isotopes of elements heavier than sulfur and chlorine. The other important factor is the kind of chemical bond in which the isotope of interest is involved within the species considered. Taking once more the example of diatomic molecules, one readily finds that the lighter isotope is enriched in the species where the vibrational force constant is the highest, i.e., where its bonding is the strongest. Schematically, the magnitude of isotope fractionation of an element between two species is therefore a relative measure of bond strengths. It is especially small between the vapor and liquid phase of the same substance. For H_2O at 298 K, D and ^{18}O have enrichments of 80 and 9‰, respectively, in (liquid) water with respect to steam.

Among light elements, only hydrogen, carbon, nitrogen, oxygen and sulfur show large isotope fractionation in nature because they form species between which there exist significant bonding differences. Of course, the contrast in possible fractionation is striking between extremes of the periodic table. Whereas D is concentrated 8 times at 190 K in ammonia with respect to hydrogen, with a suitable catalyst to ensure equilibrium, the ^{235}U and ^{238}U isotopes have so small a relative mass difference that they cannot be separated at all by chemical means. Much more expensive physical methods must thus be used to purify the fissionable, low abundance ^{235}U isotope: originally mass spectrometry during the Manhattan project, nowadays gaseous diffusion with UF_6 or ultracentrifugation.

15.3c SERIES EXPANSIONS

Against T, the internal equilibrium constant of hydrogen plots like a curve (Fig. 15.2). Against $1/T^2$, the equilibrium constant between water and hydrogen is in contrast represented by a straight line in Fig. 15.3. This apparent difference illustrates the fact that

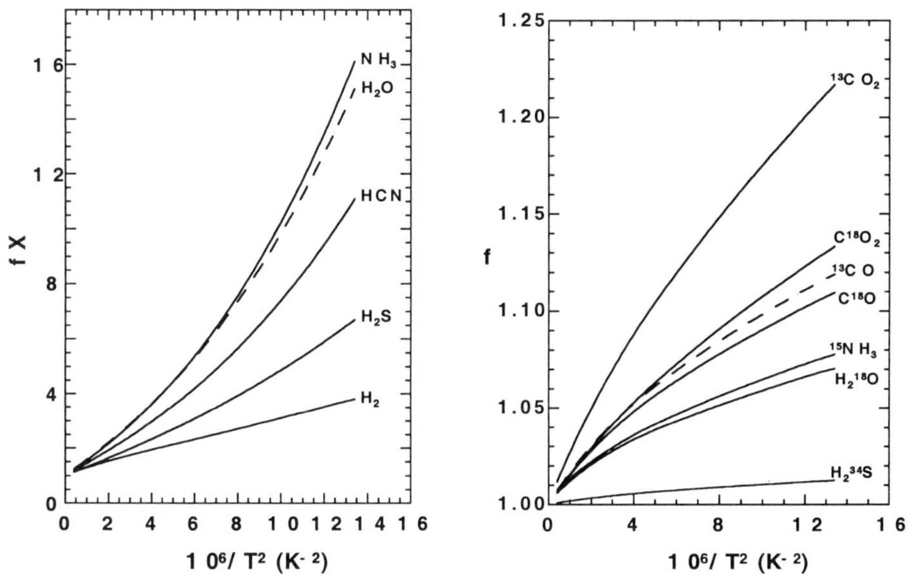

Figure 15.4. Reduced partition function ratios for hydrogen (a) and carbon, nitrogen, oxygen and sulfur isotope exchanges for some gaseous molecules (Richet et al., 1977).

isotopic equilibrium constants vary as $1/T^2$, not as T, in the high-temperature limit. To account for this law, we first note that the isotope shifts of vibrational frequencies are generally small. Writing $u = u^* + \Delta u$, with $\Delta u > 0$, we can expand (15.22) in series when Δu is small

$$(\sigma^*/\sigma)f = 1 + \sum[1/2 \; - \; 1/u_i \; + \; 1/(e^{u_i} - 1)] \; \Delta u_i. \tag{15.26}$$

With the function $G(u)$ defined as

$$G(u) = 1/2 \; - \; 1/u \; + \; 1/(e^u - 1), \tag{15.27}$$

reduced partition function ratios are written

$$(\sigma^*/\sigma)f = 1 + \sum G(u_i) \; \Delta u_i. \tag{15.28}$$

If the u_i^* are themselves small, this equation simplifies further still

$$(\sigma^*/\sigma)f = 1 + \sum u_i \Delta u_i/12 = 1 + \sum \Delta(u_i^2)/24, \tag{15.29}$$

and becomes of the form

$$(\sigma^*/\sigma)f = 1 + A/T^2. \tag{15.30}$$

Fig. 15.5 shows that eqns (15.28) and (15.29) are equivalent when all u_i are smaller than 2. Hence, (15.30) is followed beyond 700-850 K for most solids and gases whose

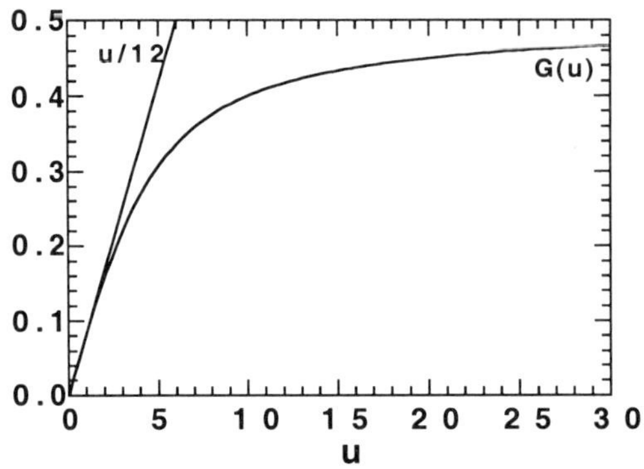

Figure 15.5. Function $G(u)$ and high-temperature limiting $1/T^2$ law.

highest vibrational wave numbers are in the range 1000-1200 cm^{-1}. Under these conditions, an isotopic equilibrium constant varies with temperature as

$$\log k = (A_i - A_j)/T^2 \tag{15.31}$$

where A_i and A_j are the A coefficients of (15.30) for the two species involved.

15.4 ISOTOPE FRACTIONATION

15.4a ISOTOPE PARTITIONING

In most applications what is actually measured is the isotope partition coefficient between different phases. For the partitioning of H and D between water and hydrogen, for instance, we write

$$\alpha_{H_2O-H_2} = [D/H]_{H_2O}/[D/H]_{H_2} \tag{15.32}$$

and call α the *isotope fractionation factor* between these two species. We will now show how these coefficients are related to the reduced partition function ratios of the two phases involved. To deal with the most general case we will again select stable hydrogen isotopes and take (15.32) as a starting point.

Neglecting the rare isotopic molecules containing tritium, the mass 3 isotope, we write the D/H ratio of hydrogen

$$(D/H)_{H_2} \sim \frac{2[D_2]+[HD]}{2[H_2]+[HD]} = X_{H_2}\left(\frac{[D_2]}{[H_2]}\right)^{1/2}, \tag{15.33}$$

where X_{H_2}, the excess factor, is given by

$$X_{H_2} = \frac{K_h^{1/2} + 2K_h^{-1/2}[HD]/[H_2]}{2+[HD]/[H_2]}. \tag{15.34}$$

Likewise, expressing the excess factor of water as

$$X_{H_2O} = \frac{K_w^{1/2} + 2K_w^{-1/2}[HDO]/[H_2O]}{2+[HDO]/[H_2O]}, \tag{15.35}$$

we write

$$\alpha_{H_2O-H_2} = \frac{X_{H_2O}}{X_{H_2}} \frac{([D_2O]/[H_2O])^{1/2}}{([D_2]/[H_2])^{1/2}} = \left(\frac{f_{H_2O}}{f_{H_2}}\right)^{1/2} \frac{X_{H_2O}}{X_{H_2}}. \tag{15.36}$$

This fractionation factor is expressed in terms of reduced partition function ratios, internal equilibrium constants and the ratios [HD]/[H_2] and [HDO]/[H_2O]. The situation is simpler than it seems, however, for we will not make any significant error by taking both of these small ratios equal to the average D/H ratio of the system (namely, D/H = 1.5 10^{-4} for terrestrial hydrogen). Note especially that both excess factors X_{H_2O} and X_{H_2} would be unity if the internal equilibrium constants of water and hydrogen were equal to 4. It is because hydrogen isotopes do not distribute randomly between isotopic molecules at not too high temperatures that the excess factors differ from unity. The excess factor is thus the analog of the fugacity coefficient in that it characterizes deviations from ideal mixing of isotopes, in this case H and D, between isotopic molecules. Without any molecular interaction non-ideality can thus result from just a mass difference.

For any fractionation factor we can write ratios analogous to (15.36). In contrast to H-D exchanges, the simplifying feature is that excess factors are all unity so that α reduces to a ratio of reduced partition function ratios whenever a random internal distribution of isotopes prevails. For instance, the relevant excess factors X_{CaCO_3} and X_{H_2O} for the $^{18}O/^{16}O$ exchange reaction between liquid water and calcium carbonate

$$1/3\ CaC^{18}O_3 + H_2^{16}O \rightleftharpoons 1/3\ CaC^{16}O_3 + H_2^{18}O, \tag{15.37}$$

are both unity. The ^{18}O fractionation factor between water and calcium carbonate

$$\alpha_{CaCO_3-H_2O} = \frac{3[CaC^{18}O_3] + 2[CaC^{18}O_2^{16}O] + [CaC^{18}O^{16}O_2]}{[H_2^{18}O]} \tag{15.38}$$

is thus simply written

$$\alpha_{CaCO_3-H_2O} = f_{CaCO_3}^{1/3} / f_{H_2O}^{1/2}. \tag{15.39}$$

As experimentally determined this fractionation is plotted against T in Fig. 15.6. Note the change in the sign of $\log\alpha_{CaCO_3-H_2O}$ at 290 K. Owing to the complex integrations

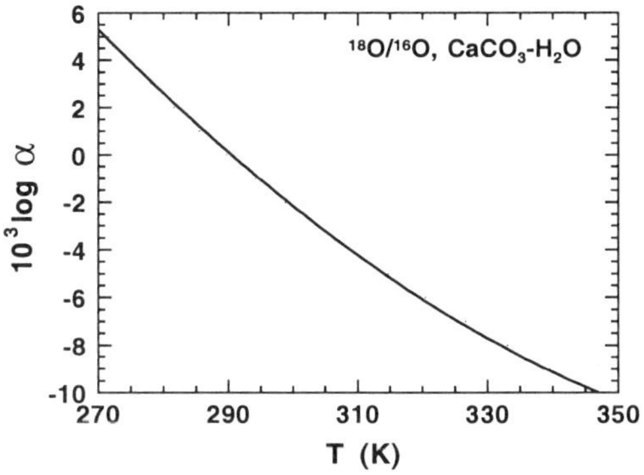

Figure 15.6. Oxygen isotope fractionation measured between water and calcium carbonate (calcite). Because α tends to 0 in the high-temperature limit, the inversion of the fractionation at 290 K implies the existence of a maximum at higher temperatures. As noted in 15.2c, fractionation factors thus do not necessarily vary monotonously with temperature. Data from O'Neil et al. (1969)

of energy levels performed in partition functions, such an *inversion* is not an uncommon feature around or below room temperature.

15.4b ENVIRONMENTAL APPLICATIONS OF ISOTOPIC EQUILIBRIA

Through innumerable practical and industrial applications thermodynamics is a tool that has been widely used for shaping the world we live in. To conclude this book, we will take two examples and show that thermodynamics can also help understanding it. This will be made with thanks to statistical mechanics.

The first application concerns palaeothermometry. This method relies on the fact that isotopic equilibrium is generally reached much faster via a chemical reaction than through isotope exchange that would take place through diffusion. As an example, $^{18}O/^{16}O$ equilibrium is reached very rapidly between carbon dioxide and water through the dissolution reaction $H_2O + CO_2 \rightleftharpoons H_2CO_3$. Likewise, the very fast kinetics of the acid-base reactions of carbonic acid ensures isotopic equilibrium between natural waters and calcium carbonate which forms either chemically or biologically. In this way the successive layers of the shells of abundant planktonic animals, the foraminifera, form in isotopic equilibrium with sea water. When sedimented after the animal's death, the shells keep a record of the mean temperature of the oceans where they grew since ^{18}O fractionation between H_2O and $CaCO_3$ depends on temperature (Fig. 15.6).

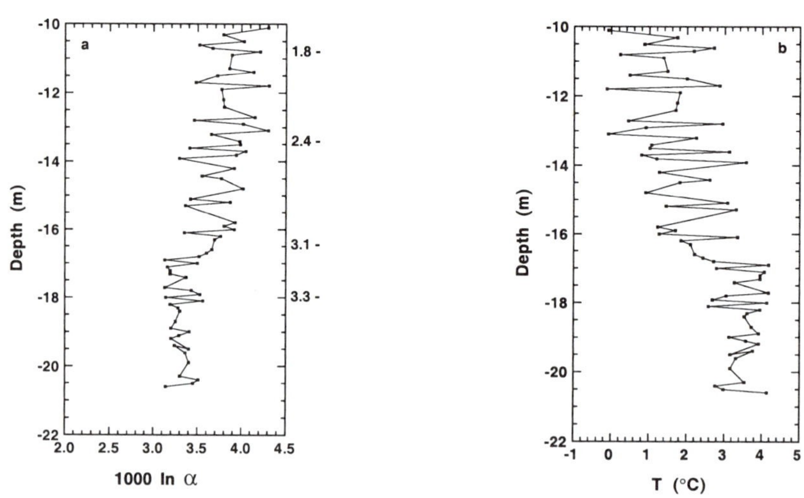

Figure 15.7. (a) Oxygen isotope composition of limestones cored in the Equatorial Pacific; (b) Ocean temperatures as calculated from the fractionation factor of Fig. 15.6. Age of the sediments (million years) indicated for 4 depths; isotope compositions expressed as 1000 times the relative deviations of the analysis from the current mean composition of oceanic waters. Isotope data from Shackelton and Opdyke (1977).

To get a glimpse at the Earth's former climate, one can thus determine palaeotemperatures from isotope analysis of marine limestones (Fig. 15.7). Such measurements point to a long period of stable sea level and low polar ice volume which ended 3.2 million years ago. Since then, climates have kept fluctuating with only relatively brief periods of pleasant climates, like our own, lasting only about 10,000 years. Determination of palaeotemperatures is made tricky, however, because the average isotopic composition of sea water also varies with the size of the ice caps due to the fact that oxygen (and hydrogen) isotopes fractionate on evaporation. This effect has been taken into account in H_2O-$CaCO_3$ palaeothermometry. It in fact offers another way of reconstructing more recent climates because the analyzed isotopic compositions of ice columns cored into the Greenland and Antarctic ice caps represent another measure of past temperatures. Hence, the study of ^{18}O distribution in ice made in 11.1a has not solely a theoretical interest!

With *mass dependent fractionation*, the second example dealt with here touches on the problem of the formation of the solar system. For an element like oxygen, which has three stable isotopes, two different fractionations can be considered: $^{17}O/^{16}O$ and $^{18}O/^{16}O$ for example between H_2O and CO_2. These fractionations are not independent, however, because their ratio is fixed by the mass differences between the isotopes of interest. Regardless of the species considered, one calculates that the $^{17}O/^{16}O$ and $^{18}O/^{16}O$ fractionations are in a ratio $0.52 \approx (17\text{-}16)/(18\text{-}16)$. Even though isotopic equilibrium is often not reached in physical, chemical or biological reactions, no exception to this rule is known for terrestrial materials.

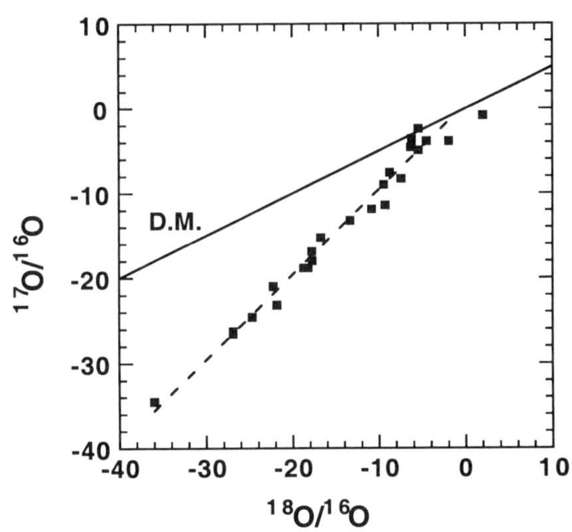

Figure 15.8. Three-isotope plot for oxygen in meteorites. Squares : meteorite (carbonaceous chondrite) data; D.M. line: fractionation depending on mass as observed for terrestrial (and lunar) materials. The $^{17}O/^{16}O$ and $^{18}O/^{16}O$ ratios represent 1000 times the relative deviations of the analyses from the values of a standard whose isotopic composition plots at the intersection of the two lines. Data of Clayton et al. (1973).

It is in meteorites, which are the remains of small planetary bodies formed 4.5 billion years ago at the same time as the Earth, from the same materials, that important violations to the mass-dependent rule have been first reported (Fig. 15.8). As it is problematic to assign the meteorite anomalies to any known reaction, the favored explanation is contamination by oxygen which had an unusual isotopic composition by being made up of almost pure ^{16}O. This oxygen could have been released during the explosion of a supernova whose shock-wave triggered the accretion of dust in what would become the solar system. Because of rapid absorption of the particles of the burst by compressed dust, contamination would have been limited to the outer part of the dust disk where the "anomalous" meteorites would have then formed. Thanks to isotope analyses and isotope fractionation theory, would some pretty mysterious stories be told more accurately?

Problems

1. One wants to produce at room temperature 1 liter of deuterium-enriched water with D/(D+H) = 0.1 from 10 m^3 of natural fresh water for which D/H = 1.5 10^{-4}. What will be the entropy change and the minimum work to be performed for this process?
2. The light isotope of an element concentrates in the species where the element is most strongly bound. Explain qualitatively this fact and demonstrate it in the case of an equilibrium between two diatomic molecules.
3. From the data of Table 14.1, calculate the internal isotope equilibrium constant of hydrogen at 100 and 700 K in the harmonic oscillator, rigid rotator approximation. Compare the results with the measurements plotted in Fig. 14.2.
4. From the vibrational frequencies of Table 14.2, calculate the isotope fractionation factor α between CO and O_2 between 300 and 1600 K in the harmonic oscillator, rigid rotator approximation. Plot the results against $1/T^2$ and comment on the variation of α at the highest temperatures.
5. For reasons of experimental accuracy, the measurements of isotopic compositions by mass spectrometry are not absolute but are made and expressed with respect to standards in δ units defined by $\delta = 1000(R_e - R_s)/R_s$, where R_e and R_s are the isotopic ratios of the sample and standard, respectively. Show that the fractionation of a given isotope between two species A and B is $\alpha_{A-B} = (1000 + \delta_A)/(1000 + \delta_B)$.
6. In volcanic gases from Mount Etna (Sicily) it has been found that carbon dioxide and water have $^{18}O/^{16}O$ ratios (expressed in δ units with respect to the same standard) δ_{H_2O} = 11.58 and δ_{CO_2} = 6.3. At 700 and 1000 K, their reduced partition function ratios are 1.0298 and 1.0155 for CO_2 and 1.0211 and 1.0124 for H_2O, respectively. Assuming a linear variation of $\alpha_{CO_2-H_2O}$ as a function of $1/T^2$, calculate at which temperature the gases were in isotopic equilibrium. The gases have in fact been emitted at around 1300 K. Why is the isotopic temperatures much lower? (Hint: what is the chemical reaction that promotes isotopic equilibrium between water and carbon dioxide; is it a fast one?)
7. Same problem for water and hydrogen sampled at around 1420 K in volcanic gases at Surtsey island (Iceland) which have D/H ratios (expressed in δ units) δ_{H_2O} = -50 and δ_{H_2} = -143.8. At 1100 and 1500 K, the reduced partition function ratios are 1.2182 and 1.1444 for H_2 and 1.4145 and 1.2634 for H_2O, respectively, and one will again assume a linear variation of $\alpha_{H_2O-H_2}$ as a function of $1/T^2$. For this equilibrium, why is the isotopic temperature consistent with that of the chemical equilibrium which was prevailing within the volcano?

Appendix A
MATHEMATICAL COMPLEMENTS

A.1 STATE FUNCTIONS AND EXACT DIFFERENTIALS

The variation of a state function depends only on the initial and final states of the system, not on the pathway followed during a transformation. If one calculates a volume variation induced by pressure and temperature changes, for instance, the order of integration of the differential $dV(P,T)$ does not matter. The differentials of functions having this property are *exact differentials*. Below we give a brief reminder of their most useful mathematical properties.

Let f be a function of two independent variables x and y; example: internal energy taken as a function of entropy and volume. If y is kept constant, $f(x,y)$ reduces to a function of a single variable and df/dx is the derivative of f with respect to x. It is denoted by $(\partial f/\partial x)_y$ to indicate clearly that this *partial* derivative refers to a constant y value (Fig. A.1). The variation df of the function f when x varies infinitesimally by dx is

$$df = (\partial f/\partial x)_y dx. \tag{A.1a}$$

What happens if both x and y vary simultaneously by infinitesimal amounts dx and dy? The variation of f is simply the sum of the individual variations resulting from the changes dx and dy

$$df = (\partial f/\partial x)_y dx + (\partial f/\partial y)_x dy. \tag{A.1b}$$

If f is a function of three independent variables, one writes in the same way

$$df = (\partial f/\partial x)_{y,z} dx + (\partial f/\partial y)_{x,z} dy + (\partial f/\partial z)_{x,y} dz, \tag{A.1c}$$

where the indices indicate the two variables kept constant during the differentiation, and so on for functions of N variables.

As for a function of a single variable, it is useful to define second derivatives which account for the variation of the first partial derivatives with respect to the variables x, y, z, etc. But the situation gets a little bit complex because, for a function of two variables, four second partial derivatives must be defined: those of $(\partial f/\partial x)_y$ with

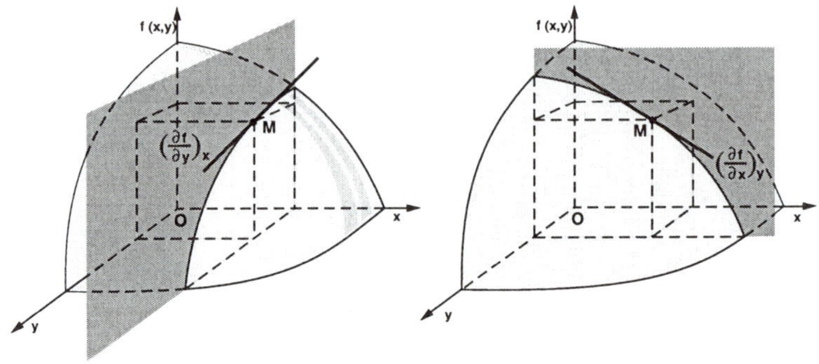

Figure A.1. Graphical representation of the partial derivatives of a function of two variables.

respect to x and y, and those of $(\partial f/\partial y)_x$ with respect to x and y. In thermodynamics, however, the number of second partial derivatives is only three because these do not depend on the order of the differentiation. In other words, the derivative of $(\partial f/\partial x)_y$ with respect to y is equal to that of $(\partial f/\partial y)_x$ with respect to x

$$\left[\frac{\partial(\partial f/\partial x)_y}{\partial y}\right]_x = \left[\frac{\partial(\partial f/\partial y)_x}{\partial x}\right]_y, \tag{A.2a}$$

which is more simply written as

$$\left(\frac{\partial^2 f}{\partial x \partial y}\right) = \left(\frac{\partial^2 f}{\partial y \partial x}\right). \tag{A.2b}$$

This equation characterizes exact differentials. It has a fundamental importance because it embodies the fact that the variations of f depend only on the changes in the variables x and y. When integrating (A.2), the same result is found whether integration is first made with respect to x or y. To avoid confusion, note finally that the notation "∂" will be reserved to partial derivatives of state functions. For heat, we will not write $(\partial Q/\partial x)_y$, but dQ/dx at constant y.

A.2 RELATIONS BETWEEN PARTIAL DERIVATIVES

Two formula are especially important for relating the partial derivatives of state functions. To illustrate the usefulness of the first one, let us consider the heat capacities at constant P and V. From (1.15), their difference is

$$C_p - C_v = T[(\partial S/\partial T)_P - (\partial S/\partial T)_V]. \tag{A.3}$$

Geometrically, this difference represents that between the slopes of the tangents to the surfaces $S = f(T,P)$ and $S = f(T,V)$ taken, at the same temperature T, in the planes P and V = constant, respectively. How are these two derivatives mutually related?

Let us begin with the exact differential of the entropy as a function of T and V

$$dS = (\partial S/\partial T)_V dT + (\partial S/\partial V)_T dV. \tag{A.4}$$

Alternatively, one can write dV as an exact differential of T and P

$$dV = (\partial V/\partial T)_P dT + (\partial V/\partial P)_T dP. \tag{A.5}$$

By replacing dV in (A.4) by its expression (A.5) we find

$$dS = \left[\left(\frac{\partial S}{\partial T}\right)_V + \left(\frac{\partial S}{\partial V}\right)_T \left(\frac{\partial V}{\partial T}\right)_P\right] dT + \left(\frac{\partial S}{\partial V}\right)_T \left(\frac{\partial V}{\partial P}\right)_T dP. \tag{A.6}$$

Because the two partial derivatives refer to the same constant variable T in the dP term, this equation simplifies to

$$dS = \left[\left(\frac{\partial S}{\partial T}\right)_V + \left(\frac{\partial S}{\partial V}\right)_T \left(\frac{\partial V}{\partial T}\right)_P\right] dT + \left(\frac{\partial S}{\partial P}\right)_T dP. \tag{A.7}$$

Of course, this differential of S as a function of T and P can be identified with

$$dS = (\partial S/\partial T)_P dT + (\partial S/\partial P)_T dP, \tag{A.8}$$

which implies

$$\left(\frac{\partial S}{\partial T}\right)_P = \left(\frac{\partial S}{\partial T}\right)_V + \left(\frac{\partial S}{\partial V}\right)_T \left(\frac{\partial V}{\partial T}\right)_P. \tag{A.9}$$

Thanks to (A.9) and the Maxwell relation (3.6c), we can thus rewrite (A.3) as

$$C_p - C_v = T\left(\frac{\partial S}{\partial V}\right)_T \left(\frac{\partial V}{\partial T}\right)_P = T\left(\frac{\partial P}{\partial T}\right)_V \left(\frac{\partial V}{\partial T}\right)_P. \tag{A.10}$$

Now, the problem lies in the fact that $(\partial P/\partial T)_V$ is not an observable property. Through the thermal expansion coefficient α and the isothermal compressibility β_T

$$\alpha = \frac{1}{V}\left(\frac{\partial V}{\partial T}\right)_P, \qquad \beta_T = -\frac{1}{V}\left(\frac{\partial V}{\partial P}\right)_T, \tag{A.11}$$

it is possible to determine the two other partial derivatives within the set of variables P, V and T. To find a relation between these three partial derivatives, we first write dP as an exact differential of T and V

$$dP = (\partial P/\partial T)_V dT + (\partial P/\partial V)_T dV. \tag{A.12}$$

Replacing dP by this expression in the differential (A.5) of V as a function of T and P, we obtain

$$dV = \left[\left(\frac{\partial V}{\partial P}\right)_T \left(\frac{\partial P}{\partial T}\right)_V + \left(\frac{\partial V}{\partial T}\right)_P\right] dT + \left(\frac{\partial V}{\partial P}\right)_T \left(\frac{\partial P}{\partial V}\right)_T dV$$

$$= \left[\left(\frac{\partial V}{\partial P}\right)_T \left(\frac{\partial P}{\partial T}\right)_V + \left(\frac{\partial V}{\partial T}\right)_P\right] dT + dV. \tag{A.13}$$

For the dT term of this equation to vanish, we must have

$$\left(\frac{\partial V}{\partial P}\right)_T \left(\frac{\partial P}{\partial T}\right)_V \left(\frac{\partial T}{\partial V}\right)_P = -1. \tag{A.14}$$

With this readily remembered formula, $(\partial P/\partial T)_V$ is replaced by two observable variables

$$(\partial P/\partial T)_V = \alpha/\beta_T. \tag{A.15}$$

Inserting this result in (A.10) we eventually find

$$C_p - C_v = TV\alpha^2/\beta_T. \tag{A.16}$$

As apparent in this example, equations of the form (A.9) and (A.14) are, along with the Maxwell relations, the most useful expressions to switch from one set of variables to another. In a more general form, we will rewrite them as

$$\left(\frac{\partial f}{\partial x}\right)_y = \left(\frac{\partial f}{\partial x}\right)_t + \left(\frac{\partial f}{\partial t}\right)_x \left(\frac{\partial t}{\partial x}\right)_y, \tag{A.17}$$

and

$$\left(\frac{\partial x}{\partial y}\right)_z \left(\frac{\partial y}{\partial z}\right)_x \left(\frac{\partial z}{\partial x}\right)_y = -1. \tag{A.18}$$

A.3 HOMOGENEOUS FUNCTIONS

By definition, a function $f(x,y,z,\ldots)$ is homogenous of degree n with respect to the variables $x,y,z,\ldots$, if it obeys the relation :

$$f(kx,ky,kz,\ldots) = k^n f(x,y,z,\ldots). \tag{A.19}$$

For such functions, the Euler theorem states that

$$x(\partial f/\partial x) + y(\partial f/\partial y) + z(\partial f/\partial z) + \ldots = nf(x,y,z). \tag{A.20}$$

This relation is obtained by differentiating A.19 with respect to k, and then taking the value $k = 1$:

$$\frac{\partial f(kx,ky,kz,\ldots)}{\partial kx} x + \frac{\partial f(kx,ky,kz,\ldots)}{\partial ky} y + \frac{\partial f(kx,ky,kz,\ldots)}{\partial kz} z + \ldots$$

$$= nk^{n-1} f(x,y,z,\ldots).$$

An intensive property is an homogenous function of degre 0 with respect to the numbers of moles:

$$f(kx, ky, kz, \ldots) = f(x, y, z, \ldots). \tag{A.21}$$

We have thus

$$x(\partial f/\partial x) + y(\partial f/\partial y) + z(\partial f/\partial z) + \ldots = 0. \tag{A.22}$$

On the other hand, extensive properties are homogenous functions of degree 1 with respect to the numbers of moles. For the internal energy, for example, we write :

$$U(T, P, kn_1, kn_2, kn_3, \ldots) = kU(T, P, n_1, n_2, n_3, \ldots). \tag{A.23}$$

which is the result derived in an elementary manner in 2.1d.

Appendix B
ADIABATIC COMPRESSIBILITIES

B.1 ACOUSTIC WAVES

Acoustic waves represent either compression movement, where matter is displaced in the direction of the wave vector, or shear movements, where matter is displaced along two different directions perpendicular to the direction of the wave vector (Fig. B.1). These longitudinal et transverse waves have very short periods compared to the timescale of heat conduction. As long as their amplitude is not too large, the periodic movements of compression and expansion can be considered as taking place under adiabatic conditions. This is why the adiabatic compressibility,

$$\beta_S = -1/V\ (\partial V/\partial P)_S \tag{B.1}$$

is directly related to acoustic wave velocities.

With regard to isothermal compressibilities, which are obtained through differentiation of volume-pressure fits, adiabatic compressibilities have thus the great advantage of being

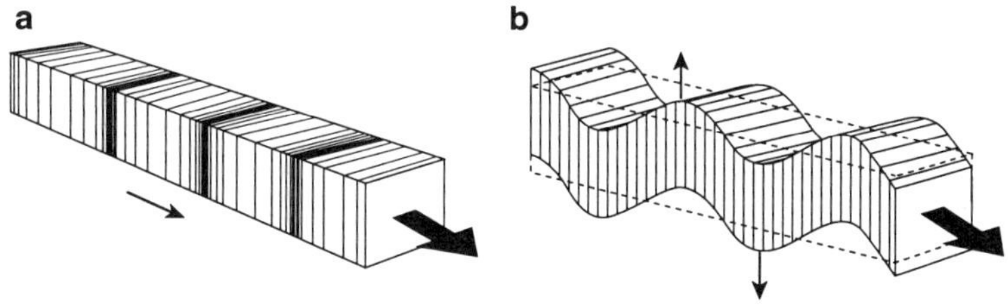

Figure B.1. Compressional (a) and shear (b) acoustic waves. The big arrows indicate the direction of the wave vector, the small ones those of the displacement of matter (shown for only one of the two transverse polarizations).

obtained accurately from direct observations. Experimentally, acoustic wave velocities are measured by various interferometry or attenuation ultrasonic methods (at MHz frequencies), or by Brillouin spectroscopy (at GHz frequencies) whereby one measures the frequency change of monochromatic (laser) electromagnetic radiation due to inelastic scattering of photons by the lattice phonons.

B.2 FLUIDS

A fluid with a low viscosity cannot be sheared significantly. Transverse acoustic waves are damped so rapidly that one can consider that their velocity is zero. Only the compressional acoustic waves have thus to be taken into account in the determination of an adiabatic compressibility. In a fluid at rest at time $t = 0$, of density ρ, at a pressure P_0, consider two points N_0 et N'_0 of coordinates x and $x+dx$. When the compressional wave is propagating, these points move to N_1 N'_1 which have, a time dt the coordinates $x+du$ and $x+dx + u+du$. The distance between the planes passing through N and N' perpendicularly to the wave vector thus changes from dx to $dx+du$. For a volume limited by two unit sections of these planes, the erlative volume change is given by

$$\frac{V - V_0}{V_0} = \frac{du}{dx} = \frac{\partial u}{\partial x} \tag{B.2}$$

From the definition of a compressibility, the pressure at point $N'1$ has varied at time dt by the amount

$$P - P_0 = dP = -\frac{1}{\beta_S}\frac{du}{dx} \tag{B.3}$$

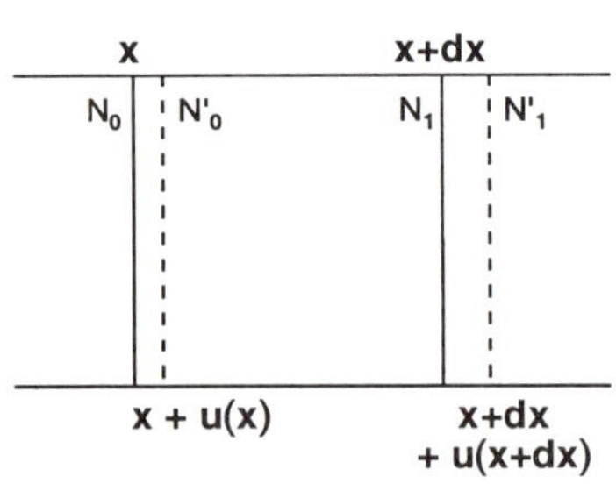

Figure B.2. Volume change due to the passage of a compressional wave.

To evaluate the force exerted on the volume element of unit cross section, les us note that the pressure change $P - P_0$ corresponds to the resultant of the forces applied (gravity excluded) on the surface elements passing through the points N'_0 and N'_1. We thus have

$$(P - P_0)_{x,t} - (P - P_0)_{x+dx,t} = -\left(\frac{\partial(P - P_0)}{\partial x}\right) dx = \frac{1}{\beta_S}\frac{\partial^2 u}{\partial x^2} dx. \tag{B.4}$$

It remains to write Newton's equation of motion. Because the mass of the volume element is ρdx, the equation of motion is

$$\frac{1}{\beta_S}\frac{\partial^2 u}{\partial x^2}dx = \rho dx \frac{\partial^2 u}{\partial t^2}. \tag{B.5}$$

Now, for a longitudinal wave with a velocity c_l the wave equation is

$$\frac{d^2 u}{dx^2} = \frac{1}{c_l^2}\frac{d^2 u}{dt^2}. \tag{B.6}$$

By identifying (B.5) and (B.6), we obtain

$$c_l = 1/(\rho\beta_S)^{1/2}, \tag{B.7}$$

which yields the desired relations

$$\beta_S = 1/\rho {c_l}^2, \qquad K_S = \rho {c_l}^2. \tag{B.8}$$

As discussed in 6.3c, the equilibrium compressibility is obtained in this way only if the pressure changes induced by the acoustic wave are short with respect to the time needed by the fluid to adjust its structure in response to this perturbation. Ultrasonic velocities, and compressibilities, are then frequency independent, which is observed for most fluids at frequencies that are typically a few MHz. The main exception is represented by viscous liquids. As shown in Fig. B.3 for diphenyl pentachloride, there exists a temperature interval where velocities are frequency dependent and where, at constant frequency, velocities diminish markedly with decreasing temperatures. At the highest temperatures, structural relaxation times in the liquid are short with respect to the period of the waves: ultrasound velocities are in this case frequency independent. At the lowest temperatures, relaxation times have

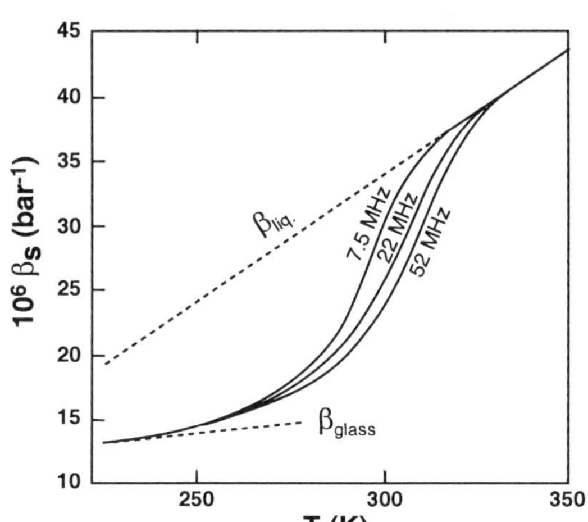

Figure B.3. Compressibility of biphenyl pentachloride as determined from ultrasonic experiments made at several frequencies (Litowitz and Lyon, 1958). The values that depends on the frequency at around 300 K signal the passage from a liquid-like to a glass-like behavior. The dashed line show the compressibility of the liquid at lower temperatures in the absence of a glass transition.

become so long that configurational changes do not take place at the timescale of acoustic waves: the liquid behaves as a solid and velocities are also frequency independent. In between, the continuous increase of relaxation times makes, at constant T, configurational changes the most complete at the lowest frequencies. This is the reason for the observed dispersion which is thus another manifestation of the glass transition. At the 10^{-6} s timescale of an ultrasonic experiment, however, the glass transition takes place at much higher temperatures than in calorimetry or dilatometry experiment whose timescale is typically of the order of 10^3 s.

B.3 SOLIDS

For a single crystal, the elastic properties are anisotropic since the compression is different along the different crystallographic orientations (see for instance the book by Nye (1985) for a complete discussion of elastic properties of crystals). They are characterized by a tensor of elastic coefficients c_{ij} the number of which depends on crystal symmetry. Only amorphous substances and polycrystals without any preferred orientation constitute isotropic materials whose compressibility and bulk modulus are related to the acoustic wave velocities by

$$\beta_S = 1/[\rho(c_l{}^2 - 4/3\, c_t{}^2)], \qquad K_S = \rho(c_l{}^2 - 4/3\, c_t{}^2) \tag{B.9}$$

where c_t is the shear wave velocity. For a fluid, c_t is zero and (B.9) reduces to (B.8).

For high-symmetry single crystals, the adiabatic modulus is related to the elastic coefficients by the following equations:

Crystals with a cubic symmetry

$$K_S = (c_{11} + 2c_{12})/3. \tag{B.10}$$

Crystals with a hexagonal, tetragonal or trigonal symmetry

$$K_S = \frac{c_{33}(c_{11} + c_{12}) - 2c_{13}{}^2}{c_{11} + c_{12} + 2c_{33} - 4c_{13}}. \tag{B.11}$$

Crystals with an orthorhombic symmetry

$$\begin{aligned} K_S = {} & [c_{11}(c_{23}{}^2 - c_{22}c_{33}) + c_{13}(c_{22}c_{13} - 2c_{12}c_{23}) + c_{33}c_{12}{}^2] \\ & /[c_{11}(2c_{23} - c_{22} - c_{33}) + c_{22}(2c_{13} - c_{33}) + 2c_{33}c_{12} + c_{23}{}^2 + c_{13}{}^2 \\ & + c_{12}^2 - 2c_{13}c_{23} - 2c_{12}c_{23} - 2c_{12}c_{13}]. \end{aligned} \tag{B.12}$$

For crystals with monoclinic or triclinic symmetry the elastic constants to be measured are too numerous and such equations too complex to be used in bulk moduli determinations.

Appendix C
SHOCK WAVES AND EQUATIONS OF STATE

C.1 SHOCK WAVES

As produced either by the explosion of supernovae or meteorite falls, shock waves have been a violent natural phenomenon for eons. It is with the invention of gunpowder that they have also become the product of human activity but the bloody art of explosives has long remained essentially empirical. Although the equations accounting for shock-wave compression of phases were obtained in the 19th century by Rankine (1820-1872) and Hugoniot (1851-1887), they were really put to use only during the Manhattan project in the second world war. The need to controll accurately the shock wave fronts created by the violent explosives used to compress plutonium in bombs resulted in the development of the techniques required to measure the propagation velocity of the shock waves. Often within the framework of military programs, since then these techniques have been used to determine up to a few Mbar pressure the equations of state of many substances, mostly solid, but also initially liquid or gaseous before being shocked.

In a shock-wave experiment, one investigates the impact of a supersonic projectile on a target initially at rest. The shock wave created during the impact propagates in the material at a supersonic speed u_s, of the order of km/s, and its front separates a compressed zone on the end of the target impacted by the projectile, which heats up on compression, from a yet unperturbed zone at the other end (Fig. C.1). The shock wave thus propagates extremely rapidly through the target before being reflected back and forth between the two ends and attenuating progressively until complete decompression. It is while the shock wave is passing for the first time that matter is accelerated at a *particle velocity*, u_p, in successive layers of the material, which is typically about half that of the shock wave. What will be described in the following is just the initial compression at a given point where the pressure, density and internal energy reach their peak values.

Experimentally, the first parameter that can be measured, by various optical or electrical methods, is the velocity of the projectile before the impact. In the simplest case where the target and projectile would have the same mass and be made up of the same material, the velocity u_p would be half the initial speed of the projectile. In general

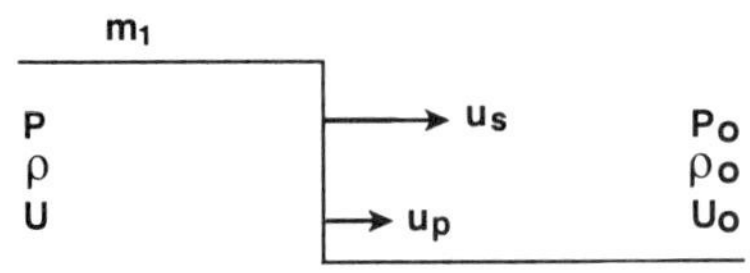

Figure C.1. Shock-wave front separating the compressed part of the target (left) from the yet uncompressed part (right).

this is not the case, but u_p can nonetheless be calculated from the projectile velocity and the material properties of both the projectile and target. On the other hand, there exists a the very large density contrast between the compressed and noncompressed zones of the target, which makes it possible to measure the velocity of the shock-wave front with high-speed X-ray cameras.

To derive an equation of state from such u_s and u_p measurements, one makes the experimentally justified assumption that hydrodynamic equilibrium is reached when the shock wave propagates through the target at a timescale of about 10^{-7} s. The reason is that the elastic limit of solids are a few tens of kbar at most. These are very small with respect to the peak pressures associated with the passage of the shock wave and makes it possible to treat the target as an ideal hydrostatic fluid. For the sake of simplicity, let us assume a time-independent, infinite planar shock wave (Fig. C.1), separating discontinuously two zones within the target where the density, pressure and internal energy are ρ, P and U, and ρ_0, P_0 and U_0, respectively. To simplify the equations, let us also consider that the shock wave propagates through a unit cross section so that a layer with a thickness dx has a volume dx and a mass ρdx.

C.2 EQUATIONS OF RANKINE-HUGONIOT AND HUGONIOT

The variables ρ, P and U are selected for describing the system because these are the properties with which one writes the mass, momentum and energy conservation equations across the shock-wave front. For a wave propagating at a velocity u_s, the amount of matter entering the front is, per unit of time, $\rho_0 u_s dx$. Because the passage of the front through a layer dx of the target results in compression and acceleration of this layer to the particle velocity u_p, the amount of matter leaving the front is $\rho dx(u_s - u_p)$. The mass conservation equation is thus

$$\rho_0 u_s = \rho(u_s - u_p). \tag{C.1}$$

The principle of momentum conservation states that the momentum variation is equal to the applied force. This force is $P - P_0$ per unit of surface area, which gives

$$P - P_0 = (\rho_0 u_s) u_p. \tag{C.2}$$

Finally, energy conservation dictates that the work done by the force $(P u_p)$ is equal to the sum of the kinetic energy gained by the target and the internal energy increase due to its compression

$$P u_p = 1/2\ \rho_0 u_s {u_p}^2 + \rho_0 u_s (U - U_0). \tag{C.3}$$

Note that compression is not isothermal. The temperature increase is actually greater for higher compressions. Although the passage of the shock wave is too rapid for heat to be exchanged between the compressed and uncompressed layers of the target, the compression is not adiabatic because part of the energy gained through compression is not converted into internal energy, but into kinetic energy. During the compression step, which is called the *Hugoniot*, dT/dP thus differs from both $(\partial T/\partial P)_T$ and $(\partial T/\partial P)_S$.

The three conservation equations (C.1)-(C.3) are the Rankine-Hugoniot equations. They mutually relate 8 variables of which 3 are known (the initial parameters), and 2 are measured (the velocities u_s and u_p). The system is thus determined analytically. Since the velocities u_s and u_p are not specific properties of the material, we must eliminate them from (C.1-3). Let us get the product $u_s u_p$ from (C.2) and the ratio u_s/u_p from (C.1) to derive from (C.3) the Hugoniot equation

$$(U - U_0) = 1/2\,(P + P_0)(V_0 - V), \tag{C.4}$$

in which we have substituted the reciprocal of the volume $1/V$ for the density ρ by taking 1 gram of matter. In a general way, this equation relates the variations of internal energy, volume and pressure. In accordance with Duhem's postulate, we check that two only out the three variables V, P and U can vary independently along a Hugoniot. Under this form, however, (C.4) cannot be used because the internal energy is not a variable whose variations can be directly measured.

C.3 PRESSURE AND VOLUME

Let us first address the simultaneous variations of pressure and volume. By changing the impact velocity of the projectile, one can make u_p vary. One then observes that u_s

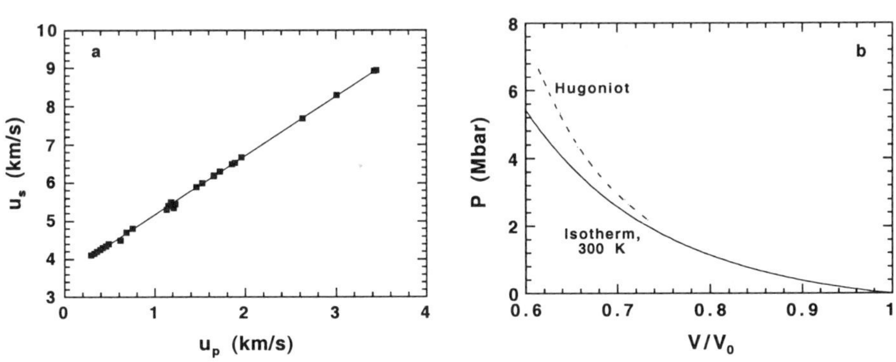

Figure C.2. (a): Relationship between the particle and shock velocities for platinum ; (b): Comparison between the Hugoniot and the 300 K isotherm calculated from the data shown in (a). Measurements and calculations of Holmes et al. (1989).

increases regularly with u_p, most often in a linear way (Fig. C.2a). Analytically, the results are reproduced by empirical polynomial expressions

$$u_s = a + bu_p + cu_p^2 + \ldots \tag{C.5}$$

where the constant a, b, etc. are obtained from fits of (C.5) to the measured velocities. When u_p is zero, the shock wave reduces to an ultrasonic wave. Hence, the parameter a is the sound velocity at room pressure along the relevant direction of the material.

With (C.5), we have an additional equation with which the internal energy can be eliminated from (C.1)-(C.3). Most often the $u_s - u_p$ relationship is observed to be linear. By taking $u_s = a + bu_p$, we obtain in this way after some rearrangement

$$P - P_0 = \frac{a^2(V_0 - V)V_0}{[V_0 - b(V_0 - V)]^2}. \tag{C.6}$$

Although this equation relates the volume to the pressure along the Hugoniot in terms of the experimentally determined parameters a and b, it does not represent an actual equation of state because the temperature is not constant (Fig. C.2b).

C.4 PRESSURE, TEMPERATURE AND VOLUME

Temperatures increase with the peak pressure of the shock wave experiments. Typically, they are of the order of 1000 K for a shock pressure of 1 Mbar. There are different ways of calculating such temperature rises from the properties of the compressed material. A simple one consists of making explicit the internal energy as given by the Rankine-Hugoniot equation

$$dU = -1/2\,(P + P_0)dV + 1/2\,(V_0 - V)dP = TdS - PdV. \tag{C.7}$$

Let us differentiate the entropy by taking T and V as independent variables

$$dS = (\partial S/\partial T)_V dT + (\partial S/\partial V)_T dV = C_v dT + \alpha K_T dV. \tag{C.8}$$

and replace in (C.7) dS by this expression after having made use of Maxwell's relation (3.6c). We obtain finally

$$C_v\left(\frac{dT}{dV}\right) = \left[\frac{1}{2}(P + P_0) - T\alpha K_T\right] + \frac{1}{2}(V_0 - V)\left(\frac{dP}{dV}\right). \tag{C.9}$$

If C_v, α and K_T were known, we could calculate the temperature by integrating (C.9) and using the slope dP/dV as given by (C.6). Of course, the problem is that α is generally not known at high pressure, and that the compressibility is the very property that one is trying to determine through shock-wave experiments! The situation is not hopeless, however, because, as already noted, the effects of dilation are relatively small with respect to those of compression. Although the empirical relationships between α, K_T and the Grüneisen parameter described in 5.5c could be used for solving (C.9), it is generally preferred to use a Mie-Grüneisen equation of state (5.51) along with

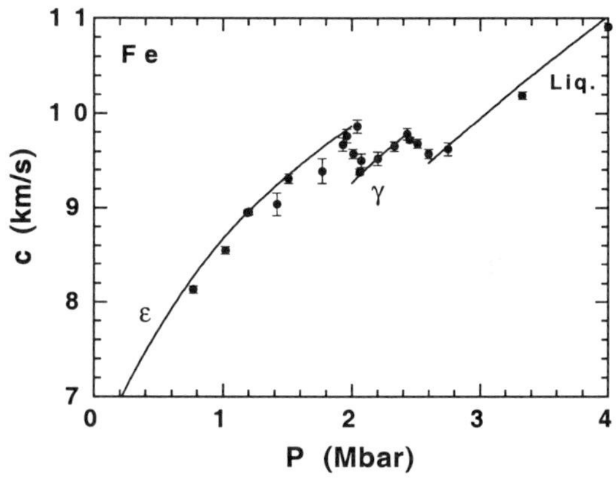

Figure C.3. Sound velocity in iron against the Hugoniot pressure. Note the changes due the transition from the ε to γ polymorphs at 2 Mbar and then to melting near 2.5 Mbar (Data of Brown and McQueen, 1986).

appropriate estimates of the Grüneisen parameter to derive an isothermal equation of state from the Hugoniot data.

A final remark. Given the very short timescale of shock-wave data, phase transitions seldom take place at the thermodynamic boundary. Over pressure intervals of the order of 100 kbar one observes instead a "mixed phase" regime where both phases coexist until the kinetics of the phase transition have become rapid enough that the transformation is complete. The $u_s - u_p$ and other relationships between properties then change progressively from the low-pressure to the high-pressure phase. For the sound velocity in iron, such effects of solid-solid and solid-liquid transitions are apparent in Fig. C.3 near 2 and 2.5 Mbar, respectively.

Appendix D
MULTICOMPONENT REGULAR AND SUBREGULAR MODELS

In this section we sketch how the activity coefficients are derived for multicomponent regular and subregular solutions. For more details, the reader is referred to the paper by Mukhopadhyay et al. (1993), or that of Cheng and Ganguly (1994) which gives an alternative formalism.

D.1 REGULAR MODEL

The activity coefficient γ_i of a component is related to the excess Gibbs-free energy of the solution by (11.68). For a regular solution, G^{xs} is

$$G^{xs} = \sum_{i<j}\sum W_{ij}x_i x_j, \tag{D.1}$$

and the simplest way of calculating γ_i is to differentiate the product $\sum n_i\ G^{xs}$ with respect to n_i. One then obtains

$$RT\ln\gamma_i = \sum_{j\neq i} 1/2\ W_{ij}x_j - G^{xs}. \tag{D.2}$$

This calculation is straightforward when one has noted that

$$\begin{aligned}\partial(n_j n_k/\textstyle\sum n_i)/\partial n_i &= -x_j x_k \text{ if } j\neq i \text{ and } k\neq i, &\text{(D.3)}\\ &= x_k - x_j x_k \text{ if } j\neq i \text{ and } k = i, &\text{(D.4)}\\ &= x_j - x_j x_k \text{ si } j = i \text{ et } k\neq i, &\text{(D.5)}\end{aligned}$$

D.2 SUBREGULAR MODEL

Activity coefficients are calculated in the same manner as for the regular model. After multiplication by $\sum n_i$, G^{xs} can be written as

$$\sum n_i G^{xs} = \sum_{j}^{n}\sum_{<k}^{n}\left(\frac{n_j n_k{}^2 W_{jk} + n_j{}^2 n_k W_{kj}}{(\sum n_i)^2}\right) + \sum_{j}^{n}\sum_{<k}^{n}\sum_{<l}^{n}\frac{n_j n_k n_l W_{jkl}}{(\sum n_i)^2} \tag{D.6}$$

For the binary part of G^{xs} the calculation involves differentials of the form

$$\frac{\partial}{\partial n_i}\left(\frac{n_j n_k^2}{(\sum n_i)^2}W_{jk}+\frac{n_j^2 n_k}{(\sum n_i)^2}W_{kj}\right)$$

$$= -(2x_jx_k^2W_{jk}+x_j^2x_kW_{kj}) \text{ si } j\neq i \text{ et } k\neq i. \quad \text{(D.7)}$$

$$= (2x_jx_i-2x_jx_i^2)W_{ji}+(x_j^2-2x_j^2x_i)W_{ij} \text{ si } j\neq i \text{ et } k=i. \quad \text{(D.8)}$$

$$= (2x_ix_k-2x_i^2x_k)W_{ki}+(x_k^2-2x_ix_k^2)W_{ik} \text{ si } j=i \text{ et } k\neq i. \quad \text{(D.9)}$$

For the ternary part, the differentials are of the form

$$\frac{\partial}{\partial n_i}\left(\frac{n_jn_kn_l}{(\sum n_i)^2}W_{jkl}\right) = -2x_jx_kx_lW_{jkl} \text{ if } j\neq i, k\neq i \text{ and } l\neq i. \quad \text{(D.10)}$$

$$= (x_kx_l-2x_ix_kx_l)W_{ikl} \text{ if } j=i,\ k\neq i \text{ and } l\neq i. \quad \text{(D.11)}$$

$$= (x_jx_l-2x_ix_jx_l)W_{jil} \text{ if } j\neq i,\ k=i \text{ and } l\neq i. \quad \text{(D.12)}$$

$$= (x_jx_k-2x_ix_jx_k)W_{jki} \text{ if } j\neq i,\ k\neq i \text{ and } l=i. \quad \text{(D.13)}$$

Noting that $W_{ijl}=W_{jil}=W_{jli}$, one eventually finds

$$RT\ln\gamma_i = 2\sum_{j\neq i}^{n}x_ix_jW_{ji}+\sum_{j\neq i}^{n}x_j^2W_{ij}+\sum_{j\neq i}^{n}\sum_{<k\neq i}^{n}x_jx_kW_{ijk}-2G^{xs}. \quad \text{(D.14)}$$

Appendix E
MOLECULAR ENERGIES

E.1 INTERNAL AND TRANSLATIONAL ENERGY

The energy of a molecule is made up of the kinetic and potential energies of its constituting particles, the latter resulting from the attractive and repulsive interactions between electrons and nuclei. In the absence of fields, the translational kinetic energy of the molecule is independent of the rest of its energy, which is termed *internal energy*. Hence, the wave function can be split into two independent parts

$$\Psi = \Psi_{trans}\Psi_{int}. \tag{E.1}$$

As indicated in 13.2d, the translation wave function can be calculated exactly. This is not the case for the internal wave function, even when considering atomic nuclei as fixed entities under the moderate energy conditions relevant to chemical thermodynamics. The difficulty results from the complex analytical form of the internal wave equation, which can be solved only through approximation methods.

E.2 BORN-OPPENHEIMER APPROXIMATION

Of foremost importance is the approximation to be made for separating first the internal wave function in two parts relative to electrons and nuclei. The mass of electrons is about 2000 times smaller than those of protons and neutrons. In view of the great difference between the mass of electrons (m) and those of nuclei (m_k), one can consider that nuclei remain at rest while electrons are moving around them. Conversely, nuclei move within a constant mean electronic environment. This is the Born-Oppenheimer approximation. Because energies are proportional to frequencies, in gaseous molecules electronic transitions (ultraviolet: $10^4 - 10^5$ cm^{-1}) are associated with much higher energies than either vibrations (infrared: from 10 to 4000 cm^{-1}) or rotations (microwaves: a few tens of cm^{-1} at most). The Born-Oppenheimer approximation thus leads to factorizing the internal wave function in two parts

$$\Psi_{int} = \Psi_{el}\Psi_{nuc}, \tag{E.2}$$

where Ψ_{el} and Ψ_{nuc} are the electronic and nuclear wave functions.

Let us make use of the indices k and l to indicate that the coordinates considered are those of nuclei or electrons, respectively, and denote by V_{kk} the potential between nuclei k, V_{ll} the potential between electrons and V_{kl} that between nuclei and electrons. With the usual notation $\Delta\Psi$ for the Laplacian of the wave function, $\partial^2\Psi/\partial x^2 + \partial^2\Psi/\partial y^2 + \partial^2\Psi/\partial z^2$, we rewrite the wave equation as

$$-\sum \frac{h^2}{8\pi^2 m_k} \Delta_k \Psi_{int} - \sum \frac{h^2}{8\pi^2 m} \Delta_l \Psi_{int} = (E_{int} - V_{kk} - V_{ll} - V_{kl})\Psi_{int}, \qquad \text{(E.3)}$$

where summations are made over all nuclei and all electrons of the molecule. Let us insert (E.2) in this equation to obtain

$$-\sum \frac{h^2}{8\pi^2 m} \Delta_k \Psi_{el} = (E_{el} - V_{ll} - V_{kl})\Psi_{el}. \qquad \text{(E.4)}$$

Mathematically, the Born-Oppenheimer approximation states that $\partial\Psi_{el}/\partial x_k$ is negligible with respect to $\partial\Psi_{nuc}/\partial x_k$. We thus have

$$-\sum \frac{h^2}{8\pi^2 m_k} \Delta_k \Psi_{nuc} = (E_{int} - E_{el} - V_{kk})\Psi_{nuc}. \qquad \text{(E.5)}$$

With this equation, it is possible to calculate the internal energy of the molecule for a given electronic state of energy E_{el} which corresponds, if this electronic state is stable, to the internuclear distances at which nuclei are in equilibrium. Therefore E_{el} represents the potential energy that determines the displacements of nuclei. For the H_2 diatomic molecule, this potential is represented as a function of the internuclear distance in Fig. E.1.

E.3 VIBRATIONAL AND ROTATIONAL ENERGIES

One cannot separate in a rigorous way the rotational and vibrational wave functions because a molecule rotates and vibrates at the same time. Vibrations cause periodic thus changes in internuclear distances, whereas the centrifugal forces induced by rotations result in increases of mean internuclear distances. Because the period of vibrations ($\sim 10^{-14}$ s) is typically 100 times shorter than that of rotations ($\sim 10^{-12}$ s), however, this influence of rotations on vibrations may be neglected to a first approximation. This one considers that the average internuclear distances are constant while a molecule is vibrating. Under these conditions, the energy of a molecule is the sum of the rotational energy of a rigid rotator and the vibrational energy of a nonrotating oscillator. More complicated treatments are needed to account for the fine structure of vibration-rotation spectra, but they are far beyond the scope of this book. We will just note that the small contribution of the energy of roatation-vibration interactions is included when the thermodynamic properties of ideal gases are calculated as sketched in 14.2c for diatomic gases.

In this section we will describe how the frequencies of the normal modes of vibration are calculated. We will adopt the usual notations whereby V and T denote the

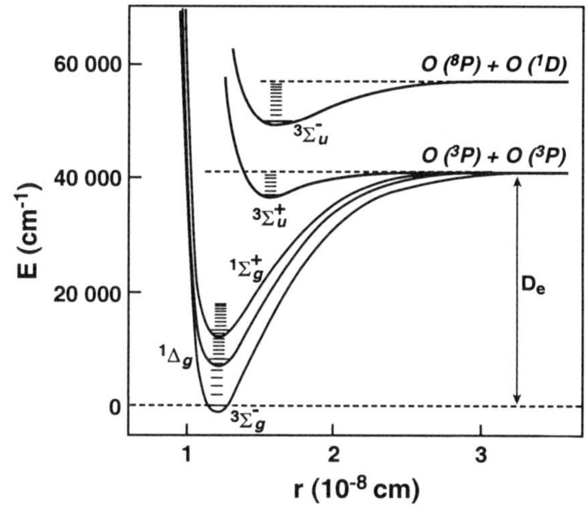

Figure E.1. Potential energy of oxygen against internuclear distance for 5 distinct electronic states of oxygen (after the book by Herzberg, 1950). The vibrational levels are indicated by the horizontal lines. Note the strong asymmetry of the potentials. Because the potential of interaction of two nuclei is zero at infinite distance, the difference between this potential and the minimum of the well represents the dissociation energy of the molecule (D_e). As shown by these plots, the dissociation energy depends on the electronic state of the molecule and on that of the oxygen atoms into which the molecule dissociates.

potential and kinetic energies, respectively, and $\dot{q}$ and $\ddot{q}$ stand for the first and second time derivatives of the coordinate q. As described in 14.2d, N-atom linear and nonlinear molecules have $3N-5$ and $3N-6$ vibrational modes, respectively. We have also seen that Cartesian coordinates are inadequate to solve the equations of motion, and that symmetry coordinates must be used instead to take advantage of the symmetry of the molecule. Expanding the interatomic potential in a Taylor series of the coordinates (14.54), one retains only the quadratic terms in the harmonic approximation, viz.

$$V = \sum_{i,j} k_{ij} q_i q_j, \tag{E.6}$$

where the summation is performed over the $3N-5$ or $3N-6$ symmetry coordinates. The kinetic energy is

$$T = \sum_{i,j} b_{ij} \dot{q}_i \dot{q}_j, \tag{E.7}$$

where, contrary to k_{ij}, the parameters b_{ij} depend on the mass of the atoms.

When solving the equations of motion one then looks for solutions of the form

$$q_i = A_i \cos(2\pi\nu t + \varepsilon), \tag{E.8}$$

where A_i is the vibrational amplitude. These equations are written for each coordinate in the generalized form of the Lagrange equation

$$\frac{\partial V}{\partial q_i} + \frac{\partial}{\partial t}\left(\frac{\partial T}{\partial \dot{q}_i}\right) = 0. \tag{E.9}$$

On the one hand we have

$$\left(\frac{\partial V}{\partial q_i}\right)_{q_j} = \Sigma_{i,j} k_{ij} q_j, \tag{E.10}$$

$$\left(\frac{\partial T}{\partial \dot{q}_i}\right)_{q_j} = \Sigma_j b_{ij} \dot{q}_j, \tag{E.11}$$

$$\frac{\partial}{\partial t}\left(\frac{\partial T}{\partial \dot{q}_i}\right) = \Sigma_j b_{ij} \ddot{q}_j. \tag{E.12}$$

On the other hand, the time derivatives of the coordinate q_i are

$$\dot{q}_i = -2\pi\nu \sin(2\pi\nu t + \varepsilon) \tag{E.13}$$

$$\ddot{q}_i = -4\pi^2\nu^2 \cos(2\pi\nu t + \varepsilon) = -4\pi^2\nu^2 q_i, \tag{E.14}$$

so that the equation of motion for q_i is

$$\sum_j (k_{ij} q_i - 4\pi^2\nu^2 b_{ij} q_i) = \sum_j (k_{ij} - 4\pi^2\nu^2 b_{ij}) A_i = 0. \tag{E.15}$$

The $3N-5$ ou $3N-6$ coordinates give rise to the same number of homogeneous linear equations (E.15). For all amplitudes A_j to differ from 0, the frequencies ν can take on only certain values that are the $3N-5$ or $3N-6$ roots of the *secular* equation, which is the determinant

$$\begin{vmatrix} k_{11} - 4\pi^2\nu^2 b_{11} & k_{12} - 4\pi^2\nu^2 b_{12} & \cdots \\ k_{21} - 4\pi^2\nu^2 b_{21} & k_{22} - 4\pi^2\nu^2 b_{ij} & \cdots \\ k_{31} - 4\pi^2\nu^2 b_{31} & \cdots & \cdots \\ \cdots & \cdots & \cdots \\ \cdots & \cdots & k_{nn} - 4\pi^2\nu^2 b_{nn} \end{vmatrix} = 0 \tag{E.16}$$

For each mode, all atoms vibrate in phase with the same frequency but with different vibrational amplitudes. The usefulness of symmetry coordinates is due to the fact that they allow the degree of the secular equation to be greatly reduced. When they are used, the determinant (E.16) is made up of determinants of smaller order along its diagonal, which concern each a limited number of modes having the same kind of symmetry (which in general number less than 4). Because the equations (E.15) are linear, any linear combination of solutions of (E.16) are also solutions to this equation.

Hence the most general solution is of the form

$$q_i = \sum_k A_{ik} \cos(2\pi\nu_k t + \varepsilon_k). \tag{E.17}$$

From the symmetry coordinates one finally derives the *normal coordinates*, whose importance stems from the fact that they allow any vibrational motion to be decomposed into normal modes, i.e., harmonic motions where all atoms oscillate at the same frequency and, in general, move in phase (Fig. 14.6). It is also for this special set of coordinates that the Schrödinger equation can be factorized into $3N-5$ or $3N-6$ independent wave functions.

Except with *ab initio* methods (see 14.4f), the force constants k_{ij} cannot be calculated. They are thus fitted to experimentally determined vibrational frequencies with (E.16). Because the number of force constants of the most general potential (E.6) is higher than the number of observed frequencies, either one uses simpler potentials, with less force constants, or one also uses the data for other isotopic molecules (for instance D_2O, HDO or $H_2^{18}O$ for water) which, according to the Born-Oppenheimer approximation, are determined by the same potential and allow one to have more observations than fitting parameters. The second method is of course to be preferred. Taking into account anharmonic and other effects in the same way, one can calculate all energy levels and the various partition functions with a high accuracy.

References

1. Quoted books and complementary reading

Allen, M.P. and D.J. Tildesley (1987) *Computer Simulation of Liquids*. Oxford University Press, Oxford.

Barthel, J.M.G., Krienke, H. and Kunz, W. (1998) *Physical Chemistry of Electrolyte Solutions. Modern aspects*. Springer, New York.

Bates, R.G. (1964) *Determination of pH, Theory and Practice*. John Wiley and Sons, New York.

Boulos, M., Fauchais, P. and Pfender, E. (1994) *Thermal Plasmas. Fundamentals and Applications*, Vol. 1, Plenum Press.

Bridgman, P.W. (1925) *A Condensed Collection of Thermodynamic Formula*. Cambridge University Press, Cambridge.

Callen, H.B. (1985) *Thermodynamics and an Introduction to Thermostatistics*. John Wiley & Sons, New York.

Cardarelli, F. (1996) *Scientific Unit Conversion*. Springer, London.

Cardwell, D.S.L. (1971) *From Watt to Clausius, the Rise of Thermodynamics in the Early Industrial Age*. Iowa State University Press, Ames.

Carnot, S. (1824) *Reflections on the motive power of heat and on machines fitted to develop this power*. T.A. Hirst, New York. (1890).

Denbigh, K. (1971) *The Principles of Chemical Equilibrium*. Cambridge University Press, Cambridge.

Dodé, M. (1965) *Le deuxième principe de la thermodynamique*. Sedes, Paris.

Durand-Vidal, S., Simonin, J.-P. and Turq, P. (2000) *Electrolytes at interfaces*. Kluwer, Dordrecht.

Einsenberg, D. and Kauzmann W. (1969) *The Structure and Properties of Water*. Oxford University Press, Oxford.

Gibbs, J.W., (1961) *The scientific papers*. Dover, New York.

Gillispie, C.C. (1970-80) *Dictionary of Scientific Biography*, 16 vols, Scribner, New York.

Guénault, A.M. (1988) *Statistical Physics*. Routledge, London.

Guggenheim, E.A. (1967) *Thermodynamics. An advanced treatment for chemists and physicists*. 5th ed., North Holland, Amsterdam.

Herzberg, G. (1950) *Molecular Spectra and Molecular Structure. I. Spectra of Diatomic Molecules*. Van Nostrand, New York.

Herzberg, G. (1955) *Molecular Spectra and Molecular Structure. II. Infrared and Raman Spectra*. Van Nostrand, New York.

Hildebrand, J.H. and Scott R.L. (1950) *The Solubility of Nonelectrolytes*. Reinhold Publishing Corporation, New York.

Kittel, C. (1996) *Introduction to solid state physics*. 7th ed, John Wiley and Sons, New York.

Landau, L. and Lifchitz, E.M. (1980) *Statistical Physics*, 3rd ed. Pergamon Press, Oxford.

Leff, H.S. and Rex, A.F. (1990) *Maxwell's Demon, Entropy, Information, Computing*. Princeton University Press, Princeton.

Lewis, G.N. and Randall M. (1961) *Thermodynamics*, revised par K.S. Pitzer et L. Brewer). Mac Graw Hill, Inc., New York.

Michard, G. (2001) *Chimie des eaux naturelles*, 2nd ed. Publisud, Paris.

Moore, W.J. (1972) *Physical Chemistry*, 5th ed. Longmans, London.

Navrotsky, A. (1994) *Physics and Chemistry of Earth Materials*. Cambridge University Press, Cambridge.

Nye, J.F. (1985) *Physical Properties of Crystals. Their Representation by Tensors and Matrices*. Oxford University Press, Oxford.

Ottonello, G. (1998) *Principles of Geochemistry*. Columbia University Press, New York.

Pauling, L. (1988) *General Chemistry*. Dover, New York.

Pitzer, K.S. (1995) *Thermodynamics*, 3rd ed. McGraw-Hill, New York.

Prausnitz, J.M. (1969) *Molecular Thermodynamics of fluid-phase equilibria*. Prentice-Hall, Englewood Cliffs, New Jersey.

Prigogine, I. and Defay R. (1954) *Chemical thermodynamics*. John Wiley & Sons, New York.

Reichl, L.E. (1980) *A modern course in statistical physics*. University of Texas Press, Austin.

Reid, R.C., Prausnitz, J.M. and Pohling, B.E. (1987) *The Properties of Gases and Liquids*, 4th ed. Mc Graw Hill, New York.

Robinson, R.A. and Stokes, R.H. (1959) *Electrolyte solutions*. Butterworths, London.

Schuster, A. (1932) *Biographical Fragments*. Macmillan, London.

Segré, E. (1984) *From Falling Bodies to Radio Waves. Classical Physicists and their Discoveries*. W.H. Freeman & Co., New York.

Stacey, F.D. (1992) *Physics of the Earth*. Brookfield Press, Brisbane.

Stanley, H.E. (1971) *Introduction to Phase Transitions and Critical Phenomena*. Oxford University Press, Oxford.

Van Zeggeren, F. and S.H. Storey (1970) *The Computation of Chemical Equilibria*. Cambridge University Press, Cambridge.

Van Holde, K.E., Johnson, W.C. and Ho, P.S. (1998) *Physical Biochemistry*. Prentice Hall, Upper Saddle River.

Yeomans, J.M. (1992) *Statistical mechanics of phase transitions*. Clarendon Press, Oxford.

Wilson, E.B., Decius J.C. and Cross P.C. (1955) *Molecular Vibrations. The Theory of Infrared and Raman Vibrational Spectra*. McGraw-Hill, New York.

2. Compilations of thermodynamic data

As examples only a few of the many compilations published by the U.S. Bureau of Mines will be quoted. Also note that the *Journal of Physical and Chemical Data* is entirely devoted to critically assessed thermodynamic data for organic and inorganic substances. For organic substances, also consult ELDATA, the *International Electronic Journal of Physical and Chemical Data* (http://www.fi-karlsruhe.de/eldata/eldata.html).

Ahrens, T.J., ed. (1995) *Mineral Physics and Crystallography, A Handbook of Physical Constants*. Amer. Geophys. Union Reference Shelf 2, AGU, Washington.

Air Liquide (1976) Encyclopédie des gaz. Elsevier, Amsterdam.

Bard, A.J., Parsons, R. and Jordan, J. (1985) *Standard Potentials in Aqueous Solutions*. Marcel Dekker, New York.

Boublik, T., Fried, V. and Hala, E. (1984) *The Vapour Pressure of Pure Substances*, 2nd ed. Physical Sciences Data 17, Elsevier, Amsterdam.

Chase, M.W., Jr., *NIST-JANAF Thermochemical Tables*, 4th ed., 2 vols. J. Phys. Chem. Ref. Data, Monograph No 9, Amer. Inst. Phys., Woodbury.

Christensen, C., Rasmussen, P., Gmehling, J. and Weidlich, T. (1984-2001) *Heats of mixing data collection*. Vol. 3, Parts 1-6, Dechema Chemistry Data Series, Dechema, Frankfurt.

Christensen J.J., Rowley, R.L. and Izatt, R.M. (1982,1988) *Handbook of Heats of Mixing, and supplement*. Wiley Interscience, New York.

Cox, J.D., Wagman, D.D., and Medvedev, V.A. (1987) *CODATA Key Values for Thermodynamics*. Hemisphere, New York.

Dawson, R.M.C., Elliott, D.C., Elliott, W.H. and Jones, K.M. (1986) *Data for Biochemical Research*. Clarendon Press, Oxford.

Domalski, E.S. and Hearing, E.D. (1996) *Heat Capacities and Entropies of Organic Compounds in the Condensed Phase*. III. J. Phys. Chem. Ref. Data, 25, 1-525.

Elliot, J.F. and Gleiser, M. (1960) *Thermochemistry for Steelmaking*. Addison-Wesley, Reading.

Gmehling, J., Schiller, M., Menke, J., Tiegs, D., Medina, A., Soares, M., Bastos, J., Alessi, P. and Kikic, I. (1986, 1994) *Activity Coefficients at Infinite Dilution*. Vol. 9, Parts 1-4, Dechema Chemistry Data Series, Dechema, Frankfurt.

Haar, L., Gallagher, J.S. and Kell, G.S. (1984) *NBS/NRC Steam Tables*. Hemisphere Publishing Corp., New York.

Handbook of Biochemistry and Molecular Biology, Physical and Chemical Data (1976) vol. 1, CRC Press, Boca Raton.

Handbook of Chemistry and Physics (1991) CRC Press, Boca Raton, Florida.

Horvath, A.L. (1985) *Handbook of aqueous electrolyte solutions. Physical properties, estimation and correlation methods*. Ellis Horwood Ltd, Chichester.

Hultgren, R., Desai, P.D., Hawkins, D.T., Gleiser, M., Kelley, K.K. and Wagman, D.D. (1973) *Selected Values of the Thermodynamic Properties of the Elements*. American Society for Metals, Metals Park.

Hultgren, R., Desai, P.D., Hawkins, D.T., Gleiser, M. and Kelley, K.K. (1973) *Selected Values of the Thermodynamic Properties of Binary Alloys*. American Society for Metals, Metals Park.

Kelley, K.K. (1960) Contribution to the Data on Theoretical Metallurgy. XIII. *High-Temperature Heat Capacity and Entropy Data for the Elements and Inorganic Compounds*. U.S. Bureau of Mines Bulletin 584.

Kelley, K.K. and King, E.G. (1962) *Contribution to the Data on Theoretical Metallurgy. XIV. Entropies of the Elements and Inorganic Compounds* U.S. Bureau of Mines Bulletin 592.

Khodakovsky, I.L., Westrum, E.F., Jr and Hemingway, B.S. (2000) *International Geothermodynamic Tables*, in press. Washington, (in press).

Kubaschevski, O. and Alcock, C.B. (1979) Metallurgical Thermochemistry. Pergamon Press, Oxford.

Lide, D.R. and Kehiaian, H.V. (1994) *CRC Handbook of Thermophysical and Thermochemical data*. CRC Press, Boca Raton.

Lobo, V.M.M. (1989) *Handbook of Electrolyte Solutions*, Parts A and B. Physical Sciences Data No 41, Elsevier, Amsterdam.

Majer, V., and Svoboda, V. (1985) *Enthalpies of Vaporisation of Organic Compounds*. IUPAC Chemical Data series No 32, Blackwell, Oxford.

Naumov, G.B., Ryzhenko, B.N. and Khodakovsky I. (1974) *Handbook of Thermodynamic Data*. N.T.I.S., U.S. Department of Commerce.

Pankratz, L.B. (1982) *Thermodynamic Properties of Elements and Oxides*. U.S. Bureau of Mines Bulletin 672.

Pankratz, L.B. (1984) *Thermodynamic Properties of Halides*. U.S. Bureau of Mines Bulletin 674.

Pankratz, L.B., Mah, A.D. and Watson, S.W. (1987) *Thermodynamic Properties of Sulfides*. U.S. Bureau of Mines Bulletin 689.

Pankratz, L.B., Stuve, J.M. and Gokcen, N.A. (1986) *Thermodynamic Data for Mineral Technology*. U.S. Bureau of Mines Bulletin 677.

Reid, R.C., Prausnitz, J.M. and Pohling, B.E. (1987) *The Properties of Gases and Liquids*. 4th ed., Mc Graw Hill, New York.

Riddick, J.A., Bunger, W.B. and Sakano, T.K. (1986) *Organic solvents, Physical Properties and Methods of Purification*. 4th ed., John Wiley & Sons, New York.

Robie, R.A., Hemingway, B.S. and Fisher, J.R. (1979) *Thermodynamic Properties of Minerals and Related Substances at 298.15 K and 1 bar (10^5 Pascals) Pressure and at Higher Temperatures*. U.S. Geological Survey Bulletin 1452.

Robie, R.A. and Hemingway, B.S. (1995) *Thermodynamic Properties of Minerals and Related Substances at 298.15 K and 1 bar (10^5 Pascals) Pressure and at Higher Temperatures*. U.S. Geological Survey Bulletin 2131.

Rosenquist, T. (1970) *Thermodynamic Data for Metallurgists*. Tapir Verlag, Trondheim.

Ruzicka, V. and Domalski, E.S. (1992) *Estimation of the Heat Capacities of Organic Liquids as a Function of Temperature Using Group Additivity*. J. Phys. Chem. Ref. Data, 22, 597-657.

Simmrock, K.H., Janowsky, R. and Ohnsorge, A. (1986) *Critical data for Pure Substances*. Vol. 2, Dechema Chemistry Data Series, Dechema, Frankfurt.

Vagarftik, N.B. (1975) *Tables on the Thermophysical Properties of Liquids and Gases in Normal and Dissociated Phases*. Hemisphere Publish. Corp., Washington, 1975.

Zabransky, M., Ruzicka, V., Jr., Majer, V. and Domalski, E.S. (1996) *Heat Capacity of Liquids: Volumes I and II. Critical Review and Recommended Values*. J. Phys. Chem. Ref. Data Monograph 6.

3. Compilations of phase equilibria data

Note that the compilations generally deal with organic substances when their titles are not explicit.

Engels, H. (1990) *Phase Equilibria and Phase Diagrams of Electrolytes*. Vol. 11, Dechema Chemistry Data Series, Dechema, Frankfurt.

Gmehling, J., Menke, J., Krafczyk, J. and Fischer, K. (1994) *Azeotropic Data*, Parts I and II, VCH, Weinheim.

Hansen, M. and K. Anderko (1958) *Constitution of Binary Alloys*, 2nd ed. and supplements, Mc Graw Hill, New York.

Knapp, H., Langhorst, R. and Teller, M. (1987) *Solid-Liquid Equilibrium Data Collection*. Vol. 8, Dechema Chemistry Data Series, Dechema, Frankfurt.

Lecat, M. (1949) *Tables azéotropiques*. Lamertin, Bruxelles.

Levin, E.M., Robbins, C.R. and McMurdie, H.F. (1964, 1969, 1975) *Phase Diagrams for Ceramists*. 2nd ed., 3 vols and supplements, American Ceramic Society, Columbus, Ohio.

Linke, W.F. (1958) *Solubilities of Inorganic and Metal-Organic Compounds*, I. Van Nostrand, Princeton.

Linke, W.F. (1965) *Solubilities of Inorganic and Metal-Organic Compounds*, II. Amer. Chem. Soc.

Liu, L.G. and Bassett, W.A. (1986) *Elements, Oxides, Silicates. High-Pressure Phases with Implications for the Earth's Interior*. Oxford University Press, Oxford.

Ohe, S. (1989) *Vapor-Liquid Equilibrium Data*. Physical Sciences Data 37, Elsevier, Amsterdam.

Ohe, S. (1991) *Vapor-Liquid Equilibrium Data. Salt Effect*. Physical Sciences Data 43, Elsevier, Amsterdam

Ohe, S. (1990) *Vapor-liquid equilibrium data at high pressure*. Physical Sciences Data 42, Elsevier, Amsterdam.

Petzow, G. and Effenberg, G. (1988-1995) *Ternary alloys. A Comprehensive Compendium of Evaluated Constitutional Data and Phase Diagrams*. 13 vols., VCH, Weinheim.

Pourbaix, M. (1963) *Atlas d'équilibres électrochimiques*. Gauthier-Villars, Paris.

Rasmussen, P., Sørensen, J.M., Arlt, W. and Macedo, M.E.A. (1979-87) *Liquid-Liquid Equilibrium Data Collection*. Vol. 5, Parts 1-4, Dechema Chemistry Data Series, Dechema, Frankfurt.

Wisniak, J. and Herskowitz, M. (1984) *Solubility of Gases and Solids*. Physical Sciences Data 18, Elsevier, Amsterdam.

Wohlfarth, C. (1994) *Vapour-Liquid Equilibrium Data of Binary Polymer Solutions*. Physical Sciences Data 44, Elsevier, Amsterdam.

4. Quoted papers

Adachi, K., Suga, H. and Seki, S. (1968) Bull. Chem. Soc. Jap., 41, 1073-1087.

Babcock, H.D. and Herzberg, L. (1948) Ap. J., 108, 167-190.

Barkhatov, L.S., Kagan, D.N., Tsytsarkin, A.F., Shpilrain, E.E. and Yakimovich, K.A. (1973) Teplofys. Vys. Temp., 11, 1188-1191.

Bilir, N. and Phillips, W.A. (1975) Phil. Mag., 32, 113-122.

Blackman, M. (1955) Handbuch der Physik, VII/I, p. 325-382.

Bloomer, O.T. and Parent, J.D. (1953) Chem. Eng. Progr. Symp., 49, 3-11 Knapp, p. 285

Boesch, J., Napolitano, A. and Macedo, P.B. (1970) J. Amer. Ceram. Soc., 53, 148-153.

Bottinga, Y. (1968) Earth Planet. Sci. Lett., 5, 301-307.

Bottinga, Y. and Richet, P. (1981) Amer. J. Sci., 281, 615.

Bowen, N.L., and Andersen, O. (1914) Amer. J. Sci., 37, 487-500.

Brown, J.M. and McQueen, R.G. (1986) J. Geophys. Res., 91, 7485-7494.

Buijs, H.L. and Gush, H.D. (1971) Can. J. Phys., 49, 2366-2375.

Cahn, J.W. (1968) Trans. Met. Soc. AIME, 242, 166-180.

Cerrai, E., Marchetti, C., Renzoni, R., Roseo, L., Silvestri, M. and Villani, S. (1954) Chem. Eng. Progr. Symp. Ser. 50, 271-280.

Chang, S.S. and Bestul, A.B. (1972) J. Chem. Phys., 56, 503-516.

Chang, S.S. and Bestul, A.B. (1974) J. Chem. Therm., 6, 325-344.

Chen, C.H. and Presnall, D.C. (1975) Amer. Mineral., 60, 398-406.

Cheng, W. and Ganguly, J. (1996) Geochim. Cosmochim. Acta, 58, 3763-3767 (1994) and 60, 3979.

Clayton, R.N., Grossman, L. and Mayeda (1973) T.K., Science, 182, 485-488.

Clyne, M.A.A. and Coxon, J.A. (1970) J. Mol. Spectr., 33, 381-406.

Connelly, D.L., Loomis, J.S. and Mapother, D.E. (1971) Phys. Rev., B3, 924-934.

Coplen, T.B., Pure Applied Chem., 68, 2839-2359 (1996)

Cox, H.L., Nelson, W.L. and Cretcher, L.H. (1927) J. Amer. Chem. Soc., 49, 1080-1083.

Debye, P. (1912) Ann. Phys., 39, 789-839.

Debye, P. and Hückel E. (1923) Phys. Z., 24, 185-206.

Debye, P. and Hückel E. (1923) Phys. Z., 24, 305-325.

Desnoyers, J.E. and Morrison, J.A. (1958) Phil. Mag., 3, 42-48.

Ditmars, D.A., Ishihara, S., Chang, S.S. and Bernstein, G. (1982) J. Res. Nat. Bur. Stand., 87, 159-163.

Dodge, B.F. (1927) Chem. Met. Eng., 10, 35, 622.

Eduljee, H.F., Newitt, D.M. and Weale, K.E. (1951) J. Chem. Soc., 3086-3091.
Edwards, M.H. and Woodbury, W.C. (1963) Phys. Rev., 129, 1911-1918.
Einstein, A. (1907) Ann. Phys., 22, 180-190.
Flubacher, P., Leadbetter, A.J., Morrison, J.A. and Stoicheff, B.P. (1959) J. Phys. Chem. Solids., 12, 53-65.
Flubacher, P., Leadbetter, A.J. and Morrison, J.A. (1960) J. Chem. Phys., 33, 1751-1755.
Giauque, W.F. and Stout, J.W. (1936) J. Amer. Chem. Soc., 58, 1144-1150.
Giauque, W.F. and Archibald, R.C. (1937) J. Amer. Chem. Soc., 59, 561-569.
Gillet, P., Fiquet, G. Daniel, I., Reynard, B. and Hanfland, M. (1999) Phys. Rev., B60, 14660-14664.
Greer, S.C., Phys. Rev. (1976) A14, 1770-1780.
Greig, J.W. (1927) Amer. J. Sci., 13, 1-44.
Grønvold, F. (1973) J. Chem. Therm., 5, 525-539.
Grønvold, F., Stølen, S. and Svendsen, S.R. (1989) Thermochim. Acta, 135, 225-243.
Guggenheim, E.A. (1945) J. Chem. Phys., 13, 253-261.
Guissani, Y. and Guillot, B. (1994) J. Chem. Phys., 101, 490-509.
Haida, O., Matsuo, T., Suga, H. and Seki, M. (1974) J. Chem. Therm., 6, 815-825.
Hamer, W.J. and WU, Y.C. (1972) J. Phys. Chem. Ref. Data, 1, 1047-1099.
Hanfland, M., Beister, H. and Syassen, K. (1989) Phys. Rev., B39, 12598-12603.
Harrison, R.J., Redfern, S.A.T. and O'Neill H.St.C. (1998) Amer. Mineral., 83, 1092-1099 .
Hemley, R.J., Jephcoat, A.P., Mao, H.K., Zha, C.S., Finger, L.W. and Cox, D.E. (1987) Nature, 330, 737-740.
Hemley, R.J., Mao, H.K., Finger, L.W., Jephcoat, A.P., Hazen, R.M. and Zha, C.S. (1990) Phys. Rev., B42, 6458-6470.
Hobson, R.W., Hartman, R.J. and Kanning, E.W. (1941) J. Amer. Chem. Soc., 63, 2094-2095.
Holden et Martin (1983) Pure Appl. Chem., 56, 653.
Holmes, N.C., Moriarty, G.R., Gathers, G.R. and Nellis, W.J. (1989) J. Appl. Phys., 66, 2962-2967.
Holste, J.C., Hall, K.R., Eubank, P.T., Esper, G., Watson, M.Q., Warowny, W., Bailey, D.M., Young, J.G. and Bellomy, M.T. (1987) J. Chem. Therm., 19, 1233-1250.
Jakli, G. and Staschewski, D. (1977) J. Chem. Soc., Farad. Trans. I, 73, 1505-1509.
Jephcoat, A., Mao, H.K. and Bell, P.M. (1986) J. Geophys. Res., 91, 4677-4684.
Kell, G.S. (1977) J. Phys. Chem. Ref. Data, 6, 1109-1131.
Kell, G.S., Mc Laurin, G.E. and Whalley, E. (1985) Phil. Trans. Roy. Soc. Lond., A315, 235-246.
Kennedy, G.C. (1954) Amer. J. Sci., 252, 225-241.
Kestin, J., Sengers, J.V., Kamgar-Parsi, B., Levelt-Sengers, J.M.H. (1984) J. Phys. Chem. Ref. Data, 13, 601-609.
Kirschenbaum, A.D., Cahill, J.A., McGonigal P.J. and Grosse, A.V. (1962) J. Inorg. Nucl. Chem., 24, 1287-1296.
Kohler, F. (1960) Monatsch. Chem., 96, 1228-1251.
Kushner, T.M., Tatsievskaia, G.I., Irzoun, L.V., Volkova, L.V., Serafimov, L.A. (1966) Zh. Fys. Khim., 40, 3010-3017.
Lane, W.H., Ind. Eng. Chem. (1946) Anal. Ed., 18, 295-296.
Levelt Sengers, J.M.H., Greer, W.L. and Sengers, J.V. (1976) J. Phys. Chem. Ref. Data, 5, 1-50.
Lewis, G.N. and Randall, M. (1921) J. Amer. Chem. Soc., 43, 1112-1154.
Litovitz, T.A. and Lyon, T. (1958) J. Acoust. Soc. Amer., 30, 856-859.
McGlashan, M.L. (1990) J. Chem. Therm., 22, 653-663.
McKinney, J.E. and Goldstein, M. (1974) J. Res. N.B.S., 78A, 331-353.
Machado, J.R.S., Street, W.B. and Delters, U. (1988) J. Chem. Eng. Data, 33, 148-152.

Mao, H.K., Wu, Y., Chen, L.C., Shu, J.F. and Jephcoat, A.P. (1990) J. Geophys. Res., 95, 21737-21742.
Marshall, W.L. and Franck, E.U. (1981) J. Phys. Chem. Ref. Data, 10, 295-304.
Metropolis, N., Rosenbluth, A.W., Rosenbluth, M.N., Teller, A.H. and Teller, E. (1953) J. Chem. Phys., 21, 1087-1092.
Michels, A., Blaisse, B. and Michels, C. (1937) Proc. Roy. Soc., A160, 358-375.
Michels, Wijker, H.U.B. and Wijker, H.K. (1949) Physica, 15, 627-633.
Michels, A., Lunbeck, R.J. and Wolkers, G.J. (1951) Physica, 17, 801-816.
Michels, A., Wassenaar, T. and Louwerse, P. (1954) Physica, 20, 99-106.
Michels, A., Levelt, J.M. and De Graaf, W. (1958) Physica (1954) 24, 659-671.
Michels, A., De Graaf, W., Wassenaar, T., Levelt, J.M.H. and Louverse P. (1959) Physica, 25, 25-42.
Michels, A., Wassenaar, T. and Louwerse, P. (1960) Physica, 26, 539-543.
Mohr, P.J. and Taylor, B.N. (1999) J. Phys. Chem. Ref. Data, 28, 1713.
Moser, H. (1936) Phys. Z., 37, 737-753.
Moynihan, C.T., Easteal, A.J., Wilder, J. and Tucker, J. (1974) J. Phys. Chem., 78, 2673-2677.
Mukhopadhyay, B., Basu, S. and Holdaway, M.J. (1993) Geochim. Cosmochim. Acta, 57, 277-283.
Nagano, Y., Miyazaki, Y., Matsuo, T. and Suga, H. (1993) J. Phys. Chem., 97, 6897-6901 .
O'Neil, J.R., Clayton, R.N. and Mayeda, T.K. (1969) J. Chem. Phys., 51, 5547-5559.
Osborne, N.S., Stimson, H.F. and Ginnings, D.C. (1939) J. Res. Natl. Bur. Stand., 23, 261-270.
Pearce, J.N. and Nelson, A.F. (1932) J. Amer. Chem. Soc., 54, 3544-3555.
Pemberton, R.C. and Mash (1978) C.J., J. Chem. Therm., 10, 867-888.
Petit, A.T. and Dulong, A.L. (1819) Ann. Chim. Phys., 10, 395-413.
Pitzer, K.S. (1973) J. Phys. Chem. 77, 268-277.
Pitzer, K.S. and Mayorga, G. (1973) J. Phys. Chem., 77, 2300-2308.
Poirier, J.P. (1999) in Microscopic Properties and Processes in Minerals, K. Wright and R. Catlow eds; Kluwer, p. 19-42.
Proks, I., Eliasova, M. and Kosa, L. (1977) Silikaty, 21, 3-11.
Pyper, J.W. and Souers, P.C. (1976) Lawrence Livermore Laboratory, University of California Report UCRL - 52104.
Redlich, O. and Kister, A.T. (1948) Ind. Eng. Chem., 40, 345-348.
Redlich, O. and Kwong, J.N.S. (1949) Chem. Rev., 44, 233-244.
Reis, J.C.R. (1982) J. Chem. Soc., Farad. Trans., 2, 1595-1608.
Richet, P., Bottinga, Y. and Javoy, M. (1977) Ann. Rev. Earth Planet. Sci., 5, 65-110.
Richet, P., Bottinga, Y., Deniélou, L., Petitet, J.P. and Téqui, C. (1982) Geochim. Cosmochim. Acta, 46, 2639-2648.
Richet, P., Mao, H.K. and Bell, P.M. (1988) Phys. Chem. Miner., 16, 207.
Richet, P., Mao, H.K. and Bell, P.M. (1988) J. Geophys. Res., 93, 15279-15288.
Rivkin, S.L., and Akhundov, T.S. (1962) Teploenergetika, No 1, 57-65.
Rivkin, S.L. and Akhundov, T.S. (1963)Teploenergetika, No 9, 66-71.
Rivkin, S.L and Troyanovskaya, G.V. (1964) Teploenergetika, No 10, 72-74.
Roder, H.M., Diller, D.E., Weber, L.A. and Goodwin, R.D. (1963) Cryogenics, 3, 16-22.
Rolston, J.H., Den Hartog, J. and Butler, J.P. (1976) J. Phys. Chem., 80, 1064-1067.
Rundman, K.B. and Hilliard, J.E. (1967) Acta Met., 15, 1025-1033.
Sasabe, H. and Moynihan, C.T. (1978) J. Polym. Sci., 16, 1447-1457.
Sato, R.K. and McMillan, P.F. (1987) J. Phys. Chem., 91, 3494-3498.
Saul, A., and Wagner, W. (1989) J. Phys. Chem. Ref. Data, 18, 1537-1564.
Shackelton, N.J. and Opdyke, N.D. (1977) Nature, 270, 216-219.
Sherman, R.H. (1965) Phys. Rev. Lett., 15, 141-142.

Simmons, R.O., and Balluffi, R.W. (1960) Phys. Rev., 117, 52-61.
Simonson, J.M., J. Chem. Therm. (1990) 22, 739-749.
Sirota, A.M., and Maltsev, B.K. (1962) Teploenergetika, No 7, 70-73.
Sirota, A.M., Maltsev, B.K. and Berkova, P.E. (1963) Teploenergetika, No 5, 64.
Speedy, R.J. (1987) J. Phys. Chem., 91, 3354-3358.
Stixrude, L., Cohen, R.E. and Singh, D.J. (1994) Phys. Rev. Lett., 50, 6442-6445.
Su, G.J., and Chang, C.H. (1946) Ind. Eng. Chem., 38, 800-806.
Trappeniers, N.J., Wassenaar, T. and Wolkers, G.J. (1956) Physica, 32, 1503-1520.
Trappeniers, N.J., Wassenaar and Abels, J.C. (1979) Physica, 98A, 289-297.
Tyler, D. (1912) J. Chem. Soc., 101, 1104-1113.
van Zandicke, F. and Verhoeye, L. (1974) J. Appl. Chem. Biotec., 24, 709-729.
Vinet, P.J., Ferrante, J.H., Rose, J.H. and Smith, J.R. (1987) J. Geophys. Res., 92, 9319-9325.
Wagner, W., Saul, A. and Pruss, A. (1994) J. Phys. Chem. Ref. Data, 23, 515-527.
Weber, H.F. (1875) Ann. Phys., 154, 367-423.
Weber, L.A. (1970) Phys. Rev., A2, 2379-2388.

Index